Calcium and Cell Physiology

Edited by Dieter Marmé

With Contributions by
L. R. Ballou, R. C. Brady, F. L. Bygrave, F. R. Cabral
W. Y. Cheung, J. R. Dedman, R. J. DeLorenzo, J. H. Exton
M. B. Feinstein, T. Godfraind, S. P. Halenda, A. Herchuelz
S. Kakiuchi, J. Kojima, L. Lagace, P. Lebrun, W. J. Malaisse
D. Marmé, S. Matzenauer, A. R. Means, A. C. Nairn
P. T. Peachell, F. L. Pearce, D. Pelzer, J. A. Putkey
H. Rasmussen, P. H. Reinhart, B. D. Roufogalis
H. J. Schatzmann, M. J. Schibler, R. C. M. Simmen
W. M. Taylor, W. Trautwein, L. J. Van Eldik, S. I. Walaas
M. P. Walsh, D. M. Watterson, G. B. Zavoico, W. Zawalich

With 94 Figures

Springer-Verlag
Berlin Heidelberg New York Tokyo

Professor Dr. DIETER MARMÉ

Gödecke Research Institute, Gödecke AG, Mooswaldallee 1–9,
7800 Freiburg, FRG

Institute of Biology III, University of Freiburg, Schänzlestr. 1,
7800 Freiburg, FRG

ISBN 3-540-13841-2 Springer-Verlag Berlin Heidelberg New York Tokyo
ISBN 0-387-13841-2 Springer-Verlag New York Heidelberg Berlin Tokyo

Typesetting and printing: Beltz, Offsetdruck, Hemsbach/Bergstr.
Bookbinding: Brühlsche Universitätsdruckerei, Giessen
2131-3130/543210

Preface

The purpose of the present volume is to give a comprehensive and up-to-date survey of the nature and role of calcium ions (Ca^{2+}) in the regulation of cellular function. Since Ca^{2+} has gained in interest over the past years as a cellular messenger in signal transduction, and since the discovery of its cellular receptor protein, calmodulin, has helped in understanding its mode of action in molecular terms, we felt that an interdisciplinary selection of topics from the calcium field could provide a good source of information for all those interested in calcium-mediated physiology.

The volume begins with an overview on the synarchic nature of the two cellular messengers, cyclic AMP and Ca^{2+}. The next three chapters deal with the various transport mechanisms for Ca^{2+}. The biochemistry and molecular biology of calmodulin, as well as the cellular localization of calmodulin and calmodulin-binding proteins, are reviewed. Calcium regulation of smooth muscle contraction introduces the pharmacology of calcium antagonists. These recently discovered drugs are known to be powerful tools in the therapy of heart diseases and hypertension. The regulation of the cytoskeleton in nonmuscular cells and the control of protein phosphorylation by Ca^{2+} are covered as important cellular processes which are involved in many physiological responses. The Ca^{2+} dependence of neurotransmitter synthesis and release and the role of Ca^{2+} in prostaglandin and thromboxane synthesis document the manifold interactions between the different signal-transducing systems. The importance of Ca^{2+} in platelet function and α-adrenergic regulation of liver function is reviewed, and the Ca^{2+}-mediated stimulus-secretion coupling is exemplified by insulin release from pancreatic β-cells and histamin release from mast cells. The volume ends with the description of a novel signal-transducing system involving protein kinase C and metabolites from (poly) phosphatidylinositides. The importance of this system is underlined by the fact that various oncogene products show enzymatic activity in the metabolism of phosphatidylinositols.

I wish to express my gratitude to all my colleagues who have spontaneously agreed to contribute to a book on Calcium and Cell Physiology. It is my hope that our book will stimulate new and productive research in the field of calcium.

Freiburg, Spring 1985   *Dieter Marmé*

Contents

List of Contributors

Ca²⁺ and cAMP in the Regulation of Cell Function

H. RASMUSSEN, W. ZAWALICH, and I. KOJIMA[1]

CONTENTS

1 Introduction

In considering the problem of how cell function is regulated by extracellular signals, it is worth making a distinction between three operationally different types of controlled processes. In some cases, when a cell is activated, all that is necessary is a brief message or trigger to initiate an already programmed sequence. This could well be the situation when cells in G_O are recruited into an active mitotic cycle, or when oocytes undergo fertilization. It is also true when an antigen-antibody reaction initiates histamine release from mast cells (Foreman et al. 1973). In other cases, there may be an inherent cyclic process which goes on in the absence of an extracellular messenger, but whose amplitude and/or frequency are changed by addition of extracellular messenger. The best-studied examples of this kind is the hormonal control of cardiac function (Tsien 1977). It is possible that mitotic cycles may also fall into this category. Finally, there are situations in which augmented cell response is initiated by addition of extracellular messenger and is sustained only as long as extracellular messenger is present; removal of messenger leads to a termination of response. This is clearly the case in the action of a variety of mammalian peptide and amine hormones ranging from insulin and growth factors to a variety of trophic hormones from the pituitary, glucagon, parathyroid hormone, angiotensin, vasopressin, and others (Rasmussen 1981).

---

[1] Departments of Internal Medicine and Cell Biology, Yale University School of Medicine, 333 Cedar Street, New Haven, CT 06510, USA

Much of our knowledge of how hormones and neurotransmitters regulate cell function has come from studies of the third type of phenomenon, in which changes in intracellular messenger concentration not only initiate an immediate cellular response, but are also involved in the more long-term trophic effects of these hormones. The present discussion will focus on the concepts that have been developed recently concerning the intracellular messenger functions of Ca^{2+} and cAMP in cells displaying sustained responses to extracellular messenger addition (Rasmussen 1981; Rasmussen and Waisman 1982; Rasmussen 1983; Takai et al. 1981). A brief presentation of our present-day views of information flow in the cAMP and Ca^{2+} messenger systems will be followed by a consideration of Ca^{2+} and cAMP as *synarchic* messengers (Rasmussen 1981). Following this discussion, the processes of *amplitude* and *sensitivity* modulation in the Ca^{2+} messenger system will be discussed in relationship to the role of changes in cytosolic calcium concentration as a means of coupling events at the plasma membrane to metabolic events within the cell (Rasmussen and Waisman 1982; Rasmussen 1983). The highly plastic behavioral characteristics of this seemingly stereotyped control device will be considered. Particular emphasis will be placed on the mechanisms which are employed by the calcium messenger system to achieve a sustained cellular response in spite of only a transient elevation of the $[Ca^{2+}]$ in the cell cytosol.

2 Intracellular Information Flow

Accepting the evidence that cAMP and Ca^{2+} are two of the major intracellular messengers in hormone and neurotransmitter action, it is useful to contrast the way in which information flows from cell surface to cell interior in these two messenger systems (Rasmussen 1981; Rasmussen and Waisman 1982; Rasmussen 1983; Takai et al. 1981; Borle 1981; Michell 1975; Robison et al. 1971; Greengard 1978).

Our present views of information flow in the cAMP system (Fig. 1) have not changed greatly since their original formulation (Robison et al. 1971); Present concepts are that cAMP is generated at a single cellular site, the plasma membrane, and by a single enzyme, adenylate cyclase. A rise in intracellular cAMP concentration leads to its binding to a single small class of receptor proteins which interact in turn with a single class of

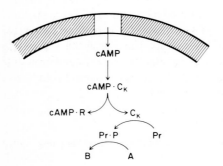

Fig. 1. A representation of the flow of information in the cAMP messenger system. Upon binding of extracellular messenger to surface receptor, adenylate cyclase is activated resulting in a rise in cAMP concentration intracellularly. The cAMP binds to a receptor (R) associated with a catalytic subunit (C_k) of one of several protein kinases. Upon binding of cAMP, C_k dissociates from R and becomes active resulting in the phosphorylation of a number of proteins (Pr). The phosphoprotein products $(P_r.P)$ catalyze various cellular reactions represented by the conversion of substrate A to product B

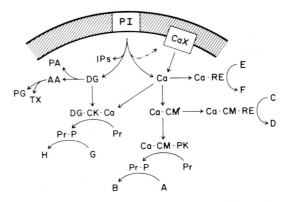

Fig. 2. A representation of the flow of information in the calcium messenger system. When extracellular messenger interacts with receptor, a phospholipase C is activated and catalyzes the hydrolysis of 4,5 diphospho-phosphatidylinositol (4,5-PIn). As a consequence, there is a rise in both the Ca^{2+} and diacyl glycerol (*DG*) concentrations in the cell. The rise in DG leads to the activation of a specific calcium-dependent protein kinase (C-kinase) which catalyzes the phosphorylation of a specific group of proteins (*Pr*.). Also, the hydrolysis of DG by a phospholipase A_2, gives rise to arachidonic acid (*AA*) and phosphatidic acid (not shown). The latter may serve to regulate cellular calcium metabolism, and the former serves as a substrate for the synthesis of prostaglandins and thromboxanes (not shown), compounds which have messenger functions in their own right. The increase in Ca^{2+} leads to the association of Ca^{2+} with receptor proteins such as calmodulin (CM. These Ca^{2+}-receptor protein complexes associate, in turn, with a variety of response elements (*RE*). Some of these response elements are protein kinases (*PK*) which catalyze the phosphorylation of a second specific group of proteins (*Pr*.). Other response elements associate directly with Ca^{2+}-calmodulin and thereby undergo a change in activity. The complexity of this messenger system is to be contrasted with the relative simplicity of the cAMP messenger system shown in Fig. 1

response elements, protein kinases, which catalyze a single type of reaction, a protein phosphorylation (Greengard 1978). Hence, specific cellular responses to cAMP are determined by the particular set of cellular proteins that serve as substrates for the cAMP-dependent protein kinase (Greengard 1978).

In contrast to this very simple and stereotyped pattern in the cAMP messenger system, events in the calcium messenger system are much more complicated and plastic (Fig. 2). The event that initiates the rise in $[Ca^{2+}]$ is probably the hydrolysis of polyphosphoinositides leading to an increase in both diacylglycerol (DG) content of the plasma membrane, and to the inositol triphosphate (InP_3) concentration of the cytosol (Takai et al. 1981; Michell 1975; Takai et al. 1979b; Downes and Michell 1982; Streb et al. 1983; Nishizuka 1983). Furthermore, the source of Ca^{2+} may be the extracellular pool and/or one or more intracellular, membrane-bound pools (Borle 1981). The rise in $[InP_3]$ may be the means by which the intracellular pool(s) of calcium is mobilized (Streb et al. 1983).

The rise in DG concentration acts as a messenger in its own right and activates a specific, calcium-activated, phospholipid-dependent protein kinase, C-kinase, which catalyzes the phosphorylation of a specific group of protein substrates (Takai et al. 1981, 1979; Nishizuka 1983). Furthermore, this DG is one of the sources of arachidonic acid which serves, in turn, as the substrate for prostaglandin, leukotriene and/or thromboxane synthesis (Takai et al. 1981, 1979; Nishizuka 1983). These latter compounds may have messenger roles themselves or more probably act as modulators of the particular response system (Takai et al. 1981; Streb et al.; Nishizuka 1983; Synder et al. 1983).

When the $[Ca^{2+}]$ in the cell cytosol increases, Ca^{2+} complexes with several different types of calcium receptor proteins, some of which have only a receptor protein function, e.g., calmodulin, whereas others are Ca^{2+}-activated enzymes (Rasmussen 1983). For example, Ca^{2+} directly activates mitochondrial glycerol oxidase (Hansford and Chappell 1967) without calmodulin involvement. Furthermore, when Ca^{2+} binds to calmodulin (CaM), this complex can either activate one of several protein kinases (Cheung 1980), or it can act directly as an allosteric modifier of enzymes such as the plasma membrane Ca^{2+} pump or the cyclic nucleotide phosphodiesterase (Cheung 1980; Means and Dedman 1980; Vincenzii et al. 1980; Waisman et al. 1981). Hence, there are multiple pathways of information flow within this branch of the calcium messenger pathway.

3 Ca^{2+} and cAMP as Synarchic Messengers

Presentation of these two major messenger systems as separate pathways of information flow from cell surface to cell interior is helpful from a didactic point of view, but misleading from a biological point of view because it has become increasingly clear that Ca^{2+} and cAMP nearly always serve as interrelated or *synarchic* messengers (Rasmussen 1981). The evidence in support of this concept has been presented in detail elsewhere (Rasmussen 1981) and will not be reiterated here, but this evidence is compelling and leads to the conclusion that events in the Ca^{2+} messenger systems are intimately related to those in the cAMP system. These interrelationships include the following: (a) cAMP regulates Ca^{2+} metabolism, (b) Ca^{2+} regulates cAMP metabolism, (c) cAMP alters the sensitivity of response elements to Ca^{2+}, (d) cAMP and Ca^{2+}-dependent protein kinases catalyze the phosphorylation of some of the same proteins, and (e) Ca^{2+} and cAMP regulate similar types of cellular processes, e.g., secretion.

A general model of synarchic regulation is presented in Fig. 3, which incorporates in a schematic way the variety and multitude of interactions possible between components in the two different messenger systems. It is apparent that in any particular cell type not all possible interactions are expressed. Reference to this model also allows one to consider the fact that there are at least five recognizable patterns of synarchic regulation. In *coordinate* regulation, both limbs of the system are turned on by the same increase in extracellular messenger concentration activating the separate limbs. In a slight variation of this pattern, *hierarchical* control is achieved by having different concentrations of the same hormone, or of different hormones activating the separate limbs to bring about a synergistic cellular response. Two different hormones, acting on one of the limbs, can produce a *redundant* pattern if each alone can produce a maximal response. Conversely, if they produce opposing effects they can interact in an *antagonistic* manner. Finally, activation of one limb can lead to the activation of the

Fig. 3. A general model of synarchic regulation illustrating the facts that: (a) in most cells flow of information from cell surface to cell interior occurs via both the cAMP and Ca²⁺ messenger systems, (b) there are multiple sites of interaction between these two limbs of information flow, (c) changes in either Ca²⁺ or cAMP concentration may lead to the phosphorylation of the same protein substrate, and (d) there are reactions which are regulated by events in one but not the other messenger system. The pattern shown is a general one, and in a real cell not all these interactions are necessarily expressed. (Rasmussen 1981)

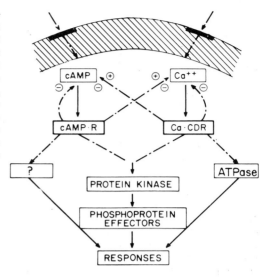

second limb as a result of the rise in messenger concentration in the first limb in a *sequential* expression of synarchic regulation.

In many cell types, it is possible to characterize the major operational events as exhibiting one of these five patterns of response, but quite often in addition to such a major theme, a supplemental minor theme displaying one of the other patterns is also evident. Aside from the merit of focusing attention upon the near universality of synarchic regulation, recognition and discussion of the patterns of synarchic regulation provide a useful introduction into the concepts of amplitude and sensitivity modulation of cell function (Rasmussen and Waisman 1982; Rasmussen 1983).

4 Amplitude and Sensitivity Modulation

In order to present these concepts, it is necessary to consider the nature of the association reactions between a messenger, Ca²⁺, a receptor protein, calmodulin, and a response element, phosphodiesterase (Fig. 4). In the generally accepted model of hormone action, the process of intracellular information flow can be characterized as exhibiting amplitude modulation. The effect of the hormone is to cause an increase in the amplitude (concentration) of the intracellular messenger, the concentration of Ca²⁺ in the cell cytosol. This rise in $[Ca^{2+}]c$ leads to the binding of 3-4 Ca²⁺ to a specific receptor protein (R), e.g., calmodulin (Cox et al. 1982; Klee 1977; Blumenthal and Stull 1980; Huang et al. 1981). The complexes thus formed, $Ca_n \cdot R$, interact with other proteins, response elements, enzymes like phosphodiesterase (Huang et al. 1981), to modulate their functions (Fig. 4).

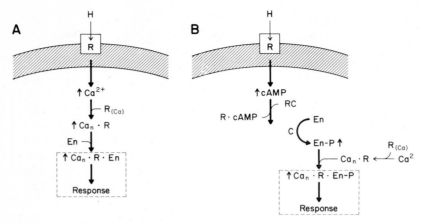

Fig. 4 A, B. The possible sequence of association reaction between Ca^{2+}, calmodulin (R), and an enzyme (En). **A** Amplitude modulation: the concentration of the activated form of the response element ($Ca_n \cdot R \cdot En$) can be increased by an increase in Ca^{2+} concentration of the cell cytosol. This leads to the formation of $Ca_n \cdot R$ which then associates with En to form the active complex, a change in the conformation of the enzyme brought about by its cAMP-dependent phosphorylation leads to an increase in the association constant of its binding to calmodulin [$R_{(Ca)}$]. **B** Sensitivity modulation: as a consequence the active form of the enzyme $Ca_n \cdot R \cdot En \text{-} P$ forms without a change in Ca^{2+} concentration in the cytosol. Nonetheless, Ca^{2+} is essential for the formation of the active forms because this form will not appear if the cytosol is depleted of Ca^{2+}, or by one of two types of sensitivity modulation. In the first type, a change in the conformation of the response element leads to an increase in the association constant of its binding to calmodulin. In the second, an increase in the concentration of either calmodulin or response element leads to an increase in formation of CM·RE and then $Ca_2 \cdot CM \cdot RE$ and finally $Ca_4 \cdot CM \cdot RE$

When such a sequence takes place in the calcium messenger system (Fig. 4A), the association reactions are ordered and cooperative so that full enzyme activation occurs over a range of $[Ca^{2+}]$ between 0.1 and 0.02 to 1 and 2 μM (Cox et al. 1982; Klee 1977; Blumenthal and Stull 1980; Huang et al. 1981). Parenthetically, when amplitude modulation occurs one of the response elements activated by Ca^{2+}-calmodulin is the calcium pump in the plasma membrane (Vincenzii et al. 1980; Waisman et al. 1981; Cox et al. 1982), so that the peak rise in $[Ca^{2+}]$ in the cytosol is brief, and soon falls to values only 1.5-2.5 times greater than basal, i.e., 0.3-0.5 μM (Charesi et al. 1983; Morgan and Morgan 1983a; Snowdowne and Borle 1984; Feinstein et al. 1983; Gershengorn and Thaw 1983; Pozzan et al. 1983). Nonetheless, cellular response continues at a high rate. One can view this situation in terms of activation calcium similar to activation energy in a chemical reaction. A sharp increase in $[Ca^{2+}]$ is necessary to switch the system on, but the system will remain on at a lower steady state $[Ca^{2+}]$.

Increasing the $[Ca^{2+}]$ in the cell cytosol is not the only way calcium-modulated response elements can be regulated. A second possible way is by the process of sensitivity modulation (Fig. 4B). By definition this is a circumstance in which there is a change in the activity of one or more of these calcium-dependent response elements under conditions where there is no change in the $[Ca^{2+}]_c$ (Rasmussen and Waisman 1982). Such a change in activity is brought about either by a change in the concentration of one of the other reactants, i.e., calmodulin or a specific response element, or by an alteration in the affinity of a response element for the various forms of calmodulin (Fig. 4B).

This second type of sensitivity modulation is of particular interest. To date most known examples are ones in which a change in response element structure brings about a change in affinity between this element and calmodulin (Rasmussen and Waisman 1982; Rasmussen 1983). A common mechanism by which sensitivity modulation of response element function in the calcium messenger system is achieved is one in which a rise in [cAMP] leads to the phosphorylation of a response element regulated by calcium-calmodulin. Two examples are phosphorylase b kinase (Walsh et al. 1980) which, when it undergoes a cAMP-dependent phosphorylation, becomes more sensitive to activation by Ca^{2+}, an example of positive sensitivity modulation; and myosin light chain kinase which, when it undergoes a cAMP-dependent phosphorylation, becomes less sensitive to activation by Ca^{2+} (Conti and Adelstein 1980), an example of negative sensitivity modulation.

The fact that cAMP-dependent phosphorylation of response elements in the calcium messenger system alters the function of these elements underscores, at the molecular level, the importance of the interactions of the Ca^{2+} and cAMP messenger systems, and provides a major validation of the synarchic messenger concept. However, sensitivity modulation can result from other kinds of interactions. Changes in response element association reactions can be brought about by: (a) a change in response element concentration, (b) a change in calmodulin concentration (Hanbauer 1980; Clayberger et al. 1981), or (c) a change in the concentration of an allosteric modifier. The calcium-activated, phospholipid-dependent protein kinase or C-kinase system is an example of a calcium-regulated enzyme which undergoes positive sensitivity modulation when it associates with diacylglycerol and membrane phospholipids (Takai 1979; Nishizuka 1983; Kishimoto et al. 1980; Takai et al. 1979a).

The C-kinase is a specific calcium-dependent protein kinase distinct from calmodulin-regulated, calcium-dependent protein kinases, and from the cAMP-dependent ones (Takai et al. 1981, 1979b). When isolated it is activated slightly by high concentrations $(10^{-4} - 10^{-5}$ M) of calcium, which means that it is virtually inactive in a nonactivated cell (Kishimoto et al. 1980; Takai et al. 1979a). However, when the cell is activated, diacylglycerol (DG) is generated in the membrane (Fig. 2), and the C-kinase then associates with the plasma membrane (Kraft et al. 1982). This association leads to a dramatic alteration in the Ca^{2+}-activation profile of the enzyme. The enzyme is now fully activated at 10^{-7} M Ca^{2+}, and its V_{max} is increased six- to ten fold. Thus, the enzyme is fully active at a calcium ion concentration thought to exist in the nonactivated cell. The enzyme in this state would be expected to be partially to fully active even at $[Ca^{2+}]_c$ similar to or slightly greater than those found in the nonactivated cell, and those found in the activated cell after the initial calcium transient.

5 Cellular Calcium Metabolism During Cell Activation

In order to appreciate the functioning of the calcium messenger system, it is necessary to understand the milieu in which it operates. The overriding fact of cellular existence is that excess intracellular Ca^{2+} leads to cell dysfunction and cell death (Schanne et al. 1979; Wrogemann and Pena 1981; Farber 1981; Fleckenstein 1983). Hence, one of the most basic cellular needs is the maintenance of cellular calcium homeostasis. This is largely achieved by a two component system of plasma and inner mitochondrial membrane (Rasmussen 1981; Rasmussen and Waisman 1982; Rasmussen 1983; Borle 1981; Kraft et al. 1982; Schanne et al. 1979). The plasma membrane is the critical first line of defense. A 10,000-fold Ca^{2+} concentration gradient exists across it $[Ca^{2+}]_e = 1000 \ \mu M$; $[Ca^{2+}]_c = 0.1 \ \mu M$), yet the maintenance of this enormous gradient is achieved by the expenditure of less than 1% of basal cellular energy utilization. This remarkable economy is brought about largely because the plasma membrane is only slightly permeable to Ca^{2+} (Borle 1981). On the other hand, there is a finite rate of calcium leak down this concentration gradient into the cell. This is pumped back out of the cell by one of two mechanisms: a Ca^{2+}:2 H^+-ATPase or calcium pump (Niggli et al. 1982; Smallwood et al. 1983), or a 3 Na^+-Ca^{2+} exchange driven by the normal inwardly directed Na^+ gradient which is maintained in turn by the Na^+-K^+-ATPase (Baker 1976; Blaustein 1974). It is possible that the Ca^{2+}:2 H^+-ATPase is responsible for the fine control of $[Ca^{2+}]_c$, and the 3 Na^+:Ca^{2+} exchange process is a less sensitive system which deals with larger calcium loads.

During cell activation, there is a four- to fivefold increase in the calcium permeability of the plasma membrane (Borle 1981). Also, any of a variety of injuries to the plasma membrane can lead to a significant increase in calcium entry. The mitochondria serve as an intracellular sink in which to sequester Ca^{2+} temporarily during times of calcium excess (Rasmussen 1981; Borle 1981; Akerman and Nicholls 1983; Carafoli and Crompton 1978; Hansford and Castro 1982). There is in the inner mitochondrial membranes a calcium uniport, and Ca^{2+} uptake is driven by the membrane potential normally existing across this membrane. Because the interior of the mitochondria has a high pH and a high inorganic phosphate concentration, $[HPO_4=]$, most of the calcium taken up by the mitochondria is stored in a biologically inactive but exchangeable pool as $(Ca)_3(HPO_4)_2$. The capacity to store calcium in this way is large but finite (Rasmussen 1981; Rasmussen and Waisman 1982; Rasmussen 1983; Borle 1981; Akerman and Nicholls 1983; Carafoli and Crompton 1978; Hansford and Castro 1982; Nicholls 1982). When this capacity is exceeded, the $[Ca^{2+}]_c$ rises to a point where it begins to exert toxic effects.

There is an additional pool of intracellular calcium: that in the plasma membrane and/or endoplasmic reticulum (Borle 1981). This pool serves as a source for activator calcium during the initial phase of cell activation (Rasmussen 1983). As a consequence of the release of Ca^{2+} from this pool and an increase in calcium entry into the cell, the $[Ca^{2+}]_c$ rises when a cell is activated by an extracellular messenger. In order for the $[Ca^{2+}]_c$ to rise from 0.1 to 1.0 μM, the total amount of extra Ca^{2+} released into and/or entering the cytosol from the extracellular space is 40-60 $\mu mol \ l^{-1}$ cell H_2O. This rather large discrepancy between the bolus of Ca^{2+} entering the cytosol and the small rise in $[Ca^{2+}]_c$ observed is a measure of both the "buffering" capacity of the mitochondria and the fact that the plasma membrane Ca^{2+} pump is activated immediately, via calcium-calmodulin, and increases the rate of Ca^{2+} efflux back out of the cell (Vincenzii et al. 1980; Waisman et al. 1981). Because of these two processes, calcium uptake by mitochondria and calcium efflux out of the cell, the $[Ca^{2+}]_c$ rises only transiently and then falls to values only 1.5 to 3-fold greater than the original basal values even though cellular response to the particular extracellular messenger is sustained (Charesi et al. 1983; Morgan and Morgan 1983a; Snowdowne and Borle 1984; Feinstein et al. 1983; Gershengorn and Thaw 1983; Pozzan et al. 1983; Tsein et al. 1982; O'Doherty et al. 1980).

These elaborate mechanisms for both regulating the $[Ca^{2+}]_c$ during the nonactivated phase of cell existence, and minimizing the changes in $[Ca^{2+}]_c$ during the activated phase of cellular response, are testaments to the essential need of the cell to continually protect itself from calcium intoxication. On the other hand, the very fact that, even in cells displaying a sustained response, the rise in $[Ca^{2+}]_c$ is transient, means that the cell has evolved a means of achieving gain control when employing the calcium messenger system to couple stimulus to response.

6 Temporal Integration of Cellular Response

One of the most interesting aspects of information flow in the calcium messenger system concerns the interactions between the two branches of this system: the calcium-calmodulin branch on the one hand, and the diacylglycerol-activated C-kinase branch on the other (Fig. 2). Kaibuchi et al. (1982) have shown that when platelets are activated by thrombin, there is a flow of information via both branches, and the effects elicited are integrated into a synergistic response. If one activates either branch maximally, only a submaximal cellular response is observed, but if both branches are activated, a maximal response is seen. These data imply that some flow of information in both branches is essential for maximal cellular response.

Our own recent experiments extend this concept and suggest that the true function of these two branches of the messenger pathway may be to achieve a temporal integration of cellular response, and provide one of the means for achieving gain control in the calcium messenger system (Kojima et al. 1983; Zawalich 1984).

In order to understand these experiments, it is necessary to understand that one can bypass receptor-mediated events and activate each branch of the calcium messenger system separately. The calcium ionophore, A23187, will increase the permeability of the plasma membrane (Reed and Lardy 1972; Prince et al. 1972). Its addition to the medium will cause the $[Ca^{2+}]_c$ to rise thereby activating the calmodulin-branch of the calcium messenger system. Addition of the phorbol ester, 12-O-tetra-decanolyl-phorbol-13-acetate (TPA) will activate the C-kinase directly (Castagna et al. 1982) and hence turn on the C-kinase branch without causing an increase in $[Ca^{2+}]_c$. Using this knowledge, the effects of these two agents, when added separately or together, upon the time courses of aldosterone secretion from adrenal glomerulosa cells and of insulin secretion from the beta cells of the Islets of Langerhans were examined (Kojima et al. 1983; Rasmussen et al. 1984).

In the case of the adrenal glomerulosa cell, the hormone angiotensin stimulates aldosterone secretion in a calcium-dependent manner (Fakunding and Catt 1980; Foster et al. 1981; Williams et al. 1981; Foster and Rasmussen 1983; Fraser et al. 1979). The hormone causes both an increase in Ca^{2+} uptake into the glomerulosa cells (Foster et al. 1981), and the release of Ca^{2+} from an intracellular pool (Kojima et al. 1984; Williams et al. 1981; Foster and Rasmussen 1983). The addition of angiotensin leads to a sustained increase in rate of aldosterone secretion (Fraser et al. 1979). Addition of A23187 also leads to a stimulation of aldosterone secretion (Fakunding and Catt 1980; Foster et al. 1981). However, this stimulation is relatively transient, and not sustained at the level seen when angiotensin is employed (Kojima et al. 1983). Conversely, addition of TPA causes a very slow and submaximal increase in aldosterone secretion (Kojima et al. 1983). However, addition of both TPA and A23187 together leads to a prompt and sustained increase in rate of aldosterone secretion qualitatively similar to that seen after angiotensin addition.

In the case of the beta cell, an increase in extracellular glucose concentration leads to a biphasic stimulation of insulin secretion: an initial transient peak followed by a slower increase to a sustained plateau (Malaisse 1973; Hedeskov 1980; Grodsky 1972). Addition of TPA induces onla a very slow rising and submaximal insulin secretory response (Zawalich et al. 1983). Addition of A23187 causes only a first phase of insulin release. Addition of A23187 and TPA together causes a secretory response qualitatively similar to that seen after an elevation of glucose concentration (Zawalich et al. 1983). Of particular interest in terms of defining the messengers involved in glucose action are more recent data showing that a combination of tolbutamide, which opens a plasma membrane calcium channel transient (Henquin 1980), forskolin, which activates adenylate cyclase (Seamon and Daly 1981), and TPA, which activates the C-kinase, can induce a biphasic pattern of insulin secretion qualitatively and quantitatively similar to that evoked by maximally stimulatory concentrations (16.7 mM) of glucose (Zawalich et al. 1984).

The biochemical basis of vascular smooth muscle contraction has been considered to involve the calcium-calmodulin-dependent activation of the enzyme myosin light chain kinase (Blumenthal and Stull 1980). The resulting phosphorylation of the myosin light chains leads to the interaction of actin with myosin, i.e., an increase in tension (Stull 1980; Kerrick et al. 1980). However, there are several lines of evidence which suggest that this model is too simple. First, when smooth muscle cells are preloaded with the protein aequorin, a calcium-sensitive phosphorescent protein, and then stimulated with either high K^+ or an agonist such as angiotensin II, the rise in the free $[Ca^{2+}]$ in the cytosol is transient even though contraction is sustained (Morgan and Morgan 1983b); second, when the degree of myosin light chain phosphorylation is followed after stimulation of smooth muscle, the extent of phosphorylation rises as contraction occurs, but then falls during the sustained phase of contraction (Aksoy et al. 1983); third addition of the phorbol ester, TPA, to smooth muscle causing a slowly developing, sustained, calcium-dependent contraction (Rasmussen et al. 1984). The rate and extent of contraction can be increased by exposure to A23187 or by repeated brief electrical stimuli, and TPA-induced contraction can be completely and rapidly reversed by addition of forskolin, an activator of adenylate cyclase (Rasmussen et al. 1984).

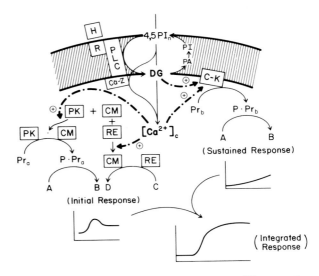

Fig. 5. Temporal integration of cellular response in the calcium messenger system. There are two branches in the calcium messenger pathway: The C-kinase (*C-K*) pathway on the *right*, and the Ca²⁺-calmodulin (*CM*) pathway on the *left*. When a hormone (*H*) interacts with its receptor (*R*), a phospholipase C (*PLC*) is activated leading to the hydrolysis of phosphatidylinositol 4,5-bisphosphate (4,5PIn). Associated with this event is a release of Ca²⁺ from an intracellular pool (not shown) and an increase in the rate of entry of Ca²⁺ into the cell cytosol. As a consequence, the calcium ion concentration, $[Ca^{2+}]_c$, rises. This leads to the amplitude modulation of a number of calcium-dependent enzymes. Some are activated directly by Ca²⁺ (not shown). Others are activated by calmodulin (*CM*). Calmodulin can alter the activity of response elements (*RE*) either by a direct allosteric activation of the enzyme (*C D*), or by activating a protein kinase (PK). This calmodulin-regulated protein kinase catalyzes the phosphorylation of a subset of cellular proteins (*Pr$_a$*). These phosphoproteins (P·Pr$_a$) regulate other cellular events (*A B*). The flow of information through this calmodulin branch leads to the initial cellular response. The activation of PLC also leads to an increase in the diacylglycerol content of the plasma membrane. This change in membrane structure leads to the binding of the C-kinase (C-K) to the membrane. The binding of C-kinase to diacylglycerol and phospholipids in the membrane brings about a positive sensitivity modulation of the calcium-activated C-kinase. This activated enzyme catalyzes the phosphorylation of a separate subset of cellular proteins (*Pr$_b$*). These phosphorylated proteins (P·Pr$_b$) regulate other cellular events (*A B*). The flow of information through this C-kinase branch is responsible for the maintenace of the sustained cellular response. *Small figures (inserts)* in the *lower part* of the figure depict the response of adrenal glomerulosa cells. The response to A23187 (a calcium-ionophore) is shown on the *left*. Addition of A23187 leads to an activation principally of the calmodulin branch, and a transient cellular response. The response to TPA (a phorbol ester) is shown on the *right*. Addition of TPA leads to an activation of the C-kinase branch, and a slowly developing and sustained but submaximal response. The response to combined TPA and A23187 is shown at the *bottom* (integrated response). Addition of the two compounds together leads to a prompt and sustained maximal response indistinguishable from that seen after the addition of the natural agonist, angiotensin II

The data in these three systems indicate that the calmodulin branch of the calcium messenger system is involved in initiating the cellular response, and the C-kinase branch in sustaining this response (Fig. 5). In other words, gain control is achieved in the calcium messenger system by having that system operate via two branches which have dif-

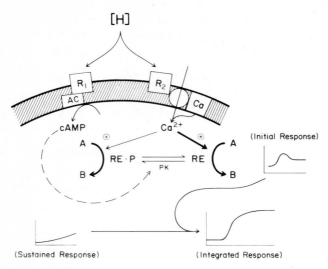

Fig. 6. An alternative means of achieving gain control and temporal integration of cellular response. In this system when the hormone (H) or intracellular messenger interacts with the cell it binds to two different surface receptors (R_1) and (R_2). The first, R_1, is coupled to adenylate cyclase. As a consequence, the [cAMP] of the cell cytosol increases. The second, R_2, is coupled to the calcium messenger system. As a consequence of its activation, the $[Ca^{2+}]_c$ rises. The rise in the $[Ca^{2+}]_c$ leads to the activation of a response element, *RE*, by the process of amplitude modulation and this event is responsible for the initial response (*right*). However, the $[Ca^{2+}]c$ soon falls so that this response would be expected to fall. This RE is a substrate for the cAMP-dependent protein kinase. When the [cAMP] increases, RE is phosphorylated so that it, *RE·P*, becomes more sensitive to activation by Ca^{2+}. As a consequence, response can be sustained (*left*) at a lower $[Ca^{2+}]$. The temporal integration of the flow of information in the two messenger systems leads to the integrated cellular response (*lower right*)

ferent properties: a calmodulin branch that is able to respond rapidly by amplitude modulation to sudden increases in cytosolic $[Ca^{2+}]$ and initiate a prompt but relatively brief response, and a C-kinase branch that is turned on by the simultaneous increase in $[Ca^{2+}]_c$ and in the diacylglycerol content of the plasma membrane, and then, because of its having undergone sensitivity modulation due to its (C-kinase) association with the plasma membrane, can operate at a $[Ca^{2+}]_c$ near or only slightly above the original basal $[Ca^{2+}]_c$.

Employment of the C-kinase pathways is not the only means by which gain control can be achieved in the calcium messenger system. In those circumstances in which the synarchic messenger, cAMP and Ca^{2+}, display either coordinate, hierarchical or a sequential pattern, a rise in cAMP concentration can, by the process of sensitivity modulation, increase the calcium sensitivity of a variety of response elements (Fig. 6).

An apparent example of this type is seen in the case of K^+-induced aldosterone secretion. When angiotensin II stimulates aldosterone secretion (see above), it activates the hydrolysis of polyphosphatidylinositides (unpublished) causes the mobilization of a dantrolene-sensitive pool of intracellular calcium (Kojima et al. 1984), but causes no change in cellular cAMP content (Fraser et al. 1979). Hence, it appears to act by causing a flow of information through both branches of the calcium messenger system. In contrast, a rise in K^+ from 4 to 8 mM causes no hydrolysis of polyphosphoinositides nor does it induce the mobilization of Ca^{2+} from an intracellular pool, hence it appears not to activate C-kinase, nonetheless it, as well as angiotensin II, causes a sustained increase in the rate of aldosterone secretion (Kojima et al. 1984). It also causes an increase in cellular cAMP content (Williams et al. 1981). It appears likely that this is sufficient to alter, by sensitivity modulation, the responsiveness of calcium-dependent response elements because the combination of A23187, a calcium ionophore which induces a transient aldosterone secretory response, and a small dose of forskolin, which by itself causes only a small increase in rate of aldosterone production, can induce a sustained increase in the rate of aldosterone secretion. Hence, in the same cell type, the adrenal glomerulosa cell, gain control in the calcium messenger system can be achieved by either of the mechanisms illustrated in Figs. 5 or 6.

7 Plasticity of Cellular Response

Reference to the six figures presented in the preceding sections of this discussion helps one to realize that, in spite of the seeming stereotyped processes by which an increase in either cAMP or Ca^{2+} is achieved, there is a remarkable plasticity in this synarchic messenger system. The fact that there are two major limbs of the information flow sequence (Fig. 3) which can interact in a coordinate, hierarchical, redundant, sequential, or anragonistic pattern already provides a considerable plasticity. Further, the dual means by which information flows from cell surface to cell interior in the calcium messenger system (Figs. 2 and 5) provides another element of plasticity. This is extended further by the multiple possible products of diacylglycerol hydrolysis, and the influences these products may have on cell function. Finally, the process of sensitivity

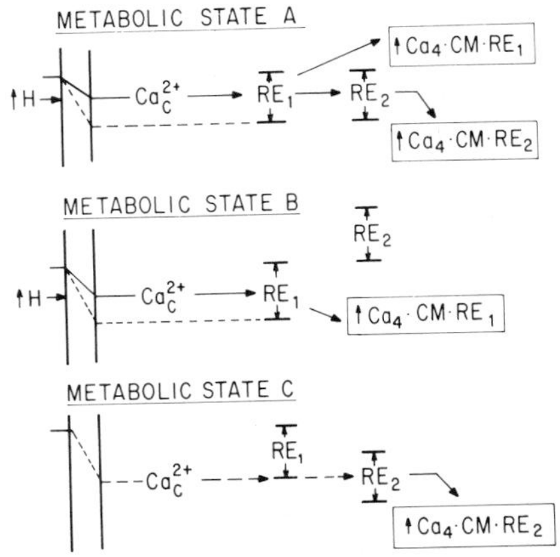

Fig. 7. The plasticity of metabolic control provided by the dual processes of amplitude and sensitivity modulation. In state A, a rise in Ca^{2+} concentration leads to the activation of two different response element, RE_1 and RE_2. In state B, a rise in Ca^{2+} causes an activation of only RE_1, because the sensitivity of RE_2 to Ca^{2+} is reduced, i.e., a higher Ca^{2+} concentration is needed to activate RE_2. In state C, the Ca^{2+} concentration does not change, but RE_2 is activated because its sensitivity to activation by Ca^{2+} has increased. There are at least two ways in which a state C type of positive sensitivity modulation can be produced. One is by activation of the C-kinase pathway by diacylglycerol. The other is by the cAMP-dependent phosphorylation of a response element in the calcium messenger system

modulation by messengers not directly involved in the calcium messenger pathway (Figs. 4 and 6) provides a means of achieving a fine control of events in either branch of the calcium messenger system by restricting or expanding the set of response elements whose function is influenced by any change in intracellular messenger concentration (Fig. 7).

8 Conclusion

As our knowledge of the roles of cAMP and Ca^{2+} as intracellular messengers has grown, our concepts as to how they regulate cell function have expanded. It is now clear that they do not usually operate independently but as dependent or synarchic messengers in the control of cell function (Rasmussen 1981). Furthermore, in the case of the calcium messenger system, a third synarchic messenger, diacylglycerol, has been identified. Along with this identification has come the recognition that events in these messenger systems are determined not only by the process of amplitude modulation, but also by the process of sensitivity modulation. All of this knowledge has led, in turn, to the recognition that the operational possibilities within this cellular control device are enormous, and provide the basis for the highly plastic character or cellular responses to multiple extracellular stimuli.

Of particular note are the recent data which show that amplitude modulation via the calmodulin branch of the calcium messenger system is responsible for the initiation of cellular response in cells displaying a sustained response to extracellular messenger addition, and sensitivity modulation via the C-kinase branch of the calcium messenger system is responsible for sustaining this cellular response.

The most important unresolved issue is whether or not extracellular messenger can interact with surface receptors in some cells in such a way that only a single branch of the calcium messenger system becomes activated. For example, if the C-kinase pathway alone were activated in some cells, then one would expect a slowly developing but sustained cellular response associated with no initial rise in the $[Ca^{2+}]_c$. Nevertheless, the cellular events underlying this response would be dependent upon intracellular Ca^{2+} because the C-kinase is a calcium-activated enzyme.

Acknowledgment. Supported by a grant from the National Institues of Health (Am 19813)

References

Akerman KEO and Nicholls DG (1983) Physiological and bioenergetic aspects of mitochondrial calcium transport. Rev Physiol Biochem Pharmacol 18149-201

Aksoy MO, Mras S, Kamm KE and Murphy RA (1983) Ca^{2+}, cAMP and change in myosin phosphorylation during contraction of smooth muscle. Am J Physiol 245 (Cell Physiol 14): C255-C270

Baker PF (1976) The regulation of intracellular Ca²⁺. SEB Symposium 30:67-88

Blaustein MP (1974) The interrelationship between sodium and calcium fluxes across cell membranes. Rev Physiol Biochem Pharmacol 70:33-82

Blumenthal DK and Stull JT (1980) Activation of skeletal muscle myosin light chain kinase by calcium and calmodulin. Biochemistry 19:5608-5614

Borle AB (1981) Control, modulation and regulation of cell calcium. Rev Physiol Biochem Pharmacol 90:13-169

Carafoli E and Crompton M (1978) The regulation of intracellular calcium by mitochondria. Ann NY Acad Sci 307:209-284

Castagna M, Takai Y, Kaibuchi K, Sano K, Kikkawa U and Nishizuka Y (1982) Direct activation of calcium-activated, phospholipid-dependent protein kinase by tumor-promoting phorbol esters. J Biol Chem 257:7847-7851

Charesi R, Blackmore PF, Berthon B and Exton JH (1983) Changes in free cytosolic Ca²⁺ in hepatocytes following β-adrenergic stimulation. J Biol Chem 258:8769-8773

Cheung WY (1980) Calmodulin plays a pivotal role in cellular regulation. Science (Wash DC) 207:19-27

Clayberger C, Goodman DBP and Rasmussen H (1981) Regulation of cyclic AMP metabolism in the rat erythrocyte during chronic β-adrenergic stimulation. J Membr Biol 58:191-201

Conti MA and Adelstein RS (1980) Phosphorylation by cyclic adenosine 3':5'-monophosphate-dependent protein kinase regulates myosin light chain kinase. Fed Proc 39:1569-1573

Cox JA, Comte M and Stein EA (1982) Activation of human erythrocyte Ca²⁺-dependent Mg²⁺-activated ATPase by calmodulin and calcium. Quantitative analysis. Proc Natl Acad Sci USA 79:4265-4269

Downes P and Michell RH (1982) Phosphatidylinositol 4-phosphate and phosphatidylinositol 4,5-bisphosphate: lipids in search of a function. Cell Calcium 3:467-502

Fakunding JL and Catt K (1980) Dependence of aldosterone stimulation in adrenal glomerulosa on calcium uptake: effects of lanthanum and verapamil. Endocrinology 107:1345-1353

Farber JL (1981) The role of calcium in cell death. Life Sci 29:1289

Feinstein MB, Egan JT, Sha'afi RI and White J (1983) The cytoplasmic concentration of free calcium in platelets is controlled by stimulators of cyclic AMP production (PGD₂, PGE, and Forskolin). Biochem Biophys Res Commun 113:598-604

Fleckenstein A (1983) Calcium antagonism in heart and smooth muscle. Wiley, New York

Foreman JC, Mongar JL and Gomperts BP (1973) Calcium ionophore and movement of calcium ions following the physiological stimulus to a secretory process. Nature (Lond) 245:249-251

Foster R, Lobo MV, Rasmussen H and Marusic ET (1981) Calcium: its role in the mechanism of action of angiotensin II and potassium in aldosterone production. Endocrinology 109:2196-2201

Foster R and Rasmussen H (1983) Angiotensin-mediated calcium efflux from adrenal glomerulosa cells. Am J Physiol 245 (Endocrinol Metab 8):E281-E288

Fraser R, Brown JJ, Lever AF, Mason PA and Robertson JIS (1979) Control of aldosterone secretion. Clin Sci (Lond) 56:389-399

Gershengorn MC and Thaw C (1983) Calcium influx is not required for TRH to elevate free cytoplasmic calcium in GH₃ cell. Endocrinology 113:1522-1524

Greengard P (1978) Cyclic nucleotides, phosphorylated proteins and neuronal function. Raven, New York

Grodsky GM (1972) A threshold distribution hypothesis for packet storage of insulin II. Effect of calcium. Diabetes 21 (Suppl):584-593

Hanbauer I, Pradhan S and Yang HYT (1980) Role of calmodulin in dopaminergic transmission. Ann NY Acad Sci 356:292-303

Hansford RF and Castro F (1982) Intramitochondrial and extramitochondrial free calcium ion concentration with very low, plausibly physiological, contents of total calcium. J Bioenerg and Biomembr 14:361-376

Hansford RG and Chappell JB (1967) The effect of Ca²⁺ on the oxidation of glycerophosphate by blowfly flight mitochondria. Biochem Biophys Res Commun 27:686-692

Hedeskov EJ (1980) Mechanism of glucose-induced insulin secretion. Physiol Rev 60:442-509

Henquin JC (1980) Tolbutamide stimulation and inhibition of insulin release: studies of the under-lying ionic mechanism in isolated rat islets. Diabetologia 18:151-160

Huang CY, Chau V, Chock PB, Wang JH and Sharma RK (1981) Mechanism of activation of cyclic nucleotides phosphodiesterase: requirement of the binding of four Ca^{2+} to calmodulin for activation. Proc Natl Acad Sci USA 78:871-874

Kaibuchi K, Sano K, Hoshijima M, Takai Y and Nishizuka Y (1982) Phosphatidylinositol turnover in platelet activation; calcium mobilization and protein phosphorylation. Cell Calcium 3:323-335

Kerrick WGL, Hoar PE and Cassidy PS (1980) Calcium-activated tension: the role of myosin light chain phosphorylation. Fed Proc 39:1558-1563

Kishimoto A, Takai Y, Mori T, Kikkawa U and Nishizuka Y (1980) Activation of calcium and phospholipid-dependent protein kinase by diacylglycerol, its relation to phosphatidylinositol turnover. J Biol Chem 255:2273-2276

Klee CB (1977) Conformational transition accompanying the binding of Ca^{2+} to the protein acti-vator of 3'-5'-cyclic adenosine monophosphate phosphodiesterase. Biochemistry 16:1017-1024

Kojima I, Lippes H, Kojima K and Rasmussen H (1983) Aldosterone production: effect of A23187 and TPA. Biochem Biophys Res Commun 116:555-562

Kojima I, Kojima K and Rasmussen H (1984) The effects of angiotensin II and K^+ on calcium efflux and aldosterone production in adrenal glomerulosa cells. Am J Physiol (in press)

Kraft AS, Anderson WB, Cooper HL and Sando JJ (1982) Decrease in cytosolic calcium/phos-pholipid-dependent protein kinase activity following phorbol ester treatment of EL4 thymoma cells. J Biol Chem 257:3193-13196

Malaisse WJ (1973) Insulin secretion: multifactorial regulation for a single process of release. Diabetalogia 9:167-173

Means AR and Dedman JR (1980) Calmodulin—an intracellular calcium receptor. Nature (Lond) 285:73-77

Michell RH (1975) Inositol phospholipids and cell surface receptor function. Biochem Biophys Acta 415:81-147

Morgan JP and Morgan KG (1983a) Vascular smooth muslce: the first recorded Ca^{2+} transient. Pflügers Arch Eur J Physiol 395:75-77

Morgan JP and Morgan KG (1983b) Vascular smooth muscle: the first recorded transients. Pflügers Arch Eur J Physiol 395:75-77

Nicholls DG (1982) Bioenergetic. Academic, London

Niggli V, Sigel E and Carafoli E (1982) The purified Ca^{2+} pump of human erythrocyte membranes catalyzes an electroneutral Ca^{2+}-H^+ exchange in reconstituted liposomal systems. J Biol Chem 257:2350-2356

Nishizuka Y (1983) A receptor-linked cascade of phospholipid turnover in hormone action. In: Shizume K, Imura H and Shimizu N (eds) Endocrinology. International Congress Series 598, Excerpta Medica, Amsterdam, pp 15-24

O'Doherty J, Youmans SJ, Armstrong W McD and Stark RJ (1980) Calcium regulation during stimulus-secretion coupling. Continuous measurement of intracellular calcium activities. Science (Wash DC) 209:510-513

Pozzan T, Lew DP, Wollheim OB and Tsein RY (1983) Is cytosolic ionized calcium regulating neutrophil activation? Science (Wash DC) 221:1413-1415

Prince WT, Berridge MJ and Rasmussen H (1972) Role of calcium and adenosine 3'-5'-cyclic monophosphate in controlling fly salivary gland secretion. Proc Natl Acad Sci 69:553-557

Rasmussen H (1981) Calcium and cAMP as synarchic messengers. Wiley, New York

Rasmussen H (1983) Pathways of amplitude and sensitivity modulation in the calcium messenger system. In: Cheung WY (ed) Calcium and cell function, vol IV. Academic, New York, pp 2-61

Rasmussen H and Waisman DM (1982) Modulation of cell function in the calcium messenger system. Rev Physiol Biochem Pharmacol 95:111-148

Rasmussen H, Forder J, Kojima I and Scriabine A (1984) TPA-induced contraction of isolated rabbit vascular smooth muscle. Biochem Biophys Res Commun 122: 776-784

Reed PW and Lardly HA (1972) A23187; a divalent cation ionophore. J Biol Chem 247:6970-6984

Robison GA, Butcher RW and Sutherland EW (1971) Cyclic AMP. Academic, New York

Schanne FAX, Kane, AB, Young EE and Farber JL (1979) Calcium-dependence of toxic cell death: a final common pathway. Science (Wash DC) 700-702

Seamon KB and Daly JW (1981) Forskolin: a unique diterpene activator of cyclic AMP generating systems. J Cyclic Nucleotide Res 7:201-224

Smallwood J, Waisman DM, Lafreniere D and Rasmussen H (1983) Evidence that the erythrocyte calcium pump catalyzes a $Ca^{2+}:nH^+$ exchange. J Biol Chem 258:11092-11097

Snowdowne KW and Borle AB (1984) Changes in cytosolic ionized calcium induced by activators of secretion in GH₃ cells. Am J Physiol 246 (Endocrinol Metab 9): E198-E201

Streb H, Irvine RF, Berridge MJ, Schulz I (1983) Release of Ca^{2+} from a non-mitochondrial store of pancreatic acinar cell by inositol-1,4,5-triphosphate. Nature (Lond) 306:67-69

Stull JT (1980) Phosphorylation of contractile proteins in relation to muscle function. Adv Cyclic Nucleotides Res 15:39-93

Synder GP, Capdevila J, Chacos N, Manna S and Falck JR (1983) Action of luteinizing hormone-releasing hormone: involvement of novel arachidonic acid metabolites. Proc Natl Acad Sci USA 90:3504-3507

Takai Y, Kishimoto A, Iwasa Y, Kawahara Y, Mori T and Nishizuka Y (1979a) Calcium-dependent activation of a multifunctional protein kinase by membrane phospholipids. J Biol Chem 254:3692-3695

Takai Y, Kishimoto A, Kikkawa U, Mori T and Nishizuka Y (1979b) Unsaturated diacylglycerol as a possible messenger for the activation of calcium-activated, phospholipid-dependent protein kinase system. Biochem Biophys Res Commun 91:1218-1224

Takai Y, Kishimoto A, Kawahara Y et al. (1981) Calcium and phosphatidylinositol turnover as signalling for transmembrane control of protein phosphorylation. Adv Cyclic Nucleotide Res 14:301-313

Tsien RW (1977) Cyclic AMP and contractile activity in the heart. Adv Cyclic Nucleotide Res 8:363-411

Tsein RY, Pozzan T and Rink TJ (1982) Calcium homeostasis in the intact lymphocytes: cytoplasmic free calcium monitored with a new intracellular trapped fluorescent indicator. J Cell Biol 94:325-334

Vincenzii FF, Hinds TR and Raess BV (1980) Calmodulin and the plasma membrane calcium pump. Ann NY Acad Sci 256:233-244

Waisman DM, Gimble JM, Goodman DBP and Rasmussen H (1981) Studies of the Ca^{2+} transport mechanism of human erythrocyte inside-out plasma membrane vesicles. I: Regulation of the Ca^{2+} pump by calmodulin. J Biol Chem 256:409-414

Walsh KX, Nallikin DM, Schendler KK and Reimann EM (1980) Stimulation of phosphorylase b kinase by the calcium-dependent regulator. J Biol Chem 255:5036-5042

Williams BB, McDougall JG, Tait JF and Tait SAS (1981) Calcium efflux and steroid output from superfused rat adrenal cells: effects of potassium, adrenocorticotrope hormone, 5-hydroxytryptamine, adenosine 3'-5'-cyclic monophosphate and angiotensin II, III. Clin Sci (Lond) 61:541-551

Wrogemann K and Pena SDJ (1981) Mitochondria calcium overload: a general mechanism for cell necrosis in muscle disease. Lancet 1:672-674

Zawalich W, Brown C and Rasmussen H (1983) Insulin secretion: combined effect of phorbol ester and A23187. Biochem Biophys Res Commun 117:448-455

Zawalich W, Zawalich K and Rasmussen H (1984) Insulin secretion: combined tolbutamide, forskolin and TPA mimic action of glucose. Cell Calcium. In Press

Calcium Extrusion Across the Plasma Membrane by the Calcium-Pump and the Ca^{2+}-Na$^+$ Exchange System

H. J. SCHATZMANN[1]

CONTENTS

---

[1] Department of Veterinary Pharmacology, University of Bern, Switzerland

1 Introduction

There is agreement that living cells keep the intracellular Ca^{2+} concentration far below that existing beyond the plasma membrane. Measuring the exceedingly low cytosolic free Ca^{2+} concentration is not an easy matter in cells surrounded by an extracellular fluid of 1 to 2 mM Ca^{2+} concentration and containing intracellular storage organelles and cation-binding structures that bristle with calcium. In the rather rare instances where reliable estimates exist, the unprepared observer finds reason to marvel: the figure is in the 10 nM range in nerve, muscle, myocardium, and human red cells. The elementary fact that cells maintain the cytosolic Ca^{2+} concentration so far below the external concentration (1.3 mM in the extracellular space of land-living mammals and 10 mM in seawater) is pertinent to most of the topics discussed in this volume. It is at the basis of the ability of cells to use calcium as a mediating link between various stimuli at the cell surface and intracellular events known to be the physiological consequences of such stimuli. The signaling by calcium is invariably achieved by an increase of the passive permeability across specific, gated channels in the plasma membrane or the membrane of intracellular organelles that contain calcium at high concentration, such as the sarcoplasmic reticulum. Not only has the calcium entering the cytosol during such activation to be removed, but there is in all cells a resting net influx of calcium that must be balanced by an equal efflux if the Ca^{2+} gradient across the plasma membrane is to be maintained. Storage organelles can remove some calcium at short notice, but it goes without saying that in the long run only calcium-exporting systems will keep the cytosol free of Ca^{2+}. In parenthesis, the reader may be reminded that the plasma membrane calcium transport also serves the purpose of transcellular uphill calcium translocation, certainly in epithelial structures and possibly in bone cells.

The two recognized mechanisms that extrude calcium from the cytosol into the surrounding medium are the Na-Ca-exchange system and the ATP-fueled calcium pump of the plasma membrane. The exchange system seems widespread in animal cells and the Ca-pump is present not only in animal cells, but in plant and prokaryotic cells as well. Since many recent reviews (for Na-Ca exchange Baker 1972; Blaustein 1974, 1977; Blaustein and Nelson 1982; Mullins 1981; Requena and Mullins 1979; Reuter 1982; for the Ca-pump Carafoli and Zurini 1982; Roufogalis 1979; Sarkadi 1980; Schatzmann 1982a, b, 1983; Schatzmann et al. 1982; general Borle 1981; Martonosi 1983; Sulakhe and St. Louis 1980) cover the field in all its details, the following narration attempts rather to convey the essence of the matter (not without adding a few new findings) than to be infallible and complete. For clarity's sake it is advisable to take cognizance of the two systems separately before making an attempt to understand why and how they cooperate in those cells where they are present together.

2 The Calcium Pump

The plasma membrane Ca-pump is an integral membrane protein of the molecular weight $\sim$ 140,000 that consumes ATP and uses the energy gained to do osmotic and electric work on Ca^{2+} ions. The maximal gradient against which the system can operate, in an idealized membrane without leaks (including slip within the pump itself), i.e., at zero net transport, is obtained by equating the work done [Eq (1)] with the free energy yielded by hydrolysis of the terminal P-0-P-bond in ATP ($\triangle$ G) [Eq (2)].

$$\text{work (J mol}^{-1}) = nRT \cdot \ln \frac{[Ca^{2+}]_o}{[Ca^{2+}]_i} + nzF \cdot V_m \tag{1}$$

$$\triangle G \text{ (J mol}^{-1}) = \triangle G° + RT \cdot \ln \frac{[ATP]}{[ADP] \cdot [PO_4]} \tag{2}$$

with z = charge of Ca^{2+} = 2, n = number of g-atoms of Ca^{2+} per mole ATP, V_m = membrane potential in volt, R = 8.31 J mol^{-1}, F = 9.65×10^4 C mol^{-1}, T = absolute temperature (K), $\triangle G°$ = standard free energy change of ATP hydrolysis in a physiological medium (30.5 kJ mol^{-1}).

Equation (2) yields, for example, with the physiological concentration in human red cells for [ATP] = 0.0015, [ADP] = 0.0003, [PO_4] = 0.0003 (M), 5.6×10^4 J mol^{-1}, and therefore from Eq. (1) with V_m = 0.01 V, $[Ca^{2+}]_o$ = 0.0013 M and T = 310 K: $[Ca^{2+}]_i^{min}$ = 5.3×10^{-8} M (for n = 2) or $[Ca^{2+}]_i^{min}$ = 1×10^{-12} M (for n = 1) if the system were fully coupled.

This system was first found in human red cells (Schatzmann 1966; Lee and Shin 1969; Olson and Cazort 1969; Schatzmann and Vincenzi 1969). They are a convenient object for its study because they do not possess the Na-Ca-exchange system (Ferreira and Lew 1977) and because 99% of their exchangeable Ca^{2+} is contained within one single compartment (Simonson et al. 1982).

2.1 The Human Red Cell System

2.1.1 The Free Ca^{2+} Concentration in Red Cells

Human red cells contain about 10 μmol calcium per liter cells (Harrison and Long 1968), most of which is associated with what is called the membrane of the cells (which includes not only the unit membrane but also the cytoskeleton).

The free Ca^{2+} concentration in the cytosol was first estimated by Simons by an indirect method (Simons 1976). Like many other cells, the red cell possesses a Ca^{2+}-sensitive potassium channel which in viable cells is definitely closed. Simons introduced calcium buffers with Ca^{2+} concentrations ranging from 10^{-8} to 10^{-4} M and observed that a measurable increase in K flux occurred at 2.5×10^{-7} M, which means that the physiological $[Ca^{2+}]_i$ must lie below this value. Later the author presented measurements obtained by a null method (Simons 1982). Cells were suspended in solutions of different $[Ca^{2+}]$ (with low concentration of Ca-buffers) monitored with a Ca^{2+}-selective electrode. Upon lysis of the cells with digitonin, the electrode gave a positive or negative signal. At zero deflection $[Ca^{2+}]_i$ was assumed to equal $[Ca^{2+}]_o$. The result was 4×10^{-7} M. Both figures are upper estimates, for obvious reasons in the first experiment and because Ca^{2+} release from internal Ca-buffering systems might occur when the cells are lysed in the second experiment.

Recently Lew et al. (1982) used benz 2 to measure $[Ca^{2+}]_i$ in intact, metabolizing red cells. Benz 2 is a Ca-chelator that has an apparent K_{Ca} of 54 nM inside red cells (Lew et al. 1982) and, like quin 2 (Tsien 1981; Tsien et al. 1982a, b), can be introduced as acetoxymethyl ester into intact cells. The ester is enzymatically hydrolyzed inside cells. Since the free chelator does not pass the plasma membrane, one can trap any desired a-mount of it inside cells by this mechanism. The authors used the chelator to set $[^{45}Ca^{2+}]_i$ at nmolar concentrations while having easily measurable μmolar total $[^{45}Ca]_i$ and found that in a medium of 1.3 mM Ca^{2+} the cells maintained a $[Ca^{2+}]_i$ of $\ll 3 \times 10^{-8}$ M. The chelator-ester liberates formaldehyde, which poisons glycolysis, but in these experiments the $[ATP]_i$ remained above 0.4 mM (Tiffert et al. 1984).

2.1.2 The Leak

Whereas earlier experiments with metabolically depleted cells yielded a Ca^{2+}-influx (from 1 mM Ca^{2+} solutions) of 10 μmol l^{-1} h^{-1} or less (Ferreira and Lew 1977), a recent study by Lew et al. in intact, metabolizing cells assessed the steady state flux at $[Ca^{2+}]_o = 1.3$ mM to be 40-50 μmol l^{-1} h^{-1} (Lew et al. 1982). This value is very similar to what was found earlier from net calcium uptake into starved but unpoisoned cells suspended in blood plasma (Schatzmann 1969). Lew et al. used benz 2 which, unlike quin 2, does not increase Ca-permeability (Tiffert et al. 1984).

At the present stage we therefore postulate that the Ca-pump must be able to transport some 50 μmol l^{-1} h^{-1} against a gradient of $\gg 10,000$ at a $[Ca^{2+}]_i < 10^{-7}$ M.

2.1.3 Biochemistry of the Red Cell Calcium Pump

Recently the Ca-pump protein has been isolated from the rest of the membrane proteins by an elegant method. The system is activated by Ca-calmodulin which binds with a K_{diss} of < 5 nM to the pump. Niggli et al. (Carafoli and Niggli 1981; Niggli et al. 1979) and Gietzen et al. (1980a) took advantage of this binding to isolate the protein. The method consists of the following: calmodulin is attached to 4B-sepharose from which a column is formed. Hemoglobin-free red cell membranes, depleted of calmodulin, are solubilized in Triton-X-100, for example, and applied in a phospholipid suspension containing free Ca^{2+} to this column. Washing with Ca^{2+} + lipid-containing solution removes all other proteins. When the effluent is protein-free, Ca^{2+} in the buffer is replaced by EGTA whereupon the pump protein dissociates from calmodulin and is eluted in more than 95% pure form as a protein-lipid complex. The preparation is very unstable (half-life a few hours at 4°C). This can be improved by adding Ca^{2+} and calmodulin ($t_{1/2}$ a few days at 4°C). At -190°C the protein is quite stable for weeks. The protein yield in the EGTA peak is about 0.1% of the total membrane proteins. The fraction of intact functional pump protein in the total amount recovered can be estimated as follows. The protein cleaves ATP under transient phosphorylation (Knauf et al. 1972, 1974; Rega and Garrahan 1975; Schatzmann and Bürgin 1978) at an acyl group (Lichtner and Wolf 1980) (the bond is acid stable and can be decomposed by hydroxylamine). By blocking dephosphorylation with lanthanum (La^{3+}) (see below) and making $[^{32}P]$-ATP saturating the amount of ^{32}P-intermediate becomes equal to the total functional protein. From such experiments with the purified protein it can be inferred that less than 10% of the yield from the column (method of Gietzen et al.) is in a functional state (Luterbacher 1982).

It is uncertain whether the protein is present within the membrane as a monomer, a dimer or an oligomer. Experiments in which azidocalmodulin was coupled to the pump are in favour of a dimer (Hinds et al. 1982). Molecular weight determination by radiation inactivation in calmodulin-depleted membranes yielded a figure of 254,000 (Cavieres 1983, 1984) or 290,000 (Minocherhomjee et al. 1983) and suggested that in the absence of calmodulin part of this functional dimer forms higher aggregates that are inert (Cavieres 1983, 1984).

2.1.4 Physiological Properties of the Red Cell Calcium Pump

The existence of the pump is not inferred from the leak and the $[Ca^{2+}]_i$ alone. It can be demonstrated by loading resealed red cells with Ca^{2+} (Schatzmann and Vincenzi 1969) or by introducing Ca^{2+} into intact cells by aid of a Ca^{2+}-ionophore (e.g., A23187) (Ferreira and Lew 1976; Sarkadi et al. 1976) or by incubation in a Na-salicylate medium (Bürgin and Schatzmann 1979; Wiley and Gill 1976). Further, inside-out vesicles made from red cells transport Ca^{2+} from the medium to the lumen if ATP is added to the medium (Sarkadi et al. 1980b), as shown in Fig. 1A. Finally, the isolated protein can be incorporated into artificial lipid vesicles (liposomes) and shown to transport calcium uphill (Carafoli and Niggli 1981; Gietzen et al. 1980b; Haaker and Racker 1979; Stieger 1982) (see Fig. 1B).

The purified protein has the properties of the $(Ca^{2+} + Mg^{2+})$-requiring ATPase, first discovered in red cell membranes by Dunham and Glynn (1961) and Wins and Schoffeniels (1966), that tally with those of Ca^{2+} transport.

The system transports Ca^{2+} at a maximal rate of 10-20 mmol 1^{-1} cells h^{-1} (Romero 1981; Schatzmann 1973) and decomposes ATP at a maximal rate of 2-4 μmol mg^{-1} membrane protein h^{-1} (Schatzmann and Roelofsen 1977). The two rates are equal if it is assumed that red cells contain 0.5% membrane protein and that the ratio Ca/ATP = 1. As the pump protein amounts to about 0.1% of the membrane proteins (Luterbacher 1982; Verma et al. 1982), the specific activity is of the order of 30 μmol mg^{-1} min^{-1} = 4000 mol/mol min^{-1}. From phosphorylation in the presence of La^{3+} the total enzyme in membranes was found to be 10^{-11} mol mg^{-1} membrane protein (Luterbacher 1982) or (with 0.5% membrane protein in red cells) 5×10^{-14} mol mg^{-1} red cells. With a volume of the erythrocyte of 90 μ^3 there are 1.1×10^7 erythrocytes per mm^{-3} packed cells and consequently 4.55×10^{-21} mol per cell. Multiplying this with Avogadro's number (6×10^{23}) yields 2700 molecules per cell. Thus there are a few thousand pumps per cell with a maximal turnover of several thousand transport cycles per minute.

The properties of the calmodulin-replete system (for the role of calmodulin see below) can be described by giving an apparent K_{Ca} of $\sim$ 1 μM (at pH 7.0) which at higher pH decreases (0.2 μM at pH 7.8) (Luterbacher, unpublished) and an apparent K_{Mg} of $\sim$ 15 μM (Schatzmann 1983). Except for Sr^{2+} no other cation has been shown to be

Fig. 1. A ATP-dependent ^{45}Ca-uptake by the Ca-pump into inside-out vesicles prepared from human red blood cells. Temp. 37°C, 0.6 mg protein ml^{-1}. Medium: (mM) Tris-Cl 20 (pH 7.0), KCl 140, MgCl$_2$ 2, CaCl$_2$ 0.2 (5x10^5 cpm ml^{-1} ^{45}Ca), tris-ATP 1. At 22 min 5 mM EGTA or 10 μM A 23187 was added to the medium. The release or the accumulated Ca^{2+} by the ionophore shows that the transport was into the water space of the vesicles and was steeply uphill. The EGTA experiment shows that the vesicles were fairly tight to Ca^{2+} (Luterbacher 1982). B ATP-dependent uptake of ^{45}Ca into liposomes into which the purified Ca-pump protein was incorporated. Liposomes of mixed soybean phospholipids (Sigma P 5638) were formed in (mM) K-MOPS 20 (pH 7.4), KCl 300, MgCl$_2$ 2, DTT 1. Medium: (mM) Imidazole 20 (pH 7.4), KCl 130, MgCl$_2$ 2, CaCl$_2$ 0.05 (5·10^5 cpm ml^{-1} ^{45}Ca). No oxalate. Start by adding 0.5 mM ATP. • with ATP; o without ATP. At 5 min EGTA (2 mM) was added. The drop in vesicular Ca^{2+} shows that the vesicles were quite leaky to Ca^{2+}. Addition of 1 μM A 23187 at 15 min shows that Ca^{2+} was accumulated into the vesicular water space against a considerable gradient. Without ATP there was no accumulation of Ca^{2+}. (Stieger 1982)

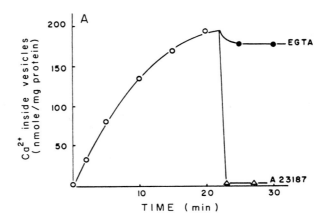

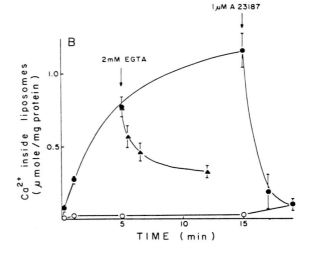

transported by the system. The Hill coefficient for Ca^{2+} is 1.3 (Schatzmann and Roelofsen 1977) -1.8 (Scharff 1978) and Ferreira and Lew have shown that 2 Ca^{2+} ions participate in the reaction cycle (Ferreira and Lew 1976), whereas there is no indication that more than one Mg^{2+} ion is involved. There are two sites for ATP with vastly differing affinity (Muallem and Karlish 1979; Richards et al. 1978; Stieger and Luterbacher 1981a). The high affinity site (K$_m$ ∼ 1-3 μM) is the enzymic site proper, because the phosphorylation reaction (see below) also displays high affinity (Luterbacher 1982; Muallem and Karlish 1980; Richards et al. 1978), and the low affinity site (K ∼ 100-300

μM) mediates a stimulatory activity of ATP which is also explained below. For phosphorylation Mg^{2+} is not required, which means that Mg-ATP is not the unique substrate, although it normally seems to be the substrate (Enyedi et al. 1982b; Penniston 1982). On the other hand, Muallem and Karlish (1980) have made it probable that, at least at $37°C$ (in the presence of calmodulin), the low affinity site requires Mg-ATP rather than free ATP. This may have obfuscated the arguments in favor of the role of Mg-ATP at the enzymic site (Enyedi et al. 1982b; Graf and Penniston 1981; Sarkadi et al. 1981).

Ca-transport is sharply discriminating potential substrates, GTP, ITP, CTP, or UTP eliciting very much less activity than ATP (Sarkadi 1980) and acylphosphates, pyrophosphate, and nitrophenylphosphate (pNPP) none (Caride et al. 1983; Olson and Cazort 1969). However, the protein is a Ca^{2+} (and K^+) stimulated pNPP-ase (Caride et al. 1982; Rega et al. 1974) if ATP is present. There is competition between ATP and pNPP (Caride et al. 1982; Luterbacher 1982) and consequently pNPP inhibits Ca^{2+} transport (Rega et al. 1973). Caride et al. (1982) have made it probable that it is the low affinity site which is responsible for pNPP-hydrolysis when the high affinity site is occupied by ATP. Interestingly, phospholipase C treatment that destroys ATPase activity enhances phosphatase activity and renders it more sensitive to Ca^{2+} (Richards et al. 1977).

Transport- and ATPase-activity in Na^+ or K^+ containing solutions are some 30-50% higher than in choline or tris (or even sucrose) media (Sarkadi et al. 1978; Schatzmann and Rossi 1971) and Rb^+ or NH_4^+, but not Li^+, mimic this action (Wolf et al. 1977). The affinities for Na^+ and K^+ are clearly different for this effect (K_{Na} = 20-30 mM, K_K = 4-6 mM) (Kratje et al. 1983, Scharff 1978; Schatzmann and Rossi 1971). Kratje et al. (1983) have recently shown quite convincingly that both K^+ and Na^+ exert their effect on the internal side of the membrane, thus ruling out a Ca-alkali-metal countertransport. They also excluded the possibility that Na^+ is cotransported by the Ca-pump.

2.1.5 The Reaction Cycle

With the present incomplete knowledge the reaction scheme may be drawn as shown in Fig. 2.

E_1 is clearly the conformation with the Ca^{2+}-binding site facing the cell interior. Phosphorylation of E_1 to $E_1 \sim P$, which can be studied using $[\gamma^{32}P]ATP$, has an absolute requirement for Ca_i^{2+} (Luterbacher 1982; Niggli et al. 1979; Rega and Garrahan 1975; Schatzmann and Bürgin 1978, Stieger and Luterbacher 1981a). Mg^{2+} ions are necessary inside for a rapid conformational change from $E_1 \sim P$ to $E_2 \sim P$. This statement is based on twofold experimental evidence. (a) Rega and Garrahan (1978) presented the following experiment (see Fig. 3A). The protein was phosphorylated in the presence of Ca^{2+}, then reaction I (Fig. 2) was stopped by addition of EGTA. The phosphoprotein decayed rapidly when ATP (at high concentration) and Mg^{2+} were added (reaction II + III), but did not with added ATP alone or Mg^{2+} alone (Fig. 3A). In a second experiment (Fig. 3B) Mg^{2+} was added during the last 5 s of phosphorylation and then Ca^{2+} and Mg^{2+} were removed by 20 mM CDTA. In this case addition of ATP was sufficient to decompose the phosphoprotein rapidly. This is interpreted by assuming that without Mg^{2+} $E_1 \sim P \rightarrow E_2 \sim P$ is slow and that $E_1 \sim P$ is not susceptible to hydrolysis by the action of high ATP. The second experiment should fail if Mg-ATP is the required species

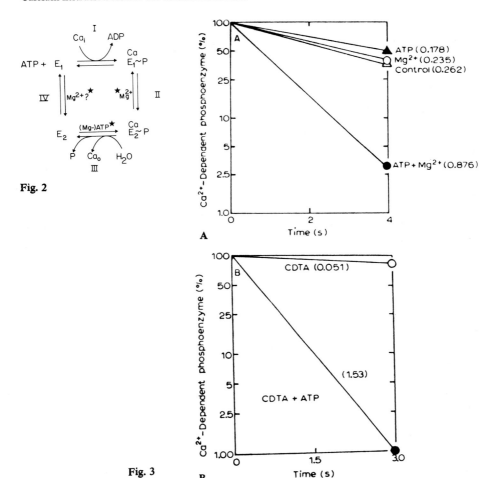

Fig. 2

Fig. 3

Fig. 2. Reaction cycle of the human red cell Ca-pump. In the normal mode the cycle turns in clockwise direction and is initiated by phosphorylation of E_1 from ATP. Ca_i means Ca^{2+} in the cytosol, Ca_0 means external Ca^{2+} and P means inorganic phosphate. E_1 and E_2 are two conformational forms of the protein; in E_1 the Ca^{2+} binding site faces the cell interior ($K_{Ca} \sim 1~\mu M$), in E_2 the Ca^{2+} site faces the exterior ($K_{Ca} \sim 10$ mM for $E_2 \sim P$ or E_2). The sequence of Ca^{2+} release (drop of Ca^{2+} affinity) and dephosphorylation is not assessed. Requirements of the reactions are indicated by *asterisks*

Fig. 3. A Fragmented membranes from human red cells. Dephosphorylation of Ca^{2+}-dependent phosphoenzyme made in the absence of $MgCl_2$ during 20 s; temp. $0°$-$3°C$. $\triangle$ control, $\blacktriangle$ medium with 1 mM ATP, $\circ$ with 0.5 mM $MgCl_2$, $\bullet$ with both (1 mM ATP + 1.5 mM $MgCl_2$). During dephosphorylation all media contained 30 mM EGTA to stop phosphorylation. **B** Effect of removal of Mg^{2+} on dephosphorylation in media with and without added ATP. Phosphoenzyme was made in the absence of Mg^{2+}. The concentration of $MgCl_2$ was raised to 0.5 mM 5 s before phosphorylation was terminated. Dephosphorylation was initiated by addition of 20 mM CDTA (trans-1, 2-diaminocyclohexanetetraacetic acid) or 20 mM CDTA plus 1 mM ATP. Figures in parentheses are rate constants (per s) calculated assuming first-order kinetics. The experiments show that Mg^{2+} and ATP are necessary for dephosphorylation, but that Mg^{2+} is required at a step preceding that accelerated by ATP. (Garrahan and Rega 1978)

in reaction III. Obviously the conditions in the experiment were such that free ATP did act. One such condition seems to be low temperature (Muallem and Karlish 1981). (b) The same conclusion is reached by experiments with La^{3+} (Luterbacher and Schatzmann 1983). Like other lanthanides, La^{3+} is inhibitory for transport and $(Ca^{2+} + Mg^{2+})$-ATPase. It has been shown (Luterbacher 1982; Schatzmann and Bürgin 1978; Szasz et al. 1978) that in the presence of La^{3+} the maximal level of phosphoenzyme (at long exposure to $[^{32}P]$ ATP) increases (to several 100% of that obtained with Ca^{2+} alone) (Luterbacher and Schatzmann 1983), and that the resulting form of the phosphoprotein can be decomposed by excess ADP, obviously in the back-reaction I, because even with Mg^{2+} and ATP dephosphorylation in the forward direction is impossible. Thus the block is at reaction II. If phosphorylation with Ca^{2+} alone proceeds for 30 s and then 200 μM La^{3+} and 5 s later 500 μM ATP + Mg^{2+} are added, the phosphoprotein decays only by about 25%; if however Mg^{2+} is already present during phosphorylation it decays by 75% in the identical dephosphorylation medium (Luterbacher and Schatzmann 1983). This again seems to show that only with Mg^{2+} present the reactions run through to the $E_2 \sim P$ form of the protein.

Reaction III seems strongly poised to the left because even under a large Ca^{2+} gradient ($\geqslant 2 \times 10^6$) in inside-out vesicles it was impossible to phosphorylate the protein from inorganic phosphate (PO_4) (Luterbacher 1982); the concentration of pump-protein was about 10^{-8} M, $^{32}PO_4$ was made 1 mM (with 2×10^7 cpm per sample), EGTA 5 mM and $MgCl_2$ 5 mM, the temperature was 0°C and the pH 7.0. From Eq. (1) (assuming zero membrane potential), it may be seen that under these conditions an easily detectable amount of 60 cpm per sample should have been incorporated, had the $\triangle G°$ of $E_2 + PO_4 \rightarrow E_2 \sim P$ (III) been = +20 kJ mol^{-1} or less. After electrophoretic separation, however, no radioactivity was detectable in the 140,000 mol wt. region. This shows that indeed the $E_2 \sim P$ is a "high energy compound"; 20 kJ mol^{-1} at 0°C corresponds to K = 6.7×10^3 M for reaction III. As the total energy drop in the cycle is equal to $\triangle G°$ for ATP hydrolysis (29.3 kJ mol^{-1} at 0°C) this means that at least 2/3 of the total energy is liberated in reactions III/IV. Since a very slow backward operation of the whole cycle (ATP synthesis from ADP and PO_4) under a large Ca^{2+}-gradient has been demonstrated (Rossi et al. 1978; Wüthrich et al. 1979), reaction III is in principle reversible. With a dodge described by de Meis et al. (de Meis and Inesi 1982; de Meis et al. 1980) Chiesi et al. (1984) have recently shown that phosphorylation from PO_4 is possible even without a Ca^{2+} gradient if the water activity is reduced by adding 30-40% dimethylsulfoxyde (DMSO) to the medium. The fact that upon restoring full H_2O activity and adding Ca^{2+} the phosphoprotein donated its PO_4 to ADP (Chiesi et al. 1984) favours the possibility that the drop in Ca^{2+} affinity is at $E_1 \sim P \rightarrow E_2 \sim P$ (II).

There is agreement that Mg^{2+} (tested in the absence of a high concentration of ATP) causes the rate constant of phosphorylation to approach that of dephosphorylation without drastically reducing the steady-state amount of phosphoprotein (Larocca et al. 1981; Luterbacher 1982) (see Fig. 4A, B). This can be explained by an accelerating effect on reaction I or reaction IV. For the time being, an action of Mg^{2+} on reaction IV which is symmetrical to that on reaction II seems the simpler assumption. This implies that $E_2 \rightleftharpoons E_1$ is poised toward E_2.

Fig. 4 A, B. Effect of Mg^{2+} on the time course of Ca^{2+}-dependent phosphorylation of the human red cell Ca-pump from $[\gamma^{32}P]$-ATP. **A** In red cell membranes. Temp. $0°-3°C$, medium: (mM) Tris-Cl 60, KCl 100, EGTA 0.1, $CaCl_2$ 0.15, $MgCl_2$ 0.5 (if present), ATP 30 μM (After Larocca et al. 1981). **B** In the isolated red cell Ca-pump protein. Temp. $0°-2°C$, protein in phosphatidylcholine (approx. 10 μg protein ml^{-1}). Medium: 2 μg ml^{-1} calmodulin; (mM) MOPS 25, KCl 120, EGTA 0.4, $CaCl_2$ 0.44 (Ca^{2+} = 45 μM), $MgCl_2$ 2 (if present), ATP 0.3 μM. Mean ± SE from three experiments. (After Luterbacher 1982). Although the rate constants are considerably higher in the isolated protein, the accelerating effect of Mg^{2+} is similar in both experiments

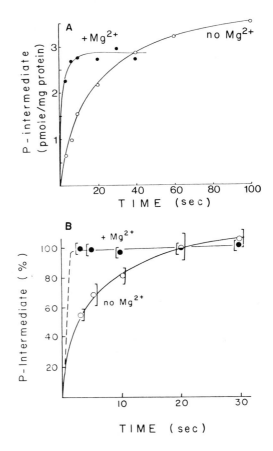

2.1.6 Stoichiometry

By stoichiometry we mean the number of g-atoms of Ca^{2+} transported per mol of ATP cleaved far from "static head" conditions. We have said above that kinetic results suggest that 2 Ca^{2+} ions are implicated. This by no means indicates that 2 Ca^{2+} ions are transferred per cycle. Direct measurement in resealed cells loaded with Ca-EGTA buffers (Schatzmann 1973; Schatzmann and Roelofsen 1977) yields a ratio Ca^{2+}/ATP = 1 after due correction for the Ca^{2+} leak and for the Mg-ATPase activity which cannot be blocked. By indirect reasoning, other authors came to the same conclusions (Ferreira and Lew 1976; Muallem and Karlish 1979).

Sarkadi (1980) reported that the Ca/ATP ratio increased from below 1 to 2 with increasing Ca^{2+} concentration up to 1 mM and in experiments by Akyempon and Roufogalis (1982) it decreased with increasing Ca^{2+} concentration. However, the most disturbing observation, with which one cannot find fault, was made by using external lanthanum (La^{3+}) (Quist and Roufogalis 1975; Sarkadi et al. 1977). Whereas La^{3+}, applied to both surfaces of membranes, blocks ATPase and Ca^{2+} transport equally, La^{3+} in the medium surrounding cells that are tight to La^{3+} (and Ca^{2+}) blocks transport nearly fully at 0.1 mM but leaves $\sim$ 50% of the (Ca^{2+} + Mg^{2+})-ATPase activity unaffected. A possible explanation of this behavior (Sarkadi et al. 1977) is that the Ca^{2+} pump is completely blocked by La^{3+} attaching to the external Ca^{2+} site, but that there is in the membrane another, nontransporting $(Ca^{2+} + Mg^{2+})$-ATPase of equal rate which is sensitive to La^{3+} present inside.

Since the only other phosphate groups in the membrane which are turned over as rapidly as that of the Ca-pump protein are those in the phosphoinositol-phosphates (Enyedi et al. 1982b), phosphorylation-dephosphorylation of phosphoinositol-phosphates catalysed by a kinase and a phosphatase has been invoked as the second $(Ca^{2+} + Mg^{2+})$-requiring ATP-consuming system.

If this were the proper explanation, the Ca/ATP ratio determined to be 1 ought to be multiplied by the factor 2. Another interpretation, however, is that external La^{3+} uncouples transport from ATPase activity leaving the pump ATPase gratuitously turning over at about half the normal rate. For this dispute see Sarkadi et al. (1979). Recently Clark and Carafoli have determined the stoichiometry in the purified protein incorporated into artificial soybean phospholipid (asolectin) vesicles and again found a Ca/ATP ratio of 1 (Clark and Carafoli 1983). It seems unlikely that a kinase and a phosphatase escaping detection in SDS gels are carried over in the calmodulin-column isolation procedure. Therefore, the uncoupling hypothesis gains some weight.

Finally, if the value of 3×10^{-8} M for $[Ca^{2+}]_i$ (Lew et al. 1982) is taken for granted, the 2:1 stoichiometry is impossible on thermodynamic grounds. The work by Eq. (1) turns out to be $58.5 \, kJ \, mol^{-1}$ (with 1.3 mM $[Ca]_o$ and a membrane potential of $-9 \, mV$), whereas the change in free energy of the ATP hydrolysis at the physiological concentration and physiological pH is taken to be $30.5 \, kJ \, mol^{-1}$ [Eq. (2)]. This means that even without any leak the concentration 3×10^{-8} M would not be reached. With a 1:1 stoichiometry, however, the work is reduced to one half ($29.25 \, kJ \, mol^{-1}$) which admits of $[Ca^{2+}]_i = 3 \times 10^{-8}$ M in the presence of a leak of $\geq 10 \, \mu mol \, l^{-1}$ cells h^{-1} (Lew et al. 1982).

2.1.7 Mode of Ca^{2+} Translocation

The question arises whether Ca^{2+} is transported as such or in exchange for another cation or as a Ca-anion pair. In the first case the system would be electrogenic, current flowing in the direction of the Ca^{2+} movement and the transport would depend on the membrane potential. The question is not easily settled because the red cell membrane is highly permeable to Cl^- and HCO_3^-, so that any pump potential is effectively shunted. Mg^{2+} as a counter-ion can be ruled out because Mg^{2+} does not move when Ca^{2+} does (Schatzmann 1975). Waisman et al. (1981a, b) have presented evidence that impermeant anions (gluconate) slow the Ca^{2+} transport, which rules out Na^+ or K^+ as counterion. Rossi and Schatzmann (1982a) showed that the system either exchanges Ca^{2+} for K^+ (in gluconate media) or transports Ca^{2+} with Cl^- (in cloride media) which rules out that either K^+-Ca^{2+}-exchange or Cl^--Ca^{2+}-cotransport through the pump mechanism is the mode of the pump. It does, however, not exclude Ca^{2+}-n H^+ exchange (contrary to what the authors say in their paper). Niggli et al. (1982) have put forward evidence to the effect that liposomes with incorporated pure pump protein extrude extra protons (over and above those produced by ATP splitting) when they accumulate Ca^{2+}. This can be explained by Ca^{2+}-n H^+ exchange or by electrogenic Ca^{2+} transport and their findings that carbonyl cyanide-m-chlorophenyl hydrazone (CCCP) + valinomycin (a protonophore and an ionophore for K) decreased the initial rate of H^+ release by 60% and that valinomycin or CCCP increased the rate of ATP splitting by 40% are also compatible with both hypotheses. Their observation that valinomycin fails to affect Ca^{2+} transport is not incontrovertible proof for Ca^{2+}-n H^+ exchange because the ion permeabilities of these vesicles are unknown (and might be high enough to balance an electrogenic Ca^{2+} transport). Unless it is shown that the permeabilities for K^+, anions, and H^+ (or OH^-) in the unpoisoned state of these vesicles are very low relative to the pump rate, these experiments do not seem to prove unequivocally the existence of the Ca^{2+}-n H^+ exchange through the Ca^{2+} pump.

Smallwood et al. (1983) blocked the anion exchange mechanism in inside-out vesicles with 10 to 20 μM DIDS and showed that acids passing the membrane in undissociated form (like acetate) accelerate Ca^{2+} uptake by the pump, while anions that pass the membrane in ionized form (like thiocyanate) do not. The first are able to collapse a proton gradient, the latter prevent the build-up of a potential gradient. This seems strong evidence in favor of a Ca^{2+}-n H^+ exchange. The authors leave open the question whether n is 1 or 2.

2.1.8 Calmodulin Dependence

Human red cells contain about 5 μM calmodulin (Foder and Scharff 1981; Jarett and Kyte 1979) or some 3×10^5 molecules per cell, which is far more than the number of pumps per cell. The discovery of the calmodulin activation of the pump is due to two groups (Gopinath and Vincenzi 1977; Jarrett and Penniston 1977, 1978; Larsen and Vincenzi 1979; Vincenzi and Hinds 1980). As in other calmodulin-dependent systems calmodulin in a Ca^{2+} form (at least 3 of the 4 calmodulin metal-binding sites occupied by Ca^{2+}) binds to the pump (Scharff 1980; Vincenzi and Hinds 1980) and thereby increases the affinity for Ca^{2+} at the transport site 30-fold, increases the V_{max} (Foder and Scharff 1981; Scharff and Foder 1978, 1982) and enhances the effect of K^+ on the rate (Scharff 1978). In experiments by Muallem and Karlish (1980), calmodulin in the presence of Mg^{2+} greatly accelerated the formation of phosphoprotein in red cell membranes without altering the steady-state amount of it. In the isolated protein (Luterbacher 1982) a similar behavior was seen in the absence of Mg^{2+}; with Mg^{2+} present the rate of phosphorylation was unmeasurably rapid even without calmodulin and calmodulin increased the steady-state amount of the phosphoprotein by about 50%. An assumption that seems to be in accord with all of this is that calmodulin accelerates the reaction $E_2 \leftrightarrows E_1$.

The K_{Ca} values of the metal binding sites of calmodulin are of the order of 1 μM. It is, therefore, possible that the second Ca^{2+} site revealed by kinetic data on transport and ATPase is on calmodulin. It is clear, however, that the transport site is distinct from the calmodulin Ca^{2+}-binding sites (see below). Cox et al. (1982) have supplied an instructive mathematical treatment of the somewhat involved problem caused by the direct and the calmodulin-mediated Ca^{2+} effects in the system.

Scharff and Foder (1982) have very thoroughly investigated the kinetics of calmodulin binding to the pump. They showed that the K_{diss} increased from 2.5 nM to 25 μM when the Ca^{2+} concentration was lowered from 20 μM to 0.1 μM. Their separate determination of the Ca^{2+} dependence of the forward and backward rate constant shows that the dissociation rate constant becomes minimal at $Ca^{2+} \sim 10^{-6}$ M and the association rate constant maximal at $\sim 10^{-5}$ M (Scharff and Foder 1982) and that both are, beyond these Ca^{2+} concentrations, of the order of 10^{-1} to 10^{-2} min^{-1} viz. nM^{-1} min^{-1}. Thus the reaction is quite slow. The practical consequence of these studies is that at physiological concentrations of calmodulin and Ca_i^{2+} (and Mg_i^{2+}) the pump is free of calmodulin (Foder and Scharff 1981; Scharff and Foder 1982). An abrupt rise of the Ca^{2+} concentration whithin a cell will induce the slow assembly of the system, resulting in an activation with a time lag such that short pulses of Ca^{2+} entry into a cell will give a prolonged $[Ca^{2+}]_i$ signal (Scharff et al. 1983). This would not be so if the system were devoid of the calmodulin mechanism and were to display the high Ca^{2+} affinity and the high pump rate at all Ca^{2+} concentrations. This reasoning provides a good raison d'être for the calmodulin dependence of the Ca^{2+} pump. It very nicely explains why $[Ca^{2+}]_i$ passes through a maximum before reaching the new steady state when a moderate Ca^{2+} leak is induced by low concentrations of the ionophore A 23187 (Ferreira and Lew 1976; Scharff et al. 1983). In addition this hysteretic behavior can bring the Ca^{2+} concentration within a cell to respond with oscillations to an increase of the Ca^{2+} permeability and in fact might be at the basis of the periodic changes of $[K^+]$ and $[Ca^{2+}]$ seen in red cells (Vestergaard-Bogind and Bennekou 1982).

2.1.9 The Endogenous Inhibitor

Membrane-free hemolysate stimulates the calmodulin-free $(Ca^{2+} + Mg^{2+})$-ATPase (Au 1978) and the Ca^{2+} transport (Sarkadi et al. 1980b) at low and inhibits at higher concentrations. The activator is of course calmodulin and the inhibitory agent is another peptide; the biphasic action of crude hemolysate rules out that the inhibitor is competitive with or binds to calmodulin. It has been isolated (Lee and Au 1981). Its physiological role is obscure, but it might be responsible for the low apparent Ca^{2+} affinity observed in intact red cells (Feirreira and Lew 1976).

2.1.10 Lipid Requirement

It has been shown that the functioning of the calmodulin replete system requires the presence of glycerophospholipids in the inner leaflet of the membrane (Roelofsen 1981; Roelofsen and Schatzmann 1977). Interestingly, function can be annihilated in membranes by enzymic degradation of glycerophospholipids and be revived by adding them again, whereas in the solubilized state the protein deteriorates irreversibly in the absence of phospholipids. Acidic lipids like phosphatidylserine, phosphatidylinositol, or cardiolipin, and even unsaturated fatty acids not only maintain function but are able to replace calmodulin (Niggli et al. 1981a, b; Stieger and Luterbacher 1981a). In their presence the Ca^{2+} affinity is high and the turnover rate larger than in phosphatidylcholine surroundings.

2.1.11 Modification of the Performance by Proteolytic Attack

The entity of molecular weight 140,000 determined by SDS gel electrophoresis under reducing conditions is in all probability the monomer of the protein which may contain considerable amounts of lipids (Verma et al. 1982). This must be kept in mind for the following discussion.

Tryptic digestion has long been known to destroy the calmodulin-dependent activity (Bond 1972; Rossi and Schatzmann 1982b) but if done judiciously (for minutes at μg ml^{-1} concentrations of trypsin) it stimulates the basal activity in a way that the major fragment becomes indistinguishable from the native enzyme coupled with calmodulin (Niggli et al. 1981a; Sarkadi et al. 1980a; Stieger and Schatzmann 1981b; Taverna and Hanahan 1980). The digested protein not only shows the high V_{max} and the high affinity for Ca^{2+}, but it also has the same affinity for Mg^{2+} and is inhibited by high Ca^{2+} concentrations as the intact Ca-pump. On the latter effect Mg^{2+} has an antagonistic competitive action as in the intact system (Stieger and Schatzmann 1981b). In membranes, activation by trypsin is only obtained if the digestion is carried out in the presence of Ca^{2+} and in fact Ca^{2+} concentrations are necessary that are in the range of the Ca^{2+} requirement of the transport site (Rossi and Schatzmann 1982b).

These experiments show clearly that the pump has the full complement of sites necessary for function in the absence of calmodulin, which rules out the possibility that Ca^{2+} bound by calmodulin is the Ca^{2+} transported by the system. Since it was shown that after trypsin activation calmodulin no longer binds to the protein (Sarkadi et al. 1980a), a tentative interpretation is that a calmodulin-binding sequence in the protein has an attenuating effect on function, which is relieved either by binding of calmodulin or by removing the sequence by trypsin.

Zurini et al. (1983) recently studied more closely the time course of the appearance of different peptides when carrying out trypsin digestion at 0°C. They conclude that first a 14,000 and a 35,000 mol. wt. peptide are removed and that the resulting 90,000 mol. wt. peptide is not only functional but still possesses the calmodulin receptor site (because it is retained on a calmodulin column). The critical step, however, seems to be the ensuing removal of a 10,000 mol. wt. peptide. The resulting 80,000-76,000 mol. wt. peptide is a species which no longer binds calmodulin but is fully active without it.

2.1.12 Studies with Antibodies

Verma et al. (1982) raised antibodies against purified human red cell Ca-pump protein in rabbits. The interesting result of these studies is that the sarcoplasmic reticulum (SR) Ca-pump did not cross-react with such antibodies, while pumps from red cells of other species and plasma membrane Ca-pump protein of tissue cells (rat corpus luteum and rat brain synaptic plasma membrane) did, although weakly. Likewise anti-SR-antibodies (raised in goats) did not react with the red cell pump protein. The antibody inhibited ATPase and Ca-transport regardless of whether calmodulin was present or not. Another study (Gietzen and Kolandt 1984) carried out with rabbit-raised antibodies against human Ca-pump protein from red cells differed in that only the calmodulin-dependent

Table 1. Propertied demonstrating that the plasma membrane and the sarcoplasmic reticulum Ca-pump are not identical

	Plasma membrane (PM)	Sarcoplasmic reticulum (SR)	References
Monomeric mol. wt.	$\sim 140{,}000$	$\sim 100{,}000$	
Calmodulin activation	+	−	
Phosphorylation from PO_4	Difficult	Easier	(Chiesi et al. 1984; Luterbacher 1982)
Antigenicity	Not reacting with SR antibodies	Not reacting with PM antibodies	(Caroni et al. 1982; Verma et al. 1982; Wuytack et al. 1983)
Substrate specificity	High (for ATP)	accepts acyl-phosphates and nitrophenylphosphate	(Olson and Cazort 1969; Rega et al. 1973; Sarkadi et al. 1980)
Proteolytic digestion	Different pattern of peptides		(Caroni et al. 1982; Zurini et al. 1983)
Phospholipase C on Ca-phosphatase activity	Activity enhanced	Activity reduced	(Richards et al. 1977)
Mg^{2+}-site of P-Intermediate	Accessible	Occluded	(Garrahan and Rega 1978; Garrahan et al. 1976)

activity was blocked by the antibody and not the basal activity. Caroni et al. (1982) prepared antibodies against the purified human red cell Ca-pump and demonstrated that bovine heart plasma membrane, but not rabbit SR Ca-ATPase, is inhibitory in the red cell radioimmuno-assay. Wuytack et al. (1983) using antibodies against the porcine stomach Ca^{2+} pump (purified on a calmodulin column) showed cross-reactivity with the red cell but not with the SR-pump protein of the pig. Their antibodies did not inhibit function.

2.2 Plasma Membrane and Sarcoplasmic Reticulum Calcium Pump

Essentially the two systems perform very similarly (Martonosi 1983). There are, however, a few differences showing that the molecular architecture cannot be identical. The divergent points are summarized in Table 1.

3 The Na-Ca-Exchange System

3.1 The Discovery of the System in Mammalian Heart and Squid Axons

Physiologists have known for a long time that Na^+ and Ca^{2+} antagonize each other in biological systems, seemingly at some common receptor. Some of these interferences are of interest for the present discussion. For instance, Na^+ in the medium antagonizes the positive inotropic effect of Ca^{2+} on the heart. Wilbrandt and Koller (1948) made the significant observation that this is not a simple 1:1 competition, but that for a constant contractile response the ratio $[Na^+]/\sqrt{[Ca^{2+}]}$ must be constant, irrespective of the absolute concentration of each. Mislead by the rapidity of the Ca^{2+} effect and the then current idea that cell membranes are impermeable to divalent cations, they assumed that Ca^{2+} had its effect at the plasma membrane outer surface, and they concluded that the phenomenon was due to a Donnan distribution between the cell surface and the bulk of the medium which indeed would result in the "quadratic law" if the Ca^{2+} concentration at the surface of the membrane were the magnitude that matters.

Twenty years later Reuter and Seitz (1968), working with guinea-pig auricles and sheep ventricular trabeculae and Baker et al. (Baker and Blaustein 1968; Baker et al. 1967, 1969), experimenting with the squid axon, made a discovery which clarified the old experiment. They observed that there is a Ca^{2+} efflux from the cell interior which is increased by external Ca^{2+} or by external Na^+. The interpretation was that this is due to a membrane protein which exchanges Ca^{2+} for Ca^{2+} or Ca^{2+} for Na^+ across the membrane, i.e., that the stimulating cation moves across the membrane in opposite direction. Reuter et al. (Glitsch et al. 1970; Reuter and Seitz 1968) noticed two important points at once: (a) between external Na^+ and external Ca^{2+} there is again a competition obeying, at least as a first approximation, the "quadratic law", $[Ca^{2+}]$ being equivalent to $[Na^+]^2$, and (b) the system has specificity for Na (over Li, Rb and K), which rules out any sort of Donnan effect.

A large number of studies (Blaustein and Nelson 1982; Reuter 1982) made it clear that the system is truly passive and symmetrical: Ca_o accelerates Na-efflux and Na_o accelerates Ca-efflux; Ca_i accelerates Na-influx and Na_i accelerates Ca-influx (Baker and Blaustein 1968; Baker et al. 1969; Blaustein 1974, 1977; Glitsch et al. 1970; Mullins and Brinley 1975; Pitts 1979; Reeves and Sutko 1979; Requena and Mullins 1979; Russel and Blaustein 1974). However, if there is a gradient for one of the two species running down across the system the partner is moved against its gradient in an uphill fashion.

Since there is a Na-K-pump in all animal cells which maintains a $[Na]_o/[Na]_i$ gradient greater than 10 but smaller than 100, the Na-Ca-exchange system constitutes a means to keep intracellular Ca^{2+} low (Jundt et al. 1975) as long as $[Na]_o$ is high. It was shown with aequorin-injected squid axons (Baker et al. 1971) that removal of external Na^+ indeed increases intracellular Ca^{2+} (see Fig. 5). However, Barry and Smith (1982), working with cultured chick ventricular cells, concluded that at physiological $[Ca^{2+}]_i$ the exchange system contributes but little to the Ca^{2+}-efflux.

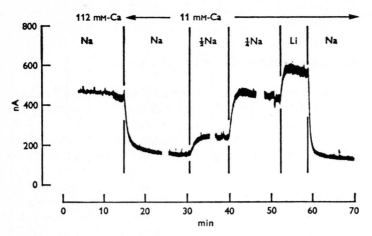

Fig. 5. Effect of different external solutions on $[Ca^{2+}]_i$ in squid axons. The axons were injected with aequorin. The ordinate is the output of the photomultiplier in nA, which is proportional to the aequorin glow, which in turn is proportional to $[Ca^{2+}]_i$. Lowering $[Ca^{2+}]_o$ lowers $[Ca^{2+}]_i$; lowering $[Na]_o$ (replacing Na by Li) increases $[Ca^{2+}]_i$. Li means zero $[Na]_o$. (Baker et al. 1971)

3.2 Stoichiometry and Voltage Dependence

For the Na-Ca exchange system it can easily be shown on purely thermodynamic grounds (Mullins 1981) that at equilibrium (when there is no net cation flux through the system in an idealized cell without leaks)

$$[Ca]_i \,/\, [Ca]_o \quad = \quad [Na]_i^2 \,/\, [Na]_o^2 \tag{3}$$

if one assumes that, in agreement with the "quadratic law", 2 Na^+ ions are transported in exchange for 1 Ca^{2+} ion (coupling factor = 2).

However, as was pointed out early (Blaustein and Hodgkin 1969), such a system will not account for the observed low value of $[Ca^{2+}]_i$; with a $[Na]_o/[Na]_i$ ratio of 20-30 a $[Ca]_o/[Ca]_i$ ratio of 400-900 results from Eq. (3), but the observed value exceeds 10,000. This may be a fallacy, because the measurement of internal Na^+ concentration may overestimate the intracellular Na^+ activity. It has been claimed that the activity coefficient for Na^+ inside cells may be as low as 0.17 (Reuter 1982). Another way out of the dilemma is to postulate a coupling factor > 2. In fact Pitts (1979), from direct measurement of initial rates in dog sarcolemmal vesicles, arrived at a 3 Na : 1 Ca stoichiometry. Flux versus concentration curves are rectangular hyperbolas for Ca^{2+} (Baker and Blaustein 1968; Baker et al. 1969; Blaustein 1977) (Fig. 6A), but have sigmoidal shape for Na^+ (Blaustein 1974, 1977; Glitsch et al. 1970; Jundt and Reuter 1977; Reuter 1982; Russel and Blaustein 1974), clearly showing that 1 Ca^{2+} ion but more than 1 Na^+ ion are involved (Fig. 6B). However, a decision between 2 and 3 Na^+ ions is not possible (Reuter 1982) by flux measurements, let alone a decision between equivalent and independent sites on one hand and positively cooperative sites on the other.

If more than 2 Na^+ ions are transferred, the system generates electrical current in the direction of the Na-movement. Such a system causes a drop in the membrane potential, and its rate and the equilibrium Ca^{2+} and Na^+ distribution which it mediates are dependent on the membrane potential. These properties have been experimentally verified (Bers et al. 1980; Brinley and Mullins 1974; Caroni et al. 1980, Lamers and Stinis 1981; Mullins and Brinley 1975; Reeves and Sutko 1980; Requena et al. 1977). The potential dependence seems to have established the fact that the number of Na^+ ions transferred is higher than 2. Yet a recent paper by Lederer and Nelson (1983) makes it doubtful whether all of these experiments are really conclusive. The authors found in internally perfused barnacle muscle cells, in which the internal pH, $[Ca^{2+}]_i$ and the membrane potential were controlled and the ATP concentration was kept zero, that Na-dependent Ca^{2+} efflux was accompanied by inward current. However, the current was so much too large that if it were ascribed solely to the exchange system, 12 Na^+ ions would move for 1 Ca^{2+} ion. It seems likely, therefore, that at elevated $[Ca^{2+}]_i$ (10 μM) a $[Ca^{2+}]_i$-sensitive (nonspecific) channel (Colquhoun et al. 1981) allows of a Na-carried inward current or else that even in these well-controlled experiments the Ca^{2+} efflux is underestimated due to dilution of $^{45}Ca^{2+}$ with unlabeled Ca^{2+} from internal stores. The conclusion is that it is difficult to attribute current generation to the exchange system as long as conductance changes in other ionic channels cannot be ruled out.

Although a stoichiometry of 3 Na : 1 Ca seems to satisfy the observed $[Ca^{2+}]_i$ in the cells, Mullins (1977, 1981) prefers the number 4, arguing that allowance has to be made for Ca^{2+} leaks that keep the $[Ca^{2+}]_i$ above the equilibrium value set by the exchange system alone.

The mathematical expression for ionic flows in the electrogenic system is rather complex (Mullins 1977). However, the equilibrium situation is easily understood if one recalls that the work done by Na^+ equals the work gained on Ca^{2+}:

$$\Delta\bar{\mu}_{Ca} = r \cdot \Delta\bar{\mu}_{Na} \tag{4}$$

$$\text{or} \quad z_{Ca} F (E_{Ca} - V_m) = r \cdot z_{Na} F (E_{Na} - V_m) \tag{5}$$

with $\Delta\mu_{Na}$, $\Delta\mu_{Ca}$ = change in electrochemical potential, E_{Na} and E_{Ca} = the equilibrium potentials for the two ions (the potential that balances the ionic gradient so that there is no movement of the ion through a diffusional pathway), V_m = membrane potential, z = the charge of the ion, r = coupling factor and F = the Faraday constant (9.65×10^4 C mol^{-1}).

The difference $(E_{Na} - V_m)$ or $(E_{Ca} - V_m)$ in Eq. (5) is the sum of all driving forces for the respective ion, translated into an electrical potential and $(r)zF$ the charge transferred. Consequently, either side of Eq. (5) is the work or energy term for the ion. At equilibrium V_m equals the potential at which there is no net ion movement through the exchange system (V_m^{eq}).

With $\quad E_{Na} \quad = \quad \dfrac{RT}{F} \quad \ln \quad \dfrac{[Na]_o}{[Na]_i}$ (6)

and $\quad E_{Ca} \quad = \quad \dfrac{RT}{2F} \quad \ln \quad \dfrac{[Ca]_o}{[Ca]_i}$ (7)

in Eq. (5) one obtains for equilibrium (with V_m^{eq} measured inside)

$$\frac{[Ca]_i}{[Ca]_o} \quad = \quad \frac{[Na]_i^r}{[Na]_o^r} \quad \cdot \quad \exp\left[(r-2)\;\frac{V_m^{eq}\cdot F}{RT}\right].$$ (8)

From Eq. (8) it is obvious that if $r > 2$, the membrane potential introduces an asymmetry into the system which can be interpreted by saying that the Ca^{2+}-binding sites are rarefied on the positive and concentrated on the negative side of the membrane.

3.3 Kinetics

Flux measurements clearly show — as one might expect — that the system exhibits saturation kinetics for Na^+ and Ca^{2+} (Fig. 6). While in crab nerve Na_o^+ is competitive with Ca_o^{2+} for Ca^{2+}-influx (Baker and Blaustein 1968) (Fig. 6A) there is no effect of $[Ca^{2+}]_i$ on the affinity for Na_o (Fig. 6B), only V_{max} of the Na_o-dependent Ca^{2+} efflux being reduced when the $[Ca^{2+}]_i$ drops from saturation to $1/10$ of saturation (in internally dialysed squid axons) (Blaustein 1977). This independence of the affinity for one ion on the cis-side from the occupancy of the site for the other ion on the trans-side speaks against a shuttling mechanism that alternatingly binds Ca^{2+} or Na^+ at the same site and is in favour of a model in which two independent sites for Ca^{2+} and Na^+ are loaded simultaneously (Blaustein 1977).

The K_{Na}^{app} (in squid axons) seems to be quite similar for sites at the external and internal surface (of the order of 10-100 mM) (Blaustein 1977; Mullins 1977), whereas it was observed that K_{Ca}^{app} is in the μM range inside and in the mM range outside (Blaustein and Russel 1975) (see Fig. 6). This behavior is not contrary to the definition of a passive system; it is in accord with the second law of thermodynamics because the asymmetry in the K values is compensated by a higher rate constant for the translocation from the side of low affinity, as was shown for the glucose transport system by Widdas (1980).

The original idea obviously was that there is one metal-binding site which accepts either 1 Ca^{2+} or 2 (or several) Na^+ ions and that only the loaded site can move across the membrane (Reuter and Seitz 1968). At present simultaneous models seem more appropriate (see above). It is assumed that the carrier molecule has two types of metal binding sites facing opposite sides of the membrane that move when the Na-site is loaded with Na^+ and the Ca-site with Ca^{2+} (Fig. 7). The model of Mullins (1977), for which the author gives a complete mathematical treatment, postulates an asymmetric protein with 4 Na^+-binding sites at one end. Upon saturation of these sites with Na^+ a Ca^{2+}-site is induced at the opposite end. The fully loaded carrier is movable, whereas that with less than 4 Na^+ is "poisoned", i.e. immobile. The free carrier is assumed to be mobile.

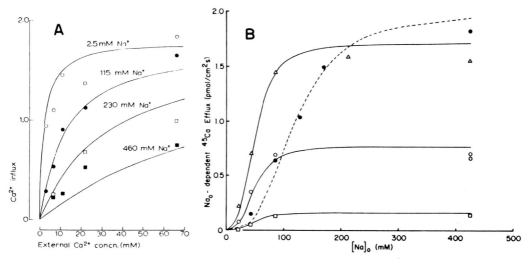

Fig. 6. A Dependence of ^{45}Ca influx into crab nerve *(Maia squinado)* on $[Ca^{2+}]_o$ at different $[Na^+]_o$. Na_o was replaced isoosmotically by Li. Ordinate: Ca influx (mmol kg^{-1} nerve during 7 min). The curves are drawn according to $J = J_{max} / [1 + K_{Ca} \cdot [Mg^{2+}] / \{K_{Mg} \cdot [Ca^{2+}]\} + \{K_{Ca} / [Ca^{2+}]\} (1 + [Na] / \overline{K}_{Na})^n]$ with $J = $ Ca-influx, $J_{max} = 1.8$, $K_{Ca} = 2$ mM, $\overline{K}_{Mg} = 20$ mM, $\overline{K}_{Na} = 75$ mM and $n = 2$. Notice that the curves are rectangular hyperbolas and that Na_o competes with Ca_o but does not affect J_{max} (Baker and Blaustein 1968). **B** Na_o-dependent ^{45}Ca efflux from internally dialyzed squid axons. Temp. $14°$-$15°$C. *Solid lines* and *open symbols* three different ATP-fuelled axons (ATP = 4 mM) of comparable diameter at -60 to -67 mV membrane potential but different $[Ca^{2+}]_i$. The mitochondria were poisoned (oligomycin, FCCP). □ $[Ca^{2+}]_i \sim 0.31$ μM; ○ $[Ca^{2+}]_i \sim 0.5$ μM; △ $[Ca^{2+}]_i \sim 2.5$ μM. *Curves* are drawn according to $J = J_{max} / [1 + (\overline{K}_{Na} / [Na]_o)^3]$ with $J = $ Na-dependent Ca-efflux and $\overline{K}_{Na} = 50$ mM. Notice that raising $[Ca^{2+}]_i$ increases J_{max} but does not affect $\overline{K}_{Na}$. *Broken line* (●) ATP-free dialyzing fluid with 100 μM Ca^{2+}. Notice that lack of ATP lowers the affinity for Na_o. For details see original. (Blaustein 1977)

Apart from the preference for 4 Na$^+$, which cannot be based on experimental facts, the model nicely predicts the alkali-metal requirement of Ca_o^{2+}-dependent Ca^{2+} efflux (Blaustein 1977). The claim that the latter is in fact through the same system is based on Reuter's early observation that the Na-Ca-exchange declines when the Ca-Ca-exchange is initiated (Blaustein 1977; Reuter and Seitz 1968) and vice versa (Baker and Blaustein 1968). Blaustein made the interesting observation that in squid axons alkali cations must be present on both sides of the membrane for Ca-Ca-exchange to occur and that Li$^+$ (that does not sustain a Li-Ca-exchange (Fig. 5)) and less effectively also K$^+$ and Rb$^+$ are suitable for this function (Blaustein 1977; Blaustein and Russell 1975). It is not clear whether in this mode a Li-Li-exchange takes place (Blaustein 1977; Blaustein et al. 1974) as shown in Fig. 7.

3.4 Role of ATP

A point that has considerably confused the issue for some time was that internal ATP has an influence on the system. It seems probable today that ATP is not hydrolyzed

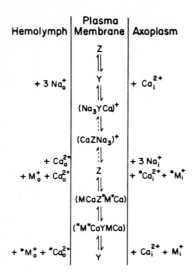

| Hemolymph | Plasma Membrane | Axoplasm |

Fig. 7. Simultaneous model for the Na-Ca/Ca-Ca-exchange system proposed by Blaustein (1977) for the squid axon. Y is the form of the protein that binds Na^+ at the outside and Ca^{2+} at the inside, Z is the form that binds Na^+ at the inside and Ca^{2+} at the outside. $Y \rightleftharpoons Z$ is reversible without cations. The upper half shows Ca-Na-exchange, the lower half Ca-Ca-exchange. The latter requires the presence or the cotransport (?) of an alkali metal ion (M^+). *Asterisks* mark the ions that were originally inside

and that it does not, therefore, impart any asymmetry to the system by supplying energy to it. Its binding seems to increase the affinity for Na^+ (Blaustein 1977; Jundt and Reuter 1977) (see Fig. 6B) and for Ca^{2+} (Baker and McNaughton 1976; Blaustein 1977). As may be said with hindsight, the difficulty was caused by the contribution of the pump to the Ca^{2+} efflux (Baker and Glitsch 1973; Baker and McNaughton 1978) (see below). What accounted for 50-90% of the Ca^{2+} efflux in unpoisoned squid axons and what Baker has termed uncoupled Ca^{2+} efflux (Baker and McNaughton 1978) probably was due to the Ca-pump. Characteristically this uncoupled Ca^{2+} efflux is inhibited by internal vanadate (Baker and Singh 1981; Di Polo et al. 1979).

3.5 Solubilization and Reconstitution

A chemical characterization of the protein is still missing. The difficulty lies in the fact that the protein can only be discerned by its function in closed membrane systems. However, solubilization and subsequent reconstitution with or without added lipids into vesicles has been successful in preserving the transport function (Miyamoto and Racker 1980; Schellenberg and Swanson 1982a). This achievement opens up the possibility of applying purification procedures to the solubilized material, so that there is hope of enriching and eventually isolating the protein.

4 Parallel Operation of Pump and Exchange

We shall see in Section 5 that in a good number of animal cell membranes the exchange system and the ATP-fueled Ca-pump have been shown to coexist. The question must be asked what might be the biological meaning of this duplication of contrivances to the same end. In at least one membrane, that of squid nerves, the maximal rate of the exchange system is considerably higher and the affinity for Ca_i lower than those of the pump (Beauge et al. 1981; Di Polo 1978; Di Polo and Beauge 1979, 1980). From this it has been inferred that the exchange system copes with heavy loads of Ca^{2+} (when $[Ca^{2+}]_i$ is high) while the pump serves the purpose of lowering intracellular $[Ca^{2+}]$ to the 10 nM range. Although correct, unless completed this statement might lead to misunderstanding. When the pump has the tendency to lower $[Ca^{2+}]_i$ to below the value which results from equilibrium distribution across the exchange system, the exchange system will start extruding Na^+ on account of incoming Ca^{2+}, i.e., the two systems antagonize each other below the equilibrium $[Ca^{2+}]_i$. This case may never occur, though. Under physiological conditions $[Ca^{2+}]_i$ might always exceed the equilibrium value of the exchange system owing to the leak fluxes. Further, it may be argued that it is important for the cell to keep $[Ca^{2+}]_i$ *and* $[Na^+]_i$ low and that the exchange system couples the Na-K-pump and the Ca-pump in such a way that either can serve to extrude Ca^{2+} *or* Na^+, resulting in a minimum of the sum of $[Ca^{2+}]_i + [Na^+]_i$. Finally, under depolarization of the membrane the exchange system will increase $[Ca^{2+}]_i$ (Mullins 1981), while the pump will not be affected by depolarization or even increase its rate.

5 Ca-Extrusion in Different Cells

In recent years reports on a Ca-pump and/or a $(Ca^{2+} + Mg^{2+})$-ATPase in the plasma membrane of different cells have appeared in rapid succession. Many cells that were thought to extrude Ca^{2+} only by the exchange system were shown to possess both systems. It must suffice to summarize the material with a few comments on the well-analyzed systems.

Because in cells of higher organization it is often difficult to decide whether a membrane preparation is pure or a mixture of plasma membrane and internal membranes (other than mitochondrial membranes which are easily recognizable by their biochemical markers), it is important to remember that only the plasma membrane is equipped with the exchange system and the Na-K-pump, and that the endoplasmic (sarcoplasmic) reticulum Ca^{2+}-pump is not calmodulin-sensitive. Further, the pump is inhibited by vanadate (Baker and Singh 1981; Di Polo and Beauge 1981; Rossi et al. 1981) and the exchange system is sensitive to amiloride (Schellenberg and Swanson 1982b; Smith et al. 1982).

5.1 Excitable Cells

5.1.1 Nerve

In the squid axon, the classical object for the study of the Na-Ca-exchange system, an uncoupled (Na_o-independent) Ca^{2+} efflux was observed in 1978 (Baker and McNaughton 1978). In the same year Di Polo (1978) demonstrated in the perfused axon of

Dorytheutis plei that addition of ATP to the perfusate increased Ca^{2+} efflux in the absence of ionic gradients and that the efflux was only partly Na_o-sensitive. In *Loligo pealei* axons (Di Polo and Beauge 1979, 1980) it was shown that the ATP driven transport saturates at a lower [ATP] (K_m = 30 μM) than what is required for half maximal effect of ATP on the exchange system (K_m = 230 μM). The maximal capacity of the ATP driven system for Ca^{2+} is about 1/10 of that of the exchange system but its affinity for Ca^{2+} ($K_{Ca} \sim 0.18$ μM) exceeds that of the exchange system. In another nerve (optic nerve of sepiotheutis sepioidea) a (Ca^{2+} + Mg^{2+})-ATPase was found which might be the biochemical correlate of the uncoupled Ca^{2+} efflux (Beauge et al. 1981).

5.1.2 Nerve Terminals, Brain

Nerve terminals can be pinched off as closed vesicles ("synaptosomes"). Both Ca-exporting systems seem to be present in these structures whose transmitter release depends on Ca^{2+} entry during the action potential (Blaustein and Oborn 1975; Kendrick et al. 1977). There is good evidence for the exchange system (Blaustein and Ector 1976, Blaustein and Oborn 1975; Blaustein and Wiesman 1970). Gill et al. (1981) have demonstrated that in synaptosomes of guinea-pig brain there is a Ca-pump with a K_{Ca} of 12 μM and an exchange system with a K_{Ca} of 40 μM. The existence of the pump is supported by the observation that proteins isolated from membranes and incorporated into soybean phospholipid vesicles imparted an ATP-dependent Ca^{2+} accumulating property to these vesicles (Papazian et al. 1979). Isolated membranes from brain may be a mixture of plasma membrane and endoplasmic reticulum material (Blaustein et al. 1978) but the presence of two (Ca^{2+} + Mg^{2+})-ATPases of mol. wt. $\sim$ 100,000 and $\sim$ 140,000 in such membranes (Blaustein et al. 1978; Hakim et al. 1982; Papazian 1979; Robinson 1976, 1978) suggests that the plasma membranes contribute their own ATPase. The calmodulin dependence, the cross-reaction with red cell Ca-pump antibodies (Hakim et al. 1982) and the association of the ATP-driven system with the Na-pump in natural membranes (Robinson 1976) are very much in favour of the idea that the 130,000-140,000 entity stems from the plasma membrane (Blaustein et al. 1978).

5.1.3 Cardiac Muscle

The myocardial cell is another classical representative with the exchange system in the plasma membrane (Bers et al. 1980; Caroni et al. 1980; Lamers and Stinis 1981; Mullins 1981; Reeves and Sutko 1980, 1983; Reuter 1982; Reuter and Seitz 1968). The technique of isolating the plasma membrane as closed vesicles has shown that the pump is also present (Caroni and Carafoli 1980, 1981a, b; Lamers and Stinis 1981; Trumble et al. 1980). As in nerve, the higher Ca^{2+} affinity of the pump suggests that the pump action is more prominent than the exchange system at low $[Ca^{2+}]_i$ (Barry and Smith 1982). An interesting observation is that the pump in myocardial cells is stimulated by phosphorylation via cAMP-dependent kinases (Caroni and Carafoli 1981b; Caroni et al. 1982; Hui et al. 1976; Sulakhe et al. 1976) (which must not be confused with the periodic phosphorylation-dephosphorylation in the transport cycle of the pump action). This seems to distinguish the heart muscle cell from the red cell.

5.1.4 Smooth Muscle

The plasma membrane of vascular, intestinal, and myometrial smooth muscle seems to be equipped with the exchange system (Kosterin et al. 1983; van Breemen et al. 1979). The prediction by Casteels et al. (1973) that the ATP-driven Ca-pump is also present has recently fully been confirmed. Wuytack et al. demonstrated a $(Ca^{2+} + Mg^{2+})$-ATPase in pig stomach muscle (Wuytack et al. 1981a) which could be purified on a calmodulin affinity column and is calmodulin-dependent (Wuytack et al. 1981b). This inspires confidence that it belongs to the plasma membrane. The protein is able, when incorporated into lipid vesicles, to transport Ca^{2+} in an uphill fashion if ATP is present in the medium (Wuytack et al. 1981a). Antibodies against this enzyme cross-reacted with the pig red cell Ca-pump (Wuytack et al. 1983). In uterine muscle plasma membranes a $(Ca^{2+} + Mg^{2+})$-ATPase was detected (Akerman and Wikstrom 1979; Popescu and Ignat 1983) that is calmodulin-sensitive (Popescu and Ignat 1983). A calmodulin-sensitive Ca-pump was also demonstrated in rat aorta plasma membranes (Morel et al. 1981).

5.1.5 Pancreas Islet Cells

The β-cells of the pancreas are excitable cells in that they respond to glucose with action potentials due to a current which seems to be carried exclusively by Ca^{2+}. A membrane preparation from pancreas islets was able to accumulate Ca^{2+} in an ATP-dependent fashion and a membrane-bound $(Ca^{2+} + Mg^{2+})$-ATPase with two different affinities for ATP and high affinity for Ca^{2+} ($K_{Ca} = 0.05$-0.1 μM) was demonstrated to exist (Colca et al. 1983; Pershadsingh et al. 1980a). If the membranes were prepared in the presence of EDTA, calmodulin had a moderate (Pershadsingh et al. 1980a) and after trifluoperazine treatment a marked (Kotagal et al. 1982) stimulatory effect on this system. It was shown that the endoplasmic reticulum Ca^{2+}-pump is distinct ($K_{Ca} \sim 2$ μM, not sensitive to calmodulin) (Colca et al. 1983).

5.2 Nonexcitable Cells

5.2.1 Epithelia

5.2.1.1 Intestine

The epithelium of the small intestine is well known to achieve transcellular Ca^{2+} transport which is controlled by the level of 1,25-dihydroxy-cholecalciferol [1,25-$(OH)_2 D_3$]. There is agreement that the active transport step takes place at the basolateral membrane. The entry of Ca^{2+} at the brush-border membrane may be the site of regulation by 1,25-$(OH)_2 D_3$.

At the basolateral membrane both Ca^{2+} transport systems exist (Ghijsen et al. 1983; Ghijsen and van Os 1979; Hildmann et al. 1979, 1982; Murer and Hildmann 1981). The separation of vesicular basolateral membranes from vesicular brush-border membranes (Mircheff and Wright 1976) has greatly clarified the situation. Ghijsen and van Os have shown that in basolateral membranes a $(Ca^{2+} + Mg^{2+})$-ATPase exists (Ghijsen and van Os 1979) which is clearly distinct from the alkaline phosphatase (Ghijsen et al. 1983). They produced unmistakable evidence for the existence of an ATP-dependent Ca^{2+} transport in basolateral membrane vesicles (Ghijsen et al. 1983; Ghijsen and van Os 1979) and a similar finding was presented by Hildmann et al. (1982). Ghijsen et al. further demonstrated that a phosphorylated intermediate can be obtained which is different from that of alkaline phosphatase (de Jonge et al. 1981).

The ATP-driven transport is stimulated by calmodulin (Nellans and Popovitch 1981). Finally, Ghijsen and van Os (1982) made it probable that $1,25$-$(OH)_2D_3$ increases the maximal activity of $(Ca^{2+} + Mg^{2+})$-ATPase in the duodenum and ileum which does of course not preclude an additional effect on the Ca^{2+} entry mechanism in the brush-border membrane. Both Ghijsen et al. (1983) and Hildmann et al. (1982) showed that a Na-gradient across basolateral membranes can induce Ca^{2+} movement.

It is not possible to assess the relative contribution of either transport system to the overall transport, because the vesicles are quite leaky to Ca^{2+} and Na^+ and because their sidedness is not ascertained (Hildmann et al. 1982). At least it seems clear the the $(Ca^{2+} + Mg^{2+})$-ATPase activity correlates with transport intensity, being highest in the duodenum and lowest in the ileum (Ghijsen and van Os 1982).

5.2.1.2 Kidney

In kidney tubule cells the situation is very similar to that in intestinal mucosa cells. A $(Ca^{2+} + Mg^{2+})$-ATPase of high Ca^{2+} affinity $(K_{Ca} \sim 5 \,\mu M)$ and a mol. wt. of 130,000 which is sensitive to calmodulin stimulation was found by De Smedt et al. (1981, 1983). Since it was difficult to find the ATPase in isolated basolateral membranes owing to a very high Mg-ATPase activity, it is gratifying that calmodulin activation was clear-cut (2.7-fold). Moore et al. (1974) and Gmaj et al. (1979) described vesicles from kidney plasma membranes that accumulated Ca^{2+} in an ATP- and Mg^{2+}-dependent manner in Na-free media. DeSmedt et al. (1983) were recently able to show in basolateral membranes a phosphoprotein of the molecular weight of the red cell pump, whereas an 85,000 phosphoprotein was attributed to the alkaline phosphatase of the brush-border membrane.

The Na-Ca-exchange revealed itself as follows: Ca-pumping into vesicles was inhibited by 62 mM Na^+ in a K^+-medium but the inhibition disappeared if the vesicles were preincubated with the Na^+-K^+-medium for 20 min (Gmaj et al. 1979), which shows that for inhibition it was the Na^+ gradient that mattered. In addition, Ca^{2+} efflux from such vesicles into EDTA containing solutions was, for a short while (1 min), faster under an inwardly directed Na^+ gradient than under a K^+ gradient (Gmaj et al. 1979).

5.2.2 Liver

Murphy et al. (1980), using a digitonin method similar to that described in the first chapter for red cells, demonstrated that free cytosolic $[Ca^{2+}]$ is 1-2×10^{-7} M in hepatocytes. As early as 1970 van Rossum (1970) had found that hepatocytes are able to extrude Ca^{2+} in a Na_o-independent fashion and that this mechanism determines the total Ca^{2+} content of the cell. A $(Ca^{2+} + Mg^{2+})$-ATPase was demonstrated (Garnett and Kemp 1975) and solubilized (Hope-Gill and Nanda 1979). Lotersztajn et al. (1981) have grappled successfully with the difficulty of distinguishing between the very active Mg-ATPase and the $(Ca^{2+} + Mg^{2+})$-ATPase in liver cell membranes purified on a concanavalin column. They found a $K_{1/2}$ for Ca^{2+} of 13 μM, a low Mg^{2+} requirement ($< 12\,\mu M$) and two sites for ATP. Curiously, the activity was not increased by calmodulin but by another protein present in liver cells. From all this the plasma membrane of liver cells appears to be devoid of the exchange system and to rely solely on the Ca-pump.

5.3 Miscellaneous Cells

Either ATP-dependent Ca^{2+} transport or a $(Ca^{2+} + Mg^{2+})$-ATPase of the red cell type or both were discovered in adipocytes (Pershadsingh et al. 1980b), macrophages (Lew and Stossel 1980), lymphocytes (Sarkadi et al. 1982), neutrophile granulocytes (Lagast et al. 1984; Volpi et al. 1982), L-cells (Lamb and Lindsay 1971), Ehrlich ascites tumor cells (Hinnen et al. 1979), fibroblasts (Katz and Ansah 1980), bone cells (Murray et al. 1983; Shen et al. 1983), corpus luteum cells (Verma and Penniston 1981) and spermatozoa (Breitbart and Rubinstein 1983). The Ca^{2+}-pump seems also to be present in bacterial membranes (Lockau and Pfeiffer 1983) and in those from cells of phanerogamic plants (Gross and Marme 1978). The exchange system was further revealed in lymphocytes (Ueda 1983), erythroleukemia cells (Smith et al. 1982) and secretory granules of lymphocytes (Saermark et al. 1983).

It is true that the evidence for the existence of the Ca^{2+}-transporting systems in the plasma membrane is not equally good in all these cells. Nevertheless, the plasma membrane Ca^{2+}-pump is apparently universal and the exchange system has been demonstrated in very many animal cells. The ability to keep Ca^{2+} from the cell interior is a fundamental property of living cells and it seems not too daring to predict that all living cells are endowed with means to this goal. Furthermore, Blaustein and Hodgkin certainly were right when they wrote in 1969 "... it would clearly be unwise ... to suppose that there is only one type of calcium pump in all cells." (Blaustein and Hodgkin 1969).

References

Akerman KEO, Wikstrom MKF (1979) $(Ca^{2+} + Mg^{2+})$-stimulated ATPase activity of rabbit myometrium plasma membrane is blocked by oxytocin. FEBS Lett 97: 283-287

Akyempon K, Roufogalis BD (1982) The stoichiometry of the Ca^{2+}-pump in human erythrocyte vesicles: modulation by Ca^{2+}, Mg^{2+} and calmodulin. Cell Calcium 3: 1-17

Au KS (1978) An endogenous inhibitor of erythrocyte membrane $(Ca^{2+} + Mg^{2+})$-ATPase. Int J Biochem 9: 477-480

Baker P (1972) Transport and metabolism of calcium ions in nerve. Prog Biophys Mol Biol 24: 177-223

Baker PF, Blaustein MP (1968) Sodium dependent uptake of calcium by crab nerve. Biochim Biophys Acta 150: 167-170

Baker PF, Glitsch HG (1973) Does metabolic energy participate directly in the Na^+-dependent extrusion of Ca^{2+} ions from squid axons. J Physiol (Lond) 233: 44 P

Baker PF, McNaughton PA (1976) Kinetics and energetics of calcium efflux from intact squid giant axons. J Physiol (Lond) 259: 103-144

Baker PF, McNaughton PA (1978) The influence of extracellular calcium binding on the calcium efflux from squid acons. J Physiol (Lond) 276: 127-150

Baker PF, Blaustein MP, Hodgkin AL, Steinhardt RA (1967) The effect of sodium concentration on calcium movements in giant axons of Loligo forbesi. J Physiol (Lond) 192: 43 P

Baker PF, Blaustein MP, Hodgkin AL, Steinhardt RA (1969) The influence of calcium on sodium efflux in squid axons. J Physiol (Lond) 200: 431-458

Barry WH, Smith TW (1982) Mechanism of transmembrane calcium movement in cultured chick embryo ventricular cells. J Physiol (Lond) 325: 243-260

Beauge L, DiPolo R, Osses L, Barnola F, Campos M (1981) A $(Ca^{2+} + Mg^{2+})$-ATPase activity in plasma membrane fragments isolated from squid nerves. Biochim Biophys Acta 644: 147-152

Bers DM, Philipson KD, Nishimoto AY (1980) Sodium-calcium exchange and sidedness of isolated sarcolemmal vesicles. Biochim Biophys Acta 601: 358-371

Blaustein MP (1974) The interrelationship between sodium and calcium fluxes across cell membranes. Rev Physiol Biochem Pharmacol 70: 33-82

Blaustein MP (1977) Effects of internal and external cations and of ATP on sodium-calcium and calcium-calcium exchange in squid axons. Biophys J 20: 79-111

Blaustein MP, Ector AC (1976) Carrier mediated sodium-dependent and calcium-dependent calcium efflux from pinched off presynaptic nerve terminals. Biochim Biophys Acta 419: 295-308

Blaustein MP, Hodgkin AL (1969) The effect of cyanide on the efflux of calcium from squid axons. J Physiol (Lond) 200: 497-527

Blaustein MP, Nelson MT (1982) Sodium-calcium exchange: its role in the regulation of cell calcium. In: Carafoli E (ed) Membrane transport of calcium. Academic London, pp 217-236

Blaustein MP, Oborn CJ (1975) The influence of sodium on calcium fluxes in pinched-off nerve terminals in vitro. J Physiol (Lond) 247: 657-686

Blaustein MP, Russell JM (1975) Sodium-calcium exchange and calcium-calcium exchange in internally dialyzed squid giant axons. J Membr Biol 22: 285-312

Blaustein MP, Wiesman WP (1970) Effect of sodium ions on calcium movements in isolated synaptic terminals. Proc Natl Acad Sci USA 66: 664-671

Blaustein MP, Russell JM, de Weer P (1974) Calcium efflux from internally dialyzed axons: the influence of external and internal cations. J Supramol Struct 2: 558-581

Blaustein MP, Ratzlaff RW, Kendrick NC, Schweitzer ES (1978) Calcium buffering in presynaptic nerve terminals. J Gen Physiol 72: 15-41

Bond GH (1972) Ligand-induced conformational changes in the $(Mg^{2+} + Ca^{2+})$-dependent ATPase of red cell membranes. Biochim Biophys Acta 288: 423-433

Borle AB (1981) Control, modulation and regulation of cell calcium. Rev Physiol Biochem Pharmacol 90: 14-153

Breitbart H, Rubinstein S (1983) Calcium transport by bull spermatozoa plasma membranes. Biochim Biophys Acta 732: 464-468

Brinley FJ Jr. Mullins LJ (1974) Effects of membrane potential on sodium and potassium fluxes in squid axons. Ann NY Acad Sci 242: 406-432

Bürgin H, Schatzmann HJ (1979) The relation between net calcium, alkalication and chloride movement in red cells exposed to salicylate. J Physiol (Lond) 287: 15-32

Carafoli E, Niggli V (1981) Purification and reconstitution of the calcium-magnesium adenosinetriphosphatase of the erythrocyte membrane. Ann NY Acad Sci 358: 159-168

Carafoli E, Zurini M (1982) The Ca^{2+} pumping ATPase of plasma membranes. Reconstitution and properties. Biochim Biophys Acta 683: 279-301

Caride AJ, Rega AF, Garrahan PJ (1982) The role of the sites for ATP of the Ca-ATPase from human red cell membranes during Ca-phosphatase activity. Biochim Biophys Acta 689: 421-428

Caride AJ, Rega AF, Garrahan PJ (1983) Effects of p-nitrophenyl-phosphate on Ca^{2+} transport in inside-out vesicles from human red cell membranes. Biochim Biophys Acta 734: 363-367

Caroni P, Carafoli E (1980) An ATP dependent Ca-pumping system in dog heart sarcolemma. Nature (Lond) 283: 765-767

Caroni P, Carafoli E (1981a) The Ca^{2+}-pumping ATPase of heart sarcolemma. J Biol Chem 256: 3267-3270

Caroni P, Carafoli E (1981b) Regulation of Ca-pumping ATPase of heart sarcolemma by a phosphorylation-dephosphorylation process. J Biol Chem 256: 9371-9373

Caroni P, Reinlib L, Carafoli E (1980) Charge movement during Na^{+}-Ca^{2+} exchange in heart sarcolemmal vesicles. Proc Natl Acad Sci USA 77: 6354-6358

Caroni P, Zurini M, Clark A (1982) The calcium-pumping ATPase of heart sarcolemma. Ann NY Acad Sci 402: 402-420

Casteels R, Goffin J, Raeymaekers L, Wuytack F (1973) Calcium-pumping in the smooth muscle cells of taenia coli. J Physiol (Lond) 231: 19 p

Cavieres JD (1983) The molecular size of the red cell calcium pump with and without calmodulin. J Physiol (Lond) 343: 96 p

Cavieres JD (1984) Calmodulin and the target size of the $(Ca^{2+} + Mg^{2+})$ ATPase of human red cell ghosts. Biochim Biophys Acta 771: 241-244

Chiesi M, Zurini M, Carafoli E (1984) ATP synthesis catalyzed by purified Ca-ATPase in the absence of Ca-gradients. Biochemistry 23: 2595-2599

Clark A, Carafoli E (1983) The stoichiometry of the Ca^{2+}-pumping ATPase of erythrocytes. Cell Calcium 4: 83-88

Colca JR, Kotagal N, Lacy PE, McDaniel M (1983) Comparison of the properties of active Ca^{2+} transport by the islet cell endoplasmic reticulum and plasma membrane. Biochim Biophys Acta 729: 176-184

Colquhoun D, Neher E, Reuter H, Stevens CF (1981) Inward current channels activated by intracellular Ca^{2+} in cultured cardiac cells. Nature (Lond) 294: 752-754

Cox JA, Comte M, Stein EA (1982) Activation of human erythrocyte Ca^{2+} dependent Mg^{2+} activated ATPase by calmodulin and calcium: quantitative analysis. Proc Natl Acad Sci USA 79: 4265-4269

de Jonge HR, Ghijsen WEJM, van Os CH (1981) Phosphorylated intermedites of Ca^{2+} ATPase and alkaline phosphatase in plasma membranes from rat duodenal epithelium. Biochim Biophys Acta 647: 140-149

de Meis L, Inesi G (1982) ATP synthesis by sarcoplasmic reticulum ATPase following Ca^{2+}, pH, temperature and water activity jumps. J Biol Chem 257: 1289-1294

de Meis L, Martins OB, Alves EW (1980) Role of water, hydrogen ion and temperature on the synthesis of ATP by the sarcoplasmic reticulum ATPase in the absence of a calcium ion gradient. Biochemistry 19: 4252-4261

de Smedt H, Parys JB, Borghgraef R, Wuytack F (1981) Calmodulin stimulation of renal $(Ca^{2+} + Mg^{2+})$-ATPase. FEBS Lett 131: 60-62

de Smedt H, Parys JB, Borghgraef R, Wuytack F (1983) Phosphorylated intermediate of $(Ca^{2+} + Mg^{2+})$-ATPase and alkaline phosphatase in renal plasma membranes. Biochim Biophys Acta 728: 409-418

di Polo R (1978) Ca^{2+}-pump driven by ATP in squid axons. Nature (Lond) 274: 390-392

di Polo R, Beauge L (1979) Physiological role of ATP-driven calcium pump in squid axon. Nature (Lond) 278: 271-273

di Polo R, Beauge L (1980) Mechanisms of calcium transport in the giant axon of the squid and their physiological role. Cell Calcium 1: 147-169

di Polo R, Beauge L (1981) The effects of vanadate on calcium transport in dialysed squid axons. Sidedness of vanadate-cation interaction. Biochim Biophys Acta 645: 229-236

di Polo R, Rojas HR, Beauge L (1979) Vanadate inhibits uncoupled Ca-efflux but not Na-Ca exchange in squid axons. Nature (Lond) 281: 228-229

Dunham ET, Glynn IM (1961) Adenosinetriphosphatase activity and the active movement of alkali metal ions. J Physiol (Lond) 156: 274-293

Enyedi A, Sarkadi B, Gardos G (1982a) On the substrate specificity of the red cell calcium pump. Biochim Biophys Acta 687: 109-112

Enyedi A, Sarkadi B, Nyers A, Gardos G (1982b) Effects of divalent metal ions on the calcium pump and membrane phosphorylation in human red cells. Biochim Biophys Acta 690: 41-49

Ferreira HG, Lew VL (1976) Use of ionophore A23187 to measure cytoplasmic Ca buffering and activation of the Ca-pump by internal calcium. Nature (Lond) 259: 47-49

Ferreira HG, Lew VL (1977) Passive Ca transport and cytoplasmic Ca buffering in intact red cells. In: Ellory JC, Lew VL (eds) Membrane transport in red cells. Academic London, pp 53-91

Foder B, Scharff O (1981) Decrease of apparent calmodulin affinity of erythrocyte ($Ca^{2+} + Mg^{2+}$)-ATPase at low Ca^{2+} concentration. Biochim Biophys Acta 649: 367-376

Garnett HM, Kemp RB (1975) ($Ca^{2+} + Mg^{2+}$)-activated ATPase in the plasma membrane of mouse liver cells. Biochim Biophys Acta 382: 526-533

Garrahan PJ, Rega AF (1978) Activation of partial reactions of the Ca^{2+}-ATPase from human red cells by Mg^{2+} and ATP. Biochim Biophys Acta 513: 59-65

Garrahan PJ, Rega AF, Alonso GL (1976) The interaction of magnesium with the calcium pump of sarcoplasmic reticulum. Biochim Biophys Acta 448: 121-132

Ghijsen WEJ, van Os CH (1979) Ca-stimulated ATPase in brush-border and basolateral membranes of rat duodenum with high affinity sites for Ca ions. Nature (Lond) 279: 802-803

Ghijsen EWJ, van Os CH (1982) 1 α-25 dihydroxy-vitamin D_3 regulates ATP dependent Ca transport in basolateral plasma membranes of rat enterocytes. Biochim Biophys Acta 689: 170-172

Ghijsen WEJ, de Jonge MD, van Os CH (1983) Kinetic properties of Na^+/Ca^{2+} exchange in basolateral plasma membranes of rat small intestine. Biochim Biophys Acta 730: 85-94

Gietzen K, Tejcka M, Wolf HU (1980a) Calmodulin affinity chromatography yields a functional purified erythrocyte ($Ca^{2+} + Mg^{2+}$)-dependent adenosine triphosphatase. Biochem J 189: 81-88

Gietzen K, Seiler S, Fleischer S, Wolf HU (1980b) Reconstitution of the Ca^{2+} transport system of human erythrocytes. Biochem J 188: 47-54

Gietzen K, Kolandt J, Konstantinova A, Bader H (1984) Antibodies against red blood cell Ca^{2+}-transporting ATPase: Studies on enzyme function and determination of antigenic sites. Hoppe-Seylers Z Physiol Chem 365: 236

Gill DL, Grollman EF, Kohn LD (1981) Calcium transport mechanisms in membrane vesicles frog pig brain synaptosomes. J Biol Chem 256: 184-192

Glitsch HG, Reuter H, Scholtz H (1970) The effect of the internal sodium concentration on calcium fluxes in isolated guinea pig auricles. J Physiol (Lond) 209: 25-43

Gmaj P, Murer H, Kinne R (1979) Calcium ion transport across plasma membranes isolated from rat kidney. Biochem J 178: 549-557

Gopinath RM, Vincenzi FF (1977) Phosphodiesterase protein activator mimics red blood cell cytoplsmic activator of ($Ca^{2+} + Mg^{2+}$)-ATPase. Biochem Biophys Res Commun 77: 1203-1209

Graf E, Penniston JT (1981) CaATP. The substrate, at low ATP concentrations, of Ca^{2+}-ATPase from human erythrocyte membranes. J Biol Chem 256: 1587-1592

Gross J, Marme D (1978) ATP-dependent Ca^{2+}-uptake into plant membrane vesicles. Proc Natl Acad Sci USA 75: 1232-1236

Haaker H, Racker E (1979) Purification and reconstituttion of the Ca^{2+}-ATPase from plasma membranes of pig erythrocytes. J Biol Chem 254: 6598-6602

Hakim G, Itano T, Verma AK, Penniston JT (1982) Purification of Ca^{2+} and Mg^{2+} requiring ATPase from rat brain synaptic plasma membrane. Biochem J 207: 225-231

Harrison DG, Long C (1968) The calcium content of human erythrocytes. J Physiol (Lond) 199: 367-381

Hildmann B, Schmidt A, Murer H (1979) Ca^{2+} transport in basal-lateral plasma membranes isolated from rat small intestinal epithelial cells. Pflügers Arch Eur J Physiol 382: R 23

Hildmann B, Schmidt A, Murer H (1982) Ca^{2+} transport across basal-lateral plasma membranes from rat small intestinal epithelial cells. J Membr Biol 65: 55-62

Hinds TR, Shattuck RR, Vincenzi FF (1982) Elucidation of a possible multimeric structure of the human RBC Ca^{2+}-pump ATPase. Acta Physiol Lat Am 32: 97-98

Hinnen R, Miyamoto H, Racker E (1979) Ca^{2+} translocation in Ehrlich ascites tumor cells. J Membr Biol 49: 309-324

Hope-Gill HF, Nanda V (1979) Stimulation of calcium ATPase by insulin, glucagon, cyclic AMP, and cyclic GMP in Triton-S-100 extracts of purified rat liver plasma membrane. Horm Metab Res 11: 698-700

Hui CW, Drummond G, Drummond I (1976) Calcium accumulation and cyclic AMP-stimulated phosphorylation in plasma membrane-enriched preparations of myocardium. Arch Biochem Biophys 173: 415-427

Jarrett HW, Kyte J (1979) Human erythrocyte calmodulin. J Biol Chem 254: 8237-8244

Jarrett HW, Penniston JT (1977) Partial purification of the $Ca^{2+} + Mg^{2+}$ ATPase activator from human erythrocytes. Its similarity to the activator of 3':5'-cyclic nucleotide phosphodiesterase. Biochem Biophys Res Commun 77: 1210-1216

Jarrett HW, Penniston JT (1978) Purification of the Ca^{2+}-stimulated ATPase activator from human erythrocytes. J Biol Chem 253: 4676-4682

Jundt H, Porzig H, Reuter H, Stucki JW (1975) The effect of substances releasing intracellular calcium ions on sodium-dependent calcium efflux from guinea pig auricles. J Physiol (Lond) 246: 229-253

Jundt H, Reuter H (1977) Is sodium actiavted calcium efflux from mammalian cardiac muscle dependent on metabolic energy? J Physiol (Lond) 266: 78 p

Katz S, Ansah TA (1980) $(Mg^{2+} + Ca^{2+})$-ATPase activity in plasma membrane enriched preparations of human skin fibroblasts: decreased activity in fibroblasts derived from cystic fibrosis patients. Clin Chim Acta 100: 245-252

Kendrick NC, Blaustein MP, Fried RC, Ratzlaff RW (1977) ATP-dependent calcium storage in presynaptic nerve terminals. Nature (Lond) 265: 246-248

Knauf PA, Proverbio F, Hoffman JF (1972) Spatial separation of Ca-ATPase from Na-K-ATPase in human red cell ghosts. 4th Int Biophys Congress Moscow 3: 104

Knauf PA, Proverbio F, Hoffman JG (1974) Electrophoretic separation of different phosphoproteins associated with Ca-ATPase and Na-K-ATPase in human red cell ghosts. J Gen Physiol 63: 324-336

Kosterin SA, Bratkova NF, Kurskii MD, Zimina VP (1983) Properties of ATP-dependent Ca^{2+} transport system in plasma membrane fraction of myometrial cells. Biochemistry (translated from Biokhimiya) 48: 212-220

Kotagal N, Patke C, Landt M, McDonald JM, Colca J, Lacy P, McDaniel ML (1982) Regulation of pancreatic islet-cell plasma membrane Ca^{2+}-Mg^{2+} ATPase by calmodulin. FEBS Lett 137: 249-252

Kratje RB, Garrahan PJ, Rega AF (1983) The effects of alkali metal ions on active Ca^{2+} transport in reconstituted ghosts from human red cells. Biochim Biophys Acta 731: 40-46

Lagast H, Lew PD, Waldvogel FA (1984) ATP-dependent Ca^{2+}-pump in the plasma membrane of guinea pig and human neutrophiles. J Clin Invest 73: 107-115

Lamb JF, Lindsay R (1971) Effect of Na, metabolic inhibitors and ATP on Ca movements in L-cells. J Physiol (Lond) 218: 691-708

Lamers JMJ, Stinis JT (1981) An electrogenic Na^+/Ca^{2+} antiporter in addition to the Ca^{2+}-pump in cardiac sarcolemma. Biochim Biophys Acta 640: 521-534

Larocca JN, Rega AF, Garrahan PJ (1981) Phosphorylation and dephosphorylation of the Ca^{2+}-pump of human red cells in the presence of monovalent cations. Biochim Biophys Acta 645: 10-16

Larsen VL, Vincenzi FF (1979) Calcium transport across the plasma membrane: Stimulation by calmodulin. Science (Wash DC) 204: 306-309

Lederer WJ, Nelson MT (1983) Effects of extracellular sodium on calcium efflux and membrane current in single muscle cells from the barnacle. J Physiol (Lond) 341: 325-339

Lee KS, Au KS (1981) Inhibitor protein of pig erythrocyte membrane $(Ca^{2+} + Mg^{2+})$-ATPase. Biochem Soc Trans 9: 132 p

Lee KS, Shin BC (1969) Studies on the active transport of calcium in human red cells. J Gen Physiol 54: 713-728

Lew PD, Stossel TP (1980) Calcium transport by macrophage plasma membranes. J Biol Chem 255: 5841-5846

Lew VL, Tsien RY, Miner C, Bookchin RM (1982) Physiological $(Ca^{2+})_i$ level and pump-leak in intact red cells measured using an incorporated Ca chelator. Nature (Lond) 298: 478-481

Lichtner R, Wolf HU (1980) Characterization of the phosphorylated intermediate of the isolated high-affinity $(Ca^{2+} + Mg^{2+})$-ATPase of human erythrocyte membranes. Biochim Biophys Acta 598: 486-493

Lockau W, Pfeiffer S (1983) ATP-dependent calcium transport in membrane vesicles of the cyano-bacterium Anabaena variabilis. Biochim Biophys Acta 733: 124-132

Lotersztajn S, Hanoune J, Pecker F (1981) A high affinity calcium-stimulated magnesium-dependent ATPase in rat liver plasma membranes. J Biol Chem 256: 11209-11215

Luterbacher S (1982) Die Teilreaktionen der ATP Spaltung durch das isolierte Protein der Ca^{2+}-Pumpe an Erythrocytenmembranen. Thesis, University of Bern

Luterbacher S, Schatzmann HJ (1983) The site of action of La^{3+} in the reaction cycle of the human red cell membrane Ca^{2+}-pump ATPase. Experientia (Basel) 39: 311-312

Martonosi AN (1983) The regulation of cytoplasmic Ca^{2+} concentration in muscle and nonmuscle cells. In: Muscle and nonmuscle motility, vol 1. Academic London, pp 233-357

Minocherhomjee AM, Beauregard G, Potier M, Roufogalis BD (1983) The molecular weight of the calcium-transport-ATPase of the human red blood cell determined by radiation inactivation. Biochem Biophys Res Commun 116: 895-900

Mircheff AK, Wright EM (1976) Analytical isolation of intestinal epithelial cells: identification of Na, K-ATPase rich membranes and the distribution of enzyme activities. J Membr Biol 28: 309-333

Miyamoto H, Racker E (1980) Solubilization and partial purification of the Ca^{2+}/Na^+ antiporter from plasma membrane of bovine heart. J Biol Chem 255: 2656-2658

Moore L, Fitzpatrick DF, Chen TS, Landon EJ (1974) Calcium pump activity of the renal plasma membrane and renal microsomes. Biochim Biophys Acta 345: 405-418

Morel N, Wibo M, Godraind T (1981) A calmodulin-stimulated Ca^{2+}-pump in rat aorta plasma membrane. Biochim Biophys Acta 644: 82-88

Muallem S, Karlish SJD (1979) Is the red cell calcium pump regulated by ATP? Nature (Lond) 277: 238-240

Muallem S, Karlish SJD (1980) Regulatory interaction between calmodulin and ATP on the red blood cell Ca^{2+}-pump. Biochim Biophys Acta 597: 631-636

Muallem S, Karlish SJD (1981) Studies on the mechanism of regulation of the red cell Ca^{2+}-pump by calmodulin and ATP. Biochim Biophys Acta 647: 73-86

Mullins LJ (1977) A mechanism for Na/Ca transport. J Gen Physiol 70: 681-695

Mullins LJ (1981) Ion transport in heart. Raven, New York

Mullins LJ, Brinley FJ (1975) Sensitivity of calcium efflux from squid axons to changes in membrane potential. J Gen Physiol 65: 135-152

Murer H, Hildmann B (1981) Transcellular transport of calcium and inorganic phosphate in the small intestinal epithelium. Am J Physiol 240: G 409-G 416

Murphy E, Coll K, Rich TL, Williamson JR (1980) Hormonal effects on calcium homeostasis in isolated hepatocytes. J Biol Chem 255: 6600-6608

Murray E, Gorski JP, Penniston JT (1983) High affinity Ca^{2+}-stimulated and Mg^{2+}-dependent ATPase from rat osteosarcoma plasma membranes. Biochem Int 6: 527-532

Nellans HN, Popovitch JE (1981) Calmodulin-regulated ATP-driven calcium transport by baso-lateral membranes of rat small intestine. J Biol Chem 256: 9932-9936

Niggli V, Penniston JT, Carafoli E (1979) Purification of the $(Ca^{2+} + Mg^{2+})$-ATPase from human erythrocyte membranes using a calmodulin affinity column. J Biol Chem 254: 9955-9958

Niggli V, Adunyah ES, Carafoli E (1981a) Acid phospholipids, unsaturated fatty acids and limited proteolysis mimic the effect of calmodulin on the purified erythrocyte Ca^{2+}-ATPase. J Biol Chem 256: 8588-8592

Niggli V, Adunyah ES, Penniston JT, Carafoli E (1981b) Purified $(Ca^{2+} + Mg^{2+})$-ATPase of the erythrocyte membrane. Reconstitution and effect of calmodulin and phospholipids. J Biol Chem 256: 395-401

Niggli V, Sigel E, Carafoli E (1982) The purified Ca^{2+}-pump of human erythrocyte membranes catalyzes an electroneutral Ca^{2+}-H^+ exchange in reconstituted liposomal systems. J Biol Chem 257: 2350-2356

Olson EJ, Cazort RJ (1969) Active calcium and strontium transport in human erythrocyte ghosts. J Gen Physiol 53: 311-322

Papazian D, Rahamimoff H, Goldin SM (1979) Reconstitution and purification by transport specific fractionation of an ATP-dependent calcium transport component from synaptosome-derived vesicles. Proc Natl Acad Sci USA 76: 3708-3712

Penniston JT (1982) Substrate specificity of erythrocyte Ca^{2+}-ATPase. Biochim Biophys Acta 688: 735-739

Pershadsingh HA, McDaniel ML, Landt M, Bry CG, Lacy PF, McDonald JM (1980a) Ca^{2+}-activated ATPase and ATP-dependent calmodulin-stimulated Ca^{2+} transport in islet cell plasma membrane. Nature (Lond) 288: 492-495

Pershadsingh HA, Landt M, McDonald JM (1980b) Calmodulin-sensitive ATP-dependent Ca^{2+} transport across adipocyte plasma membranes. J Biol Chem 255: 8983-8986

Pitts BJR (1979) Stoichiometry of sodium-calcium exchange in cardiac sarcolemmal vesicles. Coupling to the sodium pump. J Biol Chem 254: 6232-6235

Popescu LM, Ignat P (1983) Calmodulin-dependent Ca^{2+}-pump ATPase of human smooth muscle sarcolemma. Cell Calcium 4: 219-235

Quist EE, Roufogalis BD (1975) Determination of stoichiometry of the calcium pump in human erythrocytes using lanthanum as a selective inhibitor. FEBS Lett 50: 135-138

Reeves JP, Sutko JL (1979) Sodium-calcium ion exchange in cardiac membrane vesicles. Proc Natl Acad Sci USA 76: 590-594

Reeves JP, Sutko JL (1980) Sodium-calcium exchange activity generates a current in cardiac membrane vesicles. Science (Wash DC) 208: 1461-1464

Reeves JP, Sutko JL (1983) Competitive interaction of sodium and calcium with the sodium-calcium exchange system of cardiac sarcolemmal vesicles. J Biol Chem 258: 3178-3182

Rega AF, Garrahan PJ (1975) Calcium ion dependent phosphorylation of human erythrocyte membranes. J Membr Biol 22: 313-327

Rega AF, Richards DE, Garrahan PJ (1973) Calcium ion dependent p-nitrophenylphosphate phosphatase activity and calcium ion dependent ATPase activity from human erythrocyte membranes. Biochem J 136: 185-194

Rega AF, Richards DE, Garrahan PJ (1974) The effects of Ca^{2+} on ATPase and phosphatase activities of erythrocyte membranes. Ann NY Acad Sci 242: 317-323

Requena J, Mullins JL (1979) Calcium movements in nerve fibres. Rev Biophys 12: 371-460

Requena J, di Polo F, Brinley FJ Jr, Mullins LJ (1977) The control of ionized calcium in squid axons. J Gen Physiol 70: 329-353

Reuter H (1982) Na-Ca countertransport in cardiac muscle. In: Martonosi A (ed) Membranes and transport, vol 1. Plenum, New York, pp 623-631

Reuter H, Seitz N (1968) The dependence of caclium efflux from cardiac muscle on temperature and external ion composition. J Physiol (Lond) 195: 451-470

Richards DE, Vidal JC, Garrahan PJ, Rega AF (1977) ATPase and phosphatase activities from human red cell membranes: II The effects of phospholipases on Ca^{2+}-dependent enzyme activities. J Membr Biol 35: 125-136

Richards DE, Rega AF, Garrahan PJ (1978) Two classes of site for ATP in the Ca^{2+}-ATPase from human red cell membranes. Biochim Biophys Acta 511: 194-201

Robinson JD (1976) $(Ca^{2+} + Mg^{2+})$ stimulated ATPase activity of a rat brain microsomal preparation distributed with Na-K-ATPase in sucrose density gradients. Arch Biochem Biophys 176: 366-374

Robinson JD (1978) Ca stimulated phosphorylation of brain (Ca + Mg)-ATPase preparation. FEBS Lett 87: 261-264

Roelofsen B (1981) The (non)specificity in the lipid requirement of calcium- and (sodium plus potassium)- transporting ATPases. Life Sci 29: 2235-2247

Roelofsen B, Schatzmann HJ (1977) The lipid requirement of the $(Ca^{2+} + Mg^{2+})$-ATPase in the human erythrocyte membrane, as studied by various highly purified phospholipases. Biochim Biophys Acta 464: 17-36

Romero PJ (1981) Active calcium transport in red cell ghosts resealed in dextran solutions. Biochim Biophys Acta 649: 404-418

Rossi JPF, Schatzmann HJ (1982a) Is the red cell calcium pump electrogenic? J Physiol (Lond) 327: 1-15

Rossi JPF, Schatzmann HJ (1982b) Trypsin activation of the red cell Ca^{2+}-pump ATPase is calcium-sensitive. Cell Calcium 3: 583-590

Rossi JPF, Garrahan PJ, Rega AF (1978) Reversal of the calcium pump in human red cells. J Membr Biol 44: 37-46

Rossi JPF, Garrahan PJ, Rega AF (1981) Vanadate inhibition of active Ca^{2+} transport across human red cell membranes. Biochim Biophys Acta 648: 145-150

Roufogalis BD (1979) Regulation of calcium translocation across the red cell membrane. Can J Physiol Pharmacol 57: 1331-1349

Russell JM, Blaustein MP (1974) Calcium efflux from barnacle muscle fibres. J Gen Physiol 63: 144-167

Saermark T, Thorn NA, Gratzl M (1983) Calcium-sodium exchange in purified secretory vesicles from bovine lymphocytes. Cell Calcium 4: 151-170

Sarkadi B (1980) Active calcium transport in human red cells. Biochim Biophys Acta 604: 159-190

Sarkadi B, Szasz I, Gardos G (1976) The use of ionophores for rapid loading of human red cells with radioactive cations for cation pump studies. J Membr Biol 26: 357-370

Sarkadi B, Szasz I, Gerloczi A, Gardos G (1977) Transport parameters and stoichiometry of active calcium ion extrusion in intact human red cells. Biochim Biophys Acta 464: 93-107

Sarkadi B, MacIntyre JD, Gardos G (1978) Kinetics of active calcium transport in inside-out red cell membrane vesicles. FEBS Lett 89: 78-82

Sarkadi B, Szasz I, Gardos G (1979) On the red cell Ca^{2+}-pump: an estimate of stoichiometry. J Membr Biol 46: 183-184

Sarkadi B, Enyedi A, Gardos G (1980a) Molecular properties of the red cell calcium pump I/II. Cell Calcium 1: 287-310

Sarkadi B, Szasz I, Gardos G (1980b) Characteristics and regulation of active calcium transport in inside-out red cell membrane vesicles. Biochim Biophys Acta 598: 326-338

Sarkadi B, Enyedi A, Gardos G (1981) Metal-ATP complexes as substrates and free metal ions as activators of the red cell calcium pump. Cell Calcium 2: 449-458

Sarkadi B, Enyedi A, Szasz I, Gardos G (1982) Active calcium transport and calcium-dependent membrane phosphorylation in human peripheral blood lymphocytes. Cell Calcium 3: 163-182

Scharff O (1978) Stimulating effect of monovalent cations on activator-dissociated and activator-associated states of Ca^{2+}-ATPase in human erythrocytes. Biochim Biophys Acta 512: 309-317

Scharff O (1980) Kinetics of calcium-dependent membrane ATPase in human erythrocytes. In: Lassen UV, Ussing HH, Wieth JOW (eds) Membrane transport in erythrocytes. Munksgaard, Copenhagen, pp 236-254

Scharff O, Foder B (1978) Reversible shift between two states of Ca^{2+} ATPase in human erythrocytes mediated by Ca^{2+} and a membrane-bound activator. Biochim Biophys Acta 509: 67-77

Scharff O, Foder B (1982) Rate constants for calmodulin binding to Ca^{2+}-ATPase in erythrocyte membranes. Biochim Biophys Acta 691: 133-143

Scharff O, Foder B, Skibsted U (1983) Hysteretic activation of the Ca^{2+}-pump revealed by calcium transients in human red cells. Biochim Biophys Acta 730: 295-305

Schatzmann HJ (1966) ATP-dependent Ca^{2+} extrusion from human red cells. Experientia (Basel) 22: 364-368

Schatzmann HJ (1969) Transmembrane calcium movements in resealed human red cells. In: Cuthbert AW (ed) Calcium and cellular function. MacMillan, New York, pp 85-95

Schatzmann HJ (1973) Dependence on calcium concentration and stoichiometry of the calcium pump in human red cells. J Physiol (Lond) 235: 551-569

Schatzmann HJ (1975) Active calcium transport and Ca^{2+}-activated ATPase in human red cells. Current topics in membr transport, vol 6. Academic New York, pp 125-168

Schatzmann HJ (1982a) Active calcium transport in human red blood cells. In: Martonosi A (ed) Membranes and transport, vol 1. Plenum, New York, pp 601-605

Schatzmann HJ (1982b) The plasma membrane calcium pump of erythrocytes and other animal cells. In: Carafoli E (ed) Membrane transport of calcium. Academic London, pp 41-108

Schatzmann HJ (1983) The red cell calcium pump. Annu Rev Physiol 45: 303-312

Schatzmann HJ, Bürgin H (1978) Calcium in human red blood cells. Ann NY Acad Sci 307: 125-147

Schatzmann HJ, Roelofsen B (1977) Characteristics of the Ca-pump in human red blood cells (HRBC). In: Semenza G, Carafoli E (eds) Biochemistry of membrane transport. Springer, Berlin Heidelberg New York, pp 389-400

Schatzmann HJ, Rossi GL (1971) $(Ca^{2+} + Mg^{2+})$-activated membrane ATPases in human red cells and their possible relation to cation transport. Biochim Biophys Acta 241: 379-392

Schatzmann HJ, Vincenzi FF (1969) Calcium movements across the membrane of human red cells. J Physiol (Lond) 201: 369-395

Schatzmann HJ, Bürgin H, Luterbacher S, Stieger J, Wüthrich A, Rossi JP (1982) How to keep intracellular calcium low. The red cell as an example, vol 6. In: Dumont JE, Nunez J, Schultz G (Eds) Hormones and cell regulation. Elsevier Biomedical, Amsterdam, pp 13-25

Schellenberg GD, Swanson PD (1982a) Solubilization and reconstitution of membranes containing the Na^+-Ca^{2+} exchange carrier from rat brain. Biochim Biophys Acta 690: 133-144

Schellenberg GD, Swanson PD (1982b) Properties of the Na^+-Ca^{2+} exchange transport system from rat brain: inhibition by amiloride. Fed Proc 41: 673

Shen V, Kohler G, Peck WA (1983) A high affinity, calmodulin responsive $(Ca^{2+} + Mg^{2+})$-ATPase in isolated bone cells. Biochim Biophys Acta 727: 230-238

Simons TJB (1976) Calcium-dependent potassium exchange in human red cell ghosts. J Physiol (Lond) 256: 227-244

Simons TJB (1982) A method for estimating Ca^{2+} within human red blood cells, with an application to the study of their Ca-dependent K permeability. J Membr Biol 66: 235-247

Simonsen LO, Gomme J, Lew VL (1982) Uniform ionophore A23187 distribution and cytoplasmic calcium buffering in intact human red cells. Biochim Biophys Acta 692: 431-440

Smallwood JI, Waisman DM, Lafreniere D, Rasmussen H (1983) Evidence that the erythrocyte calcium pump catalyzes a Ca^{2+}:n H^+ exchange. J Biol Chem 258: 11092-11097

Smith RL, Macara IG, Levenson R, Housman D, Cantley L (1982) Evidence that Na^+/Ca^{2+} antiport system regulates murine erythroleucaemia cell differentiation. J Biol Chem 257: 773-780

Stieger J (1982) Charakterisierung und Rekonstitution der isolierten Ca^{2+}-Pumpe aus der Erythrocytenmembran. Thesis, University of Bern

Stieger J, Luterbacher S (1981a) Some properties of the purified $(Ca^{2+} + Mg^{2+})$-ATPase from human red cell membranes. Biochim Biophys Acta 641: 270-275

Stieger J, Schatzmann HJ (1981b) Metal requirement of the isolated red cell Ca^{2+}-pump ATPase after elimination of calmodulin dependence by trypsin attack. Cell Calcium 2: 601-616

Sulakhe PV, Leung N, St Louis PJ (1976) Stimulation of calcium accumulation in cardiac sarcolemma by protein kinase. Can J Biochem 54: 438-445

Sulakhe PV, St Louis PJ (1980) Passive and active calcium fluxes across plasma membranes. Prog Biophys Mol Biol 35: 135-195

Szasz I, Hasitz M, Sarkadi B, Gardos G (1978) Phopsphorylation of the Ca^{2+}-pump intermediate in intact cells, isolated membranes and inside-out vesicles. Mol Cell Biochem 22: 147-152

Taverna RD, Hanahan DJ (1980) Modulation of human erythrocyte $(Ca^{2+} + Mg^{2+})$-ATPase activity by phospholipase A and proteases. A comparison with calmodulin. Biochem Biophys Res Commun 94: 652-659

Tiffert T, Garcia-Sancho J, Lew VL (1984) Irreversible ATP depletion caused by low concentrations of formaldehyde and of Ca-chelator esters in intact human red cells. Biochim Biophys Acta. 773: 143-156

Trumble WR, Stuko JL, Reeves JP (1980) ATP-dependent calcium transport in cardiac sarcolemmal membrane vesicles. Life Sci 27: 207-214

Tsien RY (1981) A non-disruptive technique for loading calcium buffers and indicators into cells. Nature (Lond) 290: 527-528

Tsien RY, Pozzan T, Rink TJ (1982a) T-cell mitogens cause early changes in cytoplasmic free Ca^{2+} and membrane potential in lymphocytes. Nature (Lond) 295: 68-71

Tsien RY, Pozzan T, Rink TJ (1982b) Calcium homeostasis in intact lymphocytes: cytoplasmic free calcium monitored with a new intracellular trapped fluorescence indicator. J Cell Biol 94: 325-334

Ueda T (1983) Na^+-Ca^{2+} exchange activity in rabbit lymphocyte plasma membranes. Biochim Biophys Acta 734: 342-346

van Breemen C, Aaronson P, Loutzenhiser R (1979) Sodium-calcium interactions in mammalian smooth muscle. Pharmacol Rev 30: 167-208

van Rossum GDV (1970) Net movement of calcium and magnesium in slices of rat liver. J Gen Physiol 55: 18-32

Verma AK, Penniston JT (1981) A high-affinity Ca^{2+}-stimulated and Mg^{2+}-dependent ATPase in rat corpus luteum plasma membrane fractions. J Biol Chem 256: 1269-1275

Verma AK, Gorski JP, Penniston JT (1982) Antibodies directed toward human erythrocyte Ca^{2+}-ATPase: effect on enzyme function and immunoreactivity of Ca^{2+}-ATPases from other sources. Arch Biochem Biophys 215: 345-354

Vestergaard-Bogind B, Bennekou P (1982) Calcium-induced oscillations in K^+ conductance and membrane potential of human erythrocytes mediated by the ionophore A23187. Biochim Biophys Acta 688: 37-44

Vincenzi FF, Hinds TR (1980) Calmodulin and plasma membrane calcium transport. In: Cheung WY (ed) Calcium and cell function. Academic New York, pp 127-138

Volpi M, Naccache PH, Sha'afi RI (1982) Preparation of inside-out membrane vesicles from neutrophiles capable of actively transporting calcium. Biochem Biophys Res Commun 106: 123-130

Waisman DM, Gimble J, Goodman DBP, Rasmussen H (1981a) Studies of the Ca^{2+} transport mechanism of human erythrocyte inside-out plasma membrane vesicles. II Stimulation of the Ca^{2+}-pump by phosphate. J Biol Chem 256: 415-419

Waisman DM, Gimble J, Goodman DBP, Rasmussen H (1981b) III. Stimulation of the Ca^{2+}-pump by anions. J Biol Chem 256: 420-424

Widdas WF (1980) The asymmetry of the hexosetransfer system in the human red cell membrane. Current topics membr transp 14: 165-223. Academic New York

Wilbrandt W, Koller H (1948) Die Calciumwirkung am Froschherzen als Function des Ionengleichgewichts zwischen Zellmembran und Umgebung. Helv Physiol Pharmacol Acta 6: 208-221

Wiley JF, Gill FM (1976) Red cell calcium leak in congenital haemolytic anaemia with extreme microcytosis. Blood 47: 197-210

Wins T, Schoffeniels E (1966) Properties of a $Mg^{2+} + Ca^{2+}$-dependent ATPase. Biochim Biophys Acta 120: 341-350

Wolf HU, Dieckvoss G, Lichtner R (1977) Purification and properties of high-affinity Ca^{2+}-ATPase from human erythrocyte membranes. Acta Biol Med Ger 36: 847-858

Wüthrich A, Schatzmann HJ, Romero P (1979) Net ATP synthesis by running the red cell calcium pump backwards. Experientia (Basel) 35: 1589-1590

Wuytack F, de Schutter G, Casteels R (1981a) Partial purification of $(Ca^{2+} + Mg^{2+})$-dependent ATPase from pig smooth muscle and reconstitution of an ATP-dependent Ca^{2+}-transported system. Biochem J 198: 265-271

Wuytack F, de Schutter G, Casteels R (1981b) Purification of $(Ca^{2+} + Mg^{2+})$-ATPase from smooth muscle by calmodulin affinity chromatography. FEBS Lett 129: 297-300

Wuytack F, de Schutter G, Verbist J, Casteels R (1983) Antibodies to the calmodulin-binding Ca^{2+}-transport ATPase from smooth muscle. FEBS Lett 154: 191-195

Zurini M, Krebs J, Penniston JT, Carafoli E (1984) Controlled proteolysis of the purified Ca^{2+}-ATPase of the erythrocyte membrane. A correlation between structure and function of the enzyme. J Biol Chem 259: 618-627

Voltage-Dependent Gating of Single Calcium Channels in the Cardiac Cell Membrane and Its Modulation by Drugs

W. TRAUTWEIN and D. PELZER [1]

CONTENTS

---

[1] Department of Physiology II, University of the Saarland, 6650 Homburg/Saar, FRG

1 Introduction

An increase in the free calcium ion concentration, [Ca], within stimulated excitable cells underlies important processes such as muscle contraction (Ebashi and Endo 1968; Podolski 1975; Deth and Casteels 1977; Chapman 1979, 1983; Fabiato and Fabiato 1979; Fabiato 1983), the synthesis and secretion of transmitters and hormones (Katz 1969), the regulation of enzyme activities (Cohen 1982) and the control of membrane permeabilities (Meech 1978; Colquhoun et al. 1981; Maruyama and Petersen 1982). Physiologically, [Ca] within unstimulated cells is lower than 0.05 μM, whereas in the extracellular fluids it is about 2 mM. Upon electrical or chemical stimulation, the intracellular [Ca] increases by 1 to 2 orders of magnitude as a result of (1) Ca influx across the cell membrane, (2) release of Ca from intracellular stores or (3) by both mechanisms (e.g., Reuter 1983).

Ca channels seemingly play a crucial role in coupling membrane excitation to cellular responses. Thus, the purpose of this chapter is to provide an overview of Ca ion movement through the voltage-gated Ca channels in cardiac cell membranes (as one representative important example of gated Ca transport across excitable biological membranes) and its modulation by drugs. Parallels to other excitable cell membranes will be covered by recent references. The mechanisms by which the intracellular [Ca] is returned to that of the resting cell after stimulation by means of ATP-driven Ca pumps or Na-Ca countertransport as well as other Ca permeabilities (e.g. Reuter 1983) and Ca release will not be dealt with. Interested readers are referred to recent review articles (see reference list). Furthermore, we will confine ourselves mainly to information obtained with techniques which allow electrophysiological recording from single isolated adult heart cells (Powell et al. 1980; Taniguchi et al. 1981; Isenberg and Klöckner 1982a,b) and from single ion channels in heart cell membranes (Hamill et al. 1981).

Prior to 1980, the information on Ca currents in the heart was obtained from experiments on small multicellular preparations excised from SA- and AV-nodal tissue, the Purkinje fiber system, and the atrial and ventricular musculature (reviewed in McDonald 1982). Before turning to single cells and single Ca channels, however, a short summary of the information gleaned from multicellular preparations will be given. The impact of methodological developments will be reviewed in the context with the respective data.

2 The Slow Inward Calcium Current, I_{si}, in Multicellular Heart Tissue

2.1 Properties of I_{si}

Initial evidence for Ca-dependent I_{si} came from voltage-clamp results on Purkinje fibers (Reuter 1967); this current component has since been shown to occur in all cardiac tissues (Trautwein 1973; Reuter 1979; McDonald 1982). I_{si} not only determines the plateau phase of the cardiac action potential (McAllister et al. 1975; Beeler and Reuter

1977; Reuter 1979), but it is also the primary inward current involved in the spontaneous activity of SA-node (Noma and Irisawa 1976; Brown et al. 1977; Noma et al. 1980b) and AV-node (Noma et al. 1980a). When the membrane is depolarized, I_{si} supports repetitive activity in atrial (Brown et al. 1975) and ventricular fibers (Grant and Katzung 1976).

The evidence that I_{si} is an inward current distinct from the fast inward Na current, I_{Na}, has been obtained from kinetic data, from ion substitution, and from pharmacological experiments, and can be summarized as follows. (1) The voltage ranges of steady-state activation (d_∞) and inactivation (f_∞) of I_{si} are different from those of I_{Na}. For example, I_{si} can be activated from membrane potentials at which I_{Na} is completely inactivated. (2) I_{si} can still be recorded after Na removal or in the presence of TTX (Reuter 1967; Beeler and Reuter 1970a,b; New and Trautwein 1972a,b; Kohlhardt et al. 1972; Trautwein 1973; Reuter 1973; Gettes and Reuter 1974; Trautwein et al. 1975; Reuter and Scholz 1977a; McDonald and Trautwein 1978; Reuter 1979).

I_{si} can be described as $I_{si} = g_{si} \cdot d\,(V,t) \cdot f\,(V,t) \cdot (V_m - V_{si})$, where g_{si} is the limiting conductance and $(V_m - V_{si})$ the driving force (Reuter 1973; Beeler and Reuter 1977). The threshold of I_{si} lies positive to -50 mV and the steady-state activation variable, d_∞, has a sigmoidal shape with $V_{0.5}$ at about -20 mV and saturation at $+10$ mV. The relation between voltage and the steady-state inactivation variable, f_∞, also has a sigmoidal shape and is near 1.0 at -60 mV, 0.5 at -25 mV and 0 at $+10$ mV (Reuter 1973; Trautwein et al. 1975; Reuter and Scholz 1977a). During voltage-clamp steps, I_{si} reaches a peak in the range of -10 to $+10$ mV (Reuter 1973; Trautwein et al. 1975; Reuter and Scholz 1977a; McDonald and Trautwein 1978). Inactivation may not always be complete during prolonged depolarization, since there is a cross-over of d_∞ and f_∞ (Trautwein et al. 1975; Reuter and Scholz 1977a). While the time constants of activation (τ_d) reach a maximum around -20 mV, the time constants of inactivation (τ_f) seem to increase continuously in the range between -40 and $+30$ mV (Trautwein et al. 1975; Beeler and Reuter 1977; Reuter and Scholz 1977a; McDonald and Trautwein 1978). Ion-substitution studies indicate that I_{si} is largely, but not exclusively, carried by Ca ions. The apparent reversal potential, V_{si}, of I_{si} was found to be considerably more negative than the Ca equilibrium potential calculated from the Nernst equation (Trautwein 1973; Reuter 1973; Reuter and Scholz 1977a; McDonald and Trautwein 1978). However, V_{si} could be fitted by the constant field equation. The selectivities of Ca over Na and K for I_{si}, P_{Ca} / P_{Na} and P_{Ca} / P_K, have been estimated to be 100 or larger (Reuter and Scholz 1977a). Divalent cation selectivity has not been determined, but indirect evidence suggested Ba>Sr>Ca≫Mg (Reuter and Scholz 1977a). The amplitude of I_{si} increases with external [Ca] but a quantitative description of this relation is lacking. There are indications of saturation with high external [Ca] (Payet et al. 1980; Noble and Shimoni 1981) or [Sr] (Vereecke and Carmeliet 1971), a behavior that is in keeping with the saturation of Ca currents in other excitable membranes (Hagiwara and Byerly 1981).

Inorganic blockers of I_{si} include Cd, Co, La, Mn and Ni (Reuter 1973; Ochi 1976; Payet et al. 1980; McDonald et al. 1981; Maylie and Morad 1981), although Mn also has a limited ability to carry I_{si} (Ochi 1976). These cations may compete with Ca for external channel sites important for Ca permeation (Payet et al. 1980). There is a wide variety of organic blockers of I_{si}, the most intensively studied being verapamil and

D-600; micromolar concentrations block I_{si} in cardiac tissue, the mode of action being use- and voltage-dependent (McDonald et al. 1980; Pelzer et al. 1982; Trautwein et al. 1983; McDonald et al. 1984a,b).

β-adrenergic drugs increase I_{si} (Brown et al. 1975; Tsien 1977; Reuter and Scholz 1977b; Reuter 1974, 1979, 1980, 1983; Kass and Wiegers 1982), while muscarinic agonists reduce it (Giles and Noble 1976; Ten Eick et al. 1976; Hino and Ochi 1980; Reuter 1983). I_{si} inhibition by cholinergic antagonists has not been analyzed in detail, although they seem to reduce the limiting conductance, g_{si} (Hino and Ochi 1980). A detailed analysis of the epinephrine effect on I_{si} (Reuter and Scholz 1977b) indicated that neither the kinetics of I_{si} nor the selectivity of the corresponding conducting channels are altered by the drug. However, the limiting conductance, g_{si}, is greatly increased. This was interpreted as being caused by an increase in the number of functional Ca channels (Reuter and Scholz 1977b). The effects of external application of epinephrine could be mimicked by the injection of cAMP into myocardial fibers (Tsien 1973) or by the application of mono- or dibutyryl cAMP or phosphodiesterase inhibitors (Tsien et al. 1972; Tsien 1977; Reuter 1974, 1979, 1980, 1983; Morad et al. 1981).

2.2 Methodological and Analytical Problems

The data and the description of I_{si} have been used in computer models that successfully simulate electrical activity in Purkinje fibers (McAllister et al. 1975), ventricular muscle (Beeler and Reuter 1977), and SA-node (Yanagihara et al. 1980). Nevertheless, much of the outlined information on I_{si} in multicellular heart tissue may turn out to be less than definitive. Quantitative data are primarily questioned because of problems associated with voltage-clamping structures that have complicated morphological and electrical features (Johnson and Liebermann 1971; Attwell and Cohen 1977; Beeler and McGuigan 1978). Since the current flow in this cable-like arrangement of individual cells may not be spatially uniform, especially if the current is applied through a point-form current source of high resistance, membrane currents may be contaminated by clamp artifacts (Kootsey and Johnson 1972; Ramon et al. 1975). The situation is improved when the current is applied across a sucrose gap from a low-ohmic current source (New and Trautwein 1972a). Unfortunately, this method is complicated by the fact that it is difficult to achieve a reasonable ratio of transgap internal and external resistances for long periods of time (⩾1 h) (McDonald and Trautwein 1978). The core resistance often increases during long-lasting experiments or after large current flow subsequent to the loss of electrode impalement or clamp oscillations. The consequence is that the recorded membrane current becomes contaminated by local circuit and leak currents. Other problems concerning I_{si} analysis in multicellular heart tissue are the separation of I_{si} from other overlapping currents (McDonald and Trautwein 1978; Siegelbaum and Tsien 1980) and the determination of I_{si} reversal potentials (Horackova and Vassort 1976; Noble and Shimoni 1981). Therefore, voltage-clamp studies in isolated single heart cells have long been desirable (Trautwein 1973).

3 Isolated Single Cells. Preparation and Action Potentials

Most procedures for the isolation of single cells follow a common sequence of steps (Trube 1983).

1. Washout of blood and Ca. Solutions low in [Ca] help to loosen cell to cell contact.
2. Digestion by enzymes. The connective tissue is most commonly dissolved by collagenase. Purified albumin (fatty acid-free) is often added as a "protective agent".
3. Mechanical agitation. As in other tissues (Berry and Friend 1969; Amsterdam and Jamieson 1974), tight and gap junctions in heart are not or only partially cleaved by enzymes and withdrawal of Ca (Fry et al. 1979). Some mechanical agitation is needed to dissociate the cells.
4. Filtration. Remaining pieces of tissue can be separated from the single cells by filtering the suspension through gauze or by centrifugation at slow speed.

The solutions can be applied to the tissue by different means, but retrograde perfusion via the coronaries has emerged as the superior method, both in terms of time and quantity of expensive enzyme. For the preparation of cells from large animal hearts,

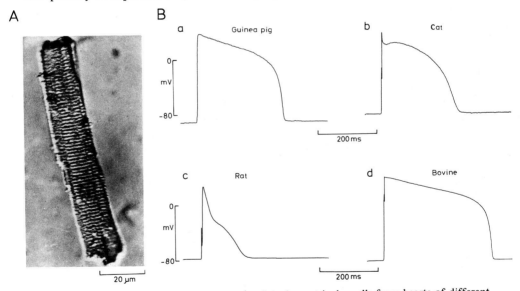

Fig. 1 A, B. Micrograph and action potentials of single ventricular cells from hearts of different adult mammals. A Photomicrograph (Nomarski optics) of a guinea pig cell taken after about 120 min perfusion with 3.6 mM Ca containing Tyrode solution ([K] 5.4 mM). Before this and other cells were exposed to Tyrode solution, they had been pre-incubated in KB-medium (Isenberg and Klöckner 1982a) for 30 min to 1h. The sarcomere lengths in these rod-shaped cells range from 1.5 to 2.1 μm. B Action protentials of isolated guinea pig (a), cat (b), rat (c) and bovine (d) ventricular myocytes obtained with the suction pipette method (Hamill et al. 1981; e.g., Fig. 2 A and B). For each species, a representative record obtained during continuous stimulation with short (0.5-1 ms) intracellular pulses through the recording pipette at 0.33 Hz for at least 15 min is presented. The temparature in parts A and B was 36° C

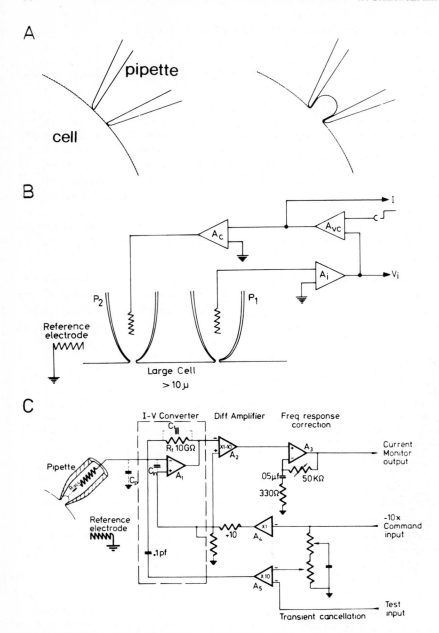

Fig. 2 A-C. Giga-seal formation between pipette tip and sarcolemma of myocytes **A**, schematic of the myocyte in the voltage-clamp circuit for whole-cell recording of membrane currents **B**, and the circuit diagram for cell-attached recording of single channel elementary currents **C. A** The left panel is the configuration of a pipette in simple mechanical contact with the cell membrane when the pipette-membrane resistance is of the order of 50-100 MOhm. Upon slight suction, a small patch of membrane (about 1.2-2.4 μm^2) is drawn into the pipette tip (right panel), forming a cell-attached patch. During the formation of this configuration the seal between membrane and pipette increases in resistance by 2-3 orders of magnitude. At this stage the omega-shaped membrane

protrusion can be destroyed by short pulses of suction or voltage applied to the pipette interior leading to the whole-cell recording configuration of action potentials (e.g., Fig. 1) and/or membrane currents (e.g., Figs. 3 and 4). Because of the large size of ventricular myocytes, a second pipette applied to the cell in the same way is required for homogeneous control of membrane potential in voltage-clamp experiments. Pipettes were made from thick-walled, hard borosilicate glass (Pyrex, Jencons) on a vertical puller in a two-step process. The pipettes were then coated with insulating varnish (Sylgard, Dow Corning) and lightly fire-polished to give a final tip diameter of about 1 μm. For recordings of action potentials and membrane currents (Figs. 1, 3, and 4), pipettes were filled either with (in mM) KCL 140, MgCl$_2$ 10, Hepes 5, pH 7.15, or glutamic acid 130, KOH 140, MgCl$_2$ 10, EGTA 5, HEPES 10, pH 7.2. When immersed in the external solution (composition in mM: NaCl 131 or 140; KCl 10.8 or 5.4, CaCl$_2$ 3.6, MgCl$_2$ 1, glucose 10, HEPES 5, pH adjusted to 7.4), pipette resistances ranged between 3 and 5 MOhm. For elementary current recordings from single Ca channels, electrodes were filled with (in mM) BaCl$_2$ 90, NaCl 2, KCl 4, HEPES 5, TTX 0.02, pH adjusted to 7.4 and resistances were between 5 and 10 MOhm. **B** Whole-cell voltage-clamp circuit. Pipette P_1 is connected to the input amplifier A_i which indicates the intracellular potential with reference to ground (V_i). Pipette P_2 injects the current delivered by the current pump. A_C. A_C converts 1 V input into 10 nA. Usually, the A_C signal comes from the clamp amplifier A_{VC}. Since the injected current is proportional to the A_C input, the A_C input voltage is an indicator of the membrane current (I). **C** A simplified diagram of the cell-attached patch-recording system taken from Hamill et al. (1981). The current-to-voltage converter is mounted on a micromanipulator, and the pipette holder (e.g., Fig. 1 of Hamill et al. 1981) plugs directly into it. Important stray capacitances (indicated by *dotted lines*) are the feedback capacitance C_f (about 0.1 pF), and the total pipette and holder capacitance C_p (about 4-7 pF). C_{in} represents the input capacitance of amplifier A_1 (for detailed circuit, see Fig. 4B of Hamill et al. 1981). With the values shown, the frequency response correction circuit compensates for time constants $R_f C_f$ up to 2.5 ms and extends the bandwidth to 10 kHz. The transient-cancellation amplifier A_5 sums two filtered signal with the time constants variable in the ranges 0.5 to 10 μs and 0.1 to 5 ms; only one filter network is shown here. The test input allows the transient response of the system to be tested: a triangle wave applied to this input should result in a square wave at the output. Amplifiers A_2, A_4 and A_5 are operational amplifiers with associated resistor networks. The operational amplifiers for A_2 and A_4 are chosen for low voltage noise; more critical for A_3 and A_5 are slew rate and bandwidth (for more details, see Hamill et al. 1981). For potential recordings from whole cells of small size (about 10 μm), a feedback amplifier is introduced between current monitor output and the voltage command input

or where the perfusion method has not been chosen, the tissue is minced and the pieces are agitated in the dispersion medium (Hume and Giles 1981; Tarr et al. 1981; Isenberg and Klöckner 1982a). Both methods can be combined (Powell et al. 1980).

The problem of most methods for the dissociation of heart muscle is the hypersensitivity of isolated cells to Ca (Dow et al. 1981; Trube 1983). At micromolar [Ca], these cells beat spontaneously; at millimolar [Ca], the cells round up and die ("Ca paradox") (Grinwald and Nayler 1981).

Ca-tolerant cells can be prepared by a number of more recent methods and the improvements have been quite diverse (Trube 1983). They include (1) minimizing the exposure of the tissue to enzymes and solution low in [Ca] (Bustamente et al. 1981; Wittenberg and Robinson 1981), (2) addition of trypsin during the final step of dissociation (Haworth et al. 1980), and (3) incubating the dispersed cells in a medium containing high potassium, low sodium, and a number of substrates to fill up intracellular energy stores ("KB medium", Isenberg and Klöckner 1982a). The latter method has been adapted by Taniguchi et al. (1981), who simplified the "KB medium" and, in addition, cannulated the aorta in situ during artificial respiration of the anesthetized animal, whereas all others excised the heart and cannulated it afterwards.

Figure 1A shows the micrograph of a guinea pig ventricular cell in 3.6 mM [Ca] solution (for composition, see legend of Fig. 2A) obtained by a method (Cavalié et al. 1983) modified from Taniguchi et al. (1981) and Isenberg and Klöckner (1982a). The cell is relaxed, quiescent and rod-shaped with a morphology similar to that of undissociated cells (Dow et al. 1981). Application of the retrograde perfusion method to hearts of other small animal species (cat, rat) or the chunk method to large animal hearts (bovine, Isenberg and Klöckner 1982a) yields similar cells. Most of them respond to intracellular pulses with action potentials whose configurations resemble those recorded from multicellular preparations excised from the hearts of these animals (Fig. 1B). Intracellular recordings can be made either by means of conventional microelectrodes or suction pipettes using part of the circuit shown in Fig. 2B (P_1). Suction pipette application to the cell is outlined in Fig. 2A; pipette-filling solutions and respective pipette resistances are given in the figure legend. A recent study suggests that the suction pipette induces less damage to single cells than the impalement of a conventional microelectrode (Pelzer et al. 1984).

4 The Calcium Current, I_{Ca}, in Isolated Ventricular Myocytes

In multicellular tissue, the Ca current has been labelled I_{si} for "slow" or "secondary" inward current, since in comparison to I_{Na}, I_{si} is secondary and has slower kinetics (see Sect. 2.1). In this section on Ca currents in isolated myocytes, I_{Ca} is preferred to I_{si} since the current seems to be carried much more selectively by Ca ions than previously recognized (Isenberg and Klöckner 1981; Isenberg 1982). The term I_{si} will only be used with reference to data from multicellular preparations.

4.1 Properties of I_{Ca}

4.1.1 Separation of I_{Na} and I_{Ca}

When a voltage-clamp test pulse depolarized the membrane to +10 mV from the resting potential of −70 mV, the cell responded with two inwardly directed current components (downward deflections), I_{Na} and I_{Ca} (Fig. 3A, left panel). I_{Ca} could be separated from I_{Na} by reducing the holding potential to −50 mV (Fig. 3A, right panel), where steady-state inactivation of I_{Na} is supposed to be virtually complete (Colatsky 1980; Brown et al. 1981a). Incomplete inactivation of I_{Na} at −50 mV was excluded for the following reasons. (1) When NaCl was replaced by tetramethylammonium chloride (TMA-Cl), the cell responded with a transient contracture and a reduction of peak I_{Ca} by about 30%, both probably related to an increase in intracellular [Ca] upon Na removal (Isenberg and Klöckner 1981). Within 20 to 30 min of Na removal, the cell had relaxed again and I_{Ca} amplitude and configuration were nearly the same as previously (Fig. 3B). (2) Records of I_{Ca} before and during 50 μM TTX application (Fig. 3C) illustrate that there was no TTX-sensitive current contributing to I_{Ca}.

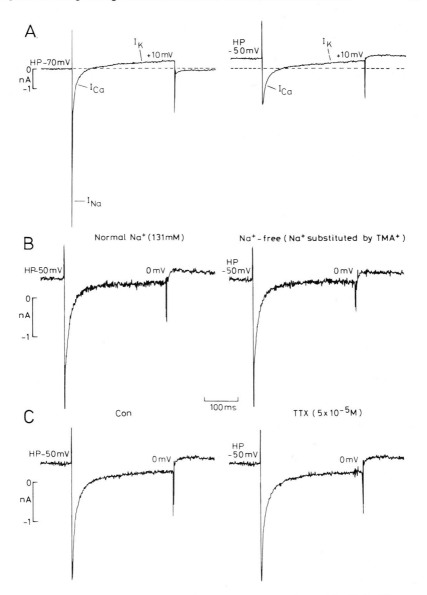

Fig. 3 A-C. Separation and selectivity of I_{Ca}. **A** Voltage-clamp pulses from −70 mV (holding potential equals about resting potential in 10.8 mM K-Tyrode) to +10 mV for 300 ms were applied at 0.33 Hz *(left panel)*. They evoked a current configuration which started with a capacitive component that was outward-directed (upward deflection). The following inward-directed current (downward deflection) comprises two components *(left panel)*: the fast inward Na current, I_{Na}, whose early phase was chopped off and the slow or secondary inward Ca current, I_{Ca}. The break in the inward-directed current waveform can be attributed to the inactivation time course which

is faster for I_{Na} than for I_{Ca}. After about 45 ms, the current became outward due to activation of the delayed K current system *(I$_K$)*. Subsequent voltage-clamp steps from −50 mV holding potential to + 10 mV for 300 ms *(right panel)* evoked a negative current deflection which we attribute to I_{Ca} alone for reasons illustrated below. *Dotted line* in both panels indicates the zero current level. Note the large outward-directed increase in the holding current due to an increased steady-state K conductance at −50 mV when compared to −70 mV, especially with respect to I_K flowing at the end of the clamp pulse. **B** Effect of Na removal on I_{Ca}. The currents were recorded during 0.33 Hz stimulation with 300 ms voltage-clamp pulses from −50 to 0 mV. *Left record* was obtained in Na-containing Tyrode solution while the right one was taken 30 min after having substituted 131 mM Na with 131 mM TMA. **C** Effect of TTX on I_{Ca}. In the presence of TTX (5 x 10^{-5} M, *right panel*), I_{Ca} evoked by voltage-clamp steps from −50 to 0 mV for 300 ms at 0.33 Hz was nearly identical to I_{Ca} on corresponding control depolarizations *(left panel)*. Parts **A-C** suggest that clamp steps starting from a holding potential of −50 mV elicit an I_{Ca} which is not significantly contaminated by TTX-sensitive I_{Na}. Temperature was 36°C

4.1.2 Effect of External [Ca] on I$_{Ca}$

When the external [Ca] was reduced from 3.6 to 0.2 mM, there was an 80 to 90% reduction in peak I_{Ca}, but little or no change in the time-to-peak or half-time for apparent inactivation (Fig. 4A). I_{Ca} is completely abolished when Ca is omitted from the external solution (Mitchell et al. 1983). Since I_{Ca} does not increase proportionally with increasing external [Ca] over the range 3.6 to 30 mM, I_{Ca} amplitude seems not to be a linear function of external [Ca].

4.1.3 Dependence of I$_{Ca}$ on Voltage

Peak I_{Ca} during a series of step depolarizations of different amplitude is shown in Fig. 4B. A depolarization to −30 mV was just suprathreshold in this cell. Larger steps produced a steep increase in I_{Ca} with a maximum between −10 and 0 mV. At more positive potentials, peak I_{Ca} declined again and extrapolation of the relation to more positive potentials defined an apparent reversal potential between +55 and +70 mV (see also Isenberg and Klöckner 1982b; Mitchell et al. 1983; Lee and Tsien 1982, 1983).

4.1.4 Inhibition and Stimulation of I$_{Ca}$

The addition of 2 μM D-600 to the external solution abolished I_{Ca} within 30 to 45 min (Fig. 4C). Intracellular injection experiments with D-600 and D-890 (the permanently charged quaternary derivative of D-600) on isolated ventricular myocytes suggest that the uncharged form of the drug crosses the membrane, is protonated, and then acts on the Ca channel from the cell interior (Hescheler et al. 1982), preferentially when the channel is open (Lee and Tsien 1983). The amplitude of I_{Ca} was increased by the addition of 0.5 μM adrenaline to the bathing solution (Fig. 4D). This augmentation of I_{Ca} can be mimicked by the intracellular injection of cAMP (Trautwein et al. 1982; Brum et al. 1983) or of the catalytic subunit of cAMP-dependent protein kinase (Osterrieder et al. 1982; Brum et al. 1983).

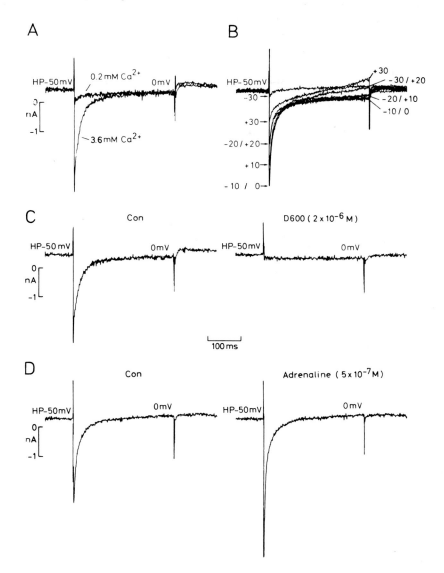

Fig. 4 A-D. Dependence of I_{Ca} on external Ca concentration, voltage and pharmacological interventions. **A** Superimposed I_{Ca} recorded in 3.6 mM and 0.2 mM external [Ca] on depolarizations from −50 to 0 mV for 300 ms at 0.33 Hz. The test current was recorded 5 min after admission of the 0.2 mM Ca solution to the bath. **B** Current-voltage relation of I_{Ca}. Superimposed I_{Ca} records on depolarizations to various potentials for 300 ms from −50 mV. Peak I_{Ca} and the current at the end of the clamp step for each potential are indicated by *arrows*. **C** I_{Ca} recorded during voltage-clamp steps from −50 to 0 mV for 300 ms at continuous stimulation (0.33 Hz) under control conditions *(left panel)* and after about 45 min perfusion with Tyrode solution containing 2×10^{-6} M D-600. **D** Adrenaline (5×10^{-7} M) increases I_{Ca} recorded during voltage-clamp pulses from −50 to 0 mV for 300 ms at 0.33 Hz *(left panel)*. The current in adrenaline *(right panel)* was recorded about 5 min after exposure to the drug. The temperature in parts **A-D** was 36°

4.2 I_{Ca} and I_{si}

I_{Ca} and I_{si} have many properties in common, for example the responses shown in Figs. 3 and 4, current-voltage relations, apparent reversal potentials and the voltage dependencies of steady-state activation and inactivation (cf., Isenberg and Klöckner 1982b; Mitchell et al. 1983 with Reuter 1979; McDonald 1982). Differences appear to exist in regard to absolute amplitudes and the time courses of activation and inactivation during voltage-clamp depolarizations. In comparison to I_{si}, I_{Ca} peaks faster (2 to 5 ms versus 10 to 20 ms) and with larger amplitude (16-42 μA cm^{-2} in 1.8-3.6 mM external [Ca] versus 5-8 μA cm^{-2}) (Isenberg and Klöckner 1981, 1982b; Mitchell et al. 1983). Voltage-clamp deficiencies, such as large series resistance resulting from endothelium and the extracellular cleft spaces, have been held responsible for the slower activation and the smaller amplitude of I_{si} (Isenberg and Klöckner 1982b; Tsien 1983). Monoexponential I_{si} inactivation at 0 mV appears to be a slow process (τ_f of 80-200 ms) compared to that observed for I_{Ca} (cf. Figs. 3 and 4 with McDonald 1982). In addition, I_{Ca} inactivation always consists of two phases (f_1 and f_2) with τ_{f_1} of about 6 ms and τ_{f_2} between 30 and 40 ms at 0 mV (Isenberg and Klöckner 1981, 1982b).

4.3 New Method — Old Problems

Hagiwara and Byerly (1981) and Tsien (1983) summarized the criteria for quantitative biophysical analysis of the Ca current in any preparation as follows: (1) The voltage clamp should be spatially uniform and rapidly settling. (2) The composition of solutions on both sides of the membrane should be under experimental control. (3) The Ca current should be separable from other ionic currents. (4) The reversal potential should be clearly defined in order to estimate the electrochemical driving force and to calculate the single channel conductance from current noise analysis. From the foregoing data and discussion it is obvious that requirements (1) and (2) are easier to fulfill in isolated myocytes than in multicellular preparations. This is especially true for the suction pipette technique which, in addition to the low-ohmic connective pathway between pipette and cell, allows relatively fast and reproducible diffusional exchange between pipette and cell interior solutions (Marty and Neher 1983). Furthermore, the probability of a shunt pathway for current flow due to membrane damage around the recording electrode is lower for suction pipettes with glass-membrane seal resistances $\geqslant$10 GOhm compared to about 500 MOhm for conventional microelectrodes (Marty and Neher 1983; Pelzer et al. 1984). However, the dissection of I_{Ca} from overlapping K outward currents is still a serious problem. In this context, measurements of the I_{Ca} reversal potential in cardiac cells have been seriously questioned, interpreted as an artifact (Isenberg and Klöckner 1981, 1982b) or remain contradictory (Lee et al. 1981; Lee and Tsien 1982).

In addition, the voltage dependence of the open-channel conductance can be confused with the voltage dependence of channel opening. To derive the probability of a channel being open from the macroscopic Ca current, the size of the elementary current through the individual open channel has to be known at each potential of interest. This is usually inferred from the "instantaneous current-voltage relation", a measurement that depends on the ability to measure current after a sudden voltage change very rapidly, before any channels open or close. Also, the instantaneous current is the sum of the open-channel current and the "leakage current" (all the conductances that we are not interested in and could not eliminate), and we usually assume that the "leakage current" is linear, which is rarely the case. However, the outlook for the Ca channel in heart and other excitable cells has improved thanks to a new approach described below.

5 Elementary Currents Through Ca Channels

5.1 Gigaseal Recording of Single Ca Channel Currents from Membrane Patches

The improved patch-clamp method of Hamill et al. (1981) has made it possible to re-cord electrical activity from individual Ca channels. Briefly, a fine-tipped, coated, fire-polished pipette is filled with a high concentration of charge carrier (see legend of Fig. 2A), and then pressed against the outside of an enzymatically-treated cell (clean cell surface). After the application of gentle suction, a "gigaseal" forms between the pipette tip and the membrane surface (e.g., Fig. 2A), so that current flow across the membrane patch and into the pipette can be recorded with high resolution under fast voltage-clamp control using the circuit schematically shown in Fig. 2C (Hamill et al. 1981; Sakmann and Neher 1983). Upon depolarization of the membrane patch by voltage-clamp pulses to potentials in the vicinity of maximal I_{Ca} (e.g. Fig. 4B), single Ca channel activity of unitary current amplitude (about 1.2 pA around 0 mV absolute membrane potential, solid lines in Fig. 5A) appears randomly as closely spaced bursts of brief channel openings separated by wider periods of inactivity (t_{ag}) (Fig. 5). Each burst of length t_b consists of n individual channel openings of length t_o ($<t_b$) and n-1 sojourns in a short-living shut state (t_s) (Fig. 5B). Per definition, the shortest burst is one channel opening (e.g., Fig. 9; Colquhoun and Hawkes 1981). The fact that no superposition of elementary currents occurs during a series of depolarizing voltage-clamp steps (>100) indicates that the patch contains only one active Ca channel. Interference from currents through other ionic channels is eliminated by the composition of the pipette solution, given in the legend of Fig. 2A. Ba, besides carrying charge through the Ca channel, abol-ishes currents through K channels (Sakmann and Trube 1984); Na and Ba fluxes through Na channels are prevented by TTX. Thus, we are left with the elementary current through the Ca channel, i.e., requirement (3) from Section 4.3 is satisfied with Ba. The recorded Ca channel currents through the membrane patch have been calculated to depolarize the rest of the cell by far less that 0.1 mV (Cavalié et al. 1983).

At present, one restriction of the patch-clamp method for the analysis of Ca chan-nels is the high concentration of the permeant divalent cation used in the recording pi-pettes, the purpose being to increase the signal-to-noise-ratio. Reduction of the Ba con-centration from 90 to 50 mM reduces the amplitude of elementary currents through the Ca channel from 1.2 to about 0.45 pA at potentials where I_{Ca} is maximum (Cavalié et al. 1983). Substitution of 50 mM [Ca] for 50 mM [Ba] further reduces the elementary

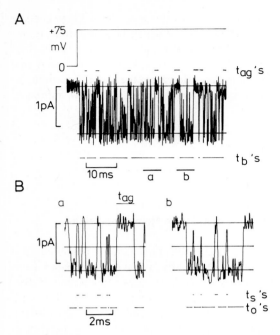

Fig. 5 A, B. Cell-attached recording of **elementary** currents through a single Ca channel (downward deflections) during 300 ms depolarizations of 75 mV from the pipette zero potential (= patch resting potential). Patch resting potentials are expected to be similar to the -65 ± 8 mV (n = 15) resting potential measured in cells (10.8 mM K Tyrode) with KCl-filled pipettes (see legend of Fig. 2). Depolarizing voltage-step command signals were generated at a rate of 0.5 Hz. The capacitive current transients and leak currents were cancelled at several stages of the recording process. The fast and part of the slower components were compensated by the analog circuit before recording (e.g., Fig. 2C). For finite capacitance correction and later digital leak subtraction, records during a pulse series that contained no channel openings (e.g., Figs. 6 C, d and 12 A, a, b) were averaged and substracted from the channel event containing records of that set. Before digitizing at a rate of 6.67 kHz by means of a Nicolet MED-80 computer, the patch currents were low-pass filtered by a 4-pole Bessel filter using various corner frequencies. **A** The baseline and average single channel amplitude levels were derived from amplitude histograms or were fitted by eye to each displayed record *(solid lines)*. Parts *a* and *b* are shown at an expanded time scale in **(B)**. *B* For analysis of open and shut time distributions (e.g., Fig. 8), an additional threshold level was installed halfway between the baseline and the average elementary current amplitude. The program detected transitions between the conducting *(lower level)* and nonconducting *(upper level)* state as crossing of this threshold level; calculated transition times were grouped in 0.15 ms intervals and displayed as time histograms for further analysis (e.g., Fig. 8). Single Ca channel traces in A and B were low-pass filtered in 3 kHz. t_o's open times; t_s's shut times within bursts; t_{ag}'s shut times between bursts; t_b's burst lengths. The temperature was 34° C

Ca channel current amplitude to about 0.35 pA (Cavalié et al. 1983). Open channel current sizes and their dependence on the species and concentration of the charge carriers in cardiac membranes agree well with findings in a wide variety of other excitable cell membranes (see Cavalié et al. 1983; Reuter 1983; Tsien 1983 for references). At physiological levels of divalent cation concentration in the recording pipettes, elementary Ca channel currents seem too small to be resolved. Nonstationary noise analysis in bovine chromaffin cells at $V_m = -12$ mV gives elementary Ca channel current amplitudes of 0.027 pA in 1 mM external [Ca] and 0.086 pA in 5 mM external [Ca], values far below those obtained in isotonic [Ba] near 0 mV (Fenwick et al. 1982).

5.2 Ensemble Analysis of Single Ca Channel Records

5.2.1 Ensemble Averaging

Ensemble statistics are basic to the study of probability (e.g. Aldrich and Yellen 1983). During depolarization, individual Ca channel opening and closing occurs on a stochastic basis (e.g., Figs. 5 and 6, parts a-d) "as if the channel repeatedly flipped a coin to decide whether or not to open, to remain open, or to close". According to the underlying probabilistic concept, a collection of single Ca channel current traces taken under identical conditions in different membrane patches (Fig. 6, A-C) during a series of identical voltage-clamp pulses (>100) forms statistical ensembles (4 examples are given in parts a-d of Fig. 6, respectively). Averaging channel behavior over such an ensemble describes the time course of Ca channel activity (parts e of Fig. 6) as does the record of macroscopic whole-cell Ca channel current through a large number of channels. However, high-frequency components of the individual signals are eliminated by ensemble averaging as if the bandwidth of the filter system had been lowered (see Fig. 6, Marshall 1978; e.g., Section 8.1).

The unitary amplitude of elementary currents indicates that Ca channels have only two distinguishable conductance levels, open and closed. Thus, the net Ca channel current (I) is given by

$$I = N \cdot p \cdot i, \tag{1}$$

where N is the total number of active channels, p the probability of a channel being open, and i the current through an open channel (Tsien 1983). In order to find p at, say, 5 ms after the onset of the voltage-clamp step in patches with only one functional channel (the simplest case to which we will confine ourselves), the fraction of records in the ensemble that have a channel open at that time are counted:

$$p\,(t = 5\text{ ms}) = \frac{(\text{number of records with channel open at } t = 5\text{ ms})}{(\text{total number of records in the ensemble})}. \tag{2}$$

If this calculation is performed over the length of the record, we obtain the probability of the Ca channel being open as a function of time. On ensemble averaging, the overall probability, p (t), of open state occupation at each particular time is determined since all channel openings in the ensemble contribute to the same current and thus serve as counters, each of which adds a unit to the total whenever the channel is open (Fig. 6, A-C). In the three examples of Fig. 6, N equals 1 and i does not change with time. Therefore, the variable behavior of p (t) results in variable I (t) in the different membrane patches (Fig. 6, A-C, parts e), the variability in channel behavior being the expected statistical deviation, if individual observations are samples from an even population of Ca channels. The mathematical extension of ensemble statistics to multichannel patches can be found elsewhere (Aldrich and Yellen 1983).

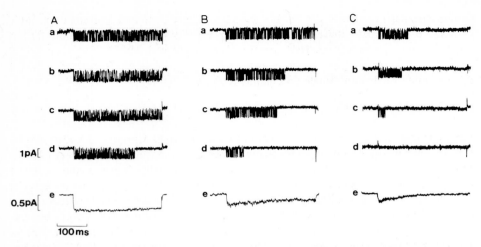

Fig. 6 A-C. Single Ca channel current records *(a-d)* and their ensemble averages (e) in response to step depolarizations of 75 mV amplitude and 300 ms duration from the pipette zero potential at 0.5 Hz **(A-C)**. Single channel current traces were low-pass filtered at 3 kHz; linear components of leakage and capacitance have been substracted. The average currents (also corrected for leakage and capacitance) were obtained from 150 averaged single channel current traces for each group, low-pass filtered at 500 Hz. The different groups classified by the degree of average current decay during depolarization are examples of the variability of the average time course of Ca channel activity seen in different patches. Average current decay was less than 10% in A, and nearly 100% in C. The fraction of blank single channel current traces (i.e., records with no channel openings; cf. part C, d) in the ensembles was 10, 23 and 44% in A, B and C, respectively. The temperature in these experiments was 35° C

5.2.2 Does Individual Ca Channel Behavior Depend on Time?

Voltage-dependent Ca channel activity lasts transiently for various durations when triggered by a sudden change in voltage (Fig. 6, A-C, parts a-d), creating a particular problem for analysis (cf. Aldrich and Yellen 1983). For a stationary process, such as a channel studied at a single constant voltage, one segment of a long record ought to be the same as another. Therefore, statistics of channel open and closed times can be pooled regardless of where they occur in the record. However, when the membrane potential is suddenly changed, the probability of the channel occupying particular conformational states can change, due to a sudden change in the rate constants for the transitions between states. In this case, the lifetime of an opening and closing event could depend on when it occurs if the rate constants are time-dependent. Such rate constants will cause the lifetime distribution to be nonexpontential, and openings occuring at different times will have different lifetimes. Other sources of "nonstationarity" and "time inhomogeneity" include a rundown of the preparation, i.e., occupancy probabilities can change gradually with time. Since we collect an ensemble of records of single channel behavior, we study the same channel over a long time and it is important to determine if the condition of long-term stability is met.

    A rundown and/or time-dependent changes of the fast gating transitions (e.g., Fig. 5) in the experiments detailed here can be ruled out for the following reasons. (1) For each ensemble, the probability that a pulse elicits channel openings was calculated as a function of the ordinal pulses number. This quantity normally converges to a characteristic value and remains constant for the rest of the experiment. (2) In each ensemble, lifetimes of opening and brief closing events do not differ significantly between the first and second halves of the clamp step (Cavalié et al. 1983). (3) Open lifetime distributions were always exponential (e.g., Fig. 8). For Ca channels, p (t) seems to represent the time course of the appearance and disappearance of Ca channel activity longer than individual openings and bursts of openings (see Sect. 8.1; Figs. 10 and 11) which reflects an "apparent open time" determined at low time resolution and bandwidth (see Sect. 5.2.1) and suggests additional slow gating transitions in the kinetics of the Ca channel (e.g., Sect. 8).

5.3 Why Use Single Channel Records?

An obvious advantage of single channel records is that we can directly observe the number of channels, N, and the current through an individual open channel, i, to obtain an approximate normalized probability function. In addition, single channel records give the channel-open probability free from all limitations of macroscopic recording. Since we can measure the elementary current through an open ionic channel, we can (besides pharmacological elimination) distinguish currents produced by different types of channels by their different open-channel current amplitudes. Potential changes caused by series resistance are negligible because of the small size of the currents through the patch. Also, we can measure any change in elementary current size produced by ion accumulation and correct for it. Furthermore, single channel records give information about channel kinetics that is impossible to estimate from macroscopic measurements. This advantage is most apparent in recordings from patches that contain only one functional channel. In this case, we can unambiguously determine the time of occurence and the duration of every opening and closing event that contributes to the total.

6 Conductance and Activation of the Single Ca Channel

The amplitude of elementary current through an open Ca channel, i, was determined from amplitude histograms and/or visually at several membrane potentials. The level of single Ca channel currents is indicated by the solid lines in Fig. 7A; i decreased the more positive the membrane potential. Over the potential range tested (about -30 to $+30$ mV absolute membrane potential), the i-V relationship was approximately ohmic with a slope conductance of 21 pS (range 15-25 pS, e.g., Reuter et al. 1982; Reuter 1983; Cavalié et al. 1983). At lower divalent cation concentrations in the recording pipette (50 mM [Ba] or 50 mM [Ca]), the single Ca channel slope conductance decreased to 9-10 pS (Cavalié et al. 1983). Clear outward Ca or Ba movement through the Ca channel was never observed under the conditions tested, even during depolarizations by 200 mV to about $+135$ mV absolute membrane potential, indicating that cardiac Ca channels are

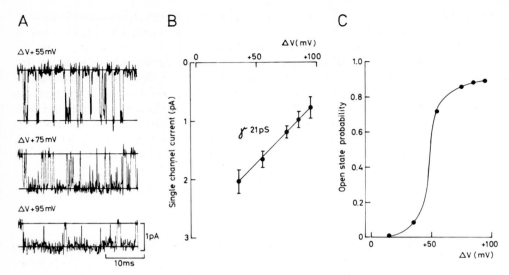

Fig. 7 A-C. Voltage dependence of single Ca channel current and open state probability. **A** Cell-attached recordings of elementary current through a single Ca channel during 300 ms depolarizations of various amplitudes from the pipette zero potential delivered at 0.5 Hz. In each record, linear components of leakage and capacitance have been subtracted; the baseline and average elementary current amplitude are indicated by the solid lines. The records were low-pass filtered at 3 kHz. **B** Current-voltage relationship of the elementary current amplitude. Each value represents the mean ± standard deviation from four experiments as shown in **A**. In each experiment, the individual single channel current amplitude was derived from amplitude histograms and/or visually. *Straight line* is drawn by eye and represents the single channel (chord) conductance (γ). **C** Open state probabilities at each membrane potential were determined between 5 and 10 ms after the onset of the depolarizing voltage-clamp pulse. They were calculated as the ratio of the number of cases, where a channel was found to be open during that time to the total number of applied voltage-clamp steps [e.g., Eq. (2)]. The temperature in parts **A-C** was 34° C

rectifying and can pass ions only into the cell. This interpretation is corroborated by reports on neuronal membranes (Kostyuk 1981; Hagiwara and Byerly 1981) where the Ca current appears to be inherently a one-way current. These considerations are based on the extreme asymmetry of the Ca ion activities across the membrane and the blocking effect of intracellular Ca (Standen 1981; Plant et al. 1983).

In dialyzed cardiac (Lee and Tsien 1982) and chromaffin (Fenwick et al. 1982) cells it has been shown that K and Cs ions can move through Ca channels in the outward direction. This may result in an apparent Ca equilibrium potential that is 50-70 mV less positive than the expected one calculated from the Nernst equation for Ca ions. The ohmic open-channel characteristic argues against supralinearity as predicted by constant field assumptions in agreement with earlier results in different tissue obtained from tail current analysis (Reuter and Scholz 1977a; Llinas et al. 1981; Byerly and Hagiwara 1982; Tsuda et al. 1982; Brown et al. 1983).

In contrast to the linear decline of i with membrane potential, the probability p that the Ca channel is open at the time of the peak of the maximal average current [e.g.,

Eqn. (2)] increased steeply with voltage in a sigmoidal manner (between -30 and $+30$ mV absolute membrane potential) (Fig. 7C). The function resembled the voltage dependence of steady-state activation (d_∞) of I_{si} and I_{Ca} (see Sects. 2.1 and 4.2). At positive potentials, p always saturates at a level lower than unity due to the occurrence of a variable number of current traces without channel openings in each ensemble (e.g., Fig. 6C, d). These empty sweeps are called "blanks" and refer to a condition or conditions where the channel is unavailable to open (e.g., Sects. 8.2 and 9.2).

7 Kinetics of Single Ca Channel Activation

As suggested in the previous section, cardiac Ca channels are mainly controlled by voltage-dependent gating. A change in membrane potential may cause a reorientation of a charged "voltage sensor" within the membrane (for example a protein group with dipole properties that may be an integral part of an ion channel), and as a consequence, a change in the ion flow through the channel (Reuter 1983). Recent evidence suggests that the type and concentration of external permeant cations moderately influence the gating of the voltage-sensitive cardiac Ca channel (Cavalié et al. 1983), a finding that is in keeping with reports on Ca channels in other excitable membranes (Saimi and Kung 1982; Hagiwara and Ohmori 1982; Wilson et al. 1983).

Quantitative kinetic information on fast voltage-dependent Ca channel gating transitions (e.g., Fig. 7 A and C) was obtained from the distribution of open and shut times at different membrane potentials (Fig. 8). The distribution of the open times could consistently be approximated by single exponentials (Fig. 8, upper panels) whose time constants, τ_0, corresponding to the average lifetime of channel openings, increased steeply with depolarization (Fig. 8, A-C, upper panels). By contrast, the distributions of the shut times corresponded to the sum of two exponentials (Fig. 8, lower panels); the short time constant, τ_s, represents the mean lifetime of the brief closing events within bursts, and the longer one, τ_{ag}, is related to the mean length of the intervals between bursts (e.g., Fig. 5). Both shut time components decreased with increasing depolarization (Fig. 8, A-C, lower panels). The average number of interruptions per burst, ν, could be derived from the shut time histograms as

$$\nu = \frac{\int_0^\infty N_s \cdot e^{-t/\tau_s}}{\int_0^\infty N_{ag} \cdot e^{-t/\tau_{ag}}} = \frac{N_s \cdot \tau_s}{N_{ag} \cdot \tau_{ag}} , \tag{3}$$

where N's are the amplitudes of the respective shut time components. Thus, it becomes obvious from Eq. (3) and Fig. 8 (A-C, lower panels) that ν increases at positive membrane potentials mainly due to the pronounced increase of the amplitude of the fast

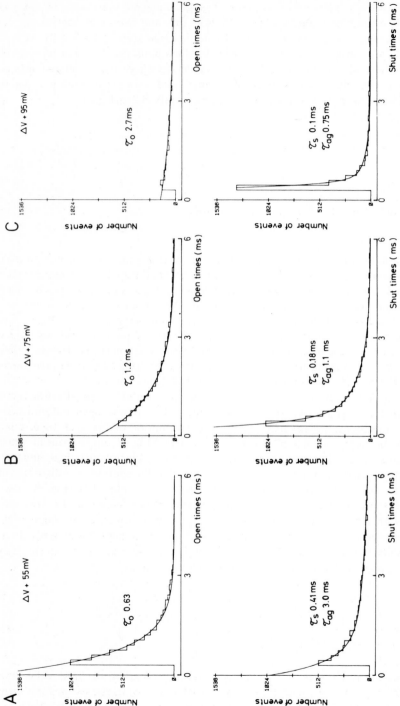

Fig. 8 A-C. Open and shut time histograms from single Ca channel current records obtained during 300 ms depolarizations of various amplitudes (A-C) delivered from the pipette zero potential at 0.5 Hz. Single channel current records were digitized at a rate of 6.67 kHz (sampling interval 0.15 ms, no filter); linear components of leakage and capacitance were digitally substracted as described in the legend of Fig. 5. The principle of analysis of time histograms can also be derived from Fig. 5B. Mono- and double-exponential curves were fitted to be displayed histograms by a nonlinear, unweighted, least-squares method. The respective time constants corresponding to the mean open and shut state lifetimes of the Ca channel at different membrane potentials are given in the inserts. Temperature during the experiment was 34°C

shut time component. As consequence, the burst length (t_b) had to increase with potential too, since t_b is related to the other quantities by

$$t_b = (\nu + 1) \cdot \tau_0 + \nu \cdot \tau_s. \tag{4}$$

Separate analysis of burst duration revealed an exponential distribution with a time constant or average length becoming longer in the positive potential range (Cavalié et al. 1983). No principle difference with respect to fast Ca channel gating was obtained among the different groups (Fig. 6A-C).

In conclusion, these data on Ca channel kinetics can well account for the voltage-dependent behavior of macroscopic peak I_{Ca} (e.g., Fig. 4B). Although the elementary current through an open channel (i) decreases linearly with increasing voltage, the macroscopic current (I) can increase within a certain voltage range because of an increasing open state probability of the individual Ca channel (p) (Fig. 7C) due to the voltage-dependent changes in channel kinetics (Fig. 8). However, on further depolarization of the membrane, the decrease in i outweighs the increase in p and I declines again [e.g., Eq. (1)].

Channel kinetics are usually described in terms of Markov processes. A lattice of states (which presumably correspond to conformational states of the channel molecule) is used to describe channel behavior, and the channel may jump from one state to another with various transition rates. For the Ca channel under the conditions tested, these rates seem most likely to be constant at a constant voltage (i.e., the test potential during the depolarizing pulses) (see Sect. 5.2.2) and independent of the channels past history (see Sect. 8.2; cf. also Fig. 12B); thus, the distribution of lifetimes of a single state is always exponential, with a characteristic lifetime determined by the reciprocal of the sum of the rate constants leaving that state. The observation that the distribution of shut times has at least two exponential components implies that there is more than one shut state (Colquhoun and Hawkes 1981, 1982). If C_1 and C_2 denote two closed states, one of very short average lifetime, and O denotes an open state, then two obvious alternative general schemes of Ca channel gating are

$$C_1 \rightleftharpoons C_2 \rightleftharpoons O$$

and

$$C_1 \rightleftharpoons O \rightleftharpoons C_2.$$

Unfortunately, it is not possible to distinguish unambiguously between these two sequences from observations of single channel records in the steady state (e.g., Colquhoun and Sakmann 1983; Sakmann and Trube 1984). However, during voltage-clamp steps, the measurement of the latencies-to-first-opening (i.e., the intervals from the onset of the voltage step to the time of the first channel opening; e.g. Fig. 5A) provides a means of discriminating between the two schemes. The distribution of latencies reached its maximum at a time considerably later than zero (Cavalié et al. 1983), thus indicating a distinct rising phase (e.g., Lux 1983). This is the expected result for a process in which multiple closed states precede the open state. The arrangement of the open state, O, between the two closed states, C_1 and C_2, would result in a distribution with a simple exponential fall. Thus, Fig. 9 illustrates the simplest possible sequence which can account for the kinetic analysis of the burst behavior of single Ca channel openings during voltage-clamp steps. In this scheme, the short closing events within bursts are due to sojourns of the channel in a short-living shut state, placed intermediately on the

Simplest plausible sequence for Ca-channel activation:

$$C_1 \underset{K_2}{\overset{K_1}{\rightleftharpoons}} C_2 \underset{K_4}{\overset{K_3}{\rightleftharpoons}} O$$

Diagrammatic representation of the possible behaviour of the single Ca channel based on the upper scheme:

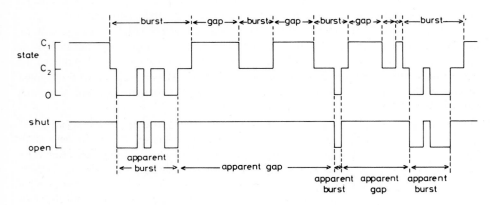

Upper part : actual behaviour of the system

Lower part : corresponding appearance of the single-channel current

Fig. 9. Schematic of Ca channel activation based on a C-C-O model. (After Colquhoun and Hawkes 1981)

opening pathway. From there it is apparent why intervals between visible bursts (apparent bursts) are labeled apparent gaps and why the shortest burst is the individual channel opening (Fig. 9, lower part). The k_i's are independent, voltage-dependent transition rates connecting the individual states. Under the restrictions of a true Markov process (see above), it can be shown (Colquhoun and Hawkes 1981; Conti and Neher 1980) that

$$\tau_0 = k_4^{-1} \tag{5}$$

and

$$\tau_S = (k_2 + k_3)^{-1}. \tag{6}$$

Furthermore, the quantity v is also related to k_2 and k_3 by the equation

$$v = k_3 \cdot k_2^{-1}. \tag{7}$$

Resolution of Eqs. (5), (6) and (7) readily yields the numerical values of k_2, k_3 and k_4. Finally, one may approximate k_1 by

$$\tau_{ag} = k_1^{-1} + k_3^{-1} + k_2 \cdot k_3^{-1} \cdot k_1^{-1}, \tag{8}$$

taking into account a couple of considerations concerning the mathematical treatment of gaps between bursts during voltage-clamp steps (Cavalié et al. 1983). It is likely that the C-C-O reaction scheme (Fig. 9) is restricted to Ca channel activation (i.e., the rise of macroscopic current to its peak upon a voltage step), since it does not consider an inactivated state.

8 Inactivation of the Single Ca Channel

8.1 Time Course of Inactivation During Depolarization

The validity of a kinetic model for channel function is tested by its ability to reproduce experimental results. The above outlined three-state sequential scheme with its fast transitions (ms and even fractions of ms), however, can neither account for the slow time-dependent decline of I during a depolarization, i.e., inactivation nor for the occurrence of "blanks" (e.g., Fig. 6). Since this sequence is based on the occurrence of Ca channel activity in bursts, it is essential at this point to specify exactly how a burst is defined. The term burst is used to refer to any group of openings that is separated by a long shut time from the next group. Channel openings will have the visual appearance of being grouped together in bursts (Figs. 5 and 10a) if the distribution of all closed times contains two exponential components, the time constant of one component being reasonably longer than the time constant of the other (Fig. 8). If we wish to actually divide our observed records (Fig. 10a) into bursts, we can define a burst as any group of openings that are separated by gaps that are less than some specified value (Fig. 10b; Colquhoun and Hawkes 1981, 1982, 1983). In order to interpret the observed bursting in terms of some mechanism, we have to define gaps within and between bursts as sojourns of the channel in a specified set of channel states as described above (see Fig. 9; Colquhoun and Hawkes 1981, 1982, 1983). To do this, the probability that a shut period of given length is part of a burst rather than separating bursts has to be known; it can be calculated if the rate constants are known [Eq. (2.50) in Colquhoun and Hawkes 1981].

The above discussion supposed that the distribution of all shut times had two well-separated time constants. During long observation periods, however, a considerable fraction of time spans of channel inactivity became apparent which were far out of even the slow exponential term of the two-exponential best fitting density function (around 2% of all closed events). The occurrence of these long-living shut times (tens to hundreds of milliseconds) indicates that bursts themselves are grouped (Fig. 10c); this type of grouping has been termed a cluster of bursts (see Neher and Steinbach 1978; Sakmann et al. 1980; Colquhoun and Hawkes 1977, 1981, 1982; Colquhoun and Sakmann 1983). Cluster durations likewise range on a tens to a hundreds of milliseconds time scale (Fig. 10c and 11). The arrangement of bursts in clusters can be attributed to an inactivated state (I) of the Ca channel occupied during the shut interval between clusters. Since this state can also be accessed without channel opening (Reuter et al. 1982; Cavalié et al. 1983; see Sect. 8.2), I seems to be coupled to the three states of the activation sequence (Fig. 9) also via C_1. Thus, the complete reaction scheme of Ca channel gating can be most generally written in a cyclic form:

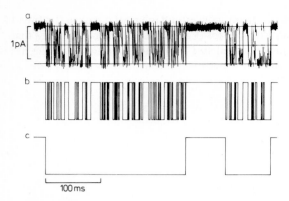

Fig. 10 a-c. Clustering of bursts of single Ca channel openings. **a** Ca channel openings occurring during 900 ms voltage-clamp steps of 75 mV amplitude delivered from the pipette zero potential at 0.5 Hz. Elementary current traces were low-pass filtered at 3 kHz; capacitance and leakage have been subtracted. *Solid lines* represent the baseline, the average elementary current amplitude and the 50% threshold level inbetween, respectively. The temperature was 35° C. **b** Idealized current trace grouping individual Ca channel openings into bursts separated by longer periods of channel inactivity (apparent gaps, e.g., Figs. 5, 8, and 9). The distribution of shut periods shorter than 20 ms can be expressed as a mixture of the distributions of *(a)* the shut periods within bursts and *(b)* the shut periods between bursts. The probability that a shut period of length t separates two bursts rather than being within a burst was determined by classifying shut periods as short/ fast (t_s) or long/slow (t_{ag}) depending on whether the fast or slow exponential term of the best fitting density function is larger for the observed lifetime of a shut interval. This classification can be performed by noting that for a time, t', defined by the relation

$$\frac{N_s \, e^{-t'/\tau_s}}{\tau_s} = \frac{N_{ag} \, e^{-t'/\tau_{ag}}}{\tau_{ag}} \qquad (9)$$

the probability densites of the short and long shut lifetimes are equal. Thus, if a shut period has a lifetime $<t'$, it has a higher probability of being a member of the fast kinetic component, i.e., a gap within a burst, whereas if a shut interval has a duration $>t'$, it is more likely to belong to the slow kinetic component, i.e., to gaps separating bursts. In this and other experiments, t' was in the order of 5 times τ_s. **c** Idealized current trace grouping bursts of Ca channel openings into clusters. Although the distribution of shut intervals separating clusters is not known, the number of apparent gaps incorrectly classified as gaps between clusters can be calculated from $N_{ag}e^{-t''/\tau_{ag}}$. Minimizing this quantity by variation of t'' makes a reliable separation of clusters possible. t'' was found to be in the order of 10-20 times τ_{ag} (20 ms in this experiment)

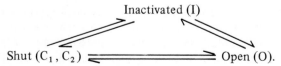

Inactivated (I)

Shut (C_1, C_2) ⇌ Open (O).

Some other examples are considered elsewhere (Colquhoun and Hawkes 1983).

Referring back to Fig. 6, there seems to exist a striking visual relation between cluster duration (i.e., the time to channel entrance into the inactivated state) and the time course of inactivation. This notion was tested during long-lasting depolarizing pulses where clustering was more evident due to the longer observation period (Fig. 11A).

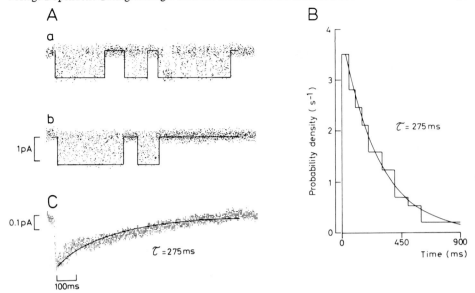

Fig. 11 A-C. Distribution of cluster lifetime and the relation to the decay of the average current during depolarization. **A** Two examples of clusters during 950-ms-long clamp pulses of 75 mV amplitude, delivered from the pipette zero potential at a rate of 0.2 Hz. In both cases, cluster reopenings occur after having entered once the longest-living shut state. The total probability to observe cluster reopenings during depolarizations of 75 ± 10 mV amplitude in all experiments ranged from about 0.05 to 0.2. Capacitive transients and leakage currents have been subtracted; the signals were low-pass filtered at 3 kHz. *Solid lines* represent the idealization of clusters according to the criteria given in the legend of Fig. 10c. **B** Distribution of cluster lifetime. For cluster durations up to 200 ms, 50 ms long bins have been used, whereas longer clusters were grouped in 100 ms long bins. The time constant corresponding to the mean cluster lifetime is given in the *inset*. **C** Ensemble average of 130 single channel current traces. *Solid line* represents the calculated time course of the average current decay using the time constant derived from the monoexponential approximation of the distribution of cluster duration in **B**. The temperature was 35° C

Clusters, in principle, have been defined in one of the three ways (see legend of Fig. 10), as discussed above for simple bursts. The distribution of all cluster lifetimes was well fitted by a single exponential with a time constant corresponding to the mean cluster lifetime (Fig. 11B). By using this τ, the time course of inactivation of the corresponding average current could be well approximated (Fig. 11C) since clusters occurred mainly once per depolarization (e.g., Fig. 6; on average 0.83 clusters per pulse) although reopenings are possible (Fig. 11A).

Figure 6 shows three different average currents obtained from ensembles of records from three different patches (A-C). The average currents (Fig. 6, parts e) illustrate that the mean cluster lifetime can vary considerably from patch to patch, ranging from less than 100 ms (C) to more than 1 s (A). The probability of observing no Ca channel activity during a voltage step (i.e., a "blank") is higher in the group with the shorter mean cluster lifetime (cf. peak amplitudes of average currents in Fig. 6). However, independent of the group, individual Ca channels examined to the present time have similar conductance and fast kinetic parameters within clusters (see above). Thus, (1) it is highly unlikely that there is more than one species of Ca channels, and (2) with respect to Section 5.2, a time-dependent behavior of p obtained by ensemble averaging [(see Sects. 5.2.1 and 5.2.2)] does not necessarily imply nonstationarity and/or time inhomogeneity for Ca channel behavior within clusters of bursts.

8.2 Steady-State Inactivation

In each ensemble of single Ca channel current records, there was a considerable fraction of "blanks" (Fig. 12A, a). This fraction increased when depolarized conditioning potentials preceded the pulses to a constant test potential (Fig. 12A,b). The probability of observing channel activity at the test potential had a sigmoidal dependence on the potential of the conditioning depolarization (Fig. 12A,c). This relation could be fitted by a Boltzmann distribution and resembled the relation between voltage and the steady-state inactivation variable, f_∞, for I_{si} and I_{Ca} (see Sects. 2.1. and 4.2.). Thus, single Ca channel inactivation occurs through a decrease in the probability that a channel is available to open, the elementary current through an open channel being independent of the conditioning voltage (Fig. 12A; e.g., Reuter et al. 1982; Cavalié et al. 1983). The channel was half-inactivated at a conditioning potential of about +35 mV (corresponding to about -30 mV absolute membrane potential), a potential just positive to the threshold for activation (e.g., Fig. 7C). There is a considerable overlap between the voltage ranges of activation and inactivation (e.g., Fig. 7C with Fig. 12A,c) predicting that the Ca channel is likely to open in the steady state within the "window" potential range (Cavalié et al. 1983). Analysis of Ca channel activity at constant potentials within this "window" range (Fig. 12A,b) revealed, in agreement with the analysis during voltage-clamp pulses (see above), that individual Ca channel openings are grouped together in bursts and bursts themselves are further grouped into clusters. If a voltage-clamp pulse was applied within a cluster, the depolarization always elicited channel openings; however, if it was applied during the shut interval between clusters, Ca channel activity was never triggered on depolarization (Fig. 12A, b).

Quantitative kinetic information on the influence of the degree of steady-state inactivation on remaining Ca channel activity during depolarization was obtained from open and shut time histograms at a constant test potential imposed from two different conditioning potentials, -60 and +40 mV (Fig. 12B). The depolarizing conditioning potential resulted in a decrease of the number of open and shut events per ensemble (smaller histograms in each panel of Fig. 12B). However, it did not affect the distributions and average lifetimes of open times (Fig. 12B, a) or shut times within or between bursts (Fig. 12B, b), the mean number of interruptions per burst [Fig. 12B, b; e.g., Eq. (3)] and thus the mean burst length [e.g., Eq. (4)]. In addition, Ca channel open and shut events analyzed at a single constant voltage within the "window" potential range (e.g., Ca channel activity between voltage steps in Fig. 12A,b) revealed, besides a lower frequency of occurrence, about the same distributions and average lifetimes as those analyzed during voltage steps to the same potential. This means that changes in the k_i's of the activation sequence due to increasing steady-state inactivation can be ruled out.

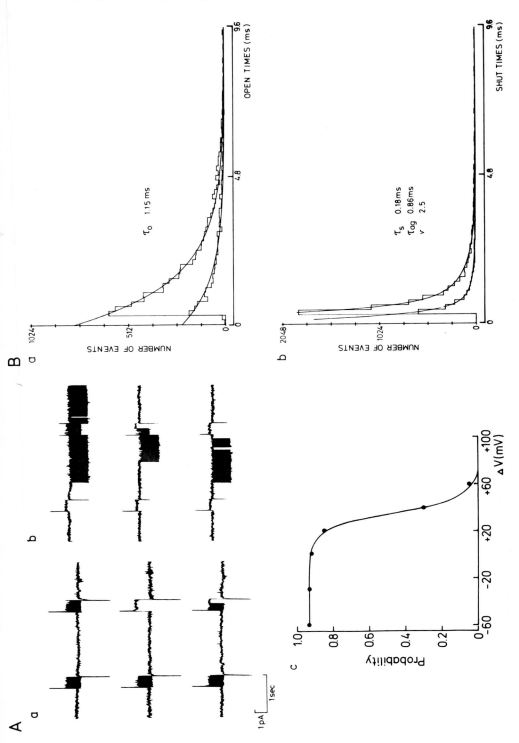

Fig. 12 A, B. (Legend see page 80)

Fig. 12. A Voltage dependence of Ca channel inactivation. *(a-b)* Elementary currents during 300 ms long depolarizing voltage-clamp pulses to a fixed potential (+75 mV) at 0.5 Hz. These test pulses were imposed from either the pipette zero potential *(a)* or a more depolarized (+60 mV) holding or conditioning potential *(b)*. Current traces (low-pass filtered at 1 kHz) are shown at a slow sweep-speed and have not been corrected for leakage and capacitance. Note (1) that the number of depolarizations eliciting channel openings at +75 mV test potential is lower stepping from +60 mV, and (2) that at the +60 mV conditioning potential, channel activity occurs in the steady state. *(c)* The probability that a depolarizing clamp pulse elicits at least one burst of Ca channel openings is plotted against the conditioning potential (abscissa). *Solid line* is a Boltzmann distribution:

$$p = 0.93 / [1 + \exp[(V - V_{0.5}) / k]]$$

with parameters chosen for the best visual fit: $V_{0.5} = + 35$ mV; $k = 7$.

B Influence of the degree of inactivation on the distribution of open times *(a)* and shut times *(b)*. Histograms depict the data obtained at a fixed potential (+75 mV), stepping for 300 ms from either −60 mV *(larger histograms)* or +40 mV *(smaller histograms)* conditioning potential. *Insets* show the time constants of the exponentials fitted to the histograms. Note that increasing inactivation decreases the absolute number of observations but leaves their distributions and average lifetimes unaffected. ν *(b)* refers to the average number of interruptions per burst, being equal to the ratio of the fast and slow components of the shut time histograms [e.g., Eq. (3)]. The temperature was 34° C

Thus, the reduced probability that a channel is available to open on pulses from depolarized conditioning potentials can be mainly attributed to an increased fraction of "blanks" in each ensemble of single Ca channel current records being related to the average lifetime of the shut intervals between clusters (e.g., Fig. 12A, b). It is suggested that at depolarized conditioning potentials the inactivated state is more probably occupied and that "blanks" may provide a measure of the steady-state occupancy of the inactivated state of the Ca channel.

The foregoing discussion attributes inactivation to a purely voltage-dependent mechanism. However, it is quite possible that an additional inactivation mechanism, Ca-dependent inactivation, exists in cardiac cells (see Fischmeister et al. 1981; Marban and Tsien 1981; reviewed in McDonald 1982; Tsien 1983) as well as in snail neurons (Brown et al. 1981b). According to Standen and Stanfield (1982), the Ca ions that enter through open Ca channels bind to a site at the inside of the membrane to cause inactivation. This site is supposed to have a lower affinity for Sr and Ba than for Ca. On the other hand, Ca channel inactivation can be produced in the absence of divalent cation influx through open Ca channels: (1) in dialyzed heart cells, outward Cs and K movements through the Ca channel show inactivation even in the absence of permeant divalent cations (Lee and Tsien 1982); (2) single Ca channels can be partially inactivated by depolarizations too weak to elicit any measurable channel opening (Reuter et al. 1982; Cavalié et al. 1983; e.g., Fig. 12A, c with Fig. 7C) and (3) single Ca channels can inactivate before opening (Cavalié et al. 1983). These observations and others (see McDonald 1982) argue against a dominant Ca-dependent inactivation mechanism.

9. Mechanisms for Ca Channel Modulation

9.1 Ca Channel Blockade

Ca channels can be blocked by various organic compounds. The mechanisms of action have recently been analyzed in multicellular (see Sect. 2.1) and single cell (Sect. 4.1.4) cardiac preparations. Verapamil and its derivatives block Ca channels predominantly from the inside surface of the membrane (Hescheler et al. 1982) by entering the channel preferentially when it is open (McDonald et al. 1980, 1984a, b; Pelzer et al. 1982; Trautwein et al. 1983; Lee and Tsien 1983). The blocking potencies of these charged tertiary amines depend on membrane potential (voltage dependence) and on the rate of stimulation (use dependence) (McDonald et al. 1980, 1984a, b; Pelzer et al. 1982, Trautwein et al. 1983; Lee and Tsien 1983). This contrasts with the 1,4-dihydropyridines (nifedipine, nitrendipine, nisoldipine) which are uncharged at physiological pH and which hardly show voltage- and use-dependent characteristics of Ca channel block (Kass 1982; Lee and Tsien 1983).

In this section, we will focus on the most widely studied organic Ca entry blocker, D-600, and how it affects single Ca channel gating. Upon bath application of D-600 (2 μM), ensemble averages of single Ca channel current traces showed a rapid decline in the average current amplitude (I) during the first 10 min, followed by a slower one during the next 20 min (Fig. 13; lower records in each panel). Distinct alterations of the average Ca channel current time course during block were not observed. Drug-induced depression of I was largely reversible upon re-introducing control solution; a washout period longer than the 60 min here is probably required for complete restoration of I (cf. Pelzer et al. 1982; McDonald et al. 1984b). The reduction in I due to D-600 apparently reflects a decrease in the probability (p) of the Ca channel to occupy the open state [e.g., Eq. (1); upper records in each panel] since the elementary Ca channel current amplitude, i, was unaffected over a wide voltage range (Figs. 13 and 14A). A complete abolition of Ca channel activity was never observed on steps from the pipette zero (= resting) or more negative conditioning potentials within 30 to 45 min of drug application.

Quantitative kinetic analysis revealed that the drug exerts a dual blocking mode of action. (1) In records with channel openings, D-600 reduces the probability of Ca channel open-state occupation by a distinct shortening of the mean lifetime of the open state (Fig. 14B) and a decreased number of openings per burst, a minute increase of the mean duration of the brief closing events within bursts (Fig. 14C), and a pronounced prolongation of the mean lifetime of the shut state between bursts (Fig. 14D). Consequently, the average burst length is reduced [e.g., Eq. (4)]. First latencies are unaffected. (2) D-600 greatly increases the fraction of "blanks" in each ensemble of single Ca channel current records, the increase being more pronounced at more positive conditioning membrane potentials (not shown). The drug-induced extra "blanks" tend to occur in groups of consecutive sweeps suggesting an additional drug modulation of Ca channel gating properties over a time scale of seconds. Both drug effects favor a nonconducting state and thereby the reduction of the total ion transfer per depolarization. In this regard, the extra "blanks" induced by D-600 seem to be more effective than the alterations in fast

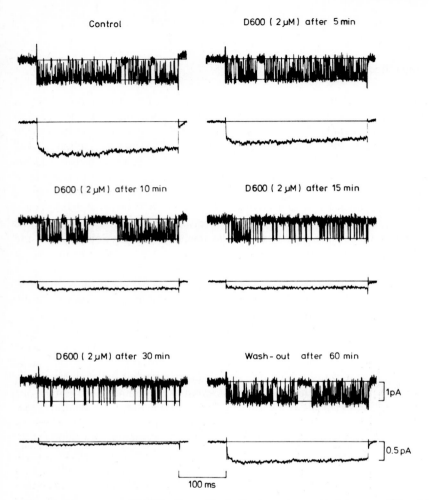

Fig. 13. Time course of D-600 effects. Single Ca channel currents *(upper panels)* and their ensemble averages *(lower panels)* at different times after bath application of 2 μM D-600 in response to step depolarizations of 75 mV amplitude and 300 ms duration at 0.5 Hz. Leakage currents and capacitive transients have been subtracted from all records. Single channel current records were lowpass filtered at 3 kHz; average currents were obtained from 150 single-channel current records, lowpass filtered at 500 Hz, for each condition; controls before D-600 application, 2.5 to 7.5 min, 7.5 to 12.5 min, 12.5 to 17.5 min and 27.5 to 32.5 min after bath application of the drug, respectively, and 57.5 to 62.5 min after washout. Analysis was restricted to experiments with only one active channel in the membrane patch and clear recovery after washout. The temperature was 35° C

kinetics. It is worth noting that the drug must have reached the Ca channel either from the lipid membrane phase or from the cytoplasm, since D-600 was only present in the bath and drug molecules cannot have diffused through the tight pipette-membrane seal which isolates the patch membrane area from the bathing solution (Hamill et al. 1981; Sakmann and Neher 1983).

Pathways which may be important for Ca channel block (and unblock) will be discussed with reference to the following channel state diagram:

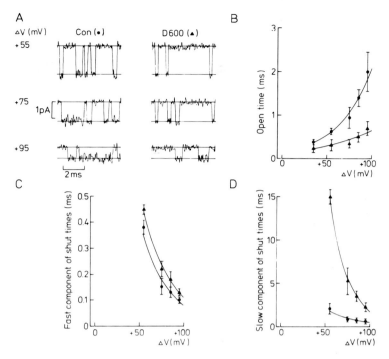

Fig. 14 A-D. Effect of D-600 (2 μM) on fast Ca channel kinetics. **A** Cell-attached recordings of elementary Ca channel current traces during 300-ms voltage-clamp pulses of various amplitudes delivered from the pipette zero potential at 0.5 Hz under control *(closed circles)* and drug conditions (30 min D-600 (2 μM) bath application; closed triangles). In each record, capacitance and leakage currents have been subtracted. *Solid lines* indicate the baseline level and the average elementary current amplitude. The records were low-pass filtered at 3 kHz. **B** Open times; **C** fast component of shut times; **D** slow component of shut times at various membrane potentials under control conditions *(closed circles)*, and 27.5 to 32.5 min after application of D-600 (2 μM) *(closed triangles)*. Data represent mean ± standard deviation from four experiments (control), and three experiments (D-600) which complied with the requirements outlined in the legend of Fig. 13. Individual values were obtained from mono- and double-exponential approximations of open and shut times histograms during 300-ms depolarizations of various amplitudes from the pipette zero potential at 0.5 Hz. *Solid lines* represent mono-exponential approximations of the mean values. The temperature was 35° C

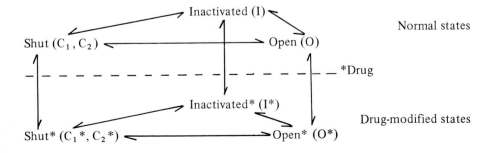

This scheme resembles the modulated receptor model previously used to describe Ca channel block and unblock by D-600 in multicellular tissue (McDonald et al. 1984b) and comprises similar assumptions. In this model (1) the drug-free channel can be found in one of the normal states, (2) the drug-bound channel can be found in one of the corresponding drug-modified states, (3) drug binding and unbinding are not restricted to a particular channel state, and (4) the drug-bound channel may have altered kinetics and equilibria. During depolarization, block can be achieved quickly by the association of drug to the open channel ($O \rightarrow O^*$) as suggested by the shortening of the average open lifetime of the Ca channel (Fig. 14B). The O^* channel will then most likely either undergo inactivation to I^* or procede towards (C_1^*, C_2^*), i.e., an easy transition from O^* to O would result in a prolongation of the mean burst length (see Neher and Steinbach 1978) rather than the observed shortening (see above). From there, the channel must have a route of unblock, since channel reopenings can occur during a single depolarization in the presence of the drug (e.g., Figs. 13 and 14A). In this regard, the likely pathway seems to be (C_1^*, C_2^*) $\rightarrow$ (C_1, C_2), since exit from the inactivated state is slow even under control conditions. The opposite transition ((C_1, C_2) $\rightarrow$ (C_1^*, C_2^*)) seems not to be a major blocking pathway, since the latency to first channel opening is not drastically affected by the drug. Thus, cycling of the Ca channel from (C_1, C_2) via O, O^*, (C_1^*, C_2^*) back to (C_1, C_2) would provide a suitable explanation for the alterations in fast channel kinetics induced by D-600. The prolongation of the shut times within and between bursts would reflect sojourns of the channel in the short-living drug-modified states. The extra "blanks" induced by D-600 point at the Ca channel getting "trapped" in the I^* state. Drug-induced stabilization of this long-living state is most likely achieved via direct slow drug association to the inactivated channel ($I \rightarrow I^*$) since any other transition ($O^* \rightarrow I^*$, (C_1^*, C_2^*) $\rightarrow I^*$) would cause an acceleration of the declining phase of the blocked ensemble average current. The data available at present fit necely into the framework of D-600 action in multicellular tissue (McDonald et al. 1984b) and single cells (Lee and Tsien 1983) although additional experiments along the lines of McDonald et al. (1984b) and Lee and Tsien (1983) are needed for further insight into the exact mechanism of Ca channel block and unblock.

9.2 Ca Channel Phosphorylation

Experiments in multicellular preparations (Sect. 2.1) and single cells (Sect. 4.1.4) have already elucidated the basic response to β-adrenergic agents such as adrenaline and isoprenaline; they have further provided evidence for the participation of cAMP, cAMP-dependent protein kinase and Ca as intracellular mediators. Until recently, it was widely believed that β-adrenergic agents, acting through cAMP, enhance Ca current by increasing the number of functional Ca channels rather than by changing the properties of individual Ca channels (Niedergerke and Page 1977; Reuter and Scholz 1977b; Tsien 1977; Reuter 1979, 1980). However, recent recordings of single Ca channel activity in cell-attached membrane patches (Reuter et al. 1982; Reuter 1983; Cachelin et al. 1983; Brum et al. 1984) suggest that this view requires modification. As illustrated in Fig. 15, bath application of adrenaline (1 μM) enhanced the average current, I, (Fig. 15b) due to an increased overall probability, p, of the Ca channel to occupy the open state. From

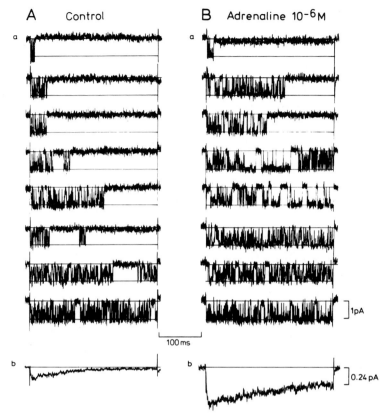

Fig. 15 A, B. Effect of β-adrenergic stimulation on the single Ca channel. Elementary Ca channel currents *(a)* (leakage and capacitance corrected; low-pass filtered at 3 kHz) during 300 ms depolarizing voltage steps of 75 mV amplitude delivered at 0.5 Hz before **(A)** and 20 min after bath application of adrenaline **(B)**. The average currents *(b)* were obtained from 150 single-channel current traces low-pass filtered at 500 Hz, respectively. In the given examples of Ca channel activity during the pulses *(a)*, "blanks" are not shown. In this experiment, 37% of all applied depolarizations did not elicit Ca channel openings under control conditions. This fraction decreased to 9% in the presence of adrenaline. The temperature was 35° C

individual elementary current records (Fig. 15a), it became evident that the Ca channel activity per depolarization was distinctly increased. The elementary Ca channel current amplitude, i, was not noticeably changed, nor was the number of active channels (i.e., N = 1; Fig. 15a). The increase in p was brought about by a slight prolongation of the mean lifetime of the open state, marked shortening of the mean lifetime of the shut state between bursts (Fig. 15a), and a reduction in the number of blanks per ensemble (not shown, see figure legend) (see Cachelin et al. 1983; Brum et al. 1984 for quantitative analysis). One hypothesis, not yet analyzed in detail, is that adrenaline seems also to increase the cluster duration and to decrease the shut intervals between clusters (e.g., Fig. 15a).

The fact that the elementary Ca channel current was unchanged (no superpositions of channel openings), and that silent (i.e., no channel activity) patches were never converted to active ones by adrenaline (Brum et al. 1984), argues strongly against a drug-induced increase in channel number. In terms of the proposed kinetic reaction scheme (see above), cAMP-dependent phosphorylation of individual Ca channels by β-adrenergic agents has been interpreted as increasing the forward rate constants, and reducing the backward ones, leading to and from the open state of the Ca channel during depolarization (Cachelin et al. 1983; Brum et al. 1984).

In contrast to other ionic channels, Ca channels do not function in isolated membrane patches (either inside-out or outside-out; see Hamill et al. 1981 for the cell-free recording techniques); Ca channel activity disappeared within 30 s to 2 min after the isolation of the membrane patch from the rest of the cell (Fenwick et al. 1982; Cavalié et al. 1983). The authors concluded that Ca channels need an intracellular substance to keep them in a functional state. Thus, it seems possible that cAMP-dependent phosphorylation will turn out to be a prerequisite for Ca channel gating.

The foregoing suggests that there could be a reserve of nonfunctional (not phosphorylated) Ca channels in the membrane. The only hint of the conversion of such channels to functional ones with β-adrenergic stimulation could have been the reduction in the number of blanks per ensemble. Much more convincing evidence seemed to emerge from recent noise-analysis experiments of frog ventricular myocytes (Bean et al. 1984): isoprenaline induced a fivefold increase in I_{Ca}, and this was mainly attributed to a threefold increase in the calculated number of functional Ca channels. On the other hand, Brum et al. (1984) show that a present there is no way to distinguish an increase in N from an increase in p in some multichannel patches. Future work will no doubt provide an explanation for the divergent results on Ca channel stimulation by β-adrenergic agents.

10 Ca Channel Density and Ca Influx Through Individual Channels

Ca channel density per unit membrane area can be estimated from the elementary current through an individual open channel and the membrane Ca current in single myocytes. Peak Ca current density has been calculated (Sect. 4.2) by estimating the average membrane surface area of a myocyte and assuming a specific membrane capacitance of $1 \mu F \, cm^{-2}$ (Isenberg and Klöckner 1982b; Mitchell et al. 1983). Furthermore, one can calculate the total number of Ca channels per cell by dividing the average peak Ca current in myocytes by the product $(p \cdot i)$ which accompanies the peak average current (I) through single channels. Relating this number of channels per cell to the cell membrane area results in estimates of channel density in the range of 0.05-0.5 per μm^2 for mammalian ventricular cells (Fabiato 1983; Reuter 1983, 1984). Knowing the channel density, one can calculate that the average distance between two Ca channels would be 1.4-4.5 μm assuming a distribution of the channels in a square array. A more likely distribution in a hexagonal array would give a slightly larger distance (Fabiato 1983).

However, Ca channels may not be homogeneously distributed over the sarcolemma (e.g., Brum et al. 1984).

Fabiato (1983) calculated the intracellular [total Ca] change associated with the Ca movement through a Ca channel assuming that the channel opening is a hemisphere. First, the change of [total Ca] at the intracellular opening of the Ca channel is 4.13 mM. This Ca then diffuses so rapidly out of the channel that the on-rates of Ca-binding sites on the inner face of the sarcolemma cannot be high enough to bind any Ca at all at this time. Thus, a Ca wave will propagate from the channel opening into the myoplasm and as the Ca diffuses away from the channel, binding occurs by several "buffers" (Fabiato 1983), the most important ones probably being the sarcoplasmic reticulum, troponin and calmodulin. In this manner, the [free Ca] will be reduced to its final cytoplasmic steady-state value during excitation.

11 Concluding Remarks

In mammalian tissue, the important cellular messengers, besides hormones, are cyclic nucleotides such as cAMP and Ca. For cAMP, the response time and the duration of action range from seconds to minutes; for Ca, they are likely to be as short as milliseconds. Both intracellular-messenger systems are interconnected and therefore overlap in time and space (Cheung 1982; Cohen 1982). Calmodulin acts as an intracellular Ca receptor and, after being activated by Ca binding, mediates important Ca-dependent cellular processes such as (1) cyclic nucleotide metabolism via activation of adenylate cyclase, guanylate cyclase and phosphodiesterase, (2) Ca metabolism via Ca-ATPase, (3) glycogen metabolism via phosphorylase kinase and glycogen synthase kinase, (4) phosphorylation of various proteins via Ca-dependent protein kinases and (5) contractile processes via myosin light-chain kinase (Cheung 1982; Cohen 1982). Furthermore, Ca itself in contractile cells regulates contraction in a beat-to-beat manner via the sarcoplasmic reticulum (Ca-induced release of Ca) and/or by binding and release to/from troponin (Fabiato 1983). These topics, as well as the control of other membrane permeabilities by Ca and the cAMP system (Meech 1978; Colquhoun et al. 1981; Maruyama and Petersen 1982; Noma 1983; Reuter 1984), are not specifically covered here since even a brief review of each of them would require at least the same space as this article. All these topics have been recently reviewed, the articles being cited in the list of references.

In general, the entry of Ca ions through Ca channels down the electrochemical gradient can be regulated by either physiological factors like transmembrane voltage and cAMP-dependent channel phosphorylation or pharmacological interventions like the application of Ca entry blockers. However, voltage-dependent gating of Ca channels and its modulation may be only one option of controlling intracellular [Ca]. Yet, the contribution of voltage-independent Ca permeabilities linked to phospholipid metabolism (Mitchell 1982) and a voltage-dependent Na-Ca-exchange (Mullins 1979) remains unclear at present in heart.

Acknowledgements. This article incorporates some data concerning Ca channel inactivation and D-600 action that have not yet been published. These were obtained in collaboration with Dr. Adolfo Cavalié in our laboratory, whose contribution we would like to acknowledge here. His experiments and suggestions were a substantial contribution to this article. We further thank Dr. T.F. McDonald for his critical review and help in preparing a final version of the manuscript. We are grateful to Mr. H. Ehrler for skillful electronics support, Mrs. D. Steimer for technical assistance, and Mrs. U. Lang for secreterial support. A major part of the Nicolet MED-80 software was developed by Mr. B. Bresser. This work was supported by the Deutsche Forschungsgemeinschaft, SFB 38, Membranforschung, project G1.

References

Aldrich RW, Yellen G (1983) Analysis of nonstationary channel kinetics. In: Sakmann B, Neher E (eds) Single-channel recording. Plenum, New York, pp 287-299

Amsterdam A, Jamieson JD (1974) Studies on dispersed pancreatic exocrine cells. J Cell Biol 63: 1037-1056

Attwell D, Cohen I (1977) The voltage clamp of multicellular preparations. Prog Biophys Mol Biol 31: 201-245

Bean BP, Nowycky MC, Tsien RW (1984) β-adrenergic modulation of calcium channels in frog ventricular heart cells. Nature (Lond) 307: 371-375

Beeler GW, Reuter H (1970a) Voltage clamp experiments on ventricular myocardium fibres. J Physiol (Lond) 207: 165-190

Beeler GW, Reuter H (1970b) Membrane calcium current in ventricular myocardium fibres. J Physiol (Lond) 207: 191-209

Beeler GW, Reuter H (1977) Reconstruction of the action potential of ventricular myocardial fibres. J Physiol (Lond) 268: 177-210

Beeler GW, McGuigan JAS (1978) Voltage clamping of multicellular preparations: capabilities and limitations of existing methods. Prog Biophys Mol Biol 34: 219-254

Berry MN, Friend DS (1969) High yield preparation of isolated rat liver parenchymal cells. A biochemical and fine structural study. J Cell Biol 43: 507-520

Brown AM, Lee KS, Powell T (1981a) Sodium current in single rat heart muscle cells. J Physiol (Lond) 318: 479-500

Brown AM, Morimoto K, Tsuda Y, Wilson DL (1981b) Calcium current-dependent and voltage-dependent inactivation of calcium channels in *Helix aspersa*. J Physiol (Lond) 320: 193-218

Brown AM, Tsuda Y, Wilson DL (1983) A description of activation and conduction in calcium channels based on tail and turn-on current measurements in the snail. J Physiol (Lond) 344: 549-583

Brown HF, McNaughton PA, Noble D, Noble SJ (1975) Adrenergic control of pacemaker currents. Philos Trans R Soc Lond B Biol Sci 270: 527-537

Brown HF, Giles W, Noble SJ (1977) Membrane currents underlying activity in frog sinus venosus. J Physiol (Lond) 271: 783-816

Brown HF, Kimura J, Noble SJ (1981) Calcium entry-dependent inactivation of the slow inward current in the rabbit sinoatrial node. J Physiol (Lond) 320: 11P

Brum G, Flockerzi V, Hofmann F, Osterrieder W, Trautwein W (1983) Injection of catalytic subunit of cAMP-dependent protein kinase into isolated cardiac myocytes. Pflügers Arch Eur J Physiol 398: 147-154

Brum G, Osterrieder W, Trautwein W (1984) β-adrenergic increase in the calcium conductance of cardiac myocytes studied with the patch clamp. Pflügers Arch Eur J Physiol 401: 111-118

Bustamente JO, Watanabe T, McDonald TF (1981) Single cells from adult mammalian heart: isolation procedure and preliminary electrophysiological studies. Can J Physiol Pharmacol 59: 907-910

Byerly L, Hagiwara S (1982) Calcium currents in internally perfused nerve cell bodies of *Limnea stagnalis.* J Physiol (Lond) 322: 503-528

Cachelin AB, dePeyer JE, Kokubun S, Reuter H (1983) Ca^{2+} channel modulation by 8-bromocyclic AMP in cultured heart cells. Nature (Lond) 304: 462-464

Cavalié A, Ochi R, Pelzer D, Trautwein W (1983) Elementary currents through Ca channels in guinea pig myocytes. Pflügers Arch Eur J Physiol 398: 284-297

Chapman RA (1979) Excitation-contraction coupling in cardiac muscle. Prog Biophys Mol Biol 35: 1-52

Chapman RA (1983) Control of cardiac contractility at the cellular level. Am J Physiol 245 (Heart Circ Physiol 14): H535-H552

Cheung WY (1982) Calmodulin. Sci Am 246: 48-56

Cohen P (1982) The role of protein phosphorylation in neural and hormonal control of cellular activity. Nature (Lond) 296: 613-620

Colatsky TJ (1980) Voltage clamp measurements of sodium channel properties in rabbit cardiac Purkinje fibres. J Physiol (Lond) 305: 215-234

Colquhoun D, Hawkes AG (1977) Relaxation and fluctuations of membrane currents that flow through drug-operated ion channels. Proc R Soc Lond B Biol Sci 199: 231-262

Colquhoun D, Hawkes AG (1981) On the stochastic properties of single ion channels. Proc R Soc Lond B Biol Sci 211: 205-235

Colquhoun D, Hawkes AG (1982) On the stochastic properties of bursts of single ion channel openings and of clusters of bursts. Philos Trans R Soc Lond B Biol Sci 300: 1-59

Colquhoun D, Hawkes AG (1983) The principles of the stochastic interpretation of ion-channel mechanisms. In: Sakmann B, Neher E (eds) Single-channel recording. Plenum, New York, pp 135-175

Colquhoun D, Neher E, Reuter H, Stevens CF (1981) Inward current channels activated by intracellular Ca in cultured cardiac cells. Nature (Lond) 294: 752-754

Colquhoun D, Sakmann B (1983) Bursts of openings in transmitter-activated ion channels. In: Sakmann B, Neher E (eds) Single-channel recording. Plenum, New York, pp 345-364

Conti F, Neher E (1980) Single channel recordings of K^+ currents in squid axons. Nature (Lond) 285: 140-143

Deth R, Casteels R (1977) A study of releasable Ca fractions in smooth muscle cells of the rabbit aorta. J Gen Physiol 69: 401-416

Dow JW, Harding NGL, Powell T (1981) Isolated cardiac myocytes. I Preparation of adult myocytes and their homology with the intact tissue. Cardiovasc Res 15: 483-514

Ebashi S, Endo M (1968) Calcium ion and muscle contraction. Prog Biophys Mol Biol 18: 125-183

Fabiato A (1983) Calcium-induced release of calcium from the cardiac sarcoplasmic reticulum. Am J Physiol 245 (Cell Physiol 14): C1-C14

Fabiato A, Fabiato F (1979) Calcium and cardiac excitation-contraction coupling. Annu Rev Physiol 41: 473-484

Fenwick EM, Marty A, Neher E (1982) Sodium and calcium channels in bovine chromaffin cells. J Physiol (Lond) 331: 599-635

Fischmeister R, Mentrard D, Vassort G (1981) Slow inward current inactivation in frog heart atrium. J Physiol (Lond) 320: 27P-28P

Fry DM, Scales D, Inesi G (1979) The ultrastructure of membrane alterations of enzymatically dissociated cardiac myocytes. J Mol Cell Cardiol 11: 1151-1163

Gettes LS, Reuter H (1974) Slow recovery from inactivation of inward currents in mammalian myocardial fibres. J Physiol (Lond) 240: 703-724

Giles W, Noble SJ (1976) Changes in membrane currents in bullfrog atrium produced by acetylcholine. J Physiol (Lond) 261: 103-123

Grant AO, Katzung BG (1976) The effects of quinidine and verapamil on electrically induced automaticity in the ventricular myocardium of guinea pig. J Pharmacol Exp Ther 196: 407-419

Grinwald PM, Nayler WG (1981) Calcium entry in the calcium paradox. J Mol Cell Cardiol 13: 867-880

Hagiwara S, Byerly L (1981) Calcium channel. Annu Rev Neurosci 4: 69-125

Hagiwara S, Ohmori H (1982) Studies of calcium channels in rat clonal pituitary cells with patch electrode voltage clamp. J Physiol (Lond) 331: 231-252

Hamill OP, Marty A, Neher E, Sakmann B, Sigworth FJ (1981) Improved patch-clamp techniques for high-resolution current recording from cells and cell-free membrane patches. Pflügers Arch Eur J Physiol 391: 85-100

Haworth RA, Hunter DR, Berkoff HA (1980) The isolation of Ca^{2+}-resistant myocytes from the adult rat. J Mol Cell Cardiol 12: 715-723

Hescheler J, Pelzer D, Trube G, Trautwein W (1982) Does the organic calcium channel blocker D-600 act from inside or outside on the cardiac cell membrane? Pflügers Arch Eur J Physiol 393: 287-291

Hino N, Ochi R (1980) Effect of acetylcholine on membrane currents in guinea-pig papillary muscle. J Physiol (Lond) 307: 183-197

Horackova M, Vassort G (1976) Calcium conductance in relation to contractility in frog myocardium. J Physiol (Lond) 259: 597-616

Hume JR, Giles W (1981) Active and passive electrical properties of single bullfrog atrial cells. J Gen Physiol 78: 19-42

Isenberg G (1982) Ca entry and contraction as studied in isolated bovine ventricular myocytes. Z Naturforsch 37c: 502-512

Isenberg G, Klöckner U (1981) Ca currents of isolated bovine ventricular myocytes. In: Ohnishi ST, Endo M (eds) The mechanism of gated calcium transport across biological membranes. Academic, New York, pp 25-33

Isenberg G, Klöckner U (1982a) Calcium tolerant ventricular myocytes prepared by preincubation in a "KB medium". Pflügers Arch Eur J Physiol 395: 6-18

Isenberg G, Klöckner U (1982b) Calcium currents of isolated bovine ventricular myocytes are fast and of large amplitude. Pflügers Arch Eur J Physiol 395: 30-41

Johnson EA, Liebermann M (1971) Heart: excitation and contraction. Annu Rev Physiol 33: 479-529

Kass RS (1982) Nisoldipine: a new, more selective calcium current blocker in cardiac Purkinje fibers. J Pharmacol Exp Ther 223: 446-456

Kass RA, Wiegers SE (1982) The ionic basis of concentration-related effects of noradrenaline on the action potential of calf cardiac Purkinje fibres. J Physiol (Lond) 322: 541-558

Katz B (1969) The release of neural transmitter substances. Liverpool University Press

Kohlhardt M, Bauer B, Krause H, Fleckenstein A (1972) Differentiation of the transmembrane Na and Ca channel in mammalian cardiac fibers by the use of specific inhibitors. Pflügers Arch Eur J Physiol 335: 309-322

Kootsey IM, Johnson EA (1972) Voltage clamp of cardiac muscle. A theoretical analysis of early currents in the single sucrose gap. Biophys J 12: 1496-1508

Kostyuk PG (1981) Calcium channel in the neuronal membrane. Biochim Biophys Acta 650: 128-150

Lee KS, Lee EW, Tsien RW (1981) Slow inward current carried by Ca^{2+} or Ba^{2+} in single isolated heart cells. Biophys J 33: 143a

Lee KS, Tsien RW (1982) Reversal of current through calcium channels in dialysed single heart cells. Nature (Lond) 297: 498-501

Lee KS, Tsien RW (1983) Mechanism of calcium channel blockade by verapamil, D-600, diltiazem and nitrendipine in single dialysed heart cells. Nature (Lond) 302: 790-794

Llinas R, Steinberg IZ, Walton K (1981) Presynaptic calcium currents in squid giant synapse. Biophys J 33: 289-322

Lux HD (1983) Observations on single calcium channels: an overview. In: Sakmann B, Neher E (eds) Single-channel recording. Plenum, New York, pp 437-449

Marban E, Tsien RW (1981) Is the slow inward calcium current of heart muscle inactivated by calcium? Biophys J 33: 143a

Marshall AG (1978) Biophysical chemistry. Principles, techniques and applications. Wiley, New York

Marty A, Neher E (1983) Tight-seal whole-cell recording. In: Sakmann B, Neher E (eds) Single-channel recording. Plenum, New York, pp 107-122

Maruyama Y, Petersen OH (1982) Cholecystokinin activation of single-channel currents is mediated by internal messenger in pancreatic acinar cells. Nature (Lond) 300: 61-63

Maylie J, Morad M (1981) Ionic characterization of pacemaker current in voltage clamped rabbit SA node. Biophys J 33: 11a

McAllister RE, Noble D, Tsien RW (1975) Reconstruction of the electrical activity of cardiac Purkinje fibres. J Physiol (Lond) 251: 1-59

McDonald TF (1982) The slow inward calcium current in the heart. Annu Rev Physiol 44: 425-434

McDonald TF, Pelzer D, Trautwein W (1980) On the mechanism of slow calcium channel block in heart. Pflügers Arch Eur J Physiol 385: 175-179

McDonald TF, Pelzer D, Trautwein W (1981) Does the calcium current modulate the contraction of the accompanying beat? A study of E-C coupling in mammalian ventricular muscle using cobalt ions. Circ Res 49: 576-583

McDonald TF, Pelzer D, Trautwein W (1984a) Cat ventricular muscle treated with D-600: effects on calcium and potassium currents. J Physiol (Lond) 352: 203-216

McDonald TF, Pelzer D, Trautwein W (1984b) Cat ventricular muscle treated with D-600: characteristics of calcium channel block and unblock. J Physiol (Lond) 352: 217-241

McDonald TF, Trautwein W (1978) Membrane currents in cat myocardium: separation of inward and outward components. J Physiol (Lond) 274: 193-216

Meech RW (1978) Calcium-dependent potassium activation in nervous tissues. Annu Rev Biophys Bioeng 7: 1-18

Mitchell RH (1982) Is phosphatidylinositol really out of the calcium gate? Nature (Lond) 296: 492-493

Mitchell MR, Powell T, Terrar DA, Twist VW (1983) Characteristics of the second inward current in cells isolated from rat ventricular muscle. Proc R Soc Lond B Biol Sci 219: 447-469

Morad M, Sanders C, Weiss J (1981) The inotropic actions of adrenaline on frog ventricular muscle: relaxing versus potentiating effects. J Physiol (Lond) 311: 585-604

Mullins JL (1979) The generation of electric currents in cardiac fibers by Na/Ca exchange. Am J Physiol 236 (3): C 103-C 110

Neher E, Steinbach JH (1978) Local anesthetics transiently block currents through single acetylcholine-receptor channels. J Physiol (Lond) 277: 153-176

New W, Trautwein W (1972a) Inward membrane currents in mammalian myocardium. Pflügers Arch Eur J Physiol 334: 1-23

New W, Trautwein W (1972b) The ionic nature of slow inward current and its relation to contraction. Pflügers Arch Eur J Physiol 334: 24-38

Niedergerke R, Page S (1977) Analysis of catecholamine effects in single atrial trabeculae of the frog heart. Proc R Soc Lond B Biol Sci 197: 333-362

Noble SJ, Shimoni Y (1981) The calcium and frequency dependence of the slow inward current "staircase" in frog atrium. J Physiol (Lond) 310: 57-75

Noma A (1983) ATP-regulated K^+ channels in cardiac muscle. Nature (Lond) 305: 147-148

Noma A, Irisawa H (1976) Membrane currents in the rabbit sinoatrial node cell as studied by the double microelectrode method. Pflügers Arch Eur J Physiol 364: 45-52

Noma A, Irisawa H, Kokubun S, Kotake H, Nishimura M, Watanabe Y (1980a) Slow current systems in the A-V node of the rabbit heart. Nature (Lond) 285: 228-229

Noma A, Kotake H, Irisawa H (1980b) Slow inward current and its role mediating the chronotropic effect of epinephrine in the rabbit sinoatrial node. Pflügers Arch Eur J Physiol 388: 1-9

Ochi R (1976) Manganese-dependent propagated action potentials and their depression by electrical stimulation in guinea-pig myocardium perfused by sodium-free media. J Physiol (Lond) 263: 139-156

Osterrieder W, Brum G, Hescheler J, Trautwein W, Hofmann F, Flockerzi V (1982) Injection of subunits of cyclic AMP-dependent protein kinase into cardiac myocytes modulates Ca^{2+} current. Nature (Lond) 298: 576-578

Payet MD, Schanne OF, Ruiz-Ceretti E (1980) Competition for slow channel of Ca^{2+}, Mn^{2+}, verapamil and D-600 in rat ventricular muscle? J Mol Cell Cardiol 12: 635-638

Pelzer D, Trautwein W, McDonald TF (1982) Calcium channel block and recovery from block in mammalian ventricular muscle treated with organic channel inhibitors. Pflügers Arch Eur J Physiol 394: 97-105

Pelzer D, Trube G, Piper HM (1984) Low resting potentials in single isolated heart cells due to membrane damage by the recording microelectrode. Pflügers Arch Eur J Physiol 400: 197-199

Plant TD, Standen NB, Ward TA (1983) The effects of injection of calcium ions and calcium chelators on calcium channel inactivation in *Helix* neurones. J Physiol (Lond) 334: 189-212

Podolsky RJ (1975) Muscle activation: the current status. Fed Proc 34: 1374-1378

Powell T, Terrar DA, Twist VW (1980) Electrical properties of individual cells isolated from adult rat ventricular myocardium. J Physiol (Lond) 302: 131-153

Ramon F, Anderson N, Joyner RW, Moore JW (1975) Axon voltage-clamp simulations. IV A multicellular preparations. Biophys J 15: 55-69

Reuter H (1967) The dependence of slow inward current in Purkinje fibres on the extracellular Ca-concentration. J Physiol (Lond) 192: 479-492

Reuter H (1973) Divalent cations as charge carriers in excitable membranes. Prog Biophys Mol Biol 26: 1-43

Reuter H (1974) Localization of beta adrenergic receptors, and effects of noradrenaline and cyclic nucleotides on action potentials, ionic currents and tension in mammalian cardiac muscle. J Physiol (Lond) 242: 429-451

Reuter H (1979) Properties of two inward membrane currents in the heart. Annu Rev Physiol 41: 413-424

Reuter H (1980) Effects of neurotransmitters on the slow inward current. In: Zipes DP, Bailey JC, Elharrar V (eds) The slow inward current and cardiac arrhythmias. Nijhoff, The Hague, pp 205-219

Reuter H (1983) Calcium channel modulation by neurotransmitters, enzymes and drugs. Nature (Lond) 301: 569-574

Reuter H (1984) Ion channels in cardiac cell membranes. Annu Rev Physiol 46: 473-484

Reuter H, Scholz H (1977a) A study of the ion selectivity and the kinetic properties of the calcium-dependent slow inward current in mammalian cardiac muscle. J Physiol (Lond) 264: 17-47

Reuter H, Scholz H (1977b) The regulation of the Ca conductance of cardiac muscle by adrenaline. J Physiol (Lond) 264: 49-62

Reuter H, Stevens CF, Tsien RW, Yellen G (1982) Properties of single calcium channels in cardiac cell culture. Nature (Lond) 297: 501-504

Saimi Y, Kung C (1982) Are ions involved in the gating of calcium channels? Science (Wash DC) 218: 153-156

Sakmann B, Neher E (1983) Single-channel recording. Plenum, New York

Sakmann B, Patlak J, Neher E (1980) Single acetylcholine-activated channels show burst-kinetics **in presence of desensitizing concentrations of agonist. Nature (Lond) 286: 71-73**

Sakmann B, Trube G (1984) Voltage-dependent inactivation of inward-rectifying single channel currents in the guinea pig heart cell membrane. J Physiol (Lond) 347: 659-683

Siegelbaum SA, Tsien RW (1980) Calcium-activated transient outward current in calf cardiac Purkinje fibres. J Physiol (Lond) 299: 485-506

Standen NB (1981) Ca^{2+} channel inactivation by intracellular Ca^{2+} injection into *Helix* neurones. Nature (Lond) 293: 158-159

Standen NB, Stanfield PR (1982) A binding-site model for calcium channel inactivation that depends on calcium entry. Proc R Soc Lond B Biol Sci 217: 101-110

Taniguchi J, Kokubun S, Noma A, Irisawa H (1981) Spontaneously active cells isolated from the sino-atrial and atrio-ventricular nodes of rabbit heart. Jpn J Physiol 31: 547-558

Tarr M, Trank JW, Leiffer P (1981) Characteristics of sarcomere shortening in single frog atrial cardiac cells during lightly loaded contractions. Circ Res 48: 189-200

Ten Eick R, Nawrath H, McDonald TF, Trautwein W (1976) On the mechanism of the negative inotropic effect of acetylcholine. Pflügers Arch Eur J Physiol 361: 207-213

Trautwein W (1973) Membrane currents in cardiac muscle fibers. Physiol Rev 53: 793-835

Trautwein W, McDonald TF, Tripathi O (1975) Calcium conductance and tension in mammalian ventricular muscle. Pflügers Arch Eur J Physiol 354: 55-74

Trautwein W, Taniguchi J, Noma A (1982) The effect of intracellular cyclic nucleotides, and calcium on the action potential and acetylcholine response of isolated cardiac cells. Pflügers Arch Eur J Physiol 392: 307-314

Trautwein W, Pelzer D, McDonald TF (1983) Interval- and voltage-dependent effects of the calcium channel-blocking agents D-600 and AQA39 on mammalian ventricular muscle. Circ Res (Suppl I) 52: I60-I68

Trube G (1983) Enzymatic dispersion of heart and other tissues. In: Sakmann B, Neher E (eds) Single-channel recording. Plenum, New York, pp 69-76

Tsien RW (1973) Adrenaline-like effects of intracellular iontophoresis of cyclic AMP in cardiac Purkinje fibres. Nature (Lond) 245: 120-122

Tsien RW (1977) Cyclic AMP and contractile activity in heart. Adv Cyclic Nucleotide Res 8: 363-420

Tsien RW (1983) Calcium channels in excitable cell membranes. Annu Rev Physiol 45: 341-358

Tsien RW, Giles W, Greengard P (1972) Cyclic AMP mediates the effect of adrenaline on cardiac Purkinje fibres. Nature New Biol 240: 181-183

Tsuda Y, Wilson DL, Brown AM (1982) Calcium tail currents in snail neurones. Biophys J 37: 181a

Vereecke J, Carmeliet E (1971) Sr action potentials in cardiac Purkinje fibres. I Evidence for a regenerative increase in Sr conductance. Pflügers Arch Eur J Physiol 322: 60-72

Wilson DL, Morimoto K, Tsuda Y, Brown AM (1983) Interaction between calcium ions and surface charge as it relates to calcium currents. J Membr Biol 72: 117-130

Wittenberg BA, Robinson TF (1981) Oxygen requirements, morphology, cell coat and membrane permeability of calcium-tolerant myocytes from hearts of adult rats. Cell Tissue Res 216: 231-251

Yanagihara K, Noma A, Irisawa H (1980) Reconstruction of sino-atrial node pacemaker potential based on the voltage clamp experiments. Jpn J Physiol 30: 841-857

Mitochondrial Calcium Fluxes and Their Role in the Regulation of Intracellular Calcium

F. L. BYGRAVE, P. H. REINHART, and W. M. TAYLOR[1]

CONTENTS

1 Introduction

In most eukaryotic cells calcium is compartmented into at least four pools located on the outer surface of the plasma membrane, in the cytoplasm, the lumen of the endoplasmic reticulum and in the mitochondrial matrix space (Fig. 1). In these compartments, as in most biological systems, calcium usually exists in two forms, either bound to membranes, macromolecules or other ions, or as the free ion, Ca^{2+} (Dawson and Hauser 1970). Although it is Ca^{2+} that plays the important role in the activation of Ca^{2+}-sensitive enzymes, only a small proportion of the total calcium content of any compartment may be in this form (Borle 1981).

This review deals with current information about the calcium compartmented in mitochondria, about the way the ion moves across the inner mitochondrial membrane and the way in which such movement might be involved in the regulation of cell calcium. To this end three general aspects will be considered, (a) a brief overview of Ca^{2+} fluxes in mitochondria from rat liver (b) the nature and content of calcium in rat liver mitochondria, and (c) some recent information about the role of mitochondria in the regulation of the cytoplasmic Ca^{2+} concentration as revealed from studies on the mechanism of action of α-adrenergic agonists in rat liver.

1 Department of Biochemistry, Faculty of Science, The Australian National University, Canberra, A.C.T. 2600, Australia

Fig. 1. Compartmentation of calcium in liver tissue. The simplest model consistent with current experimental evidence considers four pools of calcium and three Ca^{2+}-translocation cycles. Each pool consists of varying proportions of free (Ca^{2+}) and bound (Ca_b) calcium. Bound pools are represented graphically by *cross-hatching*. Show are: *pool 1* the extracellular space and binding sites on the exterior face of the plasma membrane *(PM)*; *pool 2* the cytoplasm and the interior face of the plasma membrane; *pool 3* mitochondria *(MT)*; *pool 4* the endoplasmic reticulum *(ER)*. Ca^{2+}-translocation cycles may consist of distinct Ca^{2+} uptake and Ca^{2+} efflux activities (see text). (Reinhart et al. 1984d)

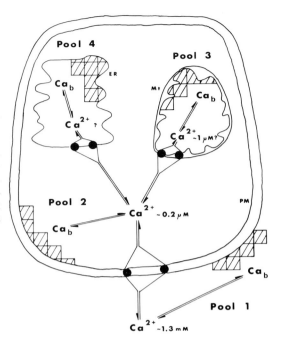

2.1 Mitochondrial Calcium Fluxes as Determined from in Vitro Studies

Most models of cellular calcium regulation, consistent with currently available data, consider the activity of Ca^{2+} transporters located in three membranes, the plasma membrane, the endoplasmic reticulum and the inner mitochondrial membrane (see Fig. 1). Movements of Ca^{2+} take place across the inner mitochondrial membrane in both directions, but employing apparently different pathways. Inwardly directed Ca^{2+} fluxes, that is from the cytoplasm into the mitochondrial matrix space, occur via an electrophoretic Ca^{2+} uniporter whose general properties are now quite well characterized. Details may be found in several recent reviews (Nicholls and Akerman 1982; Akerman and Nicholls 1983) and in the paper of Joseph et al. (1983).

Influx of Ca^{2+} occurs with a maximum velocity that can be as high as 100 nmol Ca^{2+} min^{-1} mg^{-1} of mitochondrial protein and an apparent K_m of 2-5μM that is further increased in the presence of Mg^{2+}. Ruthenium red and La^{3+} are potent inhibitors of mitochondrial Ca^{2+} influx. The organelle possesses a great capacity to take up the ion to the extent that the external free Ca^{2+} concentration can be quickly lowered to less than 1 μM. Although inward movement of the ion requires energy supplied by the mitochondrial electrochemical gradient $(\Delta\Psi)$, it can be shown that the gradient of free Ca^{2+} across the inner mitochondrial membrane never attains the predicted value of 10^{-5} to 10^{-6} if the $\Delta\Psi$ were the sole factor responsible for Ca^{2+} distribution.

This situation arises because of the existence also in the inner mitochondrial membrane of an outwardly directed, Ruthenium red-insensitive Ca^{2+} efflux system (see Nicholls and Akerman 1982). This carrier is electroneutral and in liver mitochondria

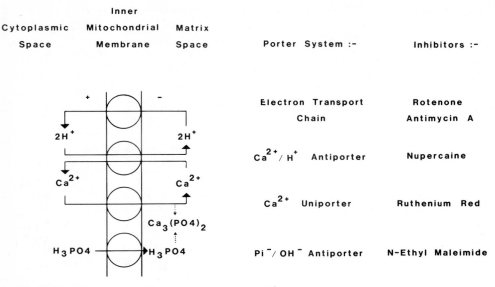

Fig. 2. Diagrammatic outline of the mitochondrial Ca^{2+} translocation cycle

catalyzes the exchange of Ca^{2+} for $2H^{+}$. The apparent K_{m} for matrix free Ca^{2+} is $10\mu M$ and the maximum velocity is of the order 5 nmol Ca^{2+} min^{-1} mg^{-1} of mitochondrial protein (Coll et al. 1982) which is much less than that of the uniporter. In mitochondria from some tissues, and to a lesser extent from liver, Na^{+} is able to induce an efflux of Ca^{2+}, a feature regarded as of potential physiological significance (Crompton et al. 1978). The local anesthetic nupercaine seems to be quite effective in inhibiting mitochondrial Ca^{2+} efflux (Dawson et al. 1979). Operation of the Ca^{2+} efflux system, in concert with that of the Ca^{2+} uniporter, leads to a situation in which the ion "cycles" rapidly between the external cytoplasmic space and the internal matrix space (see Fig. 2) thereby resulting in a steady-state concentration gradient of Ca^{2+} lower than predicted by the $\Delta\Psi$.

The interpretation of Ca^{2+} movements across the inner mitochondrial membrane in terms of a cycle has a number of important corollaries (Bygrave 1978; Akerman and Nicholls 1983).

The existence of a cycle allows for rapid control of net fluxes across the inner mitochondrial membrane.

The set of porters as shown in Fig. 2 allows for maintenance of electroneutrality in the sense that the net inward movement of Ca^{2+} is compensated by net $2H^{+}$ movement out. The loss in H^{+} from the matrix space will lead to an increase in the ΔpH that in turn compensates for the decrease in $\Delta\Psi$ following net Ca^{2+} accumulation.

The cycle provides a rationale why the addition of Ruthenium red to isolated mitochondria promotes the net efflux of Ca^{2+} from the organelle and why the addition of inorganic phosphate (Pi) inhibits Ca^{2+} efflux and is able to promote extensive accumulation of the ion (Zoccarato and Nicholls 1981).

A further important feature of Ca^{2+}-cycling in mitochondria is that only a small energy demand necessary for its operation (Stucki and Ineichen 1974).

Pi probably plays a much more important role in controlling Ca^{2+} translocation across the inner mitochondrial membrane than previously imagined. The fact that Pi is so readily able to form a complex $[Ca_3(PO_4)_2]$ within the matrix provides a means by which the anion may have a significant influence on the rates of Ca^{2+} translocation across the inner mitochondrial membrane.

Like other weak acids, Pi also has a significant influence on the value for the ΔpH as measured in vitro. In our hands for instance, the ΔpH of isolated rat liver mitochondria containing approx. 2 μM Ca^{2+} decreases markedly as the added Pi concentration is increased from 0 to 1 mM (W. M. Taylor, P. H. Reinhart and F. L. Bygrave, unpublished work). These observations are in agreement with those of Dodgson et al. (1982) who have reported values for the ΔpH of near zero when a permeant weak acid is present in vitro and raises the question of what the true value is in vivo. It also raises the more general question of how information on mitochondrial Ca^{2+} fluxes in vitro described above relates to the in vivo situation.

2.2 The Calcium Content of Rat Liver Mitochondria

The total size of the mitochondrial calcium pool has been difficult to determine in the past, with the result that a range of values have appeared in the literature (see Reinhart et al. 1984a). This largely arises because the measurements rely on the fractionation of the tissue or cells, which may produce an artificial redistribution of the ion. Previously, mitochondria were thought to contain the largest cellular pool of calcium.

A recent re-examination of the total calcium content in mitochondria (Reinhart et al. 1984a) however, has revealed that this pool of calcium may be much smaller than previous estimates of 650 nmol g^{-1} wet wt (Claret-Berthon et al. 1977). By using a rapid, single-step isolation procedure, Reinhart et al. (1982a) prepared mitochondria under a variety of conditions. It was shown that large movements of Ca^{2+} can occur during isolation of the organelle, and that inhibitors of mitochondrial Ca^{2+} fluxes, such as Ruthenium red and nupercaine are only partially effective in preventing these movements. Only when the Ca^{2+} concentration of all isolation media was adjusted close to the mitochondrial "set point", were these Ca^{2+} movements minimized (Fig. 3). Mitochondria isolated under these conditions were found to contain only approx. 2nmol calcium mg^{-1} protein, indicating that the total mitochondrial Ca^{2+} pool may be approx. 120 nmol g^{-1} wet wt (Reinhart et al. 1984a).

2.3 The Mitochondrial Ca^{2+}-Translocation Cycle and Regulation of Intra-Cellular Ca^{2+} Compartments

The relative contribution of Ca^{2+}-translocation systems located in the plasma membrane, the mitochondrial membrane, and the endoplasmic reticulum, to the regulation of intracellular Ca^{2+} has recently attracted much interest (Becker et al. 1980; Denton and McCormack 1980; Joseph et al. 1983; Nicholls and Akerman 1982; Reinhart et al.

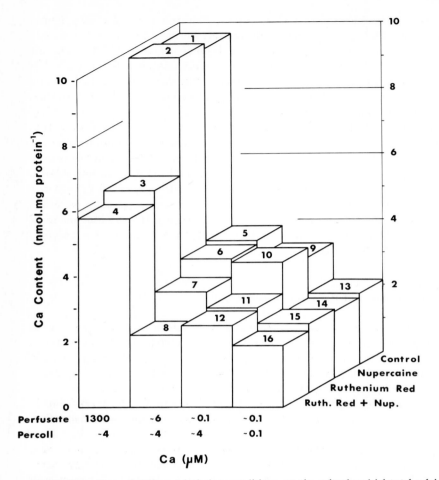

Fig. 3. The effect of different isolation conditions on the mitochondrial total calcium content. Livers of fed rats were perfused as described in Reinhart et al. (1982b). At 10 min of perfusion the perfusate free Ca^{2+} concentration was adjusted to the values shown and 5 min later the median lobe was excised and mitochondria prepared by Percoll-density-gradient centrifugation. (Reinhart et al. 1982a). The free Ca^{2+} concentration of the Percoll solutions was adjusted as shown and in some experiments nupercaine or Ruthenium red was present as indicated. For all conditions free Ca^{2+} concentrations were determined by computor analysis. Total calcium contents were determined by Atomic Absorption Spectroscopy. (Reinhart et al. 1984a)

1984 a, b, c). Because numerous Ca^{2+}-sensitive enzymes exist in the cytoplasmic compartment, considerable emphasis has been placed on defining the regulation of Ca^{2+} therein (see Denton and McCormack 1980; Borle 1981; Williamson et al. 1981; Nicholls and Akerman 1982).

A major role for mitochondria in the control of cytoplasmic Ca^{2+} was proposed by Drahota et al. (1965), and this view has been extended by numerous workers (Claret-Berthon et al. 1977; Bygrave 1978; Nicholls 1978; Borle 1981; Coll et al. 1982; Joseph et al. 1983). The operation of the mitochondrial Ca^{2+}-translocation cycle, produces a buffering of the extramitochondrial Ca^{2+} concentration at values close to those thought to exist in the cytoplasm (Nicholls 1978; Becker et al. 1980; Joseph et al. 1983). However, it seems that the ability of mitochondria to buffer cytosolic Ca^{2+} may be dependent on the total mitochondrial calcium content (Joseph et al. 1983).

Mitochondria containing less than 10 nmol calcium mg^{-1} protein were shown to buffer the extramitochondrial Ca^{2+} concentration poorly, and even between 12 and 20 nmol calcium mg^{-1} protein, less than complete buffering was observed (Joseph et al. 1983). This incomplete buffering is thought to be due to an increase in the rate of mitochondrial Ca^{2+}-efflux, in parallel with increased total calcium content (Joseph et al. 1983; Nicholls and Akerman 1983). Consistent with such a proposal is the finding that the proportion of free mitochondrial Ca^{2+} increases linearly as a function of the total calcium content in vitro (Coll et al. 1982). Half-maximal rates of efflux were observed at a matrix free Ca^{2+} concentration of 9.7 μM (equivalent to 13.8 nmol. total calcium mg^{-1} protein). To effectively buffer the cytoplasmic Ca^{2+} concentration, mitochondria in vivo therefore need to contain sufficient total calcium to maintain the rate of the Ca^{2+} efflux pathway constant for a range of total calcium contents. The recent finding that mitochondria isolated under conditions of minimal Ca^{2+} redistribution and hence approximating the in vivo state contain only 2 nmol calcium mg^{-1} protein (Reinhart et al. 1984a), is thus not consistent with the view that this organelle plays a predominant role in regulating cytoplasmic Ca^{2+}. However, it should be stressed that in vivo the relationship between the rate of efflux and the total calcium content may be quite different to what is seen in vitro. Hence endogenous Ca^{2+}-complexing ligands, such as Pi, may effectively buffer the matrix Ca^{2+} concentration, thereby making the rate of Ca^{2+} efflux less dependent on the total calcium content (Akerman and Nicholls 1983). The role of mitochondria, even containing only 2 nmol calcium mg^{-1} protein, in regulating the cytoplasmic Ca^{2+} concentration therefore remains to be established.

An alternative role for the mitochondrial Ca^{2+}-translocation cycle has been suggested by Denton and McCormack (1980). In their view this cycle may regulate the matrix free Ca^{2+} concentration, and hence the activity of numerous Ca^{2+}-sensitive, matrix-located enzymes (McCormack and Denton 1980, 1981; Hansford and Castro 1981, 1982). Although the Ca^{2+} sensitivity of such enzymes has been documented for both the isolated enzymes, and for intact, uncoupled mitochondria (McCormack and Denton 1980; Denton et al. 1980), whether or not such regulation is operative in vivo is still a controversial issue (Coll et al. 1982; Hansford 1981; Joseph et al. 1983; Nicholls and Akerman 1982; Reinhart et al. 1984a). Recent evidence showing that the ratio of mitochondrial total to free Ca^{2+} is constant at approx. 7×10^4 (i.e., more than 99.9% of mitochondrial Ca^{2+} would be in the bound form), and that mitochondria in vivo may contain only approx. 2 nmol mg^{-1} protein indicates that the free matrix Ca^{2+} concentration may approximate 1.4 μM. The presence of endogenous Ca^{2+}-complexing ligands could reduce this value even further, suggesting that the mitochondrial matrix Ca^{2+} concentration in vivo may be in a range consistent with the regulation of matrix enzymes.

Also of potential significance to intracellular Ca^{2+} homeostasis is the report that a population of vesicles of nonmitochondrial origin are able to buffer the ambient Ca^{2+}

concentration at a value lower than the "set point" attained by isolated mitochondria (Becker et al. 1980). This would indicate that under the experimental conditions employed, a nonmitochondrial Ca^{2+}-translocation cycle is the primary determinant of the final free Ca^{2+} concentration and raises the possibility that in vivo such a cycle may regulate the cytoplasmic Ca^{2+} concentration. Neither the intracellular location of this cycle nor the orientation of these vesicles has as yet, been defined. On the other hand, the finding that digitonin-treated intact hepatocytes buffer the ambient Ca^{2+} concentration at a constant value (approx. 0.1 μM) in either the presence or absence of Ruthenium red (Coll et al. 1982) indicates that the final cytoplasmic "set point" achieved, even in the absence of a functional plasma membrane Ca^{2+}-translocation cycle, is independent of the mitochondrial Ca^{2+} uptake system.

2.4 The Involvement of Mitochondria in α-Adrenergic Agonist-Induced Redistributions of Ca^{2+} in Hepatic Tissue

Some meaningful insights into the role of mitochondria in the regulation of cell Ca^{2+} have been gained from analyzing the mechanism of action of a α-adrenergic agonists in rat liver tissue. Much evidence now indicates that in rat liver α-adrenergic agonists (adrenaline, phenylephrine, noradrenaline), and the hormones vasopressin and angiotensin, stimulate the rate of glycogenolysis or gluconeogenesis without increases in the cAMP concentration, or activation of cAMP-dependent protein kinase. These findings have generated much interest about the mechanism of α-adrenergic agonist action and have led to the consideration of a role for Ca^{2+} in such a scheme. For recent reviews see Exton (1981); Williamson et al. (1981); Taylor et al. (1983a); Reinhart et al. (1984d).

Although controversy still exists with respect to the intracellular location of the adrenaline-sensitive Ca^{2+} pool, it seems evident that mitochondria are involved. Firstly, fractionation studies following the preincubation of cells or tissue with [45]Ca in either the presence or absence of α-adrenergic agonists, show an effect of the hormones on the mitochondrial [45]Ca content (e.g., Kimura et al. 1982). Secondly, fractionation studies and measurement by atomic absorption spectroscopy of the total calcium content in liver fractions isolated before or after α-adrenergic agonist challenge show again that at least a portion of the hormone-sensitive pool is mitochondrial in origin (e.g., Blackmore et al. 1979; Reinhart et al. 1982b). Indeed by employing a rapid fractionation technique and taking account of information about isolation conditions minimizing the redistribution of mitochondrial Ca^{2+}, Reinhart et al. (1982b) were able to demonstrate that approximately 50% of the total amount of Ca^{2+} mobilized by α-adrenergic agonists was correlated with a decrease in mitochondrial calcium content, while the other half was derived from a fraction enriched in both plasma membranes and endoplasmic reticulum vesicles.

A third set of general approaches has used nondisruptive techniques with intact cells and tissues and also have provided evidence for a mitochondrial location of the hormone-sensitive intracellular Ca^{2+} pools. These approaches have employed the fluorescent probe chlortetracycline as an indicator of mitochondrial membrane-bound Ca^{2+} (Babcock et al. 1979), compounds like antimycin A, 2,4-dinitrophenol and oligomycin that inhibit mitochondrial function, and measurements of rate constants from [45]Ca "wash-out" experiments (e.g., Barritt et al. 1981).

It would appear, then, that a major proportion of mitochondrial Ca^{2+} is released rapidly into the cytoplasm following the interaction of a hormone-induced signal with the organelle (see Exton 1981; Reinhart et al. 1984d). The result is an increase in the cytoplasmic Ca^{2+} concentration that then leads to the activation of the metabolic events (see Exton 1981).

In a recent series of papers in which Ca^{2+} fluxes were measured following the administration of α-adrenergic agonists to the intact perfused rat liver (reviewed in Reinhart et al. 1984d), it was possible to gain insights into the interrelation of organelle and cell plasma membrane Ca^{2+} fluxes. A summary of some of the major findings from this work is shown in Fig. 4. Mitochondria are considered a major source (Phase 1) of the intracellular Ca^{2+} that is mobilized to elevate the cytoplasmic free Ca^{2+} concentration and, once the hormone-receptor interaction is terminated (Phase 3), these organelles becomes the intracellular "sink" for the Ca^{2+} as events return to basal levels. Cycling of Ca^{2+} across the inner mitochondrial membrane is envisaged as being closely linked with that occurring across the plasma membrane and perhaps that across the endoplasmic reticular membrane. In this regard it is significant that the hormone-generated signal appears to activate plasma membrane Ca^{2+}-cycling at the same time as mobilization of mitochondrial Ca^{2+} occurs (Reinhart et al. 1984c). It is difficult from these experiments to gauge whether the Ca^{2+}-sensitive enzymes in the mitochondrial matrix are influenced by the redistribution of Ca^{2+} as depicted here, although increases tricarboxylic acid cycle activity are evident at this time (Taylor et al. 1983b).

3 Conclusions and Future Directions

It is now some 30 years since the appearance of the original paper of Slater and Cleland (1953) describing the ability of mitochondria to accumulate Ca^{2+}. Subsequent research on this topic, especially through the last two decades, has produced much basic information particularly about the mechanism of mitochondrial Ca^{2+} influx and its relation to oxidative phosphorylation. The relatively recent realization that Ca^{2+} movements are best considered in chemiosmotic terms, has led to investigations into mitochondrial Ca^{2+} efflux and into the phenomenon of mitochondrial Ca^{2+}-cycling. These in turn have greatly aided thinking about the role played by mitochondria in the regulation in intracellular Ca^{2+}.

While there still are some important outstanding questions that can be answered primarily from studying mitochondria in vitro, there is an increasing awareness that cellular Ca^{2+} regulation now should be examined where possible in situ. In this regard much more information is required about the relative contributions of the plasma membrane, the endoplasmic reticulum and the mitochondria to the regulation of cytoplasmic Ca^{2+}. Techniques are now available that give impetus to this in the intact perfused rat liver. Implementation of these techniques and the use of appropriate physiological stimuli like α-adrenergic agonists that induce a redistribution of cellular Ca^{2+}, are beginning to provide considerable insights into how mitochondria in situ contribute to the regulation of cellular Ca^{2+}

PHASE 1

PHASE 2

PHASE 3

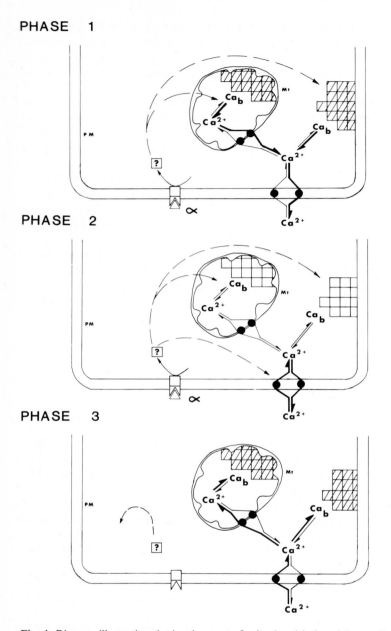

Fig. 4. Diagram illustrating the involvement of mitochondria in calcium redistribution induced by alpha-adrenergic agonists in hepatic tissue. *Phase 1* is characterized by the rapid mobilization of intracellular, presumably bound calcium (Ca_b) by one or more undefined "second messengers". Part of this calcium appears to be derived from the mitochondria, while a second site probably involves the endoplasmic reticulum and/or the plasma membrane. This mobilization is thought to elevate the cytoplasmic Ca^{2+} concentration, stimulate the rate of net Ca^{2+} efflux from the cell, and lead to the depletion of the hormone-sensitive pool of calcium. *Phase 2* occurs within 2 to 3 min of hormone administration, by which time cells have returned to a steady state with respect to net Ca^{2+} movements. During this phase, Ca^{2+}-cycling across the plasma membrane, and possible across other membranes, appears to be elevated. This cycling may allow the maintenance of an elevated cytoplasmic Ca^{2+} concentration. *Phase 3* is initiated by the removal of alpha-adrenergic agonists from the extracellular space. The discontinuation of "second messenger" generation results in the net uptake of Ca^{2+} by the cell, leading to the repletion of the hormone-sensitive pool of calcium. (Reinhart et al. 1984d)

Acknowledgement: The original work carried out by the authors was supported by a grant to F.L.B. from the National Health and Medical Research Council of Australia.

References

Akerman KEO, Nicholls, DG (1983) Physiological and bioenergetic aspects of mitochondrial calcium transport. Rev Physiol Biochem Pharmacol 95:149-201

Babcock DF, Chen J-L, Yip BP, Lardy HA (1979) Evidence for mitochondrial location of the hormone-responsive pool of calcium in isolated hepatocytes. J Biol Chem 254:8117-8120

Barritt GJ, Parker JC, Wadsworth JC (1981) A kinetic analysis of the effects of adrenaline on calcium distribution in isolated rat liver parenchymal cells. J Physiol (Lond) 312:29-55

Becker GL, Fiskum G, Lehninger AL (1980) Regulation of free calcium by liver mitochondria and the endoplasmic reticulum. J Biol Chem 255:9009-9012

Blackmore PF, Dehaye J-P, Exton JH (1979) Studies on α-adrenergic activation of hepatic glucose out put. J Biol Chem 254:6945-6950

Borle AB (1981) Control, modulation and regulation of cell calcium. Rev Physiol Biochem Pharmacol 90:13-153

Bygrave FL (1978) Mitochondria and the control of intracellular calcium. Biol Rev Camb Philos Soc 53:43-79

Claret-Berthon B, Claret M, Mazet JL (1977) Fluxes and distributions of calcium in rat liver cells: kinetic analysis and identification of pools. J Physiol (Lond) 272:529-552

Coll KE, Joseph SK, Corkey BE, Williamson JR (1982) Determination of the matrix free calcium concentration and kinetics of calcium efflux in liver and heart mitochondria. J Biol Chem 257: 8696-8704

Crompton M, Moser R, Ludi H, Carafoli E (1978) The interrelations between the transport of sodium and calcium in mitochondria of various mammalian tissues. Eur J Biochem 82:25-31

Dawson RMC, Hauser H (1970) Binding of calcium to phospholipids. In: Cuthbert AW (ed) Calcium and cellular function. McMillan, London, pp 17-41

Dawson AP, Selwyn M, Fulton DV (1979) Inhibition of calcium efflux from mitochondria by nupercaine and tetracaine. Nature (Lond) 277:484-486

Denton RM, Mc Cormack JG, Edgell NJ (1980) Role of calcium ions in the regulation of intramitochondrial metabolism. Biochem J 190:107-117

Denton RM, McCormack JG (1980) On the role of the calcium transport cycle in heart and other mammalian mitochondria. FEBS Lett 119:1-8

Denton RM, McCormack JG, Edgell NJ (1980) Role of calcium ions in the regulation of intramitochondrial metabolism. Biochem J 190:107-117

Dodgson SJ, Forster REI, Storey BT (1982) Determination of intramitochondrial pH by means of matrix carbonyic anhydrase activity measured with 180 exchange. J Biol Chem 257:1705-1711

Drahota Z, Carafoli E, Rossi CS, Gamble RL, Lehninger AL (1965) The steady-state maintenance of accumulated calcium in rat liver mitochondria. J Biol Chem 240:2712-2720

Exton JH (1981) Molecular mechanisms involved in α-adrenergic responses. Mol Cell Endocr 23: 233-264

Hansford RG (1981) Effect of micromolar concentrations of free calcium on pyruvate dehydrogenase interconversion in intact rat heart mitochondria. Biochem J 194:721-732

Hansford RG, Castro F (1981) Effect of micromolar concentrations of free calcium ions on the reduction of heart mitochondrial NAD (P) by 2-oxoglutarate. Biochem J 198:525-533

Hansford RG, Castro F (1982) Intramitochondrial and extramitochondrial free calcium concentrations of suspensions of heart mitochondria with very low, possibly physiological contents of total calcium. J Bioenerg Biomembr 14:361-376

Joseph SK, Koll KE, Cooper RH, Marks JS, Williamson JR (1983) Mechanisms underlying calcium homeostasis in isolated hepatocytes. J Biol Chem 258:731-741

Kimura S, Kugai N, Tada R, Kojima I, Abe K, Ogata E (1982) Sources of calcium mobilized by α-adrenergic stimulation in perfused rat liver. Horm Metab Res 14:133-138

McCormack JG, Denton RM (1980) Role of calcium ions in the regulation of intramitochondrial metabolism. Biochem J 190:95-105

McCormack JG, Denton RM (1981) A comparative study of the regulation of Ca^{2+} of the activities of the 2-oxoglutarate dehydrogenase complex and NAD^+-isocitrate dehydrogenase from a variety of sources. Biochem J 196:619-624

Nicholls DG (1978) The regulation of extramitochondrial free calcium ion concentration by rat liver mitochondria. Biochem J 176:463-474

Nicholls DG, Akerman KEO (1982) Mitochondrial calcium transport. Biochim Biophys Acta 683: 57-88

Reinhart PH, Taylor WM, Bygrave FL (1982a) A procedure for the rapid preparation of mitochondria from rat liver. Biochem J 204:731-735

Reinhart PH, Taylor WM, Bygrave FL (1982b) Calcium ion fluxes induced by the action of α-adrenergic agonists in perfused rat liver. Biochem J 208:619-630

Reinhart PH, van de Pol E, Taylor WM, Bygrave FL (1984a) An assessment of the calcium content of rat liver mitochondria in vivo. Biochem J 218:415-420

Reinhart PH, Taylor WM, Bygrave FL (1984b) The contribution of both extracellular and intracellular calcium to the action of α-adrenergic agonists in perfused rat liver. Biochem J 220: 35-42

Reinhart PH, Taylor WM, Bygrave FL (1984c) The action of α-adrenergic agonists on plasma membrane calcium fluxes in perfused rat liver. Biochem J 220:43-50

Reinhart PH, Taylor WM, Bygrave FL (1984d) The role of calcium in the mechanism of action of adrenaline in rat liver. Biochem J 223: 1-13

Slater EC, Cleland KW (1953) The effects of calcium on the respiratory and phosphorylation activities of heart muscle sarcosomes. Biochem J 55:566-580

Stucki JW, Ineichen A (1974) Energy dissipation by calcium recycling and the efficiency of calcium transport in rat liver mitochondria. Eur J Biochem 48:365-375

Taylor WM, Reinhart PH, Bygrave FL (1983a) On the role of calcium in the mechanism of action of α-adrenergic agonists in rat liver. Pharmacol Ther 21:125-141

Taylor WM, Reinhart PH, Bygrave FL (1983b) Stimulation by α-adrenergic agonists of calcium fluxes, mitochondrial oxidation and gluconeogenesis in perfused rat liver. Biochem J 212: 555-565

Williamson JR, Cooper RH, Hoek JB (1981) Role of calcium in the hormonal regulation of liver metabolism. Biochim Biophys Acta 639:243-295

Zoccarato F, Nicholls DG (1981) Phosphate-independent calcium efflux from liver mitochondria. FEBS Lett 128:275-277

Calmodulin Structure and Function

L. J. VAN ELDIK and D. M. WATTERSON[1]

CONTENTS

1 Introduction

Calcium is a mediator of many cellular responses and a regulator of major importance in cellular homeostasis. In order to elucidate the mechanisms of calcium control and the role of calcium in stimulus-response coupling, it is necessary to examine the cellular receptors for calcium. Thermodynamic, kinetic, and structural data strongly suggest that the cytoplasmic receptors for calcium acting as a biological messenger are a class of calcium-binding proteins referred to as calcium-modulated proteins. Calcium-modulated proteins reversibly bind calcium with dissociation constants in the micromolar range under relatively physiological conditions. Because most calcium-modulated proteins are intracellular and have dissociation constants that span the range of intracellular free calcium concentrations, they are postulated to be the major signal transducers of biological calcium signals. Most calcium-modulated proteins characterized to date are not enzymes but are effector proteins capable of transducing a calcium signal into a cellular response by their ability to regulate or modulate the activity of other macromolecules, including enzymes, in a calcium-dependent manner. Examples

[1] Howard Hughes Medical Institute and Department of Pharmacology, Vanderbilt University, Nashville, TN 37232, USA

of calcium-modulated proteins are troponin C, parvalbumin, S100, vitamin D-dependent calcium-binding protein, and calmodulin.

Over the past several years a unifying hypothesis has evolved concerning the relation of structure to calcium-binding activity in calcium-modulated proteins. Termed the EF-hand hypothesis (Kretsinger 1980a), this postulate is based on the three-dimensional structure of one calcium-modulated protein (parvalbumin) and the primary structures of a limited number of low molecular weight, acidic, calcium-modulated proteins (e.g., calmodulin and troponin C). The hypothesis attempts to correlate calcium-binding activity with a predicted three-dimensional structure based on the amino acid sequence of the protein. These predicted three-dimensional structures are called EF-hands. The essence of this hypothesis is that the targets of calcium acting as a second messenger are proteins which contain EF-hand structures. All calcium-modulated proteins whose amino acid sequences have been elucidated have putative EF-hand structures. Based on the amino acid sequence homologies among calmodulin, troponin C, and parvalbumin, a three-dimensional model of calmodulin (Fig. 1) has been proposed (Kretsinger 1980b).

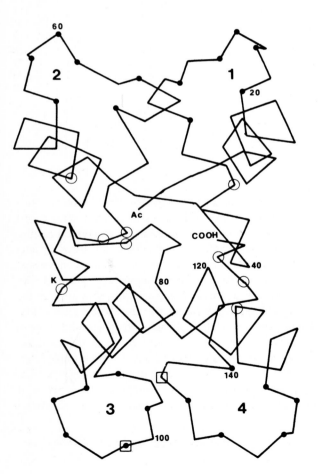

Fig. 1. Model of the three-dimensional structure and domains of calmodulin. The four structural domains, a-cetylated NH₂ terminus *(Ac)*, COOH terminus *(COOH)*,and trimethyllysine residue *(K)* are marked. *Solid circles* residues postulated to be involved in calcium binding; *open squares* tyrosinyl residues; *open circles* methionyl residues

Calmodulin is an ubiquitous calcium-modulated protein whose structure and function have been highly conserved throughout eukaryotic species. This protein has multiple in vitro activities, none of which are as primary catalyst. Calmodulin also binds to phenothiazine drugs and related compounds in a calcium-dependent manner. While the physiological or pharmacological importance of many of these activities has not been demonstrated, they do suggest that calmodulin may be a pleiotropic regulator of metabolic and mechanochemical activities in the cell. The apparent ubiquitous distribution and the highly conserved structure of calmodulin suggest that it may play a fundamental role in mediating intracellular calcium-dependent effects.

Several approaches to the study of calmodulin's role in cell function are being used with a variety of biological systems. These approaches can be grouped into five arbitrary classes: (1) pharmacology, (2) immunochemistry, (3) biochemistry, (4) mutant analysis/pathophysiology, and (5) quantitative cellular and molecular biology. The major pharmacological tools used in studies of calmodulin and cell function are phenothiazine and naphthalene sulfonic acid derivatives. Immunochemical methods have allowed the establishment of specific radioimmunoassays for calmodulin and the localization of calmodulin in various cell types by immunocytochemical procedures. Biochemical studies of the calcium-dependent activation of enzymes by calmodulin have helped clarify many points concerning calmodulin regulation, have defined what is thermodynamically and kinetically allowed in models of calmodulin regulation, and have suggested new approaches to the study of calmodulin regulation. Analysis of mutant and pathophysiological systems has demonstated that calmodulin biosynthesis is quantitatively regulated and has provided well-characterized biological systems for the further study of the role of calmodulin in cell physiology. Quantitative cellular and molecular biological studies have provided insight into the physiological importance of some of calmodulin's in vitro activities and have provided the necessary tools and information for future biological studies.

Each of these approaches to the study of calmodulin's role in cell function has its strengths and limitations. However, together they are complementary and indicate that a multidisciplinary approach to the study of this multifunctional protein may allow an accurate and incisive definition of how calmodulin is involved in cell processes. A recent comprehensive review of calmodulin structure and function (Klee and Vanaman 1982) is available and more recent updates of research on calmodulin have also appeared (Burgess et al. 1983; deBernard et al. 1983). Therefore, this chapter will concentrate on a summary of emerging trends in our knowledge of calmodulin structure and function and will discuss important biochemical and cell biological precedents that are of potential interest to calcium physiologists. The reader is referred to more general discussions (Van Eldik et al. 1982b) of calcium in biological systems for precedents outside the field of calmodulin research.

2 Chemistry of Calmodulin

2.1 Structure of Calmodulin

The amino acid sequence of calmodulin (Fig. 2) appears to be identical among various species and tissues of mammals and birds. An inferred amino acid sequence for eel calmodulin indicates at least one amino acid sequence difference (Lys/Arg at residue 74) between this phylogenetically earlier vertebrate and other vertebrates. Two co- or post-translational modifications of vertebrate calmodulin have been documented: N-acetylation of the amino terminal alanine and N-trimethyllation of lysine-115. Three other modifications of calmodulin (phosphorylation, carboxyltransmethylation, and deamidation) have been suggested (for review see Klee and Vanaman 1982), but direct proof is lacking. Historically, whenever phosphorylation of calmodulin has been reported, later, more unequivocal studies have indicated that the phosphorylated species is another macromolecule. Calmodulin will serve as a substrate for a protein carboxyl-O-methyltransferase and a slight effect of such treatment on calmodulin activator activity has been reported (Gagnon et al. 1980). However, a definitive site of stoichiometric modification has not been described yet. Data suggestive of deamidation of calmodulin have been available for several years, but only recently has a deamidation site been identified (WH Burgess, TJ Lukas, DM Watterson, unpublished).

Three amino acid sequences for invertebrate calmodulins are available (Fig. 2). These mostly differ in amide assignments which may not be real differences. The only report (Burgess 1982) of calmodulin isotypes, which appear to be in the same cell, is from the invertebrate *Arbacia punctulata* (sea urchin). The two forms of calmodulin differ in their electrophoretic mobilities, elution properties on ion exchange columns, amino acid compositions, and reactivity with site-specific antibodies raised against vertebrate calmodulin. The latter results suggest that these two calmodulins probably differ in their amino acid sequences in the COOH-terminal structural domain. No functional differences between these two forms have been reported yet. However, the feasibility of finding differences in activity between the isotypes is suggested by recent studies that indicate that the limited amino acid sequence differences among calmodulins may be reflected in quantitative functional differences (see Sect. 3.1). The possible existence of calmodulin isotypes in other tissues and species, especially gametic or embryonic tissue, is one clear question that needs to be directly addressed in future studies of calmodulin physiology.

The amino acid sequence of calmodulin from the higher plant, spinach, has recently become available (Fig. 2) as well as limited amino acid sequence analysis of calmodulin from a monocotyledon, barley (Schleicher et al. 1983). These data suggest that only minor differences in amino acid sequence, if any, exist between the dicotyledon calmodulin (spinach) and the monocotyledon calmodulin (barley). Although only 13 to 14 amino acid sequence differences between higher plant and vertebrate calmodulins were found, these do appear to result in altered physical and functional properties. Nine of the differences between the two calmodulins are found in the COOH-terminal half of the molecule. This is especially interesting physiologically in light of the suggestion that the calmodulin isotypes may differ in their COOH-terminal amino acid sequences. Two of the amino acid sequence differences between spinach and vertebrate calmodulin

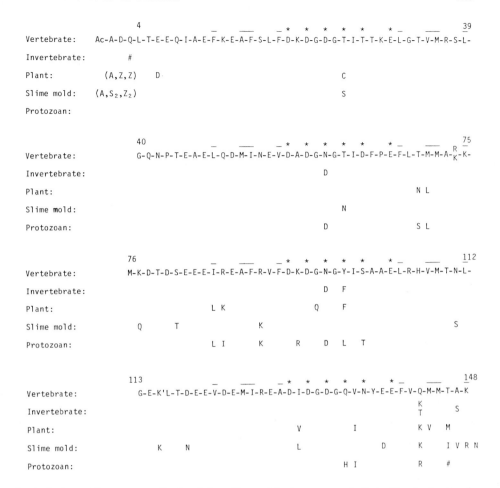

Fig. 2. Amino acid sequences of calmodulins. The vertebrate sequence is that of bovine brain calmodulin (Watterson et al. 1980; Watterson et al. 1984) and human brain calmodulin (Sasagawa et al. 1982). The known amino acid sequence differences from bovine brain calmodulin are shown for invertebrate, plant, slime mold, and protozoan calmodulins. The invertebrate sequence is a consensus sequence from that of the coelenterate *Renilla reniformis* (Jamieson et al. 1980), the sea anemone *Metridium senile* (Takagi et al. 1980), and the sea scallop *Patinopecten* (Toda et al. 1981). The plant sequence is that of spinach leaf calmodulin (Lukas et al. 1984; Watterson et al. 1984). The slime mold sequence is that of *Dictyostelium discoideum* calmodulin (Marshak et al. 1984). The protozoan sequence is that of *Tetrahymena pyriformis* calmodulin (Yazawa et al. 1981). *Asterisks* (*) indicate positions of proposed calcium ligating residues, and *bars* (−) indicate residues proposed to be on the internal aspect of α-helices (Kretsinger 1980a). The single letter abbreviations for the amino acids are: *A* alanine; *C* cysteine; *D* aspartic acid; *E* glutamic acid; *F* phenylalanine; *G* glycine; *H* histidine; *I* isoleucine; *K* lysine; *L* leucine; *M* methionine; *N* asparagine; *P* proline; *Q* glutamine; *R* arginine; *S* serine; *T* threonine; *V* valine; *X* unidentified; *Z* glutamic acid or glutamine. *K'* indicates N$^\in$-trimethyllysine. The symbol (#) indicates a gap in the sequence introduced to give maximal alignment. Residue 74 has been reported to be a lysine based on the inferred sequence of eel calmodulin (Lagace et al. 1983) and an arginine in all other vertebrate calmodulins examined. Residue 143 is a lysine in *Renilla* and *Metridium* calmodulins and a threonine in *Patinopectenn* calmodulin

are novel: a cysteine at residue 26 and a glutamine at residue 96. At the DNA level these changes require a minimum of two base changes from the vertebrate codons and both changes are in the proposed calcium binding loops. It is not clear what alterations in calcium-binding properties would result from such a change.

Another novel calmodulin is that from *Dictyostelium discoideum* (Fig. 2). The calmodulin from this slime mold has a COOH-terminal extension of one residue, asparagine, similar to that found in troponin C and lacks the post-translational modification (trimethyllation) of lysine 115. As with other calmodulins, most of the amino acid sequence differences between slime mold and vertebrate calmodulins are in the COOH-terminal half of the molecule. Although these few conservative changes do not result in alteration of phosphodiesterase activator activity under the conditions used, a quantitative difference in NAD kinase activator activity is readily detected. Similar results are obtained with *Chlamydomonas* calmodulin which also has an unmethylated lysine at residue 115 and may have a COOH-terminal extension (Schleicher et al. 1984; Lukas et al. 1985b). Altogether these results indicate that the COOH-terminal domain is a region of relatively high variability among calmodulins and that variability among COOH-terminal sequences may be a basis for some of the differences in physical characteristics among calmodulins. It is not clear what effect these COOH-terminal differences might have on function.

Protozoan (Fig. 2) and algal (Schleicher et al. 1984; Lukas et al. 1985b) calmodulins also appear to have several novel amino acid sequence differences in the COOH-terminal half of the molecule and appear to have some selectivity in their interactions with binding proteins. *Tetrahymena* calmodulin activates *Tetrahymena* guanylate cyclase but vertebrate, invertebrate, and higher plant calmodulins do not activate *Tetrahymena* guanylate cyclase (Kakiuchi et al. 1981). A calmodulin-sensitive guanylate cyclase activity in *Paramecium* is stimulated by vertebrate, higher plant, *Tetrahymena,* and *Paramecium* calmodulins but not by *Dictyostelium* calmodulin (Klumpp et al. 1983). More recently, Roberts et al. (1984) demonstrated quantitative differences in the ability of higher plant, algal, slime mold, and vertebrate calmodulins to activate plant NAD kinase. These same preparations of calmodulin were indistinguishable in their ability to activate vertebrate brain phosphodiesterase and were very similar in their activation of vertebrate myosin light chain kinase. Thus, the limited number of conservative amino acid sequence differences found among calmodulins appear to result in different spectra of activator activities and suggest that certain enzymes are selective in their interactions with calmodulins.

This limited but diverse base of structural and functional knowledge about calmodulins provides insights to the physiologist. First, although only one example of calmodulin isotypes has been documented, investigators should be continually aware of this possibility in other types of cells and under different physiological and pathological states. Phylogenetic studies have shown that a limited number of conservative amino acid sequence changes can result in dramatic effects on certain activator activities of calmodulin. Thus, the biochemical precedent exists for small changes in covalent structure, such as one might expect with isotypes, resulting in specific and quantitative changes in activity. If one thinks of calmodulin as a ligand and the various calmodulin-binding proteins as receptors, then it is simple to envision the complex number of specific and quantitative effects that could theoretically result, based on precedents in pharmacology, from small changes in either the ligand or the receptor.

Second, calmodulin is clearly a member of a class of proteins that are as structurally and functionally related to each other as the various globin chains are to each other. The direct and unequivocal demonstration of the very high structural relatedness of calmodulin and troponin C provides a firm precedent for current attempts to detect, identify, and characterize other members of this family of calcium-modulated proteins. It is clearly possible that as we increase our knowledge it might be difficult to discern a calmodulin isotype from a calmodulin-like protein; that is, there may be a continuum of calmodulin-like and troponin C-like proteins. In terms of calcium physiology, we should expect that some of these proteins will be more tissue-specific calcium signal transducers, such as troponin C, and some may be expressed only during certain stages of development or stress.

Third, the importance of well-characterized immunochemical reagents is emphasized by the demonstration that a family of closely related proteins is probably involved in the regulation of calcium homeostasis (Sect. 2.2). Fourth, the lack of absolute specificity of currently available drugs for each of these calcium-modulated proteins indicates the caution necessary in the use of such compounds in calcium physiology studies (Sect. 3.2).

2.2 Immunochemistry of Calmodulin

Immunochemical approaches have been especially useful in attempts to define the exact role of calmodulin in cell function and how these functions might be perturbed in pathophysiological states. Antibodies against calmodulin have been utilized for immunochemical measurements of calmodulin levels by radioimmunoassay, for immunocytochemical examination of the localization and distribution of calmodulin in various tissues, and in attempts to correlate calmodulin structural domains with functional domains by using immunochemical mapping approaches. It should be kept in mind that procedures using immunochemical reagents have limitations, such as "masking" of the immunoreactive site to prevent antigen-antibody interactions, cross-reactivity of antibody with other proteins, nonspecific reactivity of antigen or antibody, or artifactual localizations or measurements due to experimental conditions. However, within these limitations, immunochemical approaches have provided important information about calmodulin's role in cell physiology. Some of these immunochemical approaches are discussed in more detail in Sect. 4.

Various methods have been used to produce antibodies against calmodulin, including injection of performic acid oxidized (Van Eldik and Watterson 1981) or dinitrophenylated (Wallace and Cheung 1979) calmodulin, injection of large amounts of native calmodulin and affinity purification of the antibodies (Chafoules et al. 1979), production of monoclonal antibodies against calmodulin (Pardue et al. 1983), and engineering of site-directed anti-calmodulin antibodies by injection of synthetic peptide immunogens (Van Eldik et al. 1983a). Because calmodulin is a member of a class of structurally and functionally related proteins, a quantitative analysis of the molecular specificity of antisera against calmodulin is important for unequivocal interpretation of biological studies using immunochemical methods. In addition, new insights into the molecular mechanisms of calmodulin regulation may be obtained through immunochemical mapping of calmodulin domains using a set of site-specific antibodies. The only antibodies that have been characterized so far in terms of their molecular specificity and immunoreactive sites on calmodulin have been those elicited by performic acid oxidized calmodulin and synthetic peptides of calmodulin. Therefore, in this section, we will emphasize data on the characterization of these site-specific antibodies and discuss the utilization of these well-defined immunochemical reagents in cell physiology.

It has been previously reported (Van Eldik and Watterson 1981) that oxidation of methionine residues in calmodulin significantly enhances its immunogenicity in rabbits, and that most of the antibodies produced by injection of oxidized calmodulin react with an immunoreactive region (residues 127-144) in the COOH-terminal domain of the calmodulin molecule. The minimal immunoreactive site for one of these antisera was determined by examining the immunoreactivity of cleavage fragments and synthetic segments of calmodulin (Van Eldik et al. 1983b). These studies demonstrated that a linear segment of seven amino acid residues (residues 137-143) is the smallest calmodulin segment and the only seven-residue segment in this region that shows quantitative immunoreactivity with this anticalmodulin serum. The molecular and atomic requirements for activity were examined further by testing a series of single-substitution analogs of the heptapeptide for immunoreactivity with the anti-calmodulin serum. These studies (Watterson et al. 1984; Fok et al. 1981) demonstrated that the side chains of all seven residues of the heptapeptide are required for quantitative immunoreactivity with this anti-calmodulin serum.

The demonstration that a synthetic seven-residue peptide (residues 137-143) can show immunoreactivity similar to the 148-residue calmodulin molecule suggested the potential use of this heptapeptide as a synthetic immunogen to elicit antisera against calmodulin. The use of synthetic peptide immunogens to produce antibodies of defined specificities is an important recent advance in immunochemistry. The design of the synthetic immunogen is obviously one of the most important considerations in attempts to produce antibodies that are reactive with the native protein. The definition of an immunoreactive site on calmodulin combined with the known structural characteristics of calmodulin in the COOH-terminal region allowed a rational design of synthetic peptide immunogens for use in engineering site-directed antibodies against calmodulin (Van Eldik et al. 1983a). The affinity and specificity of these anti-peptide antisera were sufficient to allow development of radioimmunoassays using these sera, iodinated vertebrate calmodulin as tracer, and intact calmodulin as standard.

These site-specific anti-calmodulin sera have also been important tools in biological studies and in attempts to correlate structural domains with functional domains by using immunochemical mapping approaches. These antisera appear to bind to monomeric calmodulin as well as to calmodulin in supramolecular complexes such as muscle phosphorylase kinase, exhibit differential reactivity with calmodulin isotypes from sea urchin eggs, and react poorly with plant or protozoan calmodulins (Watterson et al. 1984). The affinity and specificity of these antibodies also allow their potential use as immunoadsorbents to purify calmodulin as a component of supramolecular complexes. The demonstration that calmodulin has multiple functional domains suggests that this particular immunoreactive site may not be exposed in certain other supramolecular complexes and emphasizes the utility of a library of calmodulin antibodies with different immunoreactive sites.

Table 1.   Summary of reactivity of rabbit antisera against calcium-modulated proteins[a]

Competing protein	Anti-vertebrate calmodulin (serum 1)	Anti-vertebrate calmodulin (serum 2)	Anti-plant calmodulin	Anti-calmodulin Synthetic Peptide	Anti-troponin C	Anti-S100α	Anti-S100β
Vertebrate calmodulin	+	+	±[b]	+	−[c]	−	−
Invertebrate calmodulin	±	+	±	±	ND[d]	ND	ND
Plant calmodulin	±	+	+	±	−	−	−
Protozoan calmodulin	±	+	±	±	ND	ND	ND
Troponin C	−	−	−	−	+	−	−
S100α	−	−	−	−	−	+	±
S100β	−	−	−	−	−	±	+

[a] Immunoreactivity was determined by competition radioimmunoassay as described (Van Eldik and Watterson 1981)

[b] ± = reactivity that is displaced 10- to 1000-fold from standard

[c] − = no reactivity with ≥ 1000 fold molar excess of competing protein compared to standard

[d] ND = not determined

Antisera that react with other regions of the vertebrate calmodulin molecule have also been produced, although detailed studies to define the immunoreactive sites for these antisera have not been completed. Antibodies against spinach calmodulin show good reactivity with other plant calmodulins but little, if any, reactivity with vertebrate or protozoan calmodulins (Schleicher et al. 1983; Watterson et al. 1984). Finally, rabbit antisera against rabbit skeletal muscle troponin C and bovine brain S100α and S100β are available that show no cross-reactivity with structurally related calcium modulated proteins, and mouse monoclonal antibodies against bovine brain S100β that appear to be specific for the S100β polypeptide of S100 fractions have recently been prepared (Van Eldik et al. 1984a). Thus, a library of antibodies with differing molecular specificities has been produced against several calcium-modulated proteins (Table 1). The availability of these antibodies will allow quantitative detection and measurement of and immunochemical discrimination among these structurally related proteins and may provide insight into how multiple intracellular calcium modulated proteins coordinate responses to calcium signals.

In a more general sense, these studies provide a precedent for a rational approach to the development of specific immunochemical assays for calcium-modulated proteins. The ability to design antibodies against selected regions of the calmodulin molecule will be useful in structure-function studies to provide specific molecular probes for the immunochemical mapping of functional domains on calmodulin. These approaches may allow the engineering of antisera that react specifically with regions of calmodulin other than the immunoreactive fourth domain, such as the highly conserved second domain. Hybridoma technology approaches can also be useful to yield abundant supplies of antibodies with defined chemical specificities. As our knowledge of the biochemistry of calcium-modulated proteins increases, it is evident that new immunochemical reagents will have to be developed. The potential presence in the same cell of calmodulin and other calcium modulated proteins, calmodulin isotypes, or calmodulin and calmodulin-like proteins will require specific, well-characterized antibodies that can discriminate among these highly related proteins in order to unequivocally interpret biological studies using immunochemical methods.

3 Biochemical Functions of Calmodulin

Calmodulin has been shown to have a number of in vitro biochemical activities. These activities include enzyme activity stimulation, drug-binding activity, and protein-binding activity. While some of these in vitro functions of calmodulin are appealing in terms of models of cell function, it is still not clear for many of these activities whether they reflect physiological roles for calmodulin. A recent review (Klee and Vanaman 1982) has discussed the data and biochemical basis for many of these calmodulin activities. Some of these activities are listed below and a summary of the approaches, with their strengths and limitations, used to study these calmodulin activities is given. It should be kept in mind that the very things that make calmodulin an exciting molecule to study (ubiquitous distribution among eukaryotes, highly conserved structure and function, multiple activities, and similarity to tissue-specific, calcium-modulated proteins) can also make unequivocal interpretations of biological studies difficult.

3.1 Enzyme Activation

While calmodulin has no known enzymatic activity, it will stimulate the activity of several enzymes. The enzyme activator activites of calmodulin include the calcium dependent stimulation of a cyclic nucleotide phosphodiesterase, adenylate cyclase, guanylate cyclase, actomyosin ATPase, dynein ATPase, plasma membrane ATPase, sarcolemmal ATPase, phosphorylase kinase, NAD kinase, myosin light chain kinase, glycogen synthase kinase, various protein kinase activities, a phospholipase activity, and a protein phosphatase activity. Most of these activities have been demonstrated by one or more of the following approaches; first, addition of calcium and calmodulin with resultant effects on a process or activity; second, removal of an endogenous calcium-sensitizing factor and subsequent demonstration that calcium and calmodulin will affect the measured activity; or, third, purification to homogeneity of a calcium-sensitive activity and direct demonstration that calmodulin is the calcium-binding subunit. Many of the calmodulin-regulated enzymes have not been purified to homogeneity and it is not clear if the regulation of some of the enzymes by calmodulin is a direct effect on the enzyme. Furthermore, it is not known whether calmodulin is the endogenous calcium regulatory protein for all of these activities. Other activities have been attributed to calmodulin, based on the effects of drugs that bind to calmodulin and other macromolecules. Due to the large number of these reports and the lack of a clear implication for calmodulin involvement (Sect. 3.2), these activities and approaches will not be addressed here.

The ability to demonstrate calcium-dependent calmodulin stimulation of an enzyme activity does not necessarily reflect a physiological function of calmodulin. Calcium-modulated proteins are a class of structurally and functionally related proteins. In many types of in vitro assays, multiple calcium-modulated proteins will substitute for the endogenous regulator. For example, calmodulin will substitute for troponin C in binding to troponin I and in reconstituted actomyosin ATPase assays, yet troponin C is the endogenous calcium-binding subunit of the troponin complex. Thus, it is possible that some of the in vitro activities attributed to calmodulin may be examples of calmodulin's ability to substitute for an endogenous calcium-binding protein. Some in vitro activities clearly reflect physiological roles for calmodulin. Two examples are myosin light chain kinase and muscle phosphorylase kinase regulation. In both cases the holoenzyme was purified and the calcium-binding subunit was directly shown to be calmodulin. Current physiological studies of these enzymes are concerned with defining the quantitative importance of the calcium-calmodulin regulation under a variety of physiological and disease states.

3.2 Drug Binding

The activation of various enzymes by calmodulin can be inhibited in a calcium-dependent manner by drugs of several classes including phenothiazines (Levin and Weiss 1979), naphthalene sulfonamide derivatives (Hidaka et al. 1979), diphenylbutylamine derivatives (Levin and Weiss 1979), and nitrogen mustard-based adrenergic antagonists (Lukas et al. 1983; Watterson et al. 1984). In general, there is little evidence to suggest that calmodulin is involved in the mechanism of action of these drugs, and the question of calmodulin being involved in the side effects of the drugs has not been adequately addressed. Earlier studies, especially with phenothiazines, were interpreted as showing 2 mol of drug bound per mol of calmodulin and showing specificity for calmodulin. More recent detailed analyses have shown that calmodulin binding of phenothiazines is a complex phenomenon with multiple, calcium-dependent sites (Lukas et al. 1985a; Marshak et al. 1985). In addition, a subclass of calcium-modulated proteins including S100 and troponin C have been shown to bind phenothiazines and napthalene sulfonamide derivatives in a calcium-dependent manner (Endo et al. 1981; Hidaka et al. 1980; Marshak et al. 1985).

Although these various drugs produce interesting effects on cells when used at noncytotoxic doses, it is clear that other macromolecules can interact with the drugs. In addition, little is known about the intracellular concentrations of the drugs that have been utilized in such studies or the localized concentration of the drug and the various macromolecular targets. It is also difficult to invoke calmodulin as being involved in the effects of these drugs when the effects are postulated to be on dissociated calmodulin. As discussed in more detail in Sect. 4.2, it is not known whether significant concentrations of free calmodulin exist in a eukaryotic cell.

The observations that these various classes of drugs will compete for binding to calmodulin (Weiss et al. 1980; Lukas et al. 1985a) demonstrate a functional relatedness of their binding and suggest that the drug-binding sites may be structurally similar if not overlapping. The ability of the drugs to compete for peptide-binding sites suggests that drug and peptide-binding sites may be related. Current studies using covalent cross-linking of drugs (Lukas et al. 1985a) and peptides (Giedroc et al. 1983) to calmodulin and characterization of the resultant adducts may provide insight into how calmodulin's structural domains are related to functional domains. Regardless of the pharmacological significance or limited investigational utility for calmodulin physiology, these various drugs have proven useful for mapping functional domains on calmodulin and may provide the necessary information for the future development of calmodulin-specific antagonists. The utility and limitations of the use of these various classes of drugs for designing affinity-based chromatographic purification schemes are discussed in another recent review (Marshak et al. 1984) and will not be reiterated here.

3.3 Protein Binding

In addition to calmodulin-regulated enzymes, calmodulin has been shown to bind to a variety of proteins in a calcium-dependent or independent manner. As with calmodulin-regulated enzymes, the physiological significance of this protein-protein interaction has not been shown for most of the binding proteins. Some examples of calmodulin binding proteins are histones, myelin basic protein, various microvillus proteins, spectrin, fodrin, caldesmon, gap junction protein, inhibitor protein, and calcineurin or modulator binding protein (for review see Klee and Vanaman 1982; Burgess et al. 1983). There have also been reports of several binding proteins that interact with calmodulin in the presence or absence of calcium. Some of these calmodulin-binding proteins are: gap junction proteins (Hertzberg and Gilula 1981; Welsh et al. 1982), a protein from intestinal brush border membranes (Glenney and Weber 1980), a chloroplast envelope protein (Roberts et al. 1983), several proteins from subcellular fractions of chicken embryo fibroblasts (CEF) (Van Eldik and Burgess 1983; Burgess et al. 1984), and a protein from bovine cerebral cortex membranes (Andreason et al. 1983). Many of these calmodulin-binding proteins appear to be associated with particulate structures and cell membranes. As discussed in more detail in Section 4.2, these calcium-independent interactions may have important implications for calmodulin localization and regulation of multiple cellular processes. The ability of various proteins to interact with calmodulin has usually been demonstrated by calcium-dependent interaction with calmodulin-Sepharose conjugates, by the binding of calmodulin to proteins before or after separation by polyacrylamide gel electrophoresis, or by cross-linking of calmodulin to the binding protein.

Calmodulin-Sepharose conjugates allow activity and protein characterizations to be done on the same fractions, and the method can be scaled up from analytical to preparative studies. The problems encountered in the use of calmodulin-Sepharose chromatography are those commonly encountered in flow and affinity-based chromatography procedures. Generally, the binding proteins must first be separated from endogenous calmodulin and calmodulin-like proteins before the affinity-based adsorption step. Calmodulin is an acidic protein so there can be a problem with basic polypeptides adsorbing to the column and possibly inhibiting the binding of the physiologically relevant binding protein. Because calmodulin can interact with more than one protein, there may be competition among the multiple binding proteins for the calmodulin. A calmodulin-binding protein might also bind other proteins, creating a "sandwich" effect during chromatography. Related to this latter problem is the possibility of proteolysis occurring during chromatography. When a binding protein becomes immobile while the sample is still being applied to the column, the protein becomes an immobilized substrate or "affinity column" for even trace amounts of protease that may be present. This is particularly important because the sample is applied to the column in the presence of calcium concentrations that are suficient to activate calcium-dependent proteases.

Analysis of calmodulin-binding proteins using gel methods offers some advantages over calmodulin-Sepahrose columns, but they also have limitations. The formation of complexes with labeled calmodulin before electrophoresis is limited to the use of nondenaturing gel systems and by the potential problem of dissociation of the complex during electrophoresis. These problems have been reduced with the introduction of azido derivatives of calmodulin that can be covalently cross-linked to binding proteins. However, in many cases the samples still must be depleted of endogenous calmodulin before analysis.

Several groups have recently reported the use of "gel overlay" procedures for identifying calmodulin-binding proteins that were separated by electrophoresis in the presence of sodium dodecyl sulfate (SDS) (for review see Burgess et al. 1983). There are several guidelines that are important when using gel overlay procedures. These are that (1) the chemically modified protein used as a tracer retains activity; (2) unmodified protein competes fully with the modified protein for binding; (3) known binding proteins bind in the overlay procedure; (4) for apparent calcium-dependent binding, the binding is indeed calcium-dependent, and it is reversible; (5) binding proteins in the incubation solution will compete with the same binding proteins in the gel matrix.

There are several limitations associated with gel overlay protocols. Some overlay procedures fail to detect certain known calmodulin-binding proteins. It is theoretically possible that a calmodulin-binding protein might not be identified due to insufficient renaturation of the calmodulin-binding domain on the protein. Similarly, because subunits of proteins are resolved on the SDS gels, a calmodulin-binding protein may not be identified if the binding domain is composed of portions of two or more subunits. Another problem in the use of gel overlay procedures involves the identification of calmodulin-binding proteins that are minor components of crude preparations. Major calmodulin-binding proteins that may or may not be major Coomassie blue staining proteins can be easily identified after an 18-h exposure of X-ray film. However, caution must be used in assigning calmodulin-binding activity to minor components, especially if these minor calmodulin-binding components are major Coomassie blue staining proteins. It is also possible that different calmodulin-binding proteins could co-migrate and not be distinguished from one another in the gel-binding procedure.

Within these limitations, the gel overlay procedure is an excellent method for screening protein fractions for calmodulin-binding activity and for identifying the calmodulin-binding subunit of a calmodulin-binding protein complex. In addition, the gel overlay procedure should be generally applicable to examine proteins that bind calcium-modulated proteins other than, or in addition to, calmodulin, to screen for physiologically relevant binding proteins, and to do structure-function studies on the domains involved in protein-protein interactions between calcium-modulated proteins and their targets. An example of the usefulness of the gel overlay procedure for the detection of calmodulin-binding proteins is studies on normal and transformed fibroblasts, described in Section 4.2.

4 Physiological Functions of Calmodulin

4.1 Regulation of Calmodulin Levels

The distribution and localization of calmodulin in various tissues and cells have been examined primarily by three approaches: differential centrifugation studies of tissue homogenates, biochemical cytology, and immunocytochemistry (for further details see Klee and Vanaman 1982; Van Eldik et al. 1982b). Differential centrifugation studies of tissue homogenates have shown that calmodulin is found in both soluble and particulate fractions with the relative amounts in each fraction depending on the tissue or cell type and on the isolation conditions. Quantitative biochemical cytology studies of calmodulin have demonstrated that in most cells the majority of calmodulin is found in the soluble fraction, although some calmodulin is usually present in each of the particulate fractions. Immunocytochemical studies have suggested a variety of subcellular distributions for calmodulin, depending on the cell type, cell cycle, or other conditions that affect the cell. In general, calmodulin appears to have a relatively diffuse cytoplasmic distribution in interphase cells and an association with the mitotic apparatus in mitotic cells.

Each of the three approaches described above has limitations. However, within the limitations of immunocytochemistry, differential centrifugation studies, and biochemical cytology, it appears that calmodulin can be found both freely soluble and in association with a variety of particulate structures. Although the majority of the calmodulin in many cells appears to be readily soluble, the presence of even small amounts of calmodulin associated with intracellular structures may be physiologically important. This point is discussed in more detail in Section 4.2

Although calmodulin appears to be present in all eukaryotic cells, the levels of the protein detected in tissues varies considerably. This variation could be due to the method of measurement of calmodulin, the preparation of the tissue, or the presence of interfering compounds. Within a given tissue or cell type, it appears that the levels of calmodulin may be quantitatively regulated. For example, a two to three fold increase in calmodulin concentration has been detected in many types of transformed cells and tumor tissues compared to their normal counterparts. In some cells, there may be a redistribution of calmodulin among subcellular compartments to effect local increases or decreases in the protein. There also appear to be alterations in the levels of calmodulin during the cell cycle and in rapidly growing cells, alterations which have been postulated to be involved in cell-growth regulation. Calmodulin concentration may be regulated at the genetic level or through post-translational modifications such as methylation. For further discussions of these points the reader is referred to Klee and Vanaman (1982).

It is clear that different biological systems may regulate the levels of calmodulin and calmodulin activity in different ways. In order to understand the mechanisms of calmodulin regulation, it is important to examine in detail a variety of cell types in terms of calmodulin concentration and distribution, and the subcellular array and distribution of calmodulin-binding proteins. Different cell systems are useful in attempts to answer questions about physiological roles for calmodulin. One cell system that has been particularly useful to study calmodulin regulation and its role in normal cell function and in mechanisms of pathophysiology is the well-characterized chicken embryo fibroblast (CEF)/avian sarcoma virus system.

Transformed CEF was the first biological system in which an alteration of calmodulin levels was demonstrated (Watterson et al. 1976). It is now known that the freely soluble levels of calmodulin are increased upon transformation of many types of cells by oncogenic viruses and in various tumor tissues. There is no obvious alteration in the calmodulin molecule in transformed CEF (Van Eldik and Watterson 1979), and no evidence of expression of different calmodulin isotypes upon transformation. The increase in the levels of freely soluble calmodulin in transformed CEF is not due to a subcellular redistribution of calmodulin upon transformation. Quantitative studies (Van Eldik and Burgess 1983) of the levels and subcellular distribution of calmodulin in normal and transformed CEF demonstrated that the majority of calmodulin in both normal and transformed CEF is found in the soluble fraction and that the total levels of calmodulin as measured by multiple procedures are increased in transformed CEF compared to normal CEF.

The higher levels of calmodulin found in transformed CEF appear to be the consequence of a selective increase in the rate of synthesis of calmodulin above that of total soluble or of total cellular protein with no significant difference in the rate of degradation of calmodulin between normal and transformed cells (Zendegui et al. 1984). Results similar to these have been reported (Chafouleas et al. 1981) for transformed mammalian cells. These data suggest that increases in the synthesis rate of calmodulin are a general phenomenon associated with transformation of eukaryotic cells by oncogenic viruses.

The mechanism of the increased synthesis rate of calmodulin in transformed CEF appears to be an increase in calmodulin-specific mRNA (Zendegui et al. 1984). Calmodulin consistently represents at least a two fold larger portion of the total translated protein when RNA from transformed CEF is used to direct protein synthesis. Increased activity of calmodulin mRNA could be due to its greater relative abundance in the mRNA preparations from transformed CEF or to greater translational activity per molecule of RNA. The relative abundance of calmodulin mRNA in preparations from normal and transformed CEF was measured by hybridization to a calmodulin cDNA probe in both dot-blot and Northern blot procedures. There was more hybridization to the calmodulin cDNA probe when RNA from transformed CEF was used. These dot-blot data were consistent with the results of Northern blot analysis of calmodulin mRNA. Equal amounts of poly(A)$^+$ RNA from normal and transformed CEF were subjected to electrophoresis, transferred to nitrocellulose, and hybridized to the calmodulin cDNA probe. A major species of hybridizing mRNA of approximately 2200 nucleotides was detected in RNA preparations from both normal and transformed CEF. Quantitation of the intensity of the major hybridizing mRNA bands demonstrated that approximately 1.5-fold more of the 2200 nucleotide RNA from transformed CEF hybridized to the calmodulin cDNA probe than the RNA from normal CEF. These hybridization analyses with both dot-blot and Nothern blot procedures are in agreement with data from translation experiments, and suggest that the increased translational activity of calmodulin mRNA from transformed cells is due to an increase in the quantity of calmodulin mRNA rather than a change in translational efficiency.

It is not yet possible to correlate changes in calmodulin biosynthesis in transformed cells with molecular mechanisms of transformation. Nonetheless, the potential importance of increases in calmodulin levels to the maintenance of cellular homeostasis is evident based on the multiple biochemical activities and highly conserved structure of calmodulin. For example, based on studies of calmodulin-binding proteins in CEF (see Sect. 4.2), it is now possible to suggest specific cell processes, such as cell shape and motility, that are potentially regulated by calmodulin. In addition, because calmodulin regulates multiple calcium-dependent processes and affects the activities of enzymes involved in cyclic nucleotide metabolism and action, it may coordinately modulate the two major second messenger systems: calcium and cyclic AMP. The ultimate effects of the cascades of regulation initiated by these second messengers may be influenced by the physiological regulation of intracellular calmodulin levels. Small increases in calmodulin concentration can result in 10- to 50-fold stimulation of enzyme activity (i.e., amplification). Thus, based on the amplification nature of calmodulin's enzyme activator properties and because calmodulin can interact with multiple target proteins, two fold changes in calmodulin concentration could have significant and diverse effects on multiple cellular processes. These changes could be even more significant if calmodulin exerts its regulatory effects at early steps of cascade mechanisms. Further studies of the regulation of calmodulin biosynthesis in other cell types and under different physiological states are needed in order to understand the common mechanisms of calmodulin action in eukaryotic cell function.

4.2 Calmodulin-Binding Proteins

In order to more fully understand the physiological significance of calmodulin regulation and localization, it is necessary to detect, purify, and characterize the targets of calmodulin (i.e., calmodulin-binding proteins). As described in Section 3.3, calmodulin has been shown to interact with a number of different target proteins. However, little information is available on the species, tissue, and subcellular distribution of calmodulin-regulated enzyme activities and binding proteins. The increasing availability of homogeneous preparations of calmodulin-binding proteins and the production of well-characterized antisera directed against these proteins will facilitate such studies. One biological system in which a systematic determination of the array and distribution of calmodulin-binding proteins has been done is CEF. Initial studies (Van Eldik and Burgess 1983) demonstrated the presence of calmodulin-binding proteins in both soluble and particulate subcellular fractions, showed that the majority of readily detected binding proteins are in the particulate fractions, and found no apparent qualitative differences between normal and transformed CEF with respect to the number, distribution, or apparent molecular weights of these proteins.

These studies also demonstrated that multiple classes of calmodulin-binding proteins can be identified. There are proteins that bind calmodulin only in the presence of added calcium, ones that bind calmodulin in the presence of calcium or chelator, ones that appear to bind only calmodulin, and others that bind calmodulin and other calcium-modulated proteins. These results suggest that calmodulin may contain multiple functional domains for interacting with binding proteins, that some of these domains may be shared by other calcium-modulated proteins, and that some may be unique to calmodulin.

The gel overlay procedure, calmodulin-Sepharose interactions, or crosslinking of added calmodulin do not, by themselves, provide any indication as to whether or not proteins bind calmodulin in vivo. However, it appears that many particulate calmodulin-binding proteins interact with a calmodulin domain that is found in the structurally related protein troponin C, and that several binding proteins interact with calmodulin in a relatively calcium-independent manner. In this regard, it is interesting to note that Andreasen et al. (1983) recently purified a membrane protein from bovine cerebral cortex whose apparent affinity for calmodulin is higher in the presence of calcium chelators than in the presence of calcium. Most of the soluble calmodulin-binding proteins from CEF and several particulate calmodulin-binding proteins do not bind troponin C and bind calmodulin in the presence of excess calcium but not in the presence of EDTA. Overall, these results are consistent with the possibility that calmodulin could be "compartmentalized" in the cell by virtue of its interaction with particulate-binding proteins or through the use of different binding domains on the calmodulin molecule. These interactions could occur even at very low calcium concentrations and this "compartmentalized" calmodulin could still be potentially accessible to other calmodulin-binding proteins via a different binding domain.

One implication of a compartmentalization of calmodulin includes elimination of one or possibly both of the diffusion steps that have been proposed for calcium-calmodulin regulation of various activities. For example, if calmodulin exists dissociated from its target protein, then two diffusion-limited processes are required in the calcium-dependent regulation: one for the calcium-calmodulin interactions and one for the target protein interaction with the calcium-calmodulin complex. Certain processes occur more rapidly than allowed by two diffusion-limited steps. In these cases, one possible solution is the incorporation of calmodulin as an integral subunit of the enzyme or target protein, such as in phosphorylase kinase, thus eliminating one diffusion-limited step.

However, it should be noted that the free calmodulin levels are not known for any cell type. More importantly, it is not known whether a significant amount of calmodulin ever exists in a "free" state in the cell. The concept of "soluble" levels is an operational one whose definition depends on the experimental conditions used. Although the majority of calmodulin in most cells is localized to the soluble fraction, this soluble fraction should not be equated with an intracellular, "free-floating" pool of calmodulin. Clearly, proteins localized to the soluble fraction in biochemical cytology studies could be integral components of supramolecular structures in the highly organized interior of a cell.

Further evidence for multiple functional domains on calmodulin and for potential compartmentalization of calmodulin via these multiple functional domains is provided by recent results localizing calmodulin to calmodulin-binding proteins by using calmodulin antisera (Watterson et al. 1984; Burgess WH, Watterson DM, Van Eldik LJ, unpublished). These data show that ternary complexes involving calmodulin and two calmodulin-binding proteins can be formed. Although one of the calmodulin-binding proteins is an antibody, these results demonstrate that the formation of a ternary protein complex cannot be excluded on the basis of steric arguments alone. Further, as more calmodulin antisera whose immunoreactivity is confined to specific and defined regions of the calmodulin molecule become available, further mapping studies of calmodulin and calmodulin-binding protein complexes will be possible.

The apparent molecular weights and subcellular distribution of the calmodulin-binding proteins in CEF has aided in the identification of a number of these binding proteins. Recent studies (Burgess et al. 1984) have identified six classes of calmodulin-binding proteins or calmodulin regulated activities in CEF: (1) myosin light chain kinase, (2) cyclic nucleotide phosphodiesterase, (3) calcineurin, (4) nonerythroid spectrin, (5) histones, and (6) ribosomal proteins.

Some of these proteins had been identified in many cell types other than CEF, some had been identified in CEF yet their calmodulin-binding properties had previously not been investigated, whereas others, such as ribosomal proteins, had not been reported previously to be calmodulin-binding proteins in any cell type. Myosin light chain kinase, cyclic nucleotide phosphodiesterase, and calmodulin-dependent phosphatase (calcineurin) from many cell types are calmodulin-binding proteins with calmodulin-regulated activities. The presence of myosin light chain kinase in CEF has been demonstrated by kinase activity, by immunoreactivity in immunoblots and competition radioimmunoassay, and by iodinated calmodulin binding in gel overlay. Calmodulin-binding proteins with apparent molecular weights in the range of 55,000-60,000 have been detected and calmodulin-activatable cyclic nucleotide phosphodiesterase activity and calcineurin immunoreactivity are present in CEF. Nonerythroid spectrin (fodrin) in CEF has been identified by correlation of immunoreactivity with ^{125}I-calmodulin-binding activity in CEF extracts. Similarly, DNA binding activity has been correlated with ^{125}I-calmodulin-binding activity in nuclei preparations of CEF, and these proteins co-migrate with histone standards. Finally, many of the iodinated calmodulin-binding proteins in the microsomal fraction of CEF can be identified in ribosomal protein preparation of CEF.

Thus, for CEF it is possible to postulate physiological roles for calmodulin based on the known calmodulin-binding proteins and calmodulin-regulated enzyme activities that have been identified. Overall these results suggest specific roles for calcium and calmodulin regulation of cyclic nucleotide metabolism and cell shape and motility in CEF. In order to determine the generality of these observations, systematic and quantitative examination of calmodulin and calmodulin-binding proteins in other biological systems must be done. Current studies in the CEF system are directed toward quantitative measurement of the level of the calmodulin-binding proteins in normal and transformed CEF. The fact that some of the activities change as a function of the cell cycle and upon viral transformation of CEF (Van Eldik et al. 1984b) coupled with the demonstrated changes in calmodulin levels during the cell cycle and upon viral transformation indicate future directions for a more complete understanding of the role of calmodulin in the regulation of cell function.

5 Overview

The molecular mechanisms by which calcium regulates cellular processes such as metabolic and mechanochemical events probably involve interactions with a variety of molecules. A large body of evidence suggests that the targets of calcium's regulatory effects in the cell are the class of calcium-binding proteins referred to as calcium-modulated proteins.

Calmodulin is only one of these calcium-modulated proteins, but it appears to be ubiquitous among eukaryotes, has a highly conserved structure, and has multiple in vitro biochemical activities. These characteristics suggest that calmodulin may be a multifunctional regulator in the cell and play a fundamental role in the mechanism of calcium action in cell physiology. Because calmodulin is an effector protein with no enzymatic activity, binding assays, activator assays, and immunoassays are important in investigations of the role of calmodulin in cellular processes. This chapter summarizes the various approaches, with their strengths and limitations, being used to study calmodulin structure, function, and mechanism of action. Based on current trends, a multidisciplinary approach to the study of this multifunctional protein should allow an accurate and incisive definition of calmodulin's role in the cell. Biological systems in which the expression of calmodulin can be varied have been invaluable in past investigations, and carefully designed experiments using these systems should continue to reveal information about how calmodulin and calmodulin-binding proteins are involved in biological function.

In this chapter, calmodulin structure and function have been reviewed with an emphasis on comparative studies of structure, function, and mechanism of action that provide a firm data base and precedents for physiology. Comparative sequence analyses combined with functional analyses have allowed correlation of function and structure, have suggested logical possibilities for calmodulin's mechanism of action, and have predicted the co-existence of a family of calcium-modulated proteins. Biological studies of calmodulin and calmodulin-binding proteins in cultured cell systems like chicken embryo fibroblasts have demonstrated that multiple functional classes of calmodulin-binding proteins exist and that calmodulin utilizes multiple functional domains in its interactions with target proteins. What is becoming clear is that calmodulin regulation of cell processes is not merely a simple equation of calcium and calmodulin resulting in enzyme activation. The ability of calmodulin to interact with and regulate an activity in a particular cell type is probably determined by multiple variables, including (1) the total concentration and subcellular distribution of calmodulin and calmodulin isotypes, (2) the concentrations, distributions, and properties of calmodulin-binding proteins, (3) the concentrations of calcium-modulated proteins other than calmodulin, and (4) the intracellular free calcium concentration.

Prospects for future research directions in calcium regulation require a thorough knowledge of the biological and chemical precedents which exist. Models for the role of calcium and calcium-binding proteins in cell function cannot ignore the extensive thermodynamic, kinetic, and structural data which are available. Comparative biochemistry continues to provide understanding of the possible molecular mechanisms of calcium regulation, as well as defining in chemical terms the various targets of calcium in biological systems. Therefore, current and future studies will continue to use a comparative approach to correlate structure with function in attempts to define the common and unique mechanisms used by calcium-modulated proteins in maintaining cellular homeostasis and mediating stimulus-response coupling. Elucidation of the interrelationships among calcium-binding proteins and the processes they affect is fundamental to our understanding of the role of calcium in cell physiology.

Acknowledgements. Much of the work described in this chapter is the result of collaborations with our colleagues, especially WH Burgess, BW Erickson, KF Fok, D Iverson, TJ Lukas, DR Marshak, DM Roberts, and M Schleicher, who we gratefully acknowledge. The studies were supported in part by grants from the National Institutes of Health (GM30861, GM33481, and GM30953), from the National Science Foundation (PCM8242875 and PCM8302912), and from the American Cancer Society (IN-25V).

References

Andreason TJ, Luetje CW, Heideman W, Storm DR (1983) Purification of a novel calmodulin-binding protein from bovine cerebral cortex membranes. Biochemistry 22:4615-4618

Burgess WH (1982) Characterization of calmodulin and calmodulin isotypes from sea urchin gametes. J Biol Chem 257:1800-1804

Burgess WH, Schleicher M, Van Eldik LJ, Watterson DM (1983) Comparative studies of calmodulin. Calcium Cell Function 4:209-261

Burgess WH, Watterson DM, Van Eldik LJ (1984) Identification of calmodulin-binding proteins in chicken embryo fibroblasts. J Cell Biol 99: 550-557

Chafouleas JG, Dedman JR, Munjaal RP, Means AR (1979) Calmodulin. Development and application of a sensitive radioimmunoassay. J Biol Chem 254:10262-10267

Chafouleas JG, Pardue RL, Brinkley BR, Dedman JR, Means AR (1981) Regulation of intracellular levels of calmodulin and tubulin in normal and transformed cells. Proc Natl Acad Sci USA 78:996-1000

deBernard B, Sottocasa GL, Sandri G, Carafoli E, Taylor AN, Vanaman TC, Williams RJP (eds) (1983) Calcium-binding proteins 1983. Elsevier, Amsterdam

Endo T, Tanaka T, Isobe T, Kasai H, Okuyama T, Hidaka H (1981) Calcium-dependent affinity chromatography of S100 and calmodulin on calmodulin antagonist-coupled Sepharose. J Biol Chem 256:12485-12489

Fok K-F, Van Eldik LJ, Erickson BW, Watterson DM (1981) Synthetic peptides that define an immunoreactive site of calmodulin. In: Rich DH, Gross E (eds) Peptides: synthesis-structure-function. Pierce Chem Co, pp 561-564

Gagnon C, Kelly S, Manganiello V, Vaughn M, Strittmatter W, Hoffman A, Hirata F (1980) Protein carboxyl-methylase modifies calmodulin function. Ann NY Acad Sci 356:385-386

Giedroc DP, Puett D, Ling N, Staros JV (1983) Demonstration by covalent crosslinking of a specific interaction between β-endorphin and calmodulin. J Biol Chem 258:16-19

Glenney JR, Weber K (1980) Calmodulin binding proteins of the microfilaments present in isolated brush borders and microvilli of intestinal epithelial cells. J Biol Chem 255:10551-10554

Hertzberg EL, Gilula NB (1981) Liver gap junctions and lens fiber junctions: comparative analysis and calmodulin interaction. Cold Spring Harbor Symp Quant Biol 46:639-645

Hidaka H, Yamaki T, Totsuka T, Asano M (1979) Selective inhibitors of Ca^{2+}-binding modulator of phosphodiesterase produce vascular relaxation and inhibit actin-myosin interaction. Mol Pharmacol 15:49-59

Hidaka H, Yamaki T, Naka M, Tanaka T, Hayashi H, Kobayashi R (1980) Calcium-regulated modulator protein interacting agents inhibit smooth muscle calcium-stimulated protein kinase and ATPase. Mol Pharmacol 17:66-72

Jamieson GA, Bronson DD, Schachat FH, Vanaman TC (1980) Structure and function relationships among calmodulins and troponin C-like proteins from divergent eucaryotic organisms. Ann NY Acad Sci 356:1-13

Kakiuchi S, Sobue K, Yamazaki R et al. (1981) Ca^{2+}-dependent modulator proteins from *Tetrahymena pyriformis*, sea anemone, and scallop and guanylate cyclase activation. J Biol Chem 256:19-22

Klee CB, Vanaman TC (1982) Calmodulin, Adv Protein Chem 35:213-321

Klumpp S, Kleefeld G, Schultz JE (1983) Calcium/calmodulin-regulated guanylate cyclase of the excitable ciliary membrane from *Paramecium*. J Biol Chem 258:12455-12459

Kretsinger RH (1980a) Structure and evolution of calcium-modulated proteins. CRC Crit Rev Biochem 8:119-174

Kretsinger RH (1980b) Crystallographic studies of calmodulin and homologs. Ann NY Acad Sci 356:14-19

Lagace L, Chandra T, Woo SLC, Means AR (1983) Identification of multiple species of calmodulin messenger RNA using a full length complementary DNA. J Biol Chem 258:1684-1688

Levin RM, Weiss B (1979) Selective binding of antipsychotics and other psychoactive agents to the calcium-dependent activator of cyclic nucleotide phosphodiesterase. J Pharmacol Exp Ther 208:454-459

Lukas TJ, Iverson DB, Schleicher M, Watterson DM (1984) Covalent structure of a higher plant calmodulin: *Spinacea oleracea*. Plant Physiol 75: 788-795

Lukas TJ, Marshak DR, Watterson DM (1985a) Drug-protein interactions: isolation and characterization of covalent adducts of phenoxybenzamine and calmodulin. Biochemistry 24: in press

Lukas TJ, Wiggins ME, Watterson DM (1985b) Structural characterization of a novel calmodulin from the unicellular alga *Chlamydomonas*. Submitted

Marshak DR, Lukas TJ, Watterson DM (1985) Drug-protein interactions: binding of chlorpromazine to calmodulin, calmodulin fragments, and related calcium binding proteins. Biochemistry 24: in press

Marshak DR, Lukas TJ, Cohen C, Watterson DM (1984) Calmodulin antagonists as ligands for affinity chromatography. In: Hidaka H, Hartshorne DJ (eds) Calmodulin antagonists and cellular physiology. Academic, New York. In press

Pardue RL, Brady RC, Perry GW, Dedman JR (1983) Production of monoclonal antibodies against calmodulin by in vitro immunization of spleen cells. J Cell Biol 96:1149-1154

Roberts DM, Zielinski RE, Schleicher M, Watterson DM (1983) Analysis of suborganellar fractions from spinach and pea chloroplasts for calmodulin-binding proteins. J Cell Biol 97:1644-1647

Roberts DM, Burgess WH, Watterson DM (1984) Comparison of the NAD kinase and myosin light chain kinase activator properties of vertebrate, higher plant, and algal calmodulins. Plant Physiol 75: 796-798

Sasagawa T, Ericsson LH, Walsh KA, Schreiber WE, Fisher EH, Titani K (1982) Complete amino acid sequence of human brain calmodulin. Biochemistry 21:2565-2569

Schleicher M, Lukas TJ, Watterson DM (1983) Further characterization of calmodulin from the monocotyledon barley *(Hordeum vulgare)*. Plant Physiol (Bethesda) 73:666-670

Schleicher M, Lukas TJ, Watterson DM (1984) Isolation and characterization of calmodulin from the motile green alga *Chlamydomonas reinhardtii.* Arch Biochem Biophys 228:

Takagi T, Nemoto T, Konishi KK (1980) The amino acid sequence of the calmodulin obtained from sea anemone (Metridium senile) muscle. Biochem Biophys Res Commun 96:377-381

Toda H, Yazawa M, Kondo K, Homura T, Narita K, Yagi K (1981) Amino acid sequence of calmodulin from sea scallop. J Biochem (Tokyo) 90:1493-1505

Van Eldik LJ (1984) Production and characterization of monoclonal antibodies that discriminate among individual S100 polypeptides. ICSU Short Reports 1:266-267

Van Eldik LJ, Burgess WH (1983) Analytical subcellular distribution of calmodulin and calmodulin-binding proteins in normal and virus-transformed fibroblasts. J Biol Chem 258:4539-4547

Van Eldik LJ, Watterson DM (1979) Characterization of a calcium-modulated protein from transformed chicken fibroblasts. J Biol Chem 254:10250-10255

Van Eldik LJ, Watterson DM (1981) Reproducible production of antiserum against vertebrate calmodulin and determination of the immunoreactive site. J Biol Chem 256:4205-4210

Van Eldik LJ, Watterson DM, Zendegui JG, Flockhart DA, Burgess WH (1982a) Calmodulin and calmodulin-binding proteins in normal and virus-transformed fibroblasts: levels, subcellular distribution, and regulation. In: Boynton AL, McKeehan WL, Whitfield JF (eds) Ions, cell proliferation, and cancer. Academic, New York, pp 465-487

Van Eldik LJ, Zendegui JG, Marshak DR, Watterson DM (1982b) Calcium-binding proteins and the molecular basis of calcium action. Int Rev Cytol 77:1-61

Van Eldik LJ, Fok K-F, Erickson BW, Watterson DM (1983a) Engineering of site-directed antisera against vertebrate calmodulin by using synthetic peptide immunogens containing an immunoreactive site. Proc Natl Acad Sci USA 80:6775-6779

Van Eldik LJ, Watterson DM, Fok K-F, Erickson BW (1983b) Elucidation of a minimal immunoreactive site of vertebrate calmodulin. Arch Biochem Biophys 227:522-533

Van Eldik LJ, Ehrenfried B, Jensen RA (1984a) Production and characterization of monoclonal antibodies with specificity for the S100ß polypeptide of brain S100 fractions. Proc Natl Acad Sci USA 81: 6034-6038

Van Eldik LJ, Watterson DM, Burgess WH (1984b) Immunoreactive levels of myosin light-chain kinase in normal and virus-transformed chicken embryo fibroblasts. Mol Cell Biol 4: 2224-2226

Wallace RW, Cheung WY (1979) Production of an antibody in rabbit and development of a radioimmunoassay. J Biol Chem 254:6564-6571

Watterson DM, Van Eldik LJ, Smith RE, Vanaman TC (1976) Calcium-dependent regulatory protein of cyclic nucleotide metabolism in normal and transformed chicken embryo fibroblasts. Proc Natl Acad Sci USA 73:2711-2715

Watterson DM, Sharief F, Vanaman TC (1980) The complete amino acid sequence of the Ca^{2+}-dependent modulator protein of bovine brain. J Biol Chem 255:962-975

Watterson DM, Burgess WH, Lukas TJ et al. (1984) Towards a molecular and atomic anatomy of calmodulin and calmodulin-binding proteins. Adv Cyclic Nucleotide Res 16:205-226

Weiss B, Prozialeck W, Cimino M, Barnette MS, Wallace TL (1980) Pharmacological regulation of calmodulin. Ann NY Acad Sci 356:319-345

Welsh MJ, Aster JC, Ireland M, Alcala J, Maisel H (1982) Calmodulin binds to chick lens gap junction protein in a calcium-independent manner. Science (Wash OC) 216:642-644

Yazawa M, Yagi K, Toda H et al. (1981) The amino acid sequence of the *Tetrahymena* calmodulin that specifically interacts with guanylate cyclase. Biochem Biophys Res Commun 99:1051-1057

Zendegui JG, Zielinski RE, Watterson DM, Van Eldik LJ (1984) Biosynthesis of calmodulin in normal and virus-transformed chicken embryo fibroblasts. Mol Cell Biol 4: 883-889

Calmodulin Gene Structure and Expression

A. R. MEANS, L. LAGACE, R. C. M. SIMMEN, and J. A. PUTKEY[1]

CONTENTS

1 Introduction

Calmodulin (CaM) is now recognized as a major intracellular Ca^{2+} receptor that function in a regulatory capacity for a diverse array of enzymatic reactions and biochemical pathways. The pleiotropic nature of CaM-regulated events has sparked intensive research efforts concerning the biology, structure and mechanism of action of CaM (Means et al. 1982; Klee and Vanaman 1982). This protein seems to be constitutively expressed in a number of hormonally regulated systems (Means et al. 1982; Means and Dedman 1980) but is found in elevated concentrations in virally transformed cells (Watterson et al. 1976; Chafouleas et al. 1981) and is acutely regulated at the G_1/S boundary of the cell cycle (Chafouleas et al. 1982; Chafouleas et al. 1984). As a prelude to understanding the molecular mechanism that control these changes in CaM, we have initiated a series of studies employing recombinant DNA technology to isolate and sequence the nucleic acids involved in CaM metabolism.

2 Cloning and Sequence Analysis of Calmodulin cDNAs

In an initial report, Munjaal et al. (1981) partially purified CaM mRNA from the electroplax of the eel *Electrophorous electricus* and used this material to isolate a clone that

---

[1] Department of Cell Biology, Baylor College of Medicine, Houston, TX 77030, USA

contained sequences complementary to a portion of the coding region of CaM mRNA. This cDNA clone was termed pCM109 and consisted of 340 nucleotides (NT); 168 NT coding for amino acids 93-148 of CaM and 182 NT noncoding DNA following the termination codon UGA. Lagace et al. (1983) then prepared a cDNA library from size-fractionated poly(A)⁺ RNA obtained from eel electroplax and used pCM109 to isolate a full length calmodulin cDNA clone, termed pCM116. Nucleotide sequence analysis revealed that the CaM cDNA insert from clone pCM116 contained 26 NT of 5' non-translated region, the entire coding region, and a 3' nontranslated region of 408 NT. The amino acid sequence derived from the NT sequence indicates that eel CaM is very

Table 1. Comparison of amino acid sequences for chick and eel CaM[*a] and C M-1

```
               1                                          10
Chick CaM      Ala Asp Gln Leu Thr Glu Glu Gln Ile  Ala Glu Phe Lys Glu Ala Phe Ser
Eel CaM        –   –   –   –   –   –   –   –   –    –   –   –   –   –   –   –   –
Chick C M-1    –   Glu Arg –   Ser –   –   –   –    –   –   –   –   –   –   –   –

              20                                          30
Chick CaM      Leu Phe Asp Lys Asp Gly Asp Gly Thr Ile Thr Thr Lys Glu Leu Gly Thr Val Met
Eel CaM        –   –   –   –   –   –   –   –   –   –   –   –   –   –   –   –   –   –   –
Chick C M-1    –   –   –   Arg –   –   –   –   Cys –   –   –   Met –   –   –   –   –   –

              40                                          50
Chick CaM      Arg Ser Leu Gln Gly Asn Pro Thr Glu Ala Glu Leu Gln Asp Met Ile Asn Glu Val
Eel CaM        –   –   –   –   –   –   –   –   –   –   –   –   –   –   –   –   –   –   –
Chick C M-1    –   –   –   –   –   –   –   –   –   –   –   –   –   –   –   Val Gly –   –

                        60                                70
Chick CaM      Asp Ala Asp Gly Asn Gly Thr Ile Asp Phe Pro Glu Phe Leu Thr Met Met Ala Arg
Eel CaM        –   –   –   –   –   –   –   –   –   –   –   –   –   –   –   –   –   –   Lys
Chick C M-1    –   –   –   –   Ser –   –   –   –   –   –   –   –   –   Ser Leu –   –   –

                        80                                90
Chick CaM      Lys Met Lys Asp Thr Asp Ser Glu Glu Glu Ile Arg Glu Ala Phe Arg Val Phe Asp
Eel CaM        –   –   –   –   –   –   –   –   –   –   –   –   –   –   –   –   –   –   –
Chick C M-1    –   –   Arg –   Ser –   –   –   –   –   –   –   –   –   –   –   –   –   –

                        100                               110
Chick CaM      Lys Asp Gly Asn Gly Tyr Ile Ser Ala Ala Glu Leu Arg His Val Met Thr Asn Leu
Eel CaM        –   –   –   –   –   –   –   –   –   –   –   –   –   –   –   –   –   –   –
Chick C M-1    –   –   –   –   –   –   –   –   –   –   –   –   –   –   –   –   –   –   –

                        120                               130
Chick CaM      Gly Glu Lys Leu Thr Asp Glu Glu Val Asp Glu Met Ile Arg Glu Ala Asp Ile Asp
Eel CaM        –   –   –   –   –   –   –   –   –   –   –   –   –   –   –   –   –   –   –
Chick C M-1    –   –   –   –   –   –   –   –   –   –   –   –   –   Lys –   –   –   Cys Asn

                        140                               148
Chick CaM      Gly Asp Gly Gln Val Asn Tyr Glu Glu Phe Val Gln Met Met Thr Ala Lys
Eel CaM        –   –   –   –   –   –   –   –   –   –   –   –   –   –   –   –   –
Chick C M-1    Asn –   –   –   –   –   –   –   –   –   Arg –   –   –   Glu –
```

[*a] The amino acid sequences for chick and eel CaM were derived from pCM12/pCB15 (12) and pCM116 (9) respectively

similar to that reported for CaM from other vertebrate species (Table 1). When compared to the amino acid sequence from human (Sasagawa et al. 1982) and bovine (Kasai et al. 1980), the three sequences are identical, with one exception at position 74. The substitution of an arg for lys in the eel is the first reported substitution at that position in this protein from lower invertebrates to humans. The functional implications of this conservative amino acid substitution remain unclear.

The CaM clone pCM116 was utilized to examine the size and number of CaM mRNA species in various eel, rat, chicken, and human tissues. Multiple species of cytoplasmic CaM mRNA were found in every tissue examined and ranged in size from 820 to 4200 NT. In all eel tissues examined the cytoplasmic mRNA's were 820, 1100 and 2000 NT (Lagace et al. 1983). Nuclear RNA was prepared from electroplax nuclei and hybridized to pCM116. In addition to the three mature forms of cytoplasmic CaM MRNA, nuclear RNA contained a species of 5500 NT. Sequence analysis revealed that the 5500 NT species represented the primary transcript and the three cytoplasmic forms were generated from this molecule by differential polyadenylation. pCM109 was derived from the 820 NT mRNA, whereas pCM116 represented the 1100 NT species. These data suggest that the three translatable forms of electroplax CaM mRNA are derived from a single nuclear transcript by differential polyadenylation during processing.

The next use of pCM116 was to isolate CaM cDNA clones complementary to chicken CaM mRNA. A library was prepared from poly(A)$^+$ RNA isolated from chicken brain. Screening this library with pCM116 resulted in detection of 15 positive signals among 4500 recombinants. Two overlapping clones (pCB12 and pCB15) were selected for DNA sequencing. The combined unique sequence of the two clones yielded 1395 NT and contained the entire coding region for chicken CaM together with 94 NT of 5' nontranslated region (Putkey et al. 1983). The latter sequence contained a single poly (A)$^+$ addition/termination signal as well as a terminal stretch of 28 A residues. As predicted from the coding region, the amino acid sequence of chicken CaM is identical to the mammalian protein (Sasagawa et al. 1982; Kasai et al. 1980). Compared to the eel, there is a single amino acid substitution at position 74 which is arg in chicken and lys in eel (Table 1).

The chicken cDNA's were used to construct three hybridization probes: one each from the 5 nontranslated portion, the amino acid coding region, and the 3 nontranslated region. In six chicken tissues examined, all three probes hybridized to two species of cytoplasmic RNA (Putkey et al. 1983). The strongest hybridization signal corresponds to an mRNA of 1600 NT which represents the parent mRNA for pCB12 and pCB15. The second mRNA species is 1900 NT. The relative intensities of the signals were equivalent for each probe, demonstrating that both species were CaM mRNAs. As in the case of the eel electroplax (Lagace et al. 1983), both mRNAs were found to exist on isolated polyribosomes and were derived from a common primary transcript by differential polyadenylation/termination.

Consistent with the essentially identical primary amino acid sequence of chicken and eel CaM, the NT sequence in the amino acid coding portions of the respective cDNAs are quite similar, with an overall direct homology of 79%. There are, however, regions in the chicken and eel cDNAs which exhibit either less or more homology. The 5 and 3 nontranslated regions of these cDNAs have diverged to a large extent and only reveal 21% and 29% homologies, respectively. On the other hand, those portions of the

cDNA that correspond to the four amino acid domains of the protein (Klee and Vanaman 1982) are more homologous than the overall similarity. As would be predicted
from the protein structure, the greatest inter and intraspecies similarities are seen between the NT corresponding to domains I/III and II/IV. Finally, the four putative
Ca^{2+}-binding subdomains in the chicken and eel have NT homologies of 85-95% (Putkey et al. 1983). The Ca^{2+}-binding subdomains are predicted to be comprised of 12
amino acids based on the crystalline structure of parvalbumin determined by Kretsinger
and Barry (Kretsinger 1979). These 12 amino acid subdomains would be encoded by
36 NT. We have evaluated homologies between 36 NT segments along the entire length
of the amino acid coding regions of the chicken and eel cDNAs and found that the
four Ca^{2+}-binding subdomains have considerably higher degrees of homology than all
other portions of the molecules. These data suggest that there is a greater evolutionary
pressure to conserve those NT sequences that encode the amino acids which comprise
the Ca^{2+}-binding sites.

    To assess the number of CaM genes in the chicken genome, the three chicken
cDNA restriction fragments (see Fig. 1) were used to analyze genomic complexity by
hybridization (Putkey et al. 1983). Genomic DNA was prepared from chicken liver
and digested with one of three restriction endonucleases, *Bam* HI, *Eco* RI or *Hind* III.

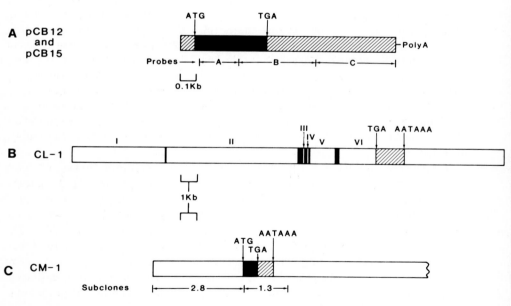

Fig. 1 A-C. Organization of chicken CaM genes. The two genomic clones, CL-1 and CM-1, are
shown together with the combined cDNA sequence from pCB12 and pCB15. *Black areas* amino
acid coding regions; *striped areas* 5 and 3-nontranslated regions; *open areas in the genomic clones*
5 and 3 flanking regions and introns I through IV. The amino acid coding and 3 nontranslated
regions in CM-1 are deduced from the open reading frame encoding the CM-1 protein and the position of the AATAAA polyadenylation/termination signal. Hybridization probes *A, B* and *C* are
shown beneath the cDNA. These probes were used for Southern analysis of chicken genomic DNA.
Shown beneath CM-1 are two subclones, *2.8* and *1.3*, which were used for DNA sequencing

The digests were resolved by electrophoresis on a 1% agarose gel, the nucleic acid fragments transferred to nitrocellulose and the fragments hybridized against the three cDNA probes. Hybridization to *Bam* HI or *Eco* RI digests revealed single positive signals at 25 Kb and 10 Kb, respectively, regardless of which cDNA probe was used. Hybridization to *Hind* III digested DNA resulted in a more complex pattern. Probe A (5) hybridized to a single 1.9 Kb fragment and probe C (3) produced a single signal at 1.7 Kb. The probe from the coding region (B) recognized both the 1.9 and 1.7 hybridization bands. The cDNA contains no cutting sites for *Bam* HI or *Eco* RI and a unique site for *Hind* III. Therefore CaM must be represented by a unique gene and this gene must be interrupted by at least one intervening sequence (intron) which contains a *Hind* III site.

3 The Chicken Calmodulin Gene, CL-1

The CaM cDNAs pCB12 and pCB15 were used as hybridization probes to screen a chicken genomic library constructed in the λ phage Charon 4A (Putkey et al. 1983). Eight positive recombinants were detected and shown to contain overlapping chicken DNA inserts by restriction enzyme mapping. One of the clones which contained the largest insert (13.5 Kb) was selected for further analysis and designated CL-1. The structural organization of the CaM gene contained in CL-1 has been determined by restriction enzyme mapping and DNA sequencing. This organization is illustrated in Fig. 1. At present we have sequenced the gene from the ATG through the transcription termination site. The gene is single copy and contains at least six intervening sequences. The first of these introns separates the ATG from the first NT of the codon which is translated into the N-terminal amino acid of CaM (ala). The remaining introns are positioned as follows: II between aa 10 and 11; III between aa 56 and 57; IV between aa 82 and 83; V between aa 100 and 101; and VI between aa 139 and 140. It is interesting to note that intervening sequences III, V, and VI interrupt the Ca^{2+}-binding sites I, II, and IV respectively, whereas intron IV occurs in the linker that serves to divide the two homologous halves of the protein. As yet we have not unambiguously identified the 5 end of the gene. We have utilized a 1.6 Kb *Taq* I fragment present at the 5 end of CL-1 to screen a cosmid library of chicken DNA. A 38 Kb molecule has been isolated and shown to contain a 9.5 Kb *Eco* RI fragment that hybridizes with a 75 NT fragment of the chicken CaM cDNA containing exclusively 5 nontranslated sequence. Restriction enzyme mapping reveals this 9.5 Kb fragment to contain 2.5 Kb of DNA 5 to the original DNA sequence present in CL-1. This additional genomic clone (CL-2) is currently being sequenced to determine whether it contains all of the remaining 5 nontranslated region of the chicken CaM gene.

4 The Chicken Calmodulin-Like Gene, CM-1

The chicken gene library was also screened with pCM109 (Munjaal et al. 1981), which contains a 350 bp insert complementary to the 820 NT eel CaM mRNA and begins with the codon for amino acid 93. The utilization of the usual stringent hybridization conditions failed to yield positive results. Empirically determined less stringent conditions resulted in eight positive hybridization signals (Stein et al. 1983). Each clone contained one large *Eco* RI fragment (10.6-11.5Kb) and one or two smaller ones (1.6-2.8 Kb). Furthermore only the large *Eco* RI fragment hybridized to the CaM cDNA. Since restriction mapping revealed the inserts to be overlapping ones, only one recombinant phage, CM-1, was selected for further analysis. The structural organization of CM-1 is also presented in Fig. 1. An *Eco* RI- *Cla* I fragment of 1.3 Kb was subcloned into pBR322 and sequenced (Stein et al. 1983). One open reading frame exists in the 1.3 Kb fragment beginning at the extreme 5 end. The NT sequence begins with amino acid 12 of CaM (phe) and extends through the termination codon at NT 412 uninterrupted by introns. A single polyadenylation/termination signal was found 486 NT from the termination codon. A 2.8 Kb fragment of CM-1, which overlaps the 5-end of the 1.3 Kb fragment, was subcloned and sequenced from its 3-end. This fragment was found to contain the remaining NT that encode CaM as well as the concensus transcription initiation sequence. Again, one open reading frame exists that begins with the initiator codon ATG.

The 1.3 Kb fragment of CM-1 was utilized as a hybridization probe to assess genomic complexity (Stein et al. 1983). Again chicken DNA was cut with *Bam* HI, *Eco* RI or *Hind* III. In all cases the size of the hybridization signal was markedly different than the signal found using the cDNA corresponding to CL-1. Thus under stringent hybridization conditions, Southern analysis of CL-1 or CM-1 yields a unique set of hybridizing bands with intensities corresponding to one copy per haploid genome. Only under conditions of reduced stringency do the two genes cross-hybridize. These observations suggest that the chicken genome contains two different CaM genes whose coding sequences are related but whose sequence organization is considerably different.

Since two genes for CaM (or CaM-like proteins) exist in the chicken genome, we examined expression of these chicken genes by Northern hybridization (Stein et al. 1983). As mentioned earlier, a cDNA probe from CL-1 hybridized to two species of cytoplasmic mRNA present in every chicken tissue examined. These species were 1600 and 1900 NT, respectively. The 1.3 Kb probe prepared from CM-1 hybridized to a single mRNA species primarily present in skeletal and cardiac muscle. This species was approximately 1000 NT and, therefore, considerably smaller than mRNAs complementary to CL-1.

The CM-1 gene encodes a 148 amino acid protein that bears extensive sequence homology with human, bovine, rat, and eel CaMs. Of the 148 residues, 129 are identical to these other vertebrate CaM's (87% homology). However, these 19 differences represent a significant departure from other CaMs, since 11 of the substitutions are of a nonconservative nature. Figure 2 shows the relationship of these amino acid substitutions to the putative calcium-binding domains in CaM. The calcium-binding domains are based on internal amino acid sequence homology (Dedman et al. 1978; Watterson et al. 1980). The assignment of residues which participate in calcium binding (6 residues/Ca^{2+}) is based on the X-ray structure of parvalbumin (Kretsinger 1979). An examination of the amino acid substitutions in the Ca^{2+}-binding subdomains of CM-1 suggests the possibility of functional changes in the CaM encoded by this gene. Sub-

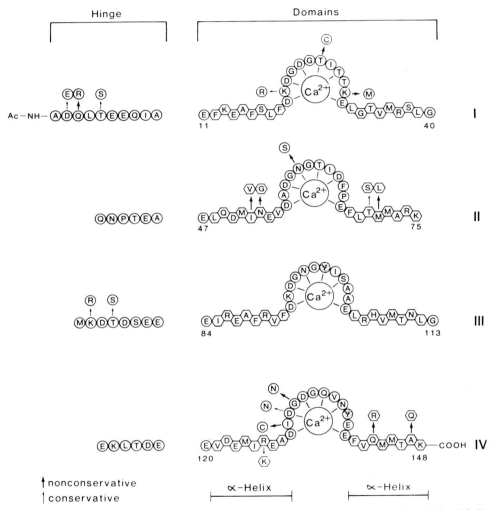

Fig. 2. Relationship of the amino acid substitution in the CM-1 protein to the calcium-binding domains and subdomains in CaM. The primary amino acid sequence for chick CaM is shown. Those amino acids which participate in α-helix formation are indicated by *hexagons*, while those amino acids in the hinge regions and the Ca^{2+}-binding subdomains are indicated by *circles*. Assignment of Ca^{2+}-binding domains is based on the maximum interdomain homology determined for rat testis CaM. The positions of nonconservative amino acid substitutions are indicated by *large arrows*, conservative amino acid substitutions are indicated by *small arrows*. The one-letter amino acid code is *A* ala, *B* Asx, *C* Cys, *D* asp, *E* glu, *F* phe, *G* gly, *H* his, *I* ile, *K* lys, *L* leu, *M* met, *N* asn, *P* pro, *Q* gln, *R* arg, *S* ser, *T* thr, *V* val, *W* trp, *Y* tyr, and *Z* glx

domain III in CM-1 is identical to that of CL-1 and to that of previously sequenced vertebrate CaMs. Subdomain II is 91% homologous with only 1 amino acid difference. However, subdomains I and IV are only 75% homologous to the corresponding Ca^{2+}-binding regions in other vertebrate CaMs including chicken CL-1. Interestingly, the single change in subdomain II is an asn, which is one of the proposed six coordination residues for Ca^{2+}, to a ser, which could also participate in ionic interaction with Ca^{2+}. Thus, subdomains II and III of CM-1 CaM could have fully functional Ca^{2+}-binding sites. The changes in subdomains I and IV are more extensive, however. Two of the three differences in subdomains IV are in critical residues implicated in Ca^{2+}-binding (asp[131] and gly[132]), and the third change is the substitution of a bulky cys residue for the adjacent ile[130]. In subdomain I, three residues are replaced by bulkier groups including the substitution of a cys for thr [26] which is located in the center of the Ca^{2+}-binding subdomain. These changes themselves would probably be sufficient to severely impede the binding of Ca^{2+} to these two subdomains. In addition, the presence of a cys residue in both subdomains suggests the possibility that a disulfide bond might be formed, which would bridge the central portions of the first and fourth Ca^{2+}-binding sites of CM-1 CaM. It thus seems likely that, for whatever reason, the protein encoded by chicken CM-1 has evolved to the point where it functions as a two-domain Ca^{2+}-binding protein. This hypothesis is supported by the nucleotide homologies of the Ca^{2+}-binding subdomains. Whereas the second and third subdomains of CM-1 are very homologous to the corresponding regions in CL-1, subdomains I and IV are considerably less so (64% and 61% respectively). If these two subdomains no longer functioned as Ca^{2+}-binding sites, then the sequence would be less constrained than subdomains II and III, and evolutionary "drift" would be more rapid.

The lack of intervening sequences in CM-1 was unexpected. The extraordinarily high sequence homology of the four domains of CaM is a cogent argument that this protein has evolved via two successive duplications of a primordial, single-domain Ca^{2+}-binding protein (Goodman 1979). In light of other evidence concerning the sequence organization of genes that have evolved by successive duplications, such as ovomucoid (Stein et al. 1980), we believe that the evolution of the CaM gene would have resulted in the presence of three introns, located between peptide domains I and II, II and III, and II and IV. Indeed six (or more) introns are present in the other CaM gene, CL-1. It is possible that CL-1 represents the primordial calmodulin gene, but the introns were subsequently lost by some mechanism of gene conversion to give rise to CM-1. Several examples of such processed genes have recently been published. A mouse pseudo α-globin gene has moved from the α-globin cluster on chromosome 11 to a position on chromosome 15, and is flanked by retrovirus-like elements (Leuders et al. 1982). Other processed genes that are flanked by short, direct repeats have been described, including human small nuclear RNA genes (Van Arsdell et al. 1981), a rat α-tubulin (Lemischka and Sharp 1982), and human immunoglobulin (Hollis et al. 1982) and β-tubulin genes (Wilde et al. 1982). All of these processed genes are pseudogenes; they are nonfunctional due to numerous point mutations and/or deletions. A different type of processed pseudogene has also been discovered — the human metallothionein II$_B$ gene (Karin and Richards 1982). Even though this gene has an intact coding sequence, it is apparently not expressed because it lacks functional promotor sequences near its 5' end. These discoveries imply that RNA transcripts of functional genes, or their DNA cognates, can be resinserted at a distant chromosomal site to generate processed gene copies. The unusual finding regarding CM-1 is that it seems to be expressed but only in a tissue-specific manner. It is tempting to hypothesize that CM-1 was orginally derived as a processed copy of CL-1. This processed copy, lacking the introns, was inserted back into the genome, and has since evolved into a functional calmodulin-like gene that codes for a Ca^{2+}-binding protein that binds only two molar equivalents of Ca^{2+}. This gene has an intact coding sequence and seems to produce a mRNA only in skeletal and cardiac muscle tissues. This hypothesis would predict that some form of repeat sequence be found flanking both ends of the gene, a poly(A)$^+$ tract exists at the 3' end of the gene, and a regulatory sequence that direct expression of this gene exist in the 5' flanking region. Indeeed preliminary sequencing experiments have revealed the presence of a 9 bp direct repeat (GTGCTTCCT) flanking both ends of CM-1. These structural features are consistent with the possibility that CM-1 represents the first example of an expressed processed gene.

5 Construction of Prokaryotic Expression Vectors

If the CM-1 gene is expressed in a selective manner in chicken striated muscle the encoded protein would be the first CaM-like protein to be demonstrated in a vertebrate species. As a prelude to answering this question, we decided to develop recombinant DNA molecules capable of synthesizing the protein encoded by CM-1. Methods developed to purify this protein from bacterial extracts could then be applied directly to chicken tissues. In addition an antibody could be produced which would selectively recognize CM-1 protein (but not authentic CaM) and used to localize the molecule in muscle. In order to accomplish this goal, expression plasmids were constructed for the bacterial synthesis of both CaM and the CM-1 protein.

5.1 CaM Expression Vector

A strategy was devised to construct an expression vector for CaM using the chicken cDNA, a pUC8 vector (Viera and Messing 1982) and a Tac promotor (Russell and Bennett 1982) as the DNA components. The strategy was targeted toward the construction of a plasmid in which there is a defined distance between the Shine-Delgarno (SD) box of the bacterial promotor and the ATG initiation codon of the CaM gene. In an initial construct, a blunt-ended *Bgl* I/*Pst* I fragment from the chicken CaM cDNA pCB12 was cloned into a blunt-ended *Sal* I/*Pst* I site of the vector pUC8 to produce pCaM4. The resulting ligation junction at the 5-end of the CaM gene contains a hybrid initiation codon in which the A is donated by the *Sal* I site of pUC8 and the TG are donated by the *Bgl* I site of pCB12. In this construct a *Bam* HI site exists 5 bp 5 to the ATG which can be used to insert a bacterial promotor. The next construct (pCaM16) involved completion of the amino acid coding region for CaM by insertion of a *Pst* I fragment from pCB15 into the *Pst* I site of pCaM4. The intact CaM gene was then transferred from pCaM16 as a *Bam* HI/*Pst* I fragment and inserted into the corresponding restriction sites of pTac2 which places the gene just 3 to the Tac promotor. Recombinant clones resulting from this last step were screened for the production of CaM by using the CaM RIA (Chafouleas et al. 1979). All colonies selected produced CaM. One recombinant, pCaM24, consistently produced 1-3% of its total protein as CaM in the presence of the inducer isopropylthio-B-galactoside (IPTG). IPTG stimulated CaM synthesis 9-12 fold. This level of CaM production yields approximaltely 2-4 mg of CaM L^{-1} of bacterial culture. The CaM produced in bacteria comigrates with rat testis CaM on SDS gels both in the absence and presence of Ca^{2+} and can be quantitatively precipitated with affinity-purified sheep antirat CaM (Chafouleas et al. 1983).

5.2 CM-1 Expression Vector

pCaM24 was used as a foundation to construct an expression vector for the CaM-like gene. A *Sma* I/*Eco* RI fragment was isolated from a subclone of CM-1 and inserted into the *Eco* RI/blunt-ended *Pst* I site of pCaM24. The resulting plasmid, pCaML1, has a CaM gene in which amino acids 1-11 are encoded by the chicken cDNA while amino acids 12-148 are encoded by CM-1. The expected protein product of this gene would have 16 of the 19 amino acid substitutions present in the CM-1 protein. The remaining three substitutions occur in amino acids 2, 3 and 5. The protein product of pCaML1 will be called CM-1 protein and has been characterized with respect to several properties. These properties are summarized in Table 2 and discussed below.

Preliminary studies indicate that CM-1 protein is recognized by the CaM antibody. Although detailed, comparative immunodilution curves have yet to be performed, the preliminary data suggest that the affinity of the polyclonal CaM antibody for the CaM-1 protein is high. The size, heat stability and calcium-binding ability of CM-1 protein were assessed by gel electrophoresis in SDS. In this experiment, cell extracts were prepared from bacteria which contained either pCaM24 or pCaML1. A sample of each extract was heat-treated at 90°C for 5 min. Aliquots of both total and heat-treated protein from each extract were then resolved on an SDS gel in the presence of 5 mM EGTA or 2 mM Ca^{2+} (Dedman et al. 1977), the gels were then stained with Coomassie blue.

Table 2. Properties of chicken CaM versus the CM-1 protein

	Protein	
Property	CaM	CM-1 Protein
Recognized by CaM Ab	Yes	Yes
M_r + EGTA	19,500	19,500
+ Ca	16,000	17,000
Heat Stable	Yes	Yes
Binds to TAPP-Sepharose	Yes	Yes
Apparent pKI	3.9	4.1
Binds to thiol-Sepharose	No	Yes

Both total and heat-treated protein extracts from bacteria containing pCaML1 exhibit a protein which comigrates, in the presence of EGTA, with authentic rat CaM and bacterially produced chick CaM at a relative M_r of 19,500. In the presence of 2 mM Ca^{2+}, CaM synthesized from pCaM24 migrates a relative M_r of 16,000. This apparent decrease in M_r is identical to that seen for rat testis CaM in response to calcium binding. The CM-1 protein produced by bacteria containing pCaML1 also undergoes an apparent molecular weight decrease in response to added Ca^{2+}, but the extent of this decrease is less than that seen for rat testis CaM or bacterially produced chicken CaM (M_r = 17,500). These data support the supposition that the CM-1 protein would bind Ca^{2+} differently than CaM.

To examine the drug-binding properties of the CM-1 protein, a heat-treated extract from bacteria containing pCaML1 was passed over phenothiazine-Sepharose in the presence of Ca^{2+}, as described by Charbonneau et al. (1983). The column was washed, and bound proteins were eluted with EGTA. Both bound and unbound proteins were analyzed by electrophoresis. The bulk of the heat-treated bacterial proteins were washed from the column, while a single major protein, which comigrated with authentic CaM on SDS gels, was retained and could be eluted by EGTA. In control experiments using heat-treated protein extracts from bacteria containing pTac 2 (a plasmid containing the bacterial promotor but no CaM gene), no bacterial proteins bound to the phenothiazine column.

The predicted primary amino acid sequence of the CM-1 protein contains two more positive charges than chicken CaM. If the CM-1 protein is faithfully synthesized in bacteria, then it should focus, on an isoelectric focusing gel, in a more basic position than CaM. Two-dimensional gel electrophoresis was performed as described by Anderson and Anderson (1978) and indeed CM-1 protein was located at a position which corresponded to the same molecular weight as CaM but at an isoelectric point which indicated it had a greater net positive charge.

6 Concluding Remarks

Thus we have succeeded in our initial aim by constructing a bacterial expression vector that harbors the CM-1 gene. In addition we have presented evidence that the CM-1 protein produced in bacteria (1) is heat-stable, (2) undergoes an aberrant Ca^{2+}-induced mobility shift on SDS gels, (3) cross-reacts with CaM in the CaM radioimmunoassay, (4) is highly acidic although differently charged than CaM, and (5) can be purified from a heat-treated bacterial extract by chromatography on sepharose to which the phenothiazine, 2-trifluoromethyl-10-aminopropyl-phenothiazine (TAPP) had been covalently linked (Charbonneau et al. 1983). The CM-1 protein eluted from this column by EGTA migrates as a single band upon analysis by both one- and two-dimensional gel electrophoresis. The final requirement before an attempt could be made to demonstrate the presence of CM-1 protein in tissue was to devise a way to purify it from CaM. This was accomplished by taking advantage of the fact that CM-1 protein contains two cysteine residues, whereas chicken CaM is devoid of this amino acid. This differnce allows the quantitative retention of CM-1 protein on a thiol-Sepharose column, whereas the contaminating CaM does not bind. CM-1 protein is then eluted from the column with L-cysteine. The amount of CM-1 protein can then be estimated utilizing the CaM radioimmunoassay, but replacing ^{125}I-CaM with CM-1 protein metabolically labeled with ^{35}S-methionine.

A preliminary experiment was conducted where chicken breast muscle was homogenized, heat-treated, and the heat-treated extract was passed over a phenothiazine-Sepharose column. Bound proteins were eluted with EGTA. The addition of a trace amount of ^{125}I-CaM allowed us to determine that over 90% of the CaM present in the muscle extract was bound to and eluted from this resin. The recovered protein was then applied to the thiol-Sepharose column. Whereas none of the calmodulin bound, as determined by recovery of the ^{125}I trace in the flow-through, protein was eluted following washing of the column with L-cysteine. Experiments are in progress to determine whether this protein represents authentic CM-1 protein. If the answer turns out to be positive then we will have obtained the first evidence for (1) the expression of an apparently processed gene, (2) presence of more than one CaM or CaM-like gene in any species, (3) existence of a novel CaM-like protein in a vertebrate, and (4) tissue-selective expression of a CaM-like protein.

Acknowledgements. The authors' work reported in this chapter was supported by grants from the American Cancer Society (BC-326) and the Robert A. Welch Foundation (Q-611) to A. R. M. Dr. Putkey was supported by cancer-directed fellowship DRG-582 of the Damon Runyon-Walter Winchell Cancer Fund. Dr. Simmen was a postdoctoral fellow of the Muscular Dystrophy Association.

References

Anderson NG, Anderson NL (1978) Analytical techniques for cell fractions XXII. two-dimensional analysis of serum and tissue proteins: multiple gradient slab electrophoresis. Anal Biochem 85:341-354

Chafouleas JG, Deman JR, Munjaal RP, Means AR (1979) Calmodulin: development and application of a sensitive radioimmunoassay. J Biol Chem 254:10262-10267

Chafouleas JG, Pardue RL, Brinkley BR, Dedman JR and Means AR (1981) Regulation of intracellular levels of calmodulin and tubulin in normal and transformed cells. Proc Natl Acad Sci USA 78:996-1000

Chafouleas JG, Bolton WE, Hidaka H, Boyd AE and Means AR (1982) Calmodulin and the cell cycle: involvement in regulation of cell cycle progression. Cell 28:41-50

Chafouleas JG, Lagace L, Bolton WE, Boyd AE and Means AR (1984) Changes in calmodulin and its mRNA accompany reentry of quiescent (G_0) cells into the cell cycle. Cell 36:73-81

Chafouleas JG, Riser ME, Lagace L, Means AR (1983) Production of polyclonal and monoclonal antibodies to calmodulin and utilization of these immunological probes. Methods Enzymol 102:104-110

Charbonneau H, Hice R, Hart RC, Cormier MJ (1983) Purification of calmodulin by Ca^{2+}- dependent affinity chromatography. Methods Enzymol 102:17-19

Dedman JR, Potter JD, Jackson RL, Johnson JD, Means AR (1977) Physicochemical properties of rat testis Ca^{2+}-dependent regulator protein of cyclic nucleotide phosphodiesterase: relationship of Ca^{2+} binding, conformational changes and phosphodiesterase activity. J Biol Chem 252:8415-8422

Dedman JR, Jackson RL, Schreiber WE, Means AR (1978) Sequence homology of the Ca^{2+}-dependent regulator of cyclic nucleotide phosphodiesterase from rat testis with other Ca^{2+} binding proteins. J Biol Chem 253:343-346

Goodman M, Pechere JF, Harech J, Demaille JG (1979) Evolutionary diversification of structure and function in the family of intracellular calcium-binding proteins. J Mol Evol 13:331-352

Hollis GF, Hieter PA, McBride OW, Swan D, Leder P (1982) Processed genes: a dispersed human immunoglobulin gene bearing evidence of RNA-type processing. Nature (Lond) 296:321-325

Karin M, Richards R (1982) Human metallothionein genes — primary structure of the metallothionein II gene and a related processed gene. Nature (Lond) 299:797-802

Kasai H, Kato Y, Isobe T, Kawasaki H, Okuyama T (1980) Determination of the complete amino acid sequence of calmodulin (phenylalanine-rich acidic protein II) from bovine brain. Biomed Res 1:248-264

Klee CB, Vanaman TC (1982) Calmodulin. Adv Protein Chem 35:213-321

Kretsinger RH (1979) The informational role of calcium in the cytosol. Adv Cyclic Nucleotide Res 11:1-26

Lagace L, Chandra T, Woo SLC and Means AR (1983) Identification of multiple species of calmodulin messenger RNA using a full length complementary DNA. J Biol Chem 258:1684-1688

Lemischka I, Sharp PA (1982) The sequence of an expressed rat α-tubulin gene and a pseudogene with an inserted repetitive element. Nature (Lond) 300:330-335

Leuders K, Leder A, Leder P, Kuff E (1982) Association between an transported α-globin pseudogene and retrovirus-like elements in the BALB/c mouse genome. Nature (Lond) 295:426-428

Means AR, Dedman JR (1980) Calmodulin — an intracellular calcium receptor. Nature (Lond) 285:73-77

Means AR, Tash JS, Chafouleas JG (1982) Physiological implications of the presence, distribution and regulation of calmodulin in eukaryotic cells. Physiol Rev 62:1-39

Munjaal RP, Chandra T, Woo SLC, Dedman JR and Means AR (1981) A cloned calmodulin structural gene probe is complementary to DNA sequences from diverse species. Proc Natl Acad Aci USA 78:2330-2334

Putkey JA, Ts'ui KF, Tanaka T, Lagace L, Stein JP, Lai EC, Means AR (1983) Chicken calmodulin genes: a species comparison of cDNA sequences and isolation of a genomic clone. J Biol Chem 258:11864-11870

Russell DR, Bennett GN (1982) Construction and analysis of an in vivo activity of *E. coli* promotor mutants that alter the −35 to −10 spacing. Gene (Amst) 20:231-243

Sasagawa T, Ericsson LH, Walsh KA, Schreiber WE, Fischer EH and Titani K (1982) Complete amino acid sequence of human brain calmodulin. Biochemistry 21:2565-2569

Stein JP, Catterall JR, Kristo P, Means AR, O'Malley BW (1980) Ovomucoid intervening sequences specify functional domains and generate protein polymorphism. Cell 21:681-687

Stein JP, Munjaal RP, Lagace L, Lai EC, O'Malley BW, Means AR (1983) Tissue-specific expression of a chicken calmodulin pseudogene lacking intervening sequences. Proc Natl Acad Sci USA 80:6485-6489

Van Arsdell SE, Denison RA, Beronstein LB, Weiner AM, Manser T and Gesteland RF (1981) Direct repeats flank three small nuclear RNA pseudogenes in the human genome. Cell 26:11-17

Viera J, Messing J (1982) The pUC plasmids, an M13mp7-derived system for insertion mutagenesis and sequencing with synthetic universal primers. Gene (Amst) 19:259-268

Watterson DM, van Eldik LJ, Smith RE, Vanaman TC (1976) Calcium-dependent regulatory protein of cyclic nucleotide metabolism in normal and transformed chick embryo fibroblasts. Proc Natl Acad Sci USA 73:2711-2715

Watterson DM, Sharief F, Vanaman TC (1980) The complete amino acid sequence of the Ca^{2+}-dependent modulator protein (calmodulin) of bovine brain. J Biol Chem 255:962-971

Wilde CD, Crowther CE, Crepe TP, Lee MG-S, Cowan NJ (1982) Evidence that a human β-tubulin pseudogene is derived from its corresponding mRNA. Nature (Lond) 297:83-84

Cellular Localization of Calmodulin and Calmodulin-Acceptor Proteins

R. C. BRADY, F. R. CABRAL, M. J. SCHIBLER and J. R. DEDMAN[1]

CONTENTS

1 Introduction

Calcium plays a pivotal role in the regulation of numerous cellular functions. Cell growth, motility, exocytosis, and endocytosis are all examples of processes which are triggered by an increase in the level of intracellular free calcium. The importance of calcium in cellular regulation was initially demonstrated in skeletal muscle. Heilbrunn and Wiercinski (1947) reported that injection of calcium into muscle fibers induced contraction of the muscle. It has since been well demonstrated that the effect of increased intracellular free calcium on skeletal muscle contraction is mediated by the calcium-binding protein troponin C. The interaction of actin with myosin and concomitant force transduction in skeletal muscle is sterically precluded unless troponin C binds calcium.

In nonmuscle cells however, there is no troponin C. The various enzymes and physiologic responses which are activated in nonmuscle cells in response to an increase in the level of cytosolic free calcium appear to be mediated by the ubiquitous, low molecular weight, heat-stable calcium-binding protein; calmodulin. While it is clear that more are yet to be identified, Table 1 lists the enzymes and processes which have, at present, been demonstrated to be activated by calmodulin. Calmodulin is a highly conserved, 17,000 molecular weight protein which is present in all eukaryotic cells and also shares a great degree of structural homology with troponin C (for reviews see Cheung 1980; Means and Dedman 1980; Klee et al. 1980). A diagrammatic represen-

---

[1] Departments of Internal Medicine and Physiology and Cell Biology, University of Texas Medical School at Houston, P.O. Box 20708, Houston, TX 77025, USA

Table 1. Enzymes and cellular processes regulated by calmodulin

1. Cyclic nucleotide phosphodiesterase	7. Phosphofructokinase
2. Adenylate cyclase	8. Plant NAD⁺ kinase
3. Guanylate cyclase	9. Ca²⁺-Dependent protein kinase
4. Erythrocyte Ca²⁺ ATPase	10. Neurotransmitter release
5. Myosin light chain kinase	11. Microtubule disassembly
6. Phosphorylase kinase	

tation of the structure of calmodulin and its amino acid composition is seen in Fig. 1. As can be seen, calmodulin contains 148 amino acids and possesses four calcium-binding domains. Upon binding calcium, calmodulin undergoes a conformational shift which results in the exposure of a hydrophobic binding site (Fig. 2). It is via this hydrophobic site that calmodulin interacts with the enzymes it activates. This hydrophobic domain is also the site with which drugs such as the phenothiazines interact to inhibit calmodulin function (Weiss and Levin 1978). The calcium-dependent interaction of calmodulin with the phenothiazines has proven useful in the rapid, large-scale purification of calmodulin. Tissue homogenates can be heat-treated and then applied, in the presence of calcium, to affinity matrices consisting of phenothiazines immobilized on Sepharose. The adsorbed calmodulin can then be eluted by washing with

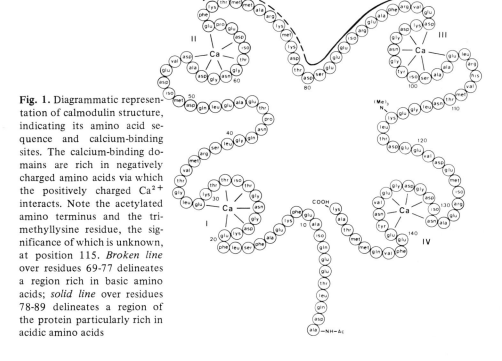

Fig. 1. Diagrammatic representation of calmodulin structure, indicating its amino acid sequence and calcium-binding sites. The calcium-binding domains are rich in negatively charged amino acids via which the positively charged Ca²⁺ interacts. Note the acetylated amino terminus and the trimethyllysine residue, the significance of which is unknown, at position 115. *Broken line* over residues 69-77 delineates a region rich in basic amino acids; *solid line* over residues 78-89 delineates a region of the protein particularly rich in acidic amino acids

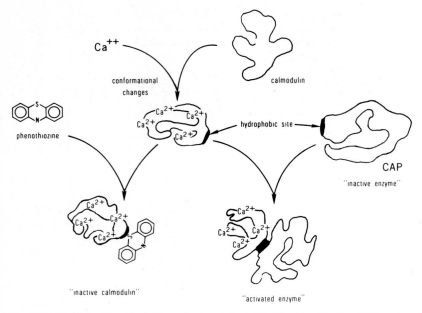

Fig. 2. Mechanism of the activation of calmodulin by Ca^{2+}. Upon binding Ca^{2+}, calmodulin undergoes a conformational shift which exposes a hydrophobic domain. This exposed site on calmodulin interacts with a complimentary hydrophobic domain present on an inactive CAP, resulting in a conformational shift and subsequent activation of the CAP. Phenothiazine antipsychotic drugs can also bind to this same hydrophobic domain, leading to the inactivation of calmodulin by precluding its interaction with various CAP's

the calcium chelator EGTA (Charbonneau and Cormier 1979; Jamieson and Vanaman 1979).

The approach taken by our laboratory in order to discern the various roles of calmodulin in the regulation of cellular function is the identification and cellular localization of the proteins with which calmodulin interacts (calmodulin-acceptor proteins; CAP's) as well as the immunocytochemical localization of calmodulin itself. Due to the fact that calmodulin is a ubiquitous protein found in all eukaryotic cells, the specific physiologic response triggered by calcium-calmodulin in a particular cell is determined by the repertoire of inherent calmodulin-acceptor proteins present within that individual cell type. Thus, the true discriminators of calmodulin function within various cells are the calmodulin-acceptor proteins. In this chapter we discuss the localization of calmodulin and the recent identification of its acceptor proteins within the mitotic spindle apparatus.

2 Localization of Calmodulin in the Mitotic Spindle Apparatus

Utilizing monospecific antibody against native calmodulin (Dedman et al. 1978), we have localized calmodulin in cultured cells via indirect immunofluorescence according

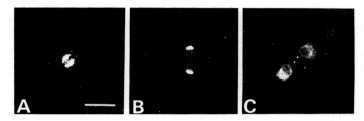

Fig. 3 A-C. Fluorescence micrograph of mitotic PTK-2 cells stained with anti-calmodulin. **A** Metaphase cell indicating calmodulin fluorescence within the spindle fibers. **B** Anaphase cell exhibiting fluorescence within the spindle poles. **C** Telophase cell exhibiting calmodulin fluorescence within the midbody. *Bar* 40 μm

to the procedure of Fuller et al. (1975). In this procedure cells grown on glass coverslips are fixed with 3% formaldehyde and then lyzed with acetone at $-20°C$. The cells are then incubated with anticalmodulin at $37°C$ (to tag the cellular calmodulin) followed by a fluorochrome-conjugated secondary antibody which recognizes only the primary, calmodulin antibody. In this manner the localization of calmodulin within the cell can be visualized microscopically by the fluorescence given off by the fluorochrome on the secondary antibody. Interphase cells subjected to this treatment routinely exhibit a diffuse pattern of calmodulin fluorescence throughout the cytoplasm which does not appear to entail any specific cellular structures. Mitotic cells, however, exhibit striking calmodulin fluorescence within the mitotic spindle apparatus (Fig. 3). In metaphase the pole-to-kinetochore fibers of the mitotic spindle are stained and in anaphase only the spindle poles (centriolar areas) are stained. In telophase, calmodulin fluorescence is observed within the midbody. These results indicate that, while homogeneously distributed throughout the cytoplasm during interphase, calmodulin becomes highly concentrated within the spindle apparatus during mitosis. Although effects of calcium on the spindle have been reported (see Dedman et al. 1979 for review), it is unclear what role is played by calmodulin in the regulation of mitotic spindle function and integrity. Marcum et al. (1978) have reported an increased sensitivity to calcium-induced depolymerization of microtubules which is conferred by calmodulin in vitro. Thus, it is possible that calmodulin functions to regulate microtubule disassembly within the mitotic spindle. This function however, remains to be defined.

3 Immunocytochemical Localization of CAP's Within the Mitotic Spindle

Calmodulin functions by binding the specific calmodulin acceptor proteins (Fig. 2). In order to localize individual CAP's immunocytochemically within the cell, pools of CAP's were isolated from Chinese hamster ovary (CHO) cells via affinity chromatography on a calmodulin-Sepharose column. Antibodies against these CAP's were produced in sheep. As can be seen in Fig. 4, cells which are subjected to indirect immunofluorescence examination with one of these antibodies exhibit fluorescence within the

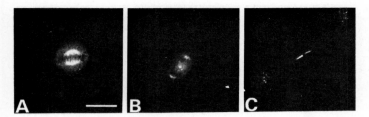

Fig. 4 A-C. Fluorescence micrograph of mitotic PTK-2 cells stained with antibody produced against a pool of CAP's isolated from CHO cells. **A** Metaphase cell exhibiting CAP fluorescence within the spindle apparatus. **B** Anaphase cell exhibiting fluorescence at the spindle poles. **C** Telophase cell exhibiting fluorescence within the midbody. Note the similarity between this pattern and the localization pattern of calmodulin during mitosis. This antibody has been demonstrated to recognize a 60,000 mol. wt. CAP present within the spindle apparatus. *Bar* 40 μm

mitotic spindle apparatus and midbody. This localization pattern correlates directly with that of calmodulin throughout the various stages of mitosis. That is, there is fluorescence within the pole-to-kinetochore fibers during metaphase, and fluorescence within the poles or centriolar areas during anaphase without any interzonal fluorescence. In telophase the midbody is stained. Interphase cells exhibit a diffuse pattern of fluorescence throughout the cytoplasm that does not appear to entail any specific cellular structures. Thus, the calmodulin-acceptor protein(s) recognized by these antibodies appear to be specifically associated with the microtubules of the spindle apparatus and not with the microtubules in the cytoplasm during interphase.

4 Identification of Mitotic Spindle CAP's by Gel Overlay Analysis

In an effort to identify the individual calmodulin-acceptor proteins present within the mitotic spindle apparatus and responsible for the fluorescence pattern observed above, we isolated spindles from CHO cells and attempted to characterize them biochemically. In order to accomplish this goal, CHO cells were synchronized by treating with thymidine and then mitotic cells were obtained by treating with nocodazole to block them in prometaphase. The block was reversed by placing the cells in drug-free medium and as the cells entered metaphase they were lyzed and the spindles harvested by differential centrifugation. Mitotic spindles isolated in this manner exhibit normal morphology, remain intact and contain chromosomes, as can be seen in Fig. 5.

Isolated spindles were then examined for their inherent content of calmodulin-acceptor proteins by gel overlay analysis (Carlin et al. 1980). In this procedure a cell or tissue sample is subjected to sodium dodecyl sulfate-polyacrylamide gel electrophoresis and its constituent proteins are separated according to their relative molecular weights. The calmodulin-acceptor proteins within the sample are then identified by incubating the gel with ^{125}I-calmodulin. The gel is then washed, dried, and subjected to autoradiography. Each calmodulin-acceptor protein within the sample appears as an

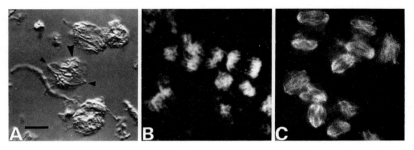

Fig. 5 A-C. Morphology of isolated CHO mitotic spindles. **A** Nomarski micrograph. *Small arrowheads* centrioles at the poles of a metaphase spindle. *Large arrowhead* chromosomes in the center of this metaphase spindle. Note the microtubule fibers that extend from the centrioles toward the central plate of chromosomes. **B** Fluorescence micrograph of isolated anaphase spindles which have been stained with Hoechst dye. This dye stains DNA and permits visualization of the chromosomes associated with the spindle apparatus. **C** Fluorescence micrograph of isolated mitotic spindles stained with anti-tubulin antibody. Microtubules within the spindles can be seen. *Bar* 10 μm

exposed band on the developed autoradiogram. With the appropriate controls one can identify the specific calmodulin-acceptor proteins present within a given sample. The results of a calmodulin overlay of isolated mitotic spindles appears in Fig. 6. Five calmodulin-acceptor proteins, specific for calmodulin, with molecular weights of 200,000, 160,000, 130,000, 60,000 and 52,000 are present. In addition, the histones and a few low molecular weight proteins can be seen to have bound ^{125}I-calmodulin. Preliminary evidence suggests, however, that these low molecular weight proteins do not represent bona fide calmodulin-acceptor proteins but rather are indicative of nonspecific interactions with ^{125}I-calmodulin, since they also bind ^{125}I-troponin C.

Furthermore, when isolated spindles are subjected to immunoblot analysis according to the procedure of Towbin et al. (1979), the anti-CAP antibodies mentioned above recognize only a single protein with a molecular weight of 60,000. This 60,000 protein and the 60,000 calmodulin-acceptor protein which is identified in the gel overlay tech-

Fig. 6. Autoradiograph of a calmodulin gel overlay of isolated mitotic spindles performed in the presence and absence of Ca^{2+}. ^{125}I-calmodulin binds specifically, in a Ca^{2+}-dependent manner, to five proteins whose molecular weights are indicated *(52K-200K) K* kilodalton. The interaction of ^{125}I-calmodulin with these proteins does not occur in the presence of 2 mM EGTA. The positively charged histones within the isolated spindles bind nonspecifically to ^{125}I-calmodulin via a charge interaction in a Ca^{2+}-independent manner. Note that one of the molecular weight standards, phosphorylase b, also binds ^{125}I-calmodulin in a Ca^{2+}-dependent manner. The physiologic significance of this interaction is not known

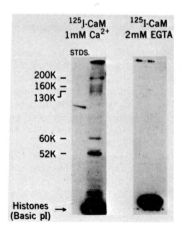

nique probably represent the same protein. The co-localization of this 60,000 CAP and calmodulin in the spindle throughout mitosis suggests that it is in part, responsible for the specific association of calmodulin with the mitotic spindle apparatus. The functions of the other four CAP's and their localization within the mitotic spindle are currently being investigated.

5 Conclusions

Information regarding the various roles of calmodulin in the regulation of cellular function can clearly be obtained by localizing calmodulin and its acceptor proteins within the cell. Calmodulin gel overlay analysis permits one to rapidly determine the total number and molecular weights of calmodulin-acceptor proteins within tissues and cells. Affinity chromatography of cell and tissue extracts on calmodulin-Sepharose matrices enables the investigator to isolate CAP's in their native state so that they may be examined for enzymatic activities of various types in order to determine their specific function. We now know that each cell type contains its own compliment of CAP's which represent the discriminators of the Ca^{2+}-calmodulin regulatory signal. Of course, once a calmodulin-acceptor protein has been identified within a particular cellular organelle engaged in a specific function, a major step toward the ultimate determination of that calmodulin-acceptor protein's function and the function of Ca^{2+}-calmodulin has been made.

References

Carlin RK, Grab DJ, Siekevitz P (1980) The binding of radioiodinated calmodulin to proteins on denaturing gels. Ann NY Acad Sci 356:73-74

Charbonneau H, Cormier MJ (1979) Purification of plant calmodulin by fluphenazine-Sepharose affinity chromatography. Biochem Biophys Res Commun 90:1039-1047

Cheung WY (1980) Calmodulin plays a pivotal role in cellular regulation. Science (Wash DC) 207:19-27

Dedman JR, Brinkley BR, Means AR (1979) Regulation of microfilaments and microtubules by calcium and cyclic AMP. Adv Cyclic Nucleotide Res 11:131-174

Dedman JR, Welsh MJ, Means AR (1978) Ca^{2+}-dependent regulator. Production and characterization of a monospecific antibody. J Biol Chem 253:7515-7521

Fuller GM, Brinkley BR, Bougher JM (1975) Immunofluorescence of mitotic spindles by using monospecific antibody against bovine brain tubulin. Science (Wash DC) 187:948-950

Heilbrunn LV, Wiercinski FJ (1947) The action of various cations on muscle protoplasm. J Cell Comp Physiol 29:15-32

Jamieson GA Jr, Vanaman TC (1979) Calcium-dependent affinity chromatography of calmodulin on an immobilized phenothiazine. Biochem Biophys Res Commun 90:1048-1056

Klee CB, Crouch TH, Richman PG (1980) Calmodulin. Annu Rev Biochem 49:489-515

Marcum JM, Dedman JR, Brinkley BR, Means AR (1978) Control of microtubule assembly-disassembly by calcium-dependent regulator protein. Proc Natl Acad Sci USA 75:3771-3775

Means AR, Dedman JR (1980) Calmodulin – an intracellular calcium receptor. Nature (Lond) 285:73-77

Towbin H, Staehelin T, Gordon J (1979) Electrophoretic transfer of proteins from polyacrylamide gels to nitrocellulose sheets: procedure and some applications. Proc Natl Acad Sci USA 76: 4350-4354

Weiss B, Levin RM (1978) Mechanism for selectively inhibiting the activation of cyclic nucleotide phosphodiesterase and adenylate cyclase by antipsychotic agents. Adv Cyclic Nucleotide Res 9:285-303

Calmodulin Antagonism

B. D. ROUFOGALIS[1]

CONTENTS

1 Introduction

Developments in the last decade formulated the concept that the Ca^{2+}-binding protein, calmodulin, mediates many of the cellular actions of Ca^{2+} as a second messenger. Attempts to elucidate the role of calmodulin in cellular regulation have led to the widespread use of calmodulin antagonists, particularly the phenothiazines, first described by Weiss et al. (1974). It has become clear, however, that the compounds presently available to antagonize calmodulin have other significant biological actions at concentrations similar to those that block calmodulin. This may seriously limit their use as

[1] Laboratory of Molecular Pharmacology, Faculty of Pharmaceutical Sciences, University of British Columbia, Vancouver, BC V6T 1WL5, Canada

tools for establishing calmodulin functions, particularly in whole cells and intact tissues. In this discussion the mode of interaction of the presently available antagonists with calmodulin, and their specificity, are reviewed. Minimum criteria are established for deciding if inhibition of cellular function is associated with calmodulin antagonism.

2 Mode of Interaction of Antagonists with Calmodulin

2.1 Kinetics of Calmodulin Antagonism

The discovery by Weiss and his colleagues (Weiss et al. 1974) of the inhibition of calmodulin-dependent forms of adenosine 3',5'-monophosphate phosphodiesterase by phenothiazine antipsychotic agents marked an important development in the availability of tools for the pharmacological modification of calmodulin-dependent cellular events. Subsequently, Levin and Weiss (1977) showed that trifluoperazine and related phenothiazines interact with calmodulin, rather than with the phosphodiesterase target of calmodulin activation (Weiss et al. 1980). Both Ca^{2+}-dependent and Ca^{2+}-independent binding of phenothiazines and related neuroleptic (antipsychotic) agents was demonstrated by equilibrium dialysis with labeled ligands (Levin and Weiss 1977, 1979). The Ca^{2+}-dependent binding occurs at a high affinity class of binding sites, while the number of molecules bound (n) varies from one to three for the various agents examined. A second class of sites binds these drugs with low affinity but with higher capacity. This binding, which occurs for example at 100 μM trifluoperazine, is Ca^{2+}-independent (Levin and Weiss 1979). Values of the affinity (K_d) and capacity per mol of calmodulin (n) for the high and low affinity sites (Levin and Weiss 1979) for phenothiazines are trifluoperazine (K_d = 1-1.5 μM, n = 2 and K_d = 5 mM, n = 24-27) and chlorpromazine (K_d = 5 μM, n = 3 and K_d = 130 μM, n = 17), while only the high affinity sites are observed for the butyrophenone, haloperidol (K_d = 9 μM, n = 2) and the diphenylbutylamine, pimozide (K_d = 0.83 μM, n = 1). There is a linear relationship between calmodulin binding and antagonism of calmodulin activation of phosphodiesterase (Levin and Weiss 1979). While there is some variation in the inhibitory concontrations between membrane preparations, inhibition of other putative calmodulin-dependent processes should occur at similar concentrations, and inhibition at concentrations either significantly greater or even lower than reported I_{50} values should be viewed cautiously. A list of drugs which inhibit calmodulin activation of phosphodiesterase and other enzymes, and the concentration at which they produce half maximal inhibition (I_{50}), is available in a recent review (Roufogalis 1982).

Divalent cations other than Ca^{2+} also promote phenothiazine binding to calmodulin. The order of effectiveness of divalent cations is $Ca^{2+} > Sr^{2+} > Ni^{2+} > Co^{2+} > Zn^{2+} > Mn^{2+}$, while Ba^{2+} and Mg^{2+} are ineffective (Levin and Weiss 1977). The binding to calmodulin is also pH-dependent, being reduced two- to threefold when the pH is increased from 7.5 to 8.5, and markedly reduced below pH_5 in the case of trifluoperazine (Weiss et al. 1980). This probably reflects an ionic contribution to binding of the ammonium head of trifluoperazine to calmodulin, in addition to the established hydrophobic interactions (Prozialeck and Weiss 1982). The Ca^{2+}-dependent high affinity binding of phenothiazines to calmodulin is reversed by Ca^{2+} chelation, and is re-established on addition of Ca^{2+} (Weiss and Levin 1978). However, ultraviolet irradiation of the phenothiazines renders their inhibition of calmodulin irreversible (Prozialeck et al. 1981). This is also true of inhibition by phenothiazines of other enzymes, including Na^+, K^+-ATPase (Akera and Brody 1969), and requires experiments with these agents to be performed in the absence of UV radiation.

2.2 The Ca^{2+}-Dependent Hydrophobic Site

The Ca^{2+}-dependence of the high affinity binding of antagonists to calmodulin appears to be associated with a conformational change in calmodulin, which exposes a hydrophobic site, as probed by hydrophobic fluorescent molecules (LaPorte et al. 1980; Tanaka and Hidaka 1980). Both calmodulin-dependent cGMP phosphodiesterase and the binding of either the fluorescent probe TNS (2-p-toluidinylnapthalene-6-sulfonate) (Tanaka and Hidaka 1980) or the napthalenesulfonamide calmodulin antagonist, W-7, are activated in a parallel fashion by Ca^{2+} from 1-10 μM (Tanaka et al. 1982). Since W-7 and the phenothiazines appear to bind at the same site (Hidaka et al. 1980), it appears that many of the calmodulin antagonists interact with the hydrophobic region on calmodulin exposed by Ca^{2+} interaction. The inhibition of calmodulin activation of various enzymes could therefore be explained by a competition between phenothiazines and these enzymes for this hydrophobic region (Tanaka et al. 1982). This interpretation is consistent with the dependence of the potency of inhibition by a variety of calmodulin antagonists on the hydrophobicity of the antagonists (Norman et al. 1979; Tanaka et al 1982; Roufogalis 1981; Prozialeck and Weiss 1982). However, a single mode of interaction may not account for the activation of all calmodulin-dependent processes, as wide differences in affinity for calmodulin and Ca^{2+} sensitivity exist between various calmodulin-sensitive systems (Tanaka et al. 1983a). It should also be noted that similar Ca^{2+}-dependent exposure of hydrophobic regions occurs in the related Ca^{2+}-binding proteins, troponin C (Tanaka and Hidaka 1981) and S-100 (Marshak et al. 1981), which may account for the relative lack of selectivity of the antagonists for calmodulin and the latter proteins (Levin and Weiss 1978; Marshak et al. 1981). On the other hand, the significance of the hydrophobic region exposed by Ca^{2+} is well illustrated by the correlation between the low potency of *Tetrahymena*-derived calmodulin for activation of phosphodiesterase and the decreased Ca^{2+}-dependent interaction with a hydrophobic fluorescent probe when compared to bovine brain calmodulin (Inagaki et al. 1983).

2.3 Localization of Antagonist-Binding Sites on Calmodulin

Considerable progress has been made in the past few years in localizing the site(s) of interaction of Ca^{2+}-binding proteins with antagonists, particularly the phenothiazines. Since these drugs are thought to sterically block interaction of Ca^{2+}-binding proteins with their effector proteins, characterization of these drug-binding sites may elucidate the nature of the sites responsible for enzyme activation. Earlier studies had implicated a major contribution of methionine and other methyl groups in the interaction of tri-fluoperazine with calmodulin, with less influence of aromatic residues (Klevit et al. 1981; Tanaka et al. 1983b). However, the interpretation of these studies may have been complicated by the high concentration of trifluoperazine (1 mM) used in NMR experiments, at which the drug may undergo aggregation (Gariépy and Hodges 1983), and by the generalized conformational changes in calmodulin induced by methionine oxidation (Walsh et al. 1978).

Recent work on troponin C and various peptide fragments derived from it and from calmodulin, have identified a trifluoperazine-binding site on troponin C to be a small helical region of around ten amino acids rich in hydrophobic aliphatic and aromatic side chains. The sequence corresponds to amino acid residues 92-102 of rabbit skeletal

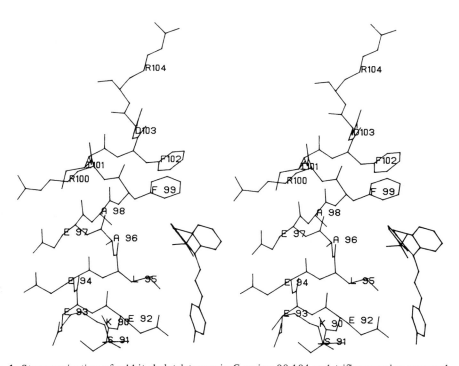

Fig. 1. Stereoprojection of rabbit skeletal troponin C region 90-104 and trifluoperazine prepared by Masao Fujinaga and Dr. R. S. Hodges, MRC Group in Protein Structure and Function, University of Alberta. This representation was constructed from the crystal coordinates of carp parvalbumin for the N-terminal region of the CD hand (residues 38-52) (Kretsinger and Nockolds 1973). The original set of coordinates for the peptide backbone and side chains was retained, but rabbit skeletal troponin C side chains of residues 90-104 were substituted in a standard conformation where different from the carp parvalbumin sequence. The crystal coordinates of trifluoperazine hydrochloride (McDowell 1980) were used. A more detailed discussion of the interaction of tri-fluoperazine and the N-terminal region of site III of troponin C has been described (Gariépy and Hodges 1983; Cachia and Hodges 1984)

muscle troponin C, and probably to residues 82-92 in the homologous sequence in calmodulin (Gariépy and Hodges 1983). This segment occurs in an N-terminal region of
the Ca^{2+}-binding domain III in tropononin C. The evidence for this assignment is as
follows. Interaction of skeletal muscle troponin I was localized to calmodulin fragments of residues 78-148 and 1-106, and a cyanogen bromide fragment of residues
84-135 in troponin C (Weeks and Perry 1978; Grabarek et al. 1981; Kuznicki et al.
1981). A cyanogen bromide-cleaved fragment (M_r 5500-6500) of calmodulin containing
this region was shown to bind to a fluphenazine-Sepharose affinity column (Head et al.
1982), while a trypsin fragment of calmodulin (TR2C, residues 78-148) was also shown
to interact with trifluoperazine (Kuznicki et al. 1981). The assignment is consistent
with the Ca^{2+}-dependent exposure of a hydrophobic drug-binding site (LaPorte et al.
1980; Tanaka and Hidaka 1980), since the binding of calmodulin fragment 77-124 to
fluphenazine is Ca^{2+}-dependent (Head et al. 1982), while the N-terminal helical region
of the troponin C-derived cyanogen bromide fragment 84-135 is induced by Ca^{2+} and
stabilized by hydrophobic patches formed by residues 95, 98, 99, 101 and 102 (Nagy
et al. 1978; Reid et al. 1981; Gariepy et al. 1982).

The interaction of trifluoperazine with its proposed binding site on troponin C is
shown in the stereochemical projection presented in Fig. 1. Gariepy and Hodges (1983)
showed large chemical shifts in the ^1H NMR spectrum of the aromatic region of a
cyanogen bromide fragment 84-135 (CB9) and trifluoperazine, and in the region of the
spectrum corresponding to methyl-protein resonances of leucine and isoleucine residues. A conformational change in this fragment accompanied trifluoperazine binding,
as demonstrated by an increase in ellipticity at 222 nm. From these results it appears
that the side chains of two phenylalanine residues (99 and 102) and the methyl residues of leucine (95) and alanine (96 and 98) create a hydrophobic cavity for interaction of the drug. The involvement of this region is further supported by the demonstrated interaction between trifluoperazine and a synthetic fragment consisting of the
region 90-104. Although the binding of trifluoperazine to this peptide, or indeed to
peptide 84-135, was Ca^{2+}-independent, the affinity of interaction is comparable to
that of the Ca^{2+}-induced site in troponin C (8 μM and 5 μM, respectively) (Gariépy
and Hodges 1983), suggesting that the fragments have sufficient flexibility to allow the
requisite conformational change on trifluoperazine binding even in the absence of
Ca^{2+}.

The ionic interaction between phenothiazines and their binding site(s) on calmodulin (Prozialeck and Weiss 1982) may occur between the positive charge of the
amino group and glutamic acid rich area, containing residues 93, 94 and 97 and particularly glutamic acid residue 92 (Fig. 1). This assignment is compatible with the broad-
ening of the β-CH_2 and γ-CH_2 resonances of glutamic acid side chains, probably due
to a reduction in their mobility (Gariépy and Hodges 1983). An ionic interaction of
this type with the flexible "piperazine-arm" of trifluoperazine, some 6-10 Å away
from its aromatic region, may account for the greater ability of the piperazine containing phenothiazines to increase the α-helical content compared to other phenothiazines (e.g., chlorpromazine) which lack the piperazine ring (Gariépy and Hodges
1983) and to the increase in potency associated with the increasing length of the
amino side chain in both phenothiazines (Prozialeck and Weiss 1982) and the naph-

thalenesulfonamide (W-7) analogs (Tanaka et al. 1982). The trifluoperazine-binding site thus appears to contain hydrophobic and negatively charged regions.

The binding of a second molecule of trifluoperazine to a high affinity, Ca^{2+}-dependent site, as reported by Levin and Weiss (1977), can occur at a similar sequence of residues on another Ca^{2+}-binding domain. Evidence has been presented for such a Ca^{2+}-dependent binding in a calmodulin fragment containing Ca^{2+}-binding sites I and II (Kuznicki et al. 1981). Similar evidence is available for a Ca^{2+}-dependent site in the N-terminal region of site I in troponin C (Johnson et al. 1978). The binding of trifluoperazine at the C-terminal of site IV was considered a less probable choice (Gariépy and Hodges 1983). The use of an antibody to the COOH-terminal portion of calmodulin (Van Eldik and Watterson 1980) may be useful for elucidation at the role of the fourth Ca^{2+}-binding domain, if any, in these interactions (Klevit et al. 1981). The possibility, therefore, that phenothiazines bind to two distinct, but structurally related, sites in different domains on calmodulin has interesting implications for a cooperative mode of interaction between calmodulin and its regulators, and the activities modulated by calmodulin (Johnson 1983; Newton et al. 1983).

3 Alternative Modes of Calmodulin Antagonism

3.1. Interaction at Calmodulin Effector-Enzyme Sites

It is possible that some drugs inhibit the actions of calmodulin either partly or entirely by interacting with the effector enzyme, thereby preventing subsequent interaction or activation by calmodulin (see Roufogalis 1982). Recent experiments have provided some experimental support for this mode of action with some calmodulin antagonists. Trifluoperazine fails to interact with phosphodiesterase (Weiss et al. 1980), nor does chlorpromazine (20-200 μM) incubation with erythrocyte $(Ca^{2+} + Mg^{2+})$-ATPase in the presence of UV irradiation prevent subsequent calmodulin activation (Roufogalis et al. 1983). Thus it appear that phenothiazines do not generally inhibit calmodulin activation by interacting with the effector enzyme. This contrasts to the $(Ca^{2+} + Mg^{2+})$-ATPase in sarcoplasmic reticulum, where phenothiazines inhibit the basal enzyme activity, and lower its Ca^{2+} sensitivity, by interacting both with the enzyme protein and at the protein-lipid interphase (Rooney and Lee 1983).

By contrast, the contribution of the interaction with the effector enzyme may be more significant with other classes of calmodulin antagonists. Vinblastine (1 μM) binds to purified human erythrocyte $(Ca^{2+} + Mg^{2+})$-ATPase (0.5 mol per mol of ATPase at 1 μM free vinblastine) as well as to calmodulin $(K_d = 2 \mu M)$ — despite the fact that vinblastine does not inhibit the basal $(Ca^{2+} + Mg^{2+})$-ATPase activity at these (or higher) concentrations (Gietzen et al. 1982b). Interaction of the antagonist with the effector enzyme, blocking subsequent calmodulin activity without blocking basal enzyme activity, as shown in Scheme 1 below, appears to account exclusively for the antagonism of calmodulin activation of erythrocyte membrane $(Ca^{2+} + Mg^{2+})$-ATPase by propranolol (Meltzer and Kassir 1983). This did not appear to be a property of the local anesthetic activity of propranol, as a related β-adrenergic blocker lacking significant local anesthetic activity (Nadolol) also appeared to inhibit by a similar mechanism (Meltzer and Kassir 1983).

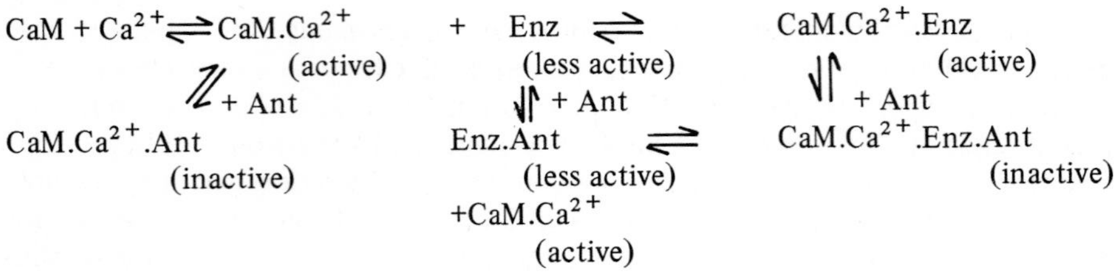

$$CaM + Ca^{2+} \rightleftharpoons CaM.Ca^{2+} \qquad + \quad Enz \rightleftharpoons \qquad CaM.Ca^{2+}.Enz$$

CaM + Ca²⁺ ⇌ CaM.Ca²⁺   +   Enz ⇌   CaM.Ca²⁺.Enz
                    (active)   (less active)   (active)
          ⇅ + Ant   ⇅ + Ant   ⇅ + Ant
CaM.Ca²⁺.Ant   Enz.Ant   ⇌   CaM.Ca²⁺.Enz.Ant
      (inactive)   (less active)   (inactive)
                                 +CaM.Ca²⁺
                                    (active)

Scheme 1

A further mode of interaction of calmodulin antagonists was revealed by the irreversible covalent agent, norchlorpromazine isothiocyanate [2-chloro-10-(3-isothiocyanatopropyl)-phenothiazine, CAPP-NCS] (Newton et al. 1983). CAPP-NCS forms a one-to-one Ca^{2+}-dependent complex with calmodulin. While this complex fails to activate phosphodiesterase, the complex does interact with the enzyme and competively inhibits its further stimulation by calmodulin. This situation can be explained by considering the two hydrophobic sites on calmodulin, one of which is free after covalent modification by CAPP-NCS to interact with phosphodiesterase, although with lower affinity and without "efficacy". The bivalent interaction could explain the enhancement of CAPP-NCS interaction with calmodulin by low concentrations of trifluoperazine, presumably acting at a second site (Newton et al. 1983).

3.2 Interaction with Activated States of Effector

While the available evidence indicates that phenothiazines and related antipsychotic agents do not interact with the unactivated state of effector enzymes, recent evidence indicates that these inhibitors directly interact with the activated state of some effector enzymes produced by limited peptide cleavage. Mild trypsin treatment activates both human erythrocyte $(Ca^{2+} + Mg^{2+})$-ATPase and brain phosphodiesterase to an extent similar to their activation by calmodulin (Cheung 1971; Wolff and Broström 1976; Taverna and Hanahan 1980; Sarkadi et al. 1980). The trypsin-activated erythrocyte $(Ca^{2+} + Mg^{2+})$-ATPase is inhibited by trifluoperazine, and related phenothiazines, at concentrations only slightly higher than those required to inhibit calmodulin activation of the enzyme (Sarkadi et al. 1982; Carafoli 1981; Roufogalis et al. 1983; Gietzen et al. 1982a). The trypsin-activated $(Ca^{2+} + Mg^{2+})$-ATPase activity was also inhibited by penfluridol and R-24571 (calmidazolium) (Gietzen et al. 1982a), but neither of the compounds inhibits the trypsin-activated state of phosphodiesterase. Activation of both $(Ca^{2+} + Mg^{2+})$-ATPase and calmodulin-deficient brain phosphodiesterase is also observed by various molecules considered to share with calmodulin an amphipathic (hydrophobic plus anionic) character (Vincenzi 1981b; Minocherhomjee et al. 1982; Gietzen et al. 1982a). These include unsaturated fatty acids (Taverna and Hanahan 1980; Wolff and Boström 1976; Al-Jobore and Roufogalis 1981), acidic phospholipids (Niggli et al. 1981; Al-Jobore and Roufogalis 1981; Wolff and Bröström 1976; Gietzen et al. 1982a) and acidic proteins or aromatic and alicyclic carboxylic and sulfonic acids (Minocherhomjee and Roufogalis 1982; Minocherhomjee et al. 1982). Activation by many of these agents is inhibited by phenothiazines and related calmodulin antagonists (Wolff and Bröström 1976; Gietzen et al. 1982a; Sarkadi et al. 1982; Vincenzi et al. 1982; Roufogalis et al. 1983). In addition, it was shown recently that phosphodiesterase is activated independently of Ca^{2+} by certain quinazolinesulfonamide derivatives, and the activation by these compounds is also inhibited by phenothiazines and napthalene-sulfonamides of the W-7 series, with an order of potency virtually indistinguishable from that for the inhibition of calmodulin activation (Tanaka et al. 1983a). A number of explanations can account for the inhibition by these antagonists of the activation induced by the diverse range of activators. These include competition between the cationic, amphipathic inhibitors and the corresponding anionic amphipathic activators (Gietzen et al. 1982a; Vincenzi 1981b), physical interactions between micellar or vesicular structures of the activator and the inhibitor (Gietzen et al. 1982a), or interaction of the inhibitors with the activated states of the effector enzymes induced by various activators, including calmodulin (Roufogalis et al. 1983). While all of these mechanisms undoubtedly contribute in different circumstances, the wide variety of activators which are antagonized by a relatively narrower spectrum of cationic amphiphiles, as discussed above, supports a mechanism whereby the inhibitors interact with an allosteric site exposed on activation of the effector enzyme. However, it is more difficult to account for why trifluoperazine antagonizes phospholipid-activated phosphodiesterase, but not the trypsin-activated enzyme (Gietzen et al. 1982a), but the finding may suggest differences in antagonist sensitivity between effector enzymes. The inhibition of the activated enzyme is included in Scheme 1, for the case of calmodulin activation.

4 Structural Specificity of Calmodulin Antagonists

4.1 Structure-Activity Relationships

Despite the early indications that antipsychotic agents showed a greater specificity for Ca^{2+}-dependent calmodulin antagonism than other psychoactive agents (Levin and Weiss 1976), it soon become clear in later work that a major determinant of this apparent specificity was the greater lipophilicity of the major tranquilizers compared to other centrally active agents or to less active analogs (see Roufogalis 1982). Thus, Norman et al. (1979) showed that the potency of a wide variety of centrally active agents, including anti-psychotic agents, could be accounted for by their relative hydrophobic

Table 1. Agents with anti-calmodulin activity

Pharmacological class	Drugs	I_{50} potency range*[a] (μM)
Phenothiazine anti-psychotics	Trifluoperazine	
	Thioridazine	6-50
	Chlorpromazine	
	Fluphenazine	
Butyrophenone anti-psychotics	Penfluridol	
	Benperidol	2.5-60
	Haloperidol	
	Spiroperidol	
Diphenylbutylamine anti-psychotics	Pimozide	0.7
Other anti-psychotics	(±)-Butaclamol	
	Clozapine	
	cis/trans-Chlorpro- thixene	2.4-19
	cis/trans-Flupen- thixol	
Muscle relaxants	W7 and analogs	23-67
Anti-depressants	Amitryptyline	
	Protriptyline	125-250
	Imipramine	
Minor tranquilizers	Diazepam	140-320
	Chlordiazepoxide	
Local anesthetics	Dibucaine	
	Tetracaine	56-310
	Mepacrine	
	Quinacrine	
	QX572	
Vinca alkaloids	Vinblastine	
	Vincrystine	
	Catharanthine	16-310
	Vindesine	
	Vindoline	
Anti-cancer agents	Adriamycin	50-85
Ca^{2+} channel blockers	Nifedipine	
	Felodipine	
	Prenylamine	8-90
	Diltiazem	
	Verapamil	
Anti-diarrheal agents	Loperamide	10-12
	Diphenyloxylate	
Insect venoms	Mellitin	0.9
Fungicides	Chloraniformethan	20-100
and herbicides	Dichlofopmethyl	
Rauwolfia alkaloids	Reserpine	50-130
Opioid peptides	β-Endorphin	3
Miscellaneous	R24571	0.35
	48/80	0.85 μg ml^{-1}

*[a] Potency represents the concentration range for antagonism of Ca^{2+}-calmodulin activation of enzymes or displacement of phenothiazine binding to calmodulin for each group of compounds

character. This could be explained by the later demonstration that the binding of these agents occurs at a hydrophobic region exposed on binding of Ca^{2+} to calmodulin (LaPorte et al. 1980; Tanaka and Hidaka 1980). Since this site is probably also the site of interaction of calmodulin with its effector enzymes, and therefore is relatively exposed (See Fig. 1), the lack of stereospecificity of the interaction (Norman et al. 1979; Raess and Vincenzi 1980; Roufogalis 1981) is not surprising. A wide range of pharmacological agents, all of which share an amphipathic (hydrophobic and cationic) character, have been shown to antagonize calmodulin-activated enzymes and cellular activities, with a potency which generally correlates with the overall hydrophobicity of the molecules. A partial list of compounds and their anti-calmodulin potencies is shown in Table 1, taken from a more detailed list in Roufogalis (1982). These include antipsychotic agents of the phenothiazine, butyrophenone, diphenylbutylamine and benzamide classes, anti-depressants (e.g., amitryptyline), muscle relaxants of the napthalenesulfonamide group (e.g., W-7), minor tranquilizers (e.g., chlordiazepoxide), local anesthetics (e.g., dibucaine), alkaloids (e.g., vinblastine, reserpine), peptides (e.g., β-endorphin) and some antimycotic agents (e.g., calmidazolinium). This list also contains some newer agents, including the anti-cancer agent, adriamycin (Katoh et al. 1981), fungicides and herbicides (Hertel and Marmé 1983), and Ca^{2+} channel blockers (Boström et al. 1981). From the discussion above, however, care should be exercised in analyzing structure activity relationships among such a diverse group of chemical structures, unless identical modes of interaction are first established, particularly for the compounds of lower potency, which may interfere with calmodulin activation by mechanisms other than (or in addition to) direct calmodulin inactivation.

Recent work has established some structural specificity of calmodulin antagonism, in addition to the contribution of hydrophobicity. It was confirmed that hydrophobicity of the aromatic ring was the major determinant of the potency of a series of phenothiazines variously substituted in the aromatic rings with chloro, thio, trifluromethyl and hydroxyl groups, in positions 2, 3, 4, 7 and 8 (Prozialeck and Weiss 1982). However, in addition, among 2-chloro-substituted phenothiazines, in which the amino substitution of the amine-containing side chain was increased or decreased by one or more methyl groups, no simple correlation was found to exist between octanol/buffer partition coefficient and calmodulin inhibitory potency (Prozialeck and Weiss 1982). Furthermore, the potency of the compounds increased as the distance between the amino group and the phenothiazine nucleus was increased from 2 to 4 atoms, although higher analogs were not tested (Prozialeck and Weiss 1982). Similar findings have been made among the 1-napthalenesulfonamide derivatives, where hydrophobicity of the aromatic ring is the major determinant of Ca^{2+}-calmodulin affinity, independently of the length of the side chain, but where increasing the length of the sulfonamido side chain from a 5-methylene bridge to a 10-methylene bridge progressively increases the potency of calmodulin interaction (Tanaka et al. 1982). The relationship of potency to hydrophobicity was less clear in the corresponding 2-substituted napthalenesulfonamido compounds. The structure activity relationships are consistent with the structure of the binding site deduced from X-ray crystal coordinates of related Ca^{2+}-binding proteins (Gariépy and Hodges 1983) (Fig. 1), which shows an interaction at both a hydrophobic area and an anionic region. The interaction of phenothiazines and a possible interaction of the napthalenesulfonamides with this site is depicted schematically in Fig. 2.

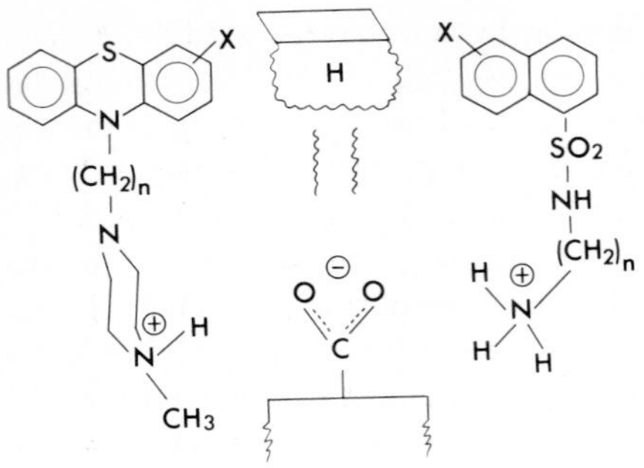

Fig. 2. Schematic representation of the calmodulin-binding site for amphipathic molecules. The binding of phenothiazine (Gariépy and Hodges 1983) and the putative binding of napthalenesulfonamides (e.g., W-7) are shown as examples. *H* hydrophobic cavity; *X* hydrophobic substituent in the aromatic ring; *n* number of methylene groups in the side chain

4.2 Relationship to Clinical Activity

4.2.1 Anti-Psychotic Activity

A number of arguments can be cited against the likelihood that anti-calmodulin potency correlates with antipsychotic activity. These include the approximately 1000-fold lower potency of the antagonism compared to other indeces of anti-psychotic activity, including dopamine receptor binding activity (Seeman 1980), lack of stereospecificity (e.g., between (+) and (−) butaclamol or cis- and trans- flupenthixol (Norman et al. 1979), lack of structural discrimination in the position of the aromatic substituent (Roufogalis 1981) compared to anti-dopamine (Roufogalis et al. 1976) or cataleptic activity in vivo (Green 1967), lack of inhibition by more polar anti-psychotic agents, including sulpiride (Norman et al. 1979) and discrepancies between relative potencies of various anti-psychotic classes (e.g., chlorpromazine vs. haloperidol) (Roufogalis 1981; Vincenzi 1981a). It thus appears that the high potency of anti-psychotic agents as calmodulin antagonists is due primarily to their higher hydrophobicity relative to other pharmacological groups of agents (Roufogalis 1975). The interaction of anti-psychotic agents with the calmodulin-binding site is depicted in Fig. 2.

4.2.2 Ca^{2+} Channel Blockers

The interesting observation that the Ca^{2+} channel blocker of the 1,4-dihydropyridine class, felodipine, interacts with calmodulin, as determined by NMR spectroscopy (Boström et al. 1981), led to the provocative possibility that calmodulin may be an intracellular site of action of the Ca^{2+} channel blockers, perhaps even accounting for their major pharmacological effects. The interaction of felodipine with calmodulin is Ca^{2+}-dependent (Johnson 1983), apparently requiring only two Ca^{2+} ions (Boström et al. 1981; Andersson et al. 1983). Other Ca^{2+} channel blockers, including diltiazem and prenylamine, also bind to calmodulin. The binding of felodipine is enhanced up to 20-fold, however, by these other Ca^{2+} channel blockers, and by calmidazolium, in a Ca^{2+}-dependent manner (Johnson 1983). These results provide evidence that various classes of calmodulin antagonists bind to distinct sites on calmodulin, and that cooper-

Table 2. Structures and potency of 1,4-dihydropyridines as calmodulin-activated phosphodiesterase inhibitors (Minocherhomjee and Roufogalis 1984)

Analog	R_1	R_2	R_3	R_4	IC_{50} (μM) (Calmodulin-activated phosphodiesterase)
Nifedipine	NO_2	H	H	CH_3	90
4-nitro Nifedipine	H	NO_2	H	CH_3	90
2-bromo Nifedipine	Br	H	H	CH_3	15
4-bromo Nifedipine	H	Br	H	CH_3	15
Felodipine	Cl	H	Cl	C_2H_5	11
(+)-Nicardipine	H	H	NO_2	X	8

$X = (CH_2)_2 \ N(CH_3)CH_2C_6H_5$

ative allosteric interactions can occur between sites. The binding of felodipine, as determined by fluorescence enhancement on interaction with dansyl-calmodulin (K_d = 2.8 μM), is similar to that of trifluoperazine (K_d = 5 μM), while verapamil and D-600 are considerably less potent (K_d = 30 μM), and diltiazem less potent again (K_d = 80 μM) (Johnson and Whitenauer 1983).

Analysis of the structural specificity of the interaction of Ca^{2+} channel blockers with calmodulin indicates, in general, a lack of correlation with the specificity observed for Ca^{2+} entry blockade (Minocherhomjee and Roufogalis 1984). The results of a structure-activity study of antagonism of calmodulin-phosphodiesterase among various 1,4-dihydropyridine Ca^{2+} channel blockers and their analogs is reproduced in Table 2. In this series, analogs containing dichloro substituents (felodipine) or bromo substituents (bromo-nifedipine) in the aromatic ring, or a large aromatic side chain in the ester group of the dihydropyridine ring (nicardipine), are six to ten times more potent than nifedipine, which has a hydrophilic NO_2 substituent in the aromatic ring. Within each group, the analogs are insensitive to the position of the aromatic substitution, whereas as Ca^{2+} entry blockers the corresponding 2-substituted compounds (R_1) are about 500-fold more potent than the 4-substituted (R_2) derivatives (see Minocherhomjee and Roufogalis 1984). Therefore, it was concluded that the interaction of 1,4-dihydropyridines occurs with relatively low affinity to structurally nonspecific sites on calmodulin, perhaps to the site depicted in Fig. 2. At the normal doses for Ca^{2+} entry blockade, it is unlikely that interactions between Ca^{2+} channel blockers and calmodulin contribute significantly to the pharmacological effects observed.

4.2.3 Anti-Cancer Activity

Microtubules are known to control various cellular functions, including cell division. The role of Ca^{2+} and calmodulin in the regulation of microtubule polymerization suggested calmodulin as a potential site of action of anti-cancer agents (Watanabe and West 1982). Vinca alkaloids, including vinblastine, were found to bind to calmodulin in a Ca^{2+}-dependent manner (Watanabe and West 1982; Gietzen et al. 1982b), at both high ($K_d = 2 \mu M$) and low ($K_d = 10 \mu M$) affinity sites (Gietzen et al. 1982b). However, the concentrations of vinca alkaloids required to block calmodulin-dependent activities is considerably higher than those required to bind to tubulin or to affect microtubular depolymerization and mitosis (see Gietzen et al. 1982b). Furthermore, in unpublished data we found that calmodulin failed to discriminate between 3,4-dehydrovinblastine and its 18-epi isomer (in which the stereochemistry between the upper and lower rings is altered) (Roufogalis 1982), whereas the 18-epi isomer has insignificant anti-tumor activity in a number of tests (J. Kutney, personal communication). This lack of stereochemical specificity suggests interaction at the hydrophobic area of the calmodulin antagonist-binding site in Fig. 2. While calmodulin antagonism could account for some of the neurotoxic properties of the vinca alkaloids, the order of potency of calmodulin antagonism (vinblastine > vincristine) (King and Boder 1979) is the reverse of the order of potency in producing peripheral neurotoxicity. It also remains to be determined if the antagonism of calmodulin activated protein kinase by adriamycin ($IC_{50} = 50\text{-}85 \mu M$) is associated with its cardiotoxicity (Katoh et al. 1981). The latter authors suggested that adriamycin may act at a hydrophobic region of calmodulin, since adriamycin also antagonizes phospholipid and Ca^{2+}-dependent protein kinase C at similar concentrations.

4.2.4 Anti-Diarrheal Activity

It has been shown that anti-calmodulin activity correlates linearly with the anti-diarrheal activity of a series of compounds, including opiates, phenothiazines, and tricyclic anti-anxiety agents (Zavecz et al. 1982). The anti-diarrheal activity was assessed by their effect on intestinal fluid secretion induced by 16,16-dimethyl prostaglandin E_2 and/or castor oil. The anti-diarrheal opiates, loperamide and diphenyloxylate, were the most potent agents, followed by chlorpromazine, promethazine, and amitryptyline. However, examination of the structure of loperamide and diphenyloxylate indicates a significant hydrophobic character, with three aromatic rings, one of which is chloro-substituted in the case of loperamide, and an ionizable amino group three atoms removed from the aromatic centers, structural features which are compatible with binding to the site depicted in Fig. 2. The less hydrophobic opiate anti-diarrheal agent, morphine, does not interact with calmodulin (Levin and Weiss 1979), raising the possibility that the in vivo anti-diarrheal potency of the compounds tested is also dependent on their hydrophobicity.

4.2.5 Opioid and Other Peptides

The neuropeptides, β-endorphin and dynorphin, inhibit activation of phosphodiesterase by calmodulin and bind to calmodulin in a Ca^{2+}-dependent manner (K_d = 4.6 μM) (Sellinger-Barnette and Weiss 1982). Cross-linking studies indicate that a maximum of two β-endorphin molecules bind to calmodulin in a Ca^{2+}-dependent manner (Giedroc et al. 1983). While β-endorphin is a basic peptide, not all basic peptides interact with calmodulin (see Sellinger-Barnette and Weiss 1982) and the binding is not greatly affected by ionic strength, suggesting that mainly hydrophobic interaction is involved in this linkage (Giedroc et al. 1983). This was supported by the strong cross-linking observed between calmodulin and the hydrophobic sequence of residues 14-23 of β-endorphin, which was shown to correspond to the region of the peptide responsible for inhibition of calmodulin-activated phosphodiesterase (Giedroc et al. 1983). The hydrophobic peptide, which contains seven nonpolar residues on the surface of a helical region, may interact with the hydrophobic region of the calmodulin-binding site depicted in Fig. 2, since β-endorphin interaction is antagonized by chlorpromazine (Sellinger-Barnette and Weiss 1982) and trifluoperazine (Giedroc et al. 1983) in a Ca^{2+}-dependent manner. The pharmacological significance of the interaction of β-endorphin with calmodulin remains to be determined, since the concentration of β-endorphin required to inhibit calmodulin is in the micromolar range, far greater than the concentration found in most brain areas (Rossier and Bloom 1979). Unlike many of the pharmacological actions produced by higher concentrations of β-endorphin the calmodulin antagonism is not blocked by naloxone, an opiate receptor antagonist (Sellinger-Barnette and Weiss 1982).

A number of other peptides, particularly those found in insect venom, antagonise calmodulin actions. Most of these, including mellitin, apear to interact as a result of their basic, positively charged groups, since acetylation markedly reduced, although it did not completely eliminate, their inhibitory actions (Barnette et al. 1983). The interaction probably occurs at sites other than the phenothiazine-binding site, and the relationship of the calmodulin antagonism to the toxicological actions of the venoms is unknown at present (Barnette et al. 1983).

4.2.6 Miscellaneous Antagonists

One of the major difficulties with the use of most, if not all, of the previously mentioned antagonists, with the possible exception of the peptides, is the inhibition of the basal, as well as the calmodulin-activated, states of the effector enzymes, although this usually occurs at somewhat higher inhibitor concentration (Hinds et al. 1981). Two recently investigated calmodulin antagonists are reported to be more selective for calmodulin than other agents examined. The antimycotic agent, R-24571, recently named calmidazolium, is reported to be the most potent inhibitor examined (Gietzen et al. 1981), with an IC_{50} = 0.35 μM. The high potency appears to be related to its hydrophobicity, as judged by the number of chlorosubstituted aromatic rings in the structure (see Roufogalis 1982). Compound 48/80, consisting of a group of cationic amphiphilic compounds of various polymerization products of N-methy-p-methoxyphenyl-

ethylamine and formaldehyde, was recently shown to be a potent calmodulin antagonist (Gietzen 1983; Gietzen et al. 1983). The latter authors propose that compound
48/80 is from 20 to 300 times more selective for calmodulin than other antagonists.
The selectivity is greater than that of calmidazolium, except in the case of the lipid
activated Ca^{2+}-transport ATPase. It was proposed that the high selectivity is determined by the polymeric structure of 48/80 (average M_r approx. 1000 (Gietzen 1983),
but the general usefulness of the inhibitor complex in other than in vitro systems remains to be assessed.

5 Conclusions and Perspective

In this discussion experimental evidence concerning the specificity of agents which inhibit calmodulin has been reviewed. A great variety of agents share the ability to inhibit calmodulin actions in vitro. Analysis of the structural requirements for this inhibition indicates that virtually all antagonists are amphipathic molecules, with both hydrophobic and cationic character. The antagonism of calmodulin is Ca^{2+}-dependent,
and as a first approximation, the potency of antagonism increases with increasing hydrophobic character of the antagonist. These facts suggest that the antagonism of calmodulin occurs at one or more hydrophobic areas exposed on binding of Ca^{2+} to
calmodulin. A binding site consisting of hydrophobic amino acid residues and negatively charged glutamic acid residues has been identified in troponin C by NMR and
circular dichroism analysis, and it was shown to be capable of accommodating the
amphipathic phenothiazine molecule. Based on this evidence, and recent structure-activity analysis of calmodulin antagonist potency, a minimal binding site topography
has been proposed, which includes hydrophobic and anionic binding residues.
   While the binding of antagonists to calmodulin is indisputable and has provided important tools for the study of Ca^{2+}-calmodulin-dependent actions, the specificity of
the commonly used calmodulin antagonists is clearly inadequate and limits the usefulness of the compounds as pharmacological antagonists. This problem is compounded
markedly in in vivo or in ex vivo cellular systems. Even for calmodulin-dependent
enzyme activation, it can be shown that antagonists may interfere with various steps in
the calmodulin-activation process, in addition to a direct effect of the antagonists on
calmodulin. For instance, many calmodulin-dependent enzymes are also activated by
acidic phospholipids, fatty acids and by proteolysis, and while the physiological significance of these activities is still largely unknown, various "calmodulin antagonists"
have been shown to antagonize these states of the enzymes, to various degrees. Furthermore, the activity of "calmodulin antagonists" is by no means restricted to calmodulin-dependent enzyme systems. These agents also inhibit other Ca^{2+}-dependent
enzymes, including the widely distributed Ca^{2+}-dependent protein kinase C, which is
thought to regulate a variety of cellular processes (Takai et al. 1979, Kuo et al. 1980),
many of which are still to be understood. The order of potency of the commonly used
antagonists is similar, although not identical, toward both calmodulin and protein
kinase C activities (Schatzman et al. 1981). Compounds considered to be among the

more selective of the calmodulin antagonists, including W-7 and calmidazolium, have now been shown to interact with Ca^{2+}-dependent protein kinase C (Schatzman et al. 1983; Wise and Kuo 1983). Furthermore, phenothiazines have been shown to interact with a wide variety of Ca^{2+} binding proteins (Moore and Dedman 1982; Head et al. 1983; Nagao et al. 1981; Levin and Weiss 1978; Van Eldik et al. 1980; McManus 1981). Phenothiazines also influence cellular Ca^{2+} levels by displacement of Ca^{2+} from membrane sites (Seeman 1972) and by blockade of both passive (Landry et al. 1981) and active (Raess and Vincenzi 1980; Levin and Weiss 1980) Ca^{2+} transport processes, including noncalmodulin regulated Ca^{2+} transport by sarcoplasmic reticulum (Chiesi and Carafoli 1982; Ho et al. 1983). These drugs, being amphipathic, can also directly affect cell shape and stability (Bereza et al. 1982; Seeman 1972; Sheetz and Singer 1974). Amphipathic drugs, including chlorpromazine and trifluoperazine, have both stimulatory and inhibitory effects on phospholipid synthesis (Michell 1975; Pelech et al. 1983). The selectivity of "calmodulin antagonists" thus is clearly dependent on dose, and will vary with the complexity of the in vitro system under examination. Effects obtained with 100 μM trifluoperazine, for example, will be extremely difficult to interpret. It should be kept in mind, however, that many of the diverse effects of the phenothiazines and related agents occur over overlapping dose ranges.

In intact cells the problem of specificity is compounded by the factors that control the distribution of the amphipathic molecules from the extracellular to the intracellular phase. Penetration of molecules across the membrane is determined by hydrophobicity and charge, precisely the structural determinants of calmodulin antagonist potency, making correlations between anti-calmodulin potency and changes in cellular activity difficult to interpret. The intracellular distribution of amphipathic molecules is also dependent on the membrane potential (Pieter Cullis, personal communication) and on phospholipid asymmetry (Sheetz and Singer 1974), since the cationic drugs will preferentially distribute at anionic, hydrophobic surfaces. Innervated cells, regulated by neurotransmitters via their actions on receptors generally located at the external cell surface, will potentially be inhibited by many of the agents used as calmodulin antagonists. Many of the "calmodulin antagonists" have even greater potency as receptor antagonists, including dopaminergic, α-adrenergic, serotonergic and muscarinic cholinergic receptors (see Roufogalis 1982). Many of the phenothiazines affect uptake and release of neurotransmitters from nerve endings at concentrations similar to or lower than those used for calmodulin antagonism (see Roufogalis 1982).

A few striking examples will serve to illustrate the limitations of the use of neuroleptic drugs as calmodulin antagonists in vivo. Incubation of C-6 glioma cells with a number of neuroleptics resulted in activation of phosphorylase B to A conversion in the absence of Ca^{2+} as well as inhibition of phosphorylase B to A conversion in the presence of the Ca^{2+} ionophore A23187 (Norman and Staehelin 1982). The activating effect was nonstereospecific and occurred only with those compounds known to be relatively potent antagonists of calmodulin, leading the authors to conclude that the activation could have been due to Ca^{2+} release, either from the membrane, by inhibition of mitochondrial Ca^{2+} uptake or by inhibition of the calmodulin-dependent Ca^{2+} pump in the plasma membrane. The drugs *decreased* the levels of cyclic AMP in the C-6 cells, possibly by inhibiting the membrane-associated Ca^{2+}-calmodulin dependent adenylate cyclase activity (Norman and Staehelin 1982) without significantly affecting the calmodulin-dependent cAMP phosphodiesterase activity. Phenothiazines may also affect the viability of cultured cells, as recently discussed by Norman and Staehelin (1982). Another limitation in the use of phenothiazines and local anesthetics in intact cells was described by Corps et al. (1982), who showed that a variety of commonly used "calmodulin antagonists", including trifluoperazine, chlorpromazine, dibucaine, and lignocaine, deplete ATP levels and increase lactate output in rat and mouse spleen cells and mouse thymocytes, at concentrations similar to those used for calmodulin antagonism. Thus, the resulting decrease in phosphorylation potential may interfere with cellular energy metabolism and cellular functions regulated by ATP (Corps et al. 1982). Although the magnitude of these effects may vary between cells and under different conditions (see Corps et al. 1982), these findings demonstrate the need for performing appropriate metabolic controls in experiments with whole cells.

The importance of dose on the pharmacological selectivity of antagonism of Ca^{2+}-dependent versus Ca^{2+}-independent events is illustrated in a study of amylase secretion from pancreatic acinar cells (Ansah and Katz 1983). Ca^{2+}-dependent secretion induced by carbachol or pancreozymin is inhibited about 50% by 10 μM chlorpromazine, whereas cyclic AMP-induced release induced by secretin is unaffected by this concentration. By contrast, release by all three secretagogs is inhibited equally by 100 μM chlorpromazine, and propranolol (10^{-7}-10^{-4} M) failed to distinguish between the secretagogs at any concentration examined. Thus, the specificity depends on both the dose and the nature of the "calmodulin antagonist" used.

In none of the groups of antagonists examined was it possible to establish a correlation between the antagonism of calmodulin and the clinical effects of the drugs examined. This included anti-psychotic activity, Ca^{2+} channel blockade, anti-cancer activity, anti-diarrheal activity, and opioid activity of peptides. A possible exception may be the neurotoxic activity of insect venom peptides, but this remains to be established. Clearly, future directions of research must be aimed at the development of specific antagonists of calmodulin, that is, drugs which specifically antagonize calmodulin at concentrations which do not effect other physiological or metabolic activities. In view of the widespread occurrence and involvement of calmodulin in Ca^{2+}-dependent cellular activities, further research should also consider the strategy by which specific calmodulin antagonists can be developed as potentially useful therapeutic agents. This possibilty will require parallel advances in the understanding of subtle differences between the modes of calmodulin activation of various calmodulin-dependent activities. Only then might it be possible to move in the direction of the development of the type of calmodulin antagonists expressed by Vincenzi (1981a), whereby antagonists can be targeted toward specific effectors, not only at the level of calmodulin, but at the sites of the calmodulin-effector interactions. It remains to be determined if specificity in new drugs can be achieved by directing them at the amphipathic binding site discussed in this review or will require the designing of new molecules directed toward other amino acid sequences on calmodulin as yet to be defined.

In the meantime, a number of criteria should be met before effects of presently available drugs can be attributed to calmodulin antagonism. The following is a list expanded from those suggested previously (Roufogalis 1982).

1. The inhibition should occur at appropriate doses for calmodulin antagonism (e.g., 5-30 μM trifluoperazine).

2. The order of inhibitory potency in a series of structurally unrelated antagonists should parallel their anticalmodulin potency. A possible series is calmidazolium > pimozide > trifluoperazine > chlorpromazine > dibucaine >>> sulpiride.

3. The inhibition should be nonstereospecific. For example, there should be no difference in potency between cis- and trans-flupenthixol, cis- and trans-chlorprothixene or 1,2,3, or 4-chloro analogs of chlorpromazine.

4. The inhibition should be Ca^{2+}-dependent. For example, the effect should occur after intracellular Ca^{2+} is increased by ionophore A-23187, but generally not in the absence of Ca^{2+} under resting conditions.

5. The inhibition should occur without general metabolic effects.

6. Inhibition at other possible sites of action should be eliminated using other agents. These may include specific receptor blockade and blockade of neurotransmitter or Ca^{2+} uptake or release.

Acknowledgements. Original research from the author's laboratory was supported by the Medical Research Council of Canada and the Canadian Heart Foundation. The generosity of Dr. R. Hodges, University of Alberta, in providing the stereoscopic drawing of trifluoperazine binding to troponin C (Fig. 1) before publication is greatly appreciated.

References

Akera T, Brody TM (1969) The interaction between chlorpromazine free radical and microsomal sodium- and potassium-activated adenosine triphosphatase from rat brain. Mol Pharmacol 5:605-614

Al-Jobore A, Roufogalis BD (1981) Phospholipid and calmodulin activation of solubilized calcium-transport ATPase from human erythrocytes: regulation by magnesium. Can J Biochem 59:880-888

Anderson A, Drakenberg T, Thulin E, Forsen S (1983) A [113]Cd and ^{1}H NMR study of the interaction of calmodulin with D600, trifluoperazine and some other hydrophobic drugs. Eur J Biochem 134:459-465

Ansah T-A, Katz S (1983) The role of calmodulin in stimulus secretion coupling in the exocrine pancreas. Proc West Pharmacol Soc 26:131-134

Barnette MS, Daly R, Weiss B (1983) Inhibition of calmodulin activity by insect venom peptides. Biochem Pharmacol 32:2929-2933

Bereza UL, Brewer GJ, Mizukami I (1982) Association of calmodulin inhibition, erythrocyte membrane stabilization and pharmacological effects of drugs. Biochim Biophys Acta 692:305-314

Boström S-L, Jung B, Mardh S, Forsen S, Thulin E (1981) Interaction of the antihypertensive drug felodipine with calmodulin. Nature (Lond) 292:777-778

Cachia PJ, Hodges RS (1984) In: Hidaka H and Hartshorne DJ (eds) Calmodulin antagonists and cellular physiology. Academic, New York (in press)

Carafoli E (1981) Calmodulin in the membrane transport of Ca^{2+}. Cell Calcium 2:353-363

Cheung WY (1971) Cyclic 3',5'-nucleotide phosphodiesterase — evidence for and properties of a protein activator. J Biol Chem 246:2859-2869

Chiesi M, Carafoli E (1982) The regulation of Ca^{2+} transport by fast skeletal muscle sarcoplasmic reticulum. J Biol Chem 257:984-991

Corps AN, Hesketh TR, Metcalfe JC (1982) Limitations of the use of phenothiazines and local anaesthetics as indicators of calmodulin function in intact cells. FEBS Lett 138:280-284

Gariépy J, Hodges RS (1983) Localization of a trifluoperazine-binding site on troponin C. Biochemistry 22:1586-1594

Gariépy J, Sykes BD, Reid RE, Hodges RS (1982) Proton nuclear magnetic-resonance investigation of synthetic calcium-binding peptides. Biochemistry 21:1506-1512

Giedroc DP, Puett D, Ling N, Staros JV (1983) Demonstration by covalent cross-linking of a specific interaction between β-endorphin and calmodulin. J Biol Chem 258:16-19

Gietzen K (1983) Comparison of the calmodulin antagonist compound 48/80 and calmidazolium. Biochem J 216:611-616

Gietzen K, Wüthrich A, Bader H (1981) R-24571: a new powerful inhibitor of red blood cell Ca^{2+}-transport ATPase and of calmodulin-regulated functions. Biochem Biophys Res Commun 101:418-425

Gietzen K, Sadorf I, Bader H (1982a) A model for the regulation of the calmodulin dependent enzymes erythrocyte Ca^{2+}-transport ATPase and brain phosphodiesterase by activators and inhibitors. Biochem J 207:541-548

Gietzen K, Wüthrich A, Bader H (1982b) Effects of microtubular inhibitors on plasma membrane dependent Ca^{2+}-transport ATPase. Mol Pharmacol 22:413-420

Gietzen K, Adamczyk-Engelmann P, Wüthrich A, Konstantinova A, Bader H (1983) Compound 48/80 is a selective and powerful inhibitor of calmodulin-regulated functions. Biochim Biophys Acta 736:109-118

Grabarek Z, Drabikowski W, Leavis PC, Rosenfeld SS, Gergely J (1981) Proteolytic fragments of troponin-C-interactions with the other troponin subunits and biological activity. J Biol Chem 256:3121-3127

Green AL (1967) Activity correlations and the mode of action of aminoalkylphenothiazine tranquillizers. J Pharm Pharmacol 19:207-208

Head JF, Masure HR, Kaminer B (1982) Identification and purification of a phenothiazine-binding fragment from bovine brain calmodulin. FEBS Lett 137:71-74

Head JF, Spielberg S, Kaminer B (1983) Two low-molecular-weight Ca^{2+}-binding proteins isolated from squid optic lobe by phenothiazine-Sepharose affinity chromatography. Biochem J 209: 797-802

Hertel H, Marmé D (1983) The fungicide chloraniformethan and the herbicide dichlofopmethyl affect calmodulin-dependent cyclic nucleotide phosphodiesterase from bovine brain. FEBS Lett 152:44-46

Hidaka H, Yamaki T, Naka M, Tanakat, Hayashi H, Kobayashi R (1980) Calcium-regulated modulator protein interacting agents inhibit smooth muscle calcium-stimulated protein kinase and ATPase. Mol Pharmacol 17:66-72

Hinds TR, Raess BU, Vincenzi FF (1981) Plasma membrane Ca^{2+} transport: antagonism by several potential inhibitors. J Membr Biol 58:57-65

Ho M-M, Scales DJ, Inesi G (1983) The effect of trifluoperazine on the sarcoplasmic reticulum membrane. Biochim Biophys Acta 730:64-70

Inagaki M, Naka M, Nozawa Y, Hidaka H (1983) Hydrophobic properties of tetrahymena calmodulin related to the phosphodiesterase activity. FEBS Lett 151:67-70

Johnson JD (1983) Allosteric interactions among drug binding sites on calmodulin. Biochem Biophys Res Commun 112:787-793

Johnson JD, Whitenauer LA (1983) A fluorescent calmodulin that reports the binding of hydrophobic inhibitory ligands. Biochem J 211:473-479

Johnson JD, Collins JH, Potter JD (1978) Dansylaziridine-labeled troponin-C. Fluorescent probe of Ca^{2+}-binding to Ca^{2+}-specific regulatory sites. J Biol Chem 253:6451-6458

Katoh N, Wise BC, Wrenn RW, Kuo JF (1981) Inhibition by adriamycin of calmodulin-sensitive and phospholipid-sensitive calcium-dependent phosphorylation of endogenous proteins from heart. Biochem J 198:199-205

King KL, Boder GB (1979) Correlation of the clinical neurotoxicity of the vinca alkaloids vincristine, vinblastine, and vindesine with their effects on cultured rat midbrain cells. Cancer Chemother Pharmacol 2:239-242

Klevit RE, Levine BA, Williams RJP (1981) A study of calmodulin and its interaction with trifluoperazine by high resolution ^{1}H NMR spectroscopy. FEBS Lett 123:25-29

Kretsinger RH, Nockolds CE (1973) Carp muscle calcium-binding protein 2. Structure determination and general description. J Biol Chem 248:3313-3326

Kuo JF, Andersson RGG, Wise BC, Mackerloua L, Salomonsson I, Brackett NL, Katoh N, Shofi M,

Wrenn RW (1980) Calcium-dependent protein kinase: widespread occurrence in various tissues and phyla of the animal kingdom and comparison of effects of phospholipid, calmodulin and trifluoperazine. Proc Natl Acad Sci USA 77:7039-7043

Kuznicki J, Grabarek Z, Brzeska H, Drabikowski W, Cohen P (1981) Stimulation of enzyme activities by fragments of calmodulin. FEBS Lett 130:141-145

Landry Y, Amellal M, Ruckstuhl M (1981) Can calmodulin inhibitors be used to probe calmodulin effects? Biochem Pharmacol 30:2031-2032

LaPorte DC, Wierman BM, Storm DR (1980) Calcium induced exposure of a hydrophobic surface on calmodulin. Biochemistry 19:3814-3819

Levin RM, Weiss B (1976) Mechanism by which psychotropic drugs inhibit adenosine cyclic 3',5'-monophosphate phosphodiesterase of brain. Mol Pharmacol 12:581-589

Levin RM, Weiss B (1977) Binding of trifluoperazine to the calcium-dependent activator of cyclic nucleotide phosphodiesterase. Mol Pharmacol 13:690-697

Levin RM, Weiss B (1978) Specificity of the binding of trifluoperazine to the calcium-dependent activator of phosphodiesterase and to a series of other calcium-binding proteins. Biochim Biophys Acta 540:197-204

Levin RM, Weiss B (1979) Selective binding of antipsychotics and other psychoactive agents to the calcium-dependent activator of cyclic nucleotide phosphodiesterase. J Pharmacol Exp Ther 208:454-459

Levin RM, Weiss B (1980) Inhibition by trifluoperazine of calmodulin-induced activation of ATPase activity of rat erythrocyte. Neuropharmacology 19:169-174

McDowell JJH (1980) Trifluoperazine hydrochloride, a phenothiazine derivative. Acta Crystallogr Sect B 36:2178-2181

McManus JP (1981) The stimulation of cyclic nucleotide phosphodiesterase by a M_r 11500 calcium-binding protein from hepatoma. FEBS Lett 126:245-249

Marshak DR, Watterson DM, Van Eldik LJ (1981) Calcium-dependent interaction of S100B, troponin-C and calmodulin with an immobilized phenothiazine. Proc Natl Acad Sci USA 78:6793-6797

Meltzer HI, Kassir S (1983) Inhibition of calmodulin-activated Ca^{2+}-ATPase by propranolol and nadolol. Biochim Biophys Acta 755:452-456

Michell RH (1975) Inositol phospholipids and cell surface receptor function. Biochim Biophys Acta 415:81-147

Minocherhomjee AM, Roufogalis BD (1982) Activation of erythrocyte Ca^{2+} and Mg^{2+}-stimulated ATPase by cyclic AMP protein kinase inhibitor: comparison with calmodulin. Biochem J 206:517-525

Minocherhomjee AM, Roufogalis BD (1984) Antagonism of calmodulin and phosphodiesterase by nifedipine and related calcium entry blockers. Cell Calcium 5:57-64

Minocherhomjee AM, Al-Jobore A, Roufogalis BD (1982) Modulation of the calcium-transport ATPase in human erythrocytes by anions. Biochim Biophys Acta 690:8-14

Moore PB, Dedman JR (1982) Calcium-dependent protein binding to phenothiazine columns. J Biol Chem 257:9663-9667

Nagao S, Kudo S, Nozawa Y (1981) Effects of phenothiazines on the membrane-bound guanylate and adenylate cyclase in tetrahymena pyriformis. Biochem Pharmacol 30:2709-2712

Nagy B, Potter JD, Gergely J (1978) Calcium-induced conformational changes in a cyanogen-bromide fragment of troponin-C that contains one of the binding sites. J Biol Chem 253:5971-5974

Newton DL, Burke TR, Rice KC, Klee CB B (1983) Calcium ion-dependent covalent modification of calmodulin with norchlorpromazine isothiocyanate. Biochemistry 22:5412-5416

Niggli V, Adunyah ES, Carafoli E (1981) Acidic phospholipids, unsaturated fatty acids and limited proteolysis mimic the effect of calmodulin on the purified erythrocyte Ca^{2+}-ATPase. J Biol Chem 256:8588-8592

Norman JA, Staehelin M (1982) Calmodulin inhibitors activate glycogen phosphorylase B to A conversion in C6 glioma cells. Mol Pharmacol 22:395-402

Norman JA, Drummond AH, Moser P (1979) Inhibition of calcium-dependent regulator-stimulated phosphodiesterase activity by neuroleptic drugs is unrelated to their clinical efficacy. Mol Pharmacol 16:1089-1094

Pelech SL, Jetha F, Vance DE (1983) Trifluoperazine and other anaesthetics inhibit rat liver CTP: phosphocholine cytidylyltransferase. FEBS Lett 158:89-92

Prozialeck WC, Weiss B (1982) Inhibition of calmodulin by phenothiazines and related drugs: structure-activity relationships. J Pharmacol Exp Ther 222:509-516

Prozialeck WC, Cimino M, Weiss B (1981) Photoaffinity labeling of calmodulin by phenothiazine antipsychotics. Mol Pharmacol 19:264-269

Raess BU, Vincenzi FF (1980) Calmodulin activation of red blood cell $(Ca^{2+} + Mg^{2+})$-ATPase and its antagonism by phenothiazines. Mol Pharmacol 18:253-258

Reid RE, Gariépy J, Saund AK, Hodges RS (1981) Calcium-induced protein folding-structure affinity relationships in synthetic analogs of the helix-loop-helix calcium-binding unit. J Biol Chem 256:2742-2751

Rooney EK, Lee AG (1983) Binding of hydrophobic drugs to lipid bilayers and to the $(Ca^{2+} + Mg^{2+})$-ATPase. Biochim Biophys Acta 732:428-440

Rossier J, Bloom F (1979) Central pharmacology of endorphins. Adv Biochem Psychopharmacol 20:165-185

Roufogalis BD (1975) Comparative studies on the membrane actions of depressant drugs: the role of lipophilicity in inhibition of brain sodium and potassium-stimulated ATPase. J Neurochem 24:51-61

Roufogalis BD (1981) Phenothiazine antagonism of calmodulin: a structurally nonspecific interaction. Biochem Biophys Res Commun 98:607-613

Roufogalis BD (1982) Specificity of trifluoperazine and related phenothiazines for calcium-binding proteins. In: Cheung WY (ed) Calcium cell function, vol 3. Academic Press, New York, pp 129-159

Roufogalis BD, Thornton M, Wade DN (1976) Specificity of the dopamine sensitive adenylate cyclase for antipsychotic antagonists. Life Sci 19:927-934

Roufogalis BD, Minocherhomjee AM, Al-Jobore A (1983) Pharmacological antagonism of calmodulin. Can J Biochem Cell Biol 61:927-933

Sarkadi B, Enyedi Á, Gárdós G (1980) Molecular properties of the red cell calcium pump. 1. Effects of calmodulin, proteolytic digestion and drugs on the kinetics of active calcium uptake in inside-out red cell membrane vesicles. Cell Calcium 1:287-297

Sarkadi B, Enyedi A, Nyers A, Gardos G (1982) The function and regulation of the calcium pump in the erythrocyte membrane. Ann NY Acad Sci 402:329-348

Seeman P (1972) The membrane actions of anesthetics and tranquilizers. Pharmacol Rev 24:583-655

Seeman P (1980) Brain dopamine receptors. Pharmacol Rev 32:229-313

Sellinger-Barnette M, Weiss B (1982) Interaction of β-endorphin and other opioid peptides with calmodulin. Mol Pharmacol 21:86-91

Schatzman RC, Wise BC, Kuo JF (1981) Phospholipid-sensitive calcium-dependent protein kinase: inhibition by antipsychotic drugs. Biochem Biophys Res Commun 98:669-676

Schatzman RC, Raynor RL, Kuo JF (1983) N-(6-aminohexyl)-5-chloro-1-napthalene-sulfonamide (W-7), a calmodulin antagonist, also inhibits phospholipid-sensitive calcium-dependent protein kinase. Biochim Biophys Acta 755:144-147

Sheetz MP, Singer SJ (1974) Biological membranes as bilayer couples. A molecular mechanism of drug-erythrocyte interactions. Proc Natl Acad Sci USA 71:4457-4461

Takai Y, Kishimoto A, Iwasa Y, Kawahara Y, Mori T, Nishizuka Y (1979) Calcium-dependent activation of a multifunctional protein kinase by membrane phospholipids. J Biol Chem 254:3692-3695

Tanaka T, Hidaka H (1980) Hydrophobic regions function in calmodulin-enzyme(s) interaction. J Biol Chem 255:11078-11080

Tanaka T, Hidaka H (1981) Hydrophobic regions of calcium-binding proteins exposed by calcium. Biochem Int 2:71-75

Tanaka T, Ohmura T, Hidaka H (1982) Hydrophobic interaction of the Ca^{2+}-calmodulin complex with calmodulin antagonists. Mol Pharmacol 22:403-407

Tanaka T, Yamada E, Sone T, Hidaka H (1983a) Calcium-independent activation of calcium ion-dependent cyclic nucleotide phosphodiesterase by synthetic compounds: Quinazolinesulfonamide derivatives. Biochemistry 22:1030-1034

Tanaka T, Ohmura T, Hidaka H (1983b) Calmodulin antagonists' binding sites on calmodulin. Pharmacology (Basel) 26:249-257

Taverna DR, Hanahan JD (1980) Modulation of human erythrocyte Ca^{2+}/Mg^{2+} ATPase activity by phospholipase A_2 and proteases. A comparison with calmodulin. Biochem Biophys Res Commun 94:652-659

Van Eldik LJ, Watterson DM (1980) Reproducible production and characterization of anti-calmodulin anti-sera. Ann NY Acad Sci USA 356:437-438

Van Eldik LJ, Piperno G, Watterson DM (1980) Similarities and dissimilarities between calmodulin and a *Chlamydomonas* flagellar protein. Proc Natl Acad Sci USA 77:4779-4783

Vincenzi FF (1981a) Calmodulin pharmacology. Cell Calcium 2:387-409

Vincenzi FF (1981b) Pharmacological differentiation of calmodulin receptors? Proc West Pharmacol Soc 24:193-196

Vincenzi FF, Adunyah ES, Niggli V, Carafoli E (1982) Purified red blood cell Ca^{2+}-pump ATPase: evidence for direct inhibition by presumed anti-calmodulin drugs in the absence of calmodulin. Cell Calcium 3:545-559

Walsh M, Stevens FC, Oikawa K, Kay CM (1978) Chemical modification studies on Ca^{2+}-dependent protein modulator-role of methionine residues in activation of cyclic nucleotide phosphodiesterase. Biochemistry 17:3924-3930

Watanabe K, West WL (1982) Calmodulin, activated cyclic nucleotide phosphodiesterase, microtubules, and vinca alkaloids. Fed Proc 41:2292-2299

Weeks RA, Perry SV (1978) Characterization of a region of primary sequence of troponin-C involved in calcium ion-dependent interactions with Troponin-I. Biochem J 173:449-457

Weiss B, Levin RM (1978) Mechanism for selectively inhibiting the activation of cyclic nucleotide phosphodiesterase and adenylate cyclase by antipsychotic agents. Adv Cyclic Nucleotide Res 9:285-303

Weiss B, Fertel R, Giglin R, Uzunov P (1974) Selective alteration of the activity of the multiple forms of adenosine 3',5'-monophosphate phosphodiesterase of rat cerebrum. Mol Pharmacol 10:615-625

Weiss B, Prozialeck W, Cimino M, Barnette MS, Wallace TL (1980) Pharmacological regulation of calmodulin. Ann NY Acad Sci USA 356:319-345

Wise BC, Kuo JF (1983) Modes of inhibition by acylcarnitines, adriamycin and trifluoperazine of cardiac phospholipid-sensitive calcium-dependent protein kinase. Biochem Pharmacol 32:1259-1265

Wolff DJ, Broström CO (1976) Calcium-dependent cyclic nucleotide phosphodiesterase from brain: identification of phospholipids as calcium-independent activators. Arch Biochem Biophys 173:720-731

Zavecz JH, Jackson TE, Limp GL, Yellin TO (1982) Relationship between anti-diarrheal activity and binding to calmodulin. Eur J Pharmacol 78:375-377

Calcium Regulation of Smooth Muscle Contraction

M. P. WALSH[1]

CONTENTS

Abbreviations: ATPγS, adenosine 5'-0(3-thiotriphosphate); CaD, caldesmon; CaM, calmodulin; cAMP, cyclic adenosine 3':5'-monophosphate; EGTA, ethylene glycol bis(β-aminoethylether)-N,N, N',N'-tetraacetic acid; GTPγS, guanosine 5'-0-(3-thiotriphosphate); ITPγS, inosine 5'-0-(3-thiotriphosphate); LC_{20}, M_r 20,000 light chain of smooth muscle myosin; MLCK, myosin light chain kinase; MLCP, myosin light chain phosphatase; SDS, sodium dodecyl sulfate.

[1] Department of Medical Biochemistry, Faculty of Medicine, University of Calgary, Calgary, Alberta T2N 1N4, Canada

1 Introduction

In recent years, our knowledge of the ultrastructure of smooth muscle and of the mechanisms of regulation of smooth muscle contraction has undergone exponential growth. This is due primarily to differences from the striated muscle systems, differences which have attracted widespread interest and attention. This chapter is concerned with the mechanism whereby Ca^{2+} ions control the contraction of smooth muscle. Most attention will be focused on the myosin phosphorylation theory which is widely believed to represent the Ca^{2+} regulatory system responsible at least for switching on the contractile process. Other Ca^{2+}-dependent mechanisms which may play a role in regulating smooth muscle contraction will also be discussed. Finally, some recent interesting observations relating, at least potentially, to the control of smooth muscle contraction will be summarized and future prospects analyzed.

2 The Contractile Apparatus

The contractile machinery of smooth muscle consists of myosin, actin, tropomyosin, and their associated regulatory proteins. The properties of the contractile proteins of smooth and skeletal muscle have been compared in some detail recently (Walsh and Hartshorne 1982). As in the case of striated muscles, the major contractile proteins are present in filamentous form: myosin forms the thick filaments and actin the thin filaments. Tropomyosin is located on the thin filaments. The location of regulatory protein components will be discussed later.

It is now believed, largely on the basis of ultrastructural studies of a number of different smooth muscles (see Murphy 1979; Somlyo 1980; Small and Sobieszek 1980 for reviews), that the mechanism of contraction of smooth muscle is fundamentally similar to the sliding filament-cross-bridge cycling model of skeletal muscle contraction (Huxley and Niedergerke 1954; Huxley and Hanson 1954). Cycles of attachment and detachment of myosin cross-bridges to actin filaments occur at the expense of ATP hydrolysis, the site of ATPase activity being located in the globular head region of the myosin molecule. Measurement of the actin-activated myosin Mg^{2+}-ATPase activity is widely used as a quantitative in vitro correlate of muscle contraction. Another frequently measured parameter in vitro is superprecipitation, the increased turbidity of actomyosin upon activation. Superprecipitation is commonly measured by the increase in absorbance at 660 nm and again is believed to be an in vitro measure of muscle contraction.

Electron microscopic studies of smooth muscles from which the soluble proteins have been removed by saponin treatment have demonstrated clearly that the thin filaments are attached at their ends to structures called dense bodies, which are distributed throughout the smooth muscle cell, some being cytoplasmic and others attached to the cell membrane (Somlyo 1980). Thick filaments are also clearly visible in such preparations. The dense bodies appear to be structurally and functionally analogous to the Z lines of striated muscle. Smooth muscle cells appear, therefore, to contain a sarcomere-like unit, delineated by the dense bodies, and both thick and thin filaments which potentially operate via a sliding filament-cross-bridge cycling mechanism. However, a significant body of evidence is consistent with a more labile state of organization of smooth muscle myosin (Kelly and Rice 1969; Fay and Cooke 1973; Shoenberg 1973; Ohashi and Nonomura 1979; Kendrick-Jones and Scholey 1981; Cooke 1982; Cande et al. 1983). This possibility will be discussed in more detail in Section 3.1.5.

As in the case of skeletal muscle, the regulation of smooth muscle contraction by Ca^{2+} is at the level of the myosin cross-bridge-actin interaction. The mechanism(s) whereby this regulation is achieved in smooth muscle, however, differs significantly from that operative in striated muscles.

3 Regulation of Smooth Muscle Contraction

In all muscle types the contractile state is controlled by the sarcoplasmic $[Ca^{2+}]$. The $[Ca^{2+}]$ transient leading to contraction is thought to occur over the range 10^{-7} to 10^{-5} M based on studies of glycerinated or otherwise chemically skinned smooth muscle (Filo et al. 1965; Endo et al. 1977; Ebashi 1980) in response to depolarization of the sarcolemma. In skeletal muscle, the additional Ca^{2+} comes predominantly from the sarcoplasmic reticulum (SR). In smooth muscle, it is unclear whether the SR can provide all the activating Ca^{2+}, or whether Ca^{2+} entry from the extracellular space may also be involved (Somlyo 1980). Mitochondria do not appear to be a source of activating Ca^{2+} in smooth muscle (Somlyo et al. 1979, 1982). As in numerous other Ca^{2+}-regulated physiological processes, a sensor(s) detects the rise in sarcoplasmic $[Ca^{2+}]$ and initiates a sequence of biochemical events which ultimately triggers cross-bridge cycling, i.e., contraction. In striated muscles the Ca^{2+} sensor is troponin C, a Ca^{2+}-binding subunit of the troponin complex associated with the thin filaments (Greaser and Gergely 1971, 1973). In smooth muscle the primary Ca^{2+} sensor is calmodulin, the ubiquitous multifunctional Ca^{2+}-dependent regulator protein (see Cheung 1980; Klee and Vanaman 1982; Walsh and Hartshorne 1983 for reviews). Ca^{2+} binding to these sensory proteins induces conformational changes which affect interactions between the Ca^{2+}-binding protein and target protein(s). In striated muscle the conformational change induced in troponin C by the binding of Ca^{2+} induces conformational changes in the neighbouring troponin subunits (T and I) and tropomyosin. Consequently, tropomyosin shifts its position on the thin filament thereby permitting cross-bridge-actin interaction. In smooth muscle the conformational change induced by the binding of Ca^{2+} to calmodulin permits its interaction with and activation of myosin light chain kinase (see Sect. 3.1.2). Myosin phosphorylation results, allowing rapid cross-bridge cycling to occur.

The detailed biochemical events occurring during relaxation are less well understood. It is likely, however, both for striated and smooth muscles, that repolarization of the sarcolemma results in Ca^{2+} sequestration within the SR, thereby decreasing the sarcoplasmic $[Ca^{2+}]$ to resting levels ($\sim 10^{-7}$ M). As a consequence of the Ca^{2+}-binding affinities of the Ca^{2+} sensors, these Ca^{2+}-binding proteins release their bound Ca^{2+},

Fig. 1. The reaction catalyzed by Ca^{2+}, calmodulin-dependent myosin light chain kinase. Smooth muscle myosin is a hexamer composed of a pair of heavy chains (M_r 200,000 each) and two pairs of light chains (M_r 20,000 and M_r 17,000). The enzyme MLCK, which requires Ca^{2+} and calmodulin for activity, catalyzes the transfer of the terminal phosphoryl group of MgATP to a specific serine residue (ser-19) on each of the two M_r 20,000 light chains of myosin. The myosin molecule is depicted diagrammatically to illustrate the globular head with its associated light chains and the long rod-like tail. The exact location of the light chains in the globular head region is unknown, but is probably the neck region (head-tail junction)

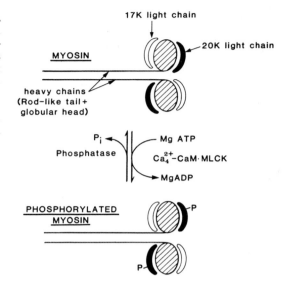

CaM = calmodulin
MLCK = myosin light chain kinase

revert to the Ca^{2+}-free conformation so that interaction with target protein(s) is broken and cross-bridge cycling ceases.

As mentioned earlier, it is widely believed that Ca^{2+} turns on smooth muscle contraction by activating the enzyme myosin light chain kinase. This regulatory system will now be considered in detail.

3.1 Myosin Phosphorylation

Sobieszek (1977) originally reported that gizzard myosin could be phosphorylated with concomitant actin-activation of the myosin Mg^{2+}-ATPase activity. Similar observations made with a wide variety of smooth muscles led to formulation of the phosphorylation theory of the regulation of smooth muscle contraction. According to this theory the mechanism of Ca^{2+} regulation of smooth muscle contraction involves 6 stages:

1. The sarcoplasmic $[Ca^{2+}]$ rises following excitation of the muscle cell.
2. This results in formation of the Ca_4^{2+}-calmodulin complex with a change in conformation of the calmodulin (CaM) molecule.
3. As a result of this Ca^{2+}-induced conformation change, CaM can interact with the inactive apoenzyme of myosin light chain kinase (MLCK) to form an active ternary complex, $Ca_4^{2+} \cdot CaM \cdot MLCK$.
4. The active kinase catalyzes phosphorylation of serine-19 in each of the two M_r 20,000 light chains (LC_{20}) of myosin, the so-called regulatory light chains (Fig. 1).

5. This phosphorylation reaction apparently affects the conformation of the myosin molecule in such a way as to permit actin-activation of the myosin Mg^{2+}-ATPase, i.e., the cross-bridge cycling rate is enhanced.
6. Relaxation occurs when the sarcoplasmic $[Ca^{2+}]$ returns to resting levels as a consequence, primarily, of sequestration by the SR. The fall in $[Ca^{2+}]$ induces dissociation of CaM from MLCK, thereby regenerating the inactive apoenzyme of MLCK. Myosin phosphorylation ceases and dephosphorylation occurs catalyzed by one or more phosphatases.

This scheme is probably an oversimplification. Recent observations leading to modifications of the basic theory will be discussed later. At this point some of the important properties of the protein components of this regulatory system will be examined.

3.1.1 Calmodulin

Calmodulin was discovered independently by Cheung and Kakiuchi et al. in 1970 and was known for several years as a Ca^{2+}-dependent activator of cyclic nucleotide phosphodiesterase. In recent years, however, this ubiquitous Ca^{2+}-binding protein has been implicated in the control by Ca^{2+} of numerous enzymatic activities including skeletal muscle phosphorylase kinase (Cohen et al. 1978), adenylate cyclase of brain (Brostrom et al. 1975; Cheung et al. 1975) and smooth muscle (Piascik et al. 1983), membrane Ca^{2+}, Mg^{2+}-ATPases (Gopinath and Vincenzi 1977; Jarrett and Penniston 1977), phospholamban kinase (Le Peuch et al. 1979), calcineurin (a protein phosphatase) (Wang and Desai 1977; Stewart et al. 1982), and plant NAD kinase (Anderson and Cormier 1978). Calmodulin also appears to be involved in Ca^{2+}-mediated regulation of various cellular processes such as microtubule assembly-dissembly (Marcum et al. 1978), but the mechanisms whereby such control is achieved are, as yet, unknown.

Calmodulin fulfills its role in Ca^{2+}-dependent regulation as a consequence of its Ca^{2+}-binding properties and the conformational changes which accompany Ca^{2+}-binding. While the actual values of the binding constants for each of the four Ca^{2+}-binding sites of calmodulin determined in different laboratories vary quite considerably (see Walsh and Hartshorne 1983 for review), it is clear that, at resting $[Ca^{2+}]$ and physiological $[Mg^{2+}]$, calmodulin is in the Ca^{2+}-free state and exhibits a rather loose conformation. The increase in $[Ca^{2+}]$ following excitation of the cell exceeds the binding constants of CaM for Ca^{2+} so that Ca^{2+} is bound. It is not clear what the exact state of CaM is under these conditions, but it is probably $Ca_4^{2+} \cdot CaM$. The binding of Ca^{2+} to calmodulin induces a conformational change in the calciprotein with exposure of a hydrophobic region which has been implicated in the binding of CaM to several of its target proteins (La Porte et al. 1980; Tanaka and Hidaka 1980). This region appears to be located between the second and third Ca^{2+}-binding sites of calmodulin (Walsh and Stevens 1978). In this new conformation CaM has a high affinity for its target proteins.

The dependence of MLCK activity on CaM was first demonstrated by Dabrowska et al. (1977) for the chicken gizzard enzyme and has since been shown for MLCK from numerous tissues and species of vertebrates. Indeed, it is now apparent that all vertebrate MLCKs require Ca^{2+} and calmodulin for activity. The requirement of smooth muscle (specifically bovine stomach) MLCK for Ca^{2+} and calmodulin is illustrated in Fig. 2.

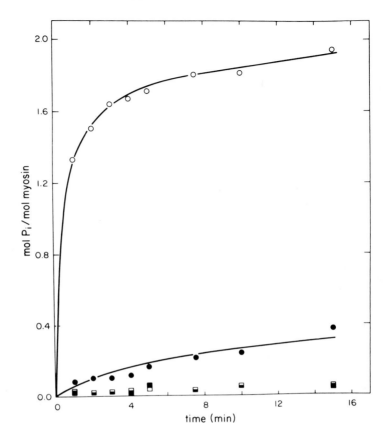

Fig. 2. Ca^{2+}- and calmodulin-dependence of bovine stomach MLCK. Purified bovine stomach MLCK (25.8 nM) was incubated at 25° in 25 mM Tris-HCl (pH 7.5), 4 mM $MgCl_2$, 60 mM KCl, 0.74 mM $[\gamma^{-32}P]$ATP (2500 cpm $nmol^{-1}$), and 0.5 mg $m\ell^{-1}$ turkey gizzard myosin, with (○, □) or without (●, ■) 15 μg $m\ell^{-1}$ bovine brain calmodulin, in the presence of 0.1 mM $CaCl_2$ (○, ●) or 1 mM EGTA (□, ■) in a total reaction volume of 4 $m\ell$. Aliquots (0.48 $m\ell$) of reaction mixtures were withdrawn at the indicated times and quenched by addition to 25% trichloroacetic acid, 2% sodium pyrophosphate (0.5 $m\ell$). Protein-bound ^{32}P was quantitated as described by Walsh et al. (1983b). The very low rate of myosin phosphorylation observed in the presence of Ca^{2+} but absence of added calmodulin (●) is due to a trace of contaminating calmodulin in the myosin preparation. (Walsh et al. 1982c)

Based on studies of Ca^{2+} and CaM activation of skeletal muscle MLCK (Blumenthal and Stull 1980) and measurements of the stoichiometry of CaM binding to gizzard MLCK (Hartshorne and Mrwa 1980; Adelstein and Klee 1981), the mechanism of activation of MLCK by Ca^{2+}-calmodulin may be summarized as follows:

$$4\,Ca^{2+} + CaM \rightleftharpoons Ca_4^{2+} \cdot CaM$$

$$Ca_4^{2+} \cdot CaM \rightleftharpoons Ca_4^{2+} \cdot CaM*$$

$$Ca_4^{2+} \cdot CaM* + MLCK \rightleftharpoons Ca_4^{2+} \cdot CaM* \cdot MLCK$$
          (inactive
          apoenzyme)

$$Ca_4^{2+} \cdot CaM* \cdot MLCK \rightleftharpoons Ca_4^{2+} \cdot CaM* \cdot MLCK*$$
                                    (active ternary complex)

where (*) denotes the active conformation of the protein.

3.1.2 Myosin Light Chain Kinase

Myosin light chain kinase (MLCK) of smooth muscle catalyzes the reaction shown in
Fig. 1. The properties of MLCK's isolated from numerous sources have been reviewed
recently (Walsh and Hartshorne 1982). The enzyme has been isolated or partially purified from several smooth muscles: chicken gizzard (Dabrowska et al. 1977; Uchiwa et
al. 1982; Ngai et al. 1984), turkey gizzard (Adelstein and Klee 1981; Walsh et al. 1983b),
bovine aorta (Vallet et al. 1981), bovine stomach (Walsh et al. 1982c), porcine uterus
(Higashi et al. 1983), and bovine carotid artery (Bhalla et al. 1982). Some of the more
important properties of these enzymes are summarized in Table 1. Amino acid compositions have been determined for the native forms of MLCK from chicken gizzard (M_r
136,000) and bovine stomach (M_r 155,000) (Table 2) and are very similar. Some differences are apparent, however, notably in the lower methionine and cysteine contents of
the bovine stomach enzyme and the higher contents of aspartic acid and the uncharged
residues threonine, proline, glycine, and leucine in bovine stomach MLCK. Sequence
differences are reflected in the facts that these two MLCK's exhibit completely different chymotryptic peptide maps (Ngai PK and Walsh MP, unpublished result, 1983) and
non-cross-reactivity of bovine stomach MLCK with a monoclonal antibody to the gizzard
enzyme (Adachi et al. 1983).

MLCK exhibits a high degree of substrate specificity, data from several laboratories
indicating that this enzyme is specific for phosphorylation of the regulatory light chain
of myosin. While MLCK's will generally phosphorylate myosins obtained from other
sources, they prefer myosin of their own tissue. The high degree of specificity of MLCK
indicates rigid structural requirements in the substrate. Comparison of the amino acid
sequences around the phosphorylated serine residue of the regulatory light chains of
rabbit and chicken skeletal and chicken gizzard myosins indicates obvious sequence
homologies, particularly at the C-terminal side of the phosphorylated serine (shown in
capitals):

Rabbit skeletal: X-pro-lys-lys-ala-lys-arg-arg-ala-ala-ala-
Chicken skeletal: (X-pro-lys-lys-ala-lys-arg-arg-ala-ala- —)
Chicken gizzard: X-ser-ser-lys-arg-ala-lys-ala-lys-thr-thr-lys-lys-arg-pro-

Rabbit skeletal: -glu-gly-gly-ser-SER-asn-val-phe-ser-met-phe-asp-gln-thr-
Chicken skeletal: (-glu-gly-) -ser-SER-asn-val-phe-ser-met-phe-asp-gln-thr-
Chicken gizzard: (-gln-arg-ala-thr-SER-asn-val-phe-ser-)

Table 1. Properties of smooth muscle myosin light chain kinases

Smooth muscle	M_r	V_{max}	K_d CaM (nM)	K_m ATP (μM)	K_m LC$_{20}$ (μM)	Reference
Chicken gizzard	105,000[a]	5-15[b]	ND[d]	60-70	ND	(Dabrowska et al. 1977; Mrwa and Hartshorne 1980)
Turkey gizzard	125,000-130,000[a]	10-30[b]	1-2	50	5	(Adelstein and Klee 1981; Walsh et al. 1983b; Adelstein et al. 1978)
Bovine stomach	155,000	8.9[c]	1.3	ND	ND	(Walsh et al. 1982c)
Pig uterus	130,000	16.1[b]	0.6	23	20	(Higashi et al. 1983)
Bovine aorta[e]	142,000	2-6[b]	6.3	ND	ND	(Vallet et al. 1981)

[a] Recent evidence (Adachi et al. 1983) indicates that these represent proteolytic fragments of the native enzyme which exhibits M_r = 136,000

[b] μmol phosphate incorporated min^{-1} mg^{-1} using isolated myosin light chains as substrate. Values given for 25°; adjusted to this temperature, if necessary, assuming a Q_{10} of 2

[c] μmol phosphate incorporated min^{-1} mg^{-1} using intact bovine stomach myosin as substrate at 25°

[d] ND, not determined

[e] Using the partially purified enzyme

Table 2. Amino acid composition of chicken gizzard and bovine stomach myosin light chain kinase[a]

Residue	Chicken gizzard[b]	Bovine stomach[c]
Lysine	119.4	127.6
Histidine	14.9	18.8
Arginine	47.2	46.9
Aspartic acid	119.4	139.8
Threonine[d]	75.5	98.2
Serine[d]	108.7	102.8
Glutamic acid	167.5	169.6
Proline	64.5	108.6
Glycine	81.6	100.3
Alanine	104.8	98.0
Cysteine	29.0	22.8
Valine	74.6	77.2
Methionine	26.8	17.1
Isoleucine	45.7	54.6
Leucine	71.0	99.0
Tyrosine	28.9	27.3
Phenylalanine	34.4	38.7
Tryptophan	20.1	ND[e]

[a]Values expressed as residues per mole. Measurements were made in triplicate at three times of hydrolysis (24, 48, and 72 h)
[b]Ngai PK and Walsh MP, unpublished results (1983). This enzyme is the native enzyme of $M_r =$ 136,000 identified in our laboratory (Adachi et al. 1983)
[c]Walsh et al. (1982c)
[d]Values obtained after extrapolation to zero-time hydrolysis
[e]ND, not determined

Sequences are aligned to give the maximum degree of homology. This necessitates two deletions in the N-terminal sequence of chicken skeletal light chain and an additional four residues at the amino-terminus of the chicken gizzard protein. The sequences are taken from the following sources: rabbit skeletal (Perrie et al. 1973; Collins 1976; Jakes et al. 1976), chicken skeletal (Matsuda et al. 1977), and chicken gizzard (Jakes R and Kendrick-Jones J, personal communication, 1983).

Recent studies with synthetic peptides have shed more light on the structural requirements for substrates of MLCK. Kemp et al. (1982, 1983) have studied the kinetics of phosphorylation of a series of synthetic peptide substrate analogs of chicken gizzard LC_{20} catalyzed by gizzard MLCK. The best peptide substrate corresponded to the N-terminal 23 residues as determined by Jakes and Kendrick-Jones (see above) and exhibited apparent K_m (6.9 μM) similar to that of the intact LC_{20} (8.6 μM) and a lower V_{max} (3.9 μmol min^{-1} mg^{-1} for the peptide and 36.2 μmol min^{-1} mg^{-1} for intact LC_{20}). Reducing the length of this peptide up to lys-11 had relatively little effect on the phosphorylation of ser-19. Large increases in apparent K_m were observed, however, when the basic residues, lys-11, lys-12 and arg-13, were successively substituted by ala, when lys-11 was substituted by thr, or when two threonine residues were inserted between lys-11 and lys-12, suggesting this trio of basic residues to be important specificity determinants.

Other protein kinases, e.g., cAMP-dependent protein kinase and phosphorylase b kinase, also require basic residues at the N-terminal side of the sites of phosphorylation in their substrates. Nevertheless, MLCK and these other kinases clearly have different specificity determinants since neither a synthetic dodecapeptide substrate of phosphorylase b kinase nor a synthetic heptapeptide substrate of cAMP-dependent protein kinase was phosphorylated by MLCK (Kemp et al. 1983). It is probably significant that, while the apparently essential basic residues in LC_{20} are separated by five residues from the site of phosphorylation, the essential basic residues in substrates of cAMP-dependent protein kinase are located closer to the site of phosphorylation (Zetterquist and Ragnarsson 1982). In this regard it is interesting to note that LC_{20} itself is also a substrate of cAMP-dependent protein kinase, phosphorylation occurring on the same serine as with MLCK (Noiman 1980). This may be attributed to the proximity of arg-16 rather than the group lys-11, lys-12, arg-13. In support of this conclusion, the skeletal myosin regulatory light chain was not phosphorylated by cAMP-dependent protein kinase, consistent with the absence of a basic residue this close to the phosphorylated serine (Noiman 1980). The observation that the isolated LC_{20} of gizzard myosin was a substrate of cAMP-dependent protein kinase raised the possibility that this reaction may be of physiological significance. We confirmed Noiman's observation with isolated LC_{20}, but also showed that intact gizzard myosin was not phosphorylated by cAMP-dependent protein kinase, indicating that this phosphorylation is not physiological (Walsh et al. 1981).

Singh et al. (1983) demonstrated stoichiometric phosphorylation of the isolated LC_{20} by phosphorylase kinase, casein kinase I (a cyclic nucleotide- and Ca^{2+}-independent glycogen synthase kinase) and casein kinase II in addition to MLCK and cAMP-dependent kinase. However, only MLCK was capable of phosphorylating intact myosin. Similarly, it was shown recently (Gallis et al. 1983) that isolated chicken gizzard LC_{20} was phosphorylated stoichiometrically on two of its three tyrosine residues (tyr-142 and tyr-155) by the epidermal growth factor receptor kinase of human epidermoid carcinoma (A431) cells. Until further studies are performed, including the use of intact gizzard myosin as a potential substrate, the physiological significance of tyrosine phosphorylation in LC_{20} will remain in doubt. As discussed in detail in Section 4.2, both isolated LC_{20} and intact smooth muscle myosin are substrates of a Ca^{2+}- and phospholipid-dependent protein kinase.

An important feature of smooth muscle MLCK to be considered is its localization. An early clue that it may be bound to the contractile machinery came from the method developed to purify the enzyme from turkey gizzard. Adelstein and Klee (1981) purified MLCK from an insoluble fraction corresponding to myofibrils of striated muscle. They found that treatment of the "myofibrils" with high $[Mg^{2+}]$ resulted in extraction of the kinase along with actin, filamin and several additional minor protein components. Immunofluorescence studies indicated that MLCK is localized mainly along the actin-containing microfilament bundles (stress fibers) of various nonmuscle cells (chicken embryo, human fibroblasts and gerbil fibroma cells (de Lanerolle et al. 1981). In similar studies, anti-gizzard MLCK has been used to show that MLCK is located along the thin filaments of cardiac and skeletal muscle (Guerriero et al. 1981; Cavadore et al. 1982). Subsequently, in vitro binding studies indicated that gizzard MLCK binds to skeletal actin with high affinity, this interaction being unaffected by gizzard tropomyosin, Ca^{2+}-CaM or Mg^{2+}-ATP (Dabrowska et al. 1982). Similar high-affinity binding of MLCK to smooth muscle myosin was not observed. These observations are all consistent with localization of MLCK on the thin filament. This is of importance in considering the mechanics of myosin phosphorylation. If tightly bound to the thin filament, a given MLCK molecule will have access to a limited number of myosin molecules. An important consideration in this regard is the concentration of MLCK in the tissue. Recent estimates in our laboratory put the MLCK concentration in chicken gizzard at ~4 μM (Ngai PK, Walsh MP, unpublished result, 1983). This corresponds to ~2 molecules of MLCK per 1 μm length of thin filament. The full significance of the association of MLCK with the thin filament must await further investigation.

3.1.3 Myosin Light Chain Phosphatase

As indicated earlier, according to the phosphorylation theory myosin dephosphorylation will occur when the sarcoplasmic $[Ca^{2+}]$ returns to resting levels, thereby permitting relaxation to occur. Before the importance of myosin phosphorylation in the regulation of smooth muscle contraction had been recognized, myosin light chain phosphatase (MLCP) activity was identified in extracts of uterine smooth muscle, cardiac muscle and both fast and slow skeletal muscles (Frearson et al. 1976). The fast skeletal muscle phosphatase (M_r 70,000) was subsequently isolated (Morgan et al. 1976) and found to be specific for dephosphorylation of the regulatory light chain of myosin.

Since myosin phosphorylation was implicated in the Ca^{2+}-mediated control of smooth muscle actin-myosin interaction, attention has been focused on MLCP of smooth muscle. Advances in understanding of the mechanism and possible control of myosin dephosphorylation have lagged behind those relating to the myosin kinase system, largely as a result of numerous difficulties associated with studying phosphatases (reviewed by diSalvo et al. 1983a). However, several phosphatases have recently been isolated which dephosphorylate LC_{20}. Pato and Adelstein (1980) isolated 2 MLCP's from turkey gizzard: MLCP I (consisting of 3 subunits of M_r 60,000, 55,000 and 38,000 present in equimolar amounts) was independent of Mg^{2+} and relatively nonspecific; MLCP II (consisting of a single polypeptide chain of M_r 43,000) required Mg^{2+} for activity and appeared to be specific for LC_{20}. While MLCP I dephosphorylated isolated LC_{20}, it did not accept intact phosphorylated myosin as a substrate, suggesting that this phosphatase does not function in vivo in the dephosphorylation of myosin (Pato and Adelstein 1983a). MLCP II also did not dephosphorylate intact myosin (Pato and Adelstein 1983b). Two forms of MLCP which did dephosphorylate intact gizzard myosin were, however, identified following gel filtration of a gizzard extract (Pato and Adelstein 1983b). These enzymes have not, as yet, been isolated or characterized.

Werth et al. (1982) isolated a MLCP from bovine aorta which contained two subunits (M_r 67,000 and 38,000) in equimolar amounts and which exhibited a high level of activity with intact myosin as substrate. This phosphatase also dephosphorylated the isolated LC_{20} and histone IIA, albeit at a slower rate, and exhibited significant p-nitro-phenyl phosphatase activity. This MLCP required Mn^{2+} or Co^{2+} for activity; these cations could not be replaced by Mg^{2+}, Ca^{2+} or Mg^{2+}-ATP. This enzyme may represent the true myosin phosphatase, since its specific activity was comparable to that of MLCK using intact myosin as substrate. Alternatively, it may be derived from MLCP I of Pato and Adelstein (1983a, b, see above) by dissociation of the M_r 55,000 subunit. A similar enzyme was isolated by Onishi et al. (1982) from chicken gizzard. Like MLCP I, this phosphatase contained three subunits (M_r 67,000, 54,000 and 34,000), although not in equimolar amounts (1:1.8:0.6) indicating possible subunit dissociation; it dephos-phorylated intact myosin. It is conceivable that dissociation of the M_r 55,000 subunit of MLCP I may alter the substrate specificity of the phosphatase so that it becomes active towards intact myosin. In support of this idea, eIF-2 phosphatase, which consists of two subunits (M_r 60,000 and 38,000), dephosphorylated intact myosin, and the free M_r 38,000 subunit of MLCP I also dephosphorylated intact myosin (Pato and Adelstein 1983a). The interesting possibility exists, of course, that subunit dissociation may be a physiological mechanism for controlling phosphatase activity by altering its substrate specificity.

Preparations of MLCP's have been used in experiments involving ATPase and super-precipitation measurements in simplified systems and tension measurements in skinned smooth muscle fiber bundles to assess the importance of myosin phosphorylation in the contractile response in smooth muscle (see Sect. 3.1.4.4).

3.1.4 Evidence in Support of the Phosphorylation Theory

3.1.4.1 Correlation Between Myosin Phosphorylation and Actin-Activated Myosin Mg^{2+}-ATPase, Superprecipitation and Tension Development

As mentioned earlier, Sobieszek (1977) first observed a positive correlation between my-osin phosphorylation and the actin-activated myosin Mg^{2+}-ATPase of smooth muscle, suggesting the possibility of regulation of actin-myosin interaction by phosphorylation of LC_{20}. Several investigators have since confirmed this correlation using various smooth muscle actomyosin preparations (see Hartshorne and Mrwa 1982 for review). For example, Fig. 3 demonstrates the necessity of myosin phosphorylation for superprecipita-tion and actin activation of myosin Mg^{2+}-ATPase to occur in a system reconstituted from purified actin, myosin, tropomyosin, calmodulin, and MLCK (Ngai PK and Walsh MP, unpublished data, 1983).

More detailed examination revealed that the correlation between myosin phosphory-lation and actin-activated myosin Mg^{2+}-ATPase is not a simple linear relationship (Per-sechini and Hartshorne 1981, 1982; Ikebe et al. 1982). Below $\sim 50\%$ maximal phos-phorylation of myosin, very little actin activation of the myosin Mg^{2+}-ATPase was ob-served. Above 50% phosphorylation, the ATPase activity was markedly enhanced. On the basis of kinetic studies of myosin phosphorylation and actin-activated myosin Mg^{2+}-ATPase activity, Persechini and Hartshorne (1981) concluded that myosin phosphory-lation is an ordered (sequential) process, i.e., at 50% phosphorylation the myosin popu-lation consists entirely of molecules with one head phosphorylated, and myosin with a

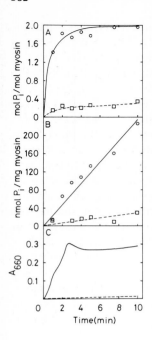

Fig. 3 A–C. The correlation between Ca^{2+}-dependent phosphorylation of smooth muscle myosin and actin-activated myosin Mg^{2+}-ATPase activity on the one hand and superprecipitation on the other. Chicken gizzard myosin (0.44 mg mℓ^{-1}) was incubated at 30° in 25 mM Tris-HCl (pH 7.5), 4 mM $MgCl_2$, 60 mM KCl, 5 μg mℓ^{-1} MLCK, 10 μg mℓ^{-1} CaM, 50 μg mℓ^{-1} chicken gizzard tropomyosin, and 0.25 mg mℓ^{-1} rabbit skeletal muscle actin in the presence of 0.1 mM $CaCl_2$ (0 in **A** and **B**, *solid line* in **C**) or 1 mM EGTA (□ in **A** and **B**, *dashed line* in **C**). Reactions were started by addition of [γ-^{32}P]ATP (2500 cpm nmol^{-1}) (**A** and **B**) or unlabeled ATP (**C**) to a final concentration of 1 mM. **A** and **B** Aliquots (0.5 mℓ) of reaction mixtures were withdrawn at the indicated times for quantitation of protein-bound phosphate (**A**) as described by Walsh et al. (1983b) and ATPase activity (**B**) as described by Ferenczi et al. (1978). ATPase activity is expressed as nmol P_i released mg^{-1} myosin. Superprecipitation (**C**) was monitored by recording the absorbance at 660 nm in a Bausch and Lomb Spectronic 2000

single head phosphorylated exhibits only a fraction of the actin-activated ATPase of fully phosphorylated myosin. The phosphorylation of the first head of myosin occurs relatively easily, while the second head is phosphorylated with more difficulty. This could be due to negative cooperativity between the myosin heads (phosphorylation of the first head hindering phosphorylation of the second head) or pre-existing asymmetry in the myosin molecule (the two heads of unphosphorylated myosin not being equivalent). Recent kinetic evidence (Persechini and Hartshorne 1983) suggests pre-existing asymmetry in the myosin molecule, with the possibility of additional negative cooperativity between the phosphorylated and nonphosphorylated heads. In contrast to intact myosin, smooth muscle heavy meromyosin phosphorylation is a random rather than ordered process (Ikebe et al. 1982; Sellers and Adelstein 1982). Numerous studies have indicated that myosin phosphorylation, in addition to showing a correlation with the actin-activated myosin Mg^{2+}-ATPase and superprecipitation, also correlates with active stress development in skinned smooth muscle fibers (Cassidy et al. 1980; Kerrick et al. 1980; Walsh et al. 1982a; Chatterjee and Murphy 1983). Furthermore, myosin phosphorylation precedes or coincides with force development in stimulated intact smooth muscles (Barron et al. 1979; Janis et al. 1980; Barron et al. 1980; de Lanerolle and Stull 1980; Dillon et al. 1981; Aksoy et al. 1982; Butler and Siegman 1982; Silver and Stull 1982; Gerthoffer and Murphy 1983).

3.1.4.2 Use of ATP Analogs

The ATP analog, adenosine 5'-0-(3-thiotriphosphate), (abbreviated ATPγS) has been widely used to test the phosphorylation theory in a variety of smooth muscle systems.

Its usefulness is based on the fact that ATPγS is commonly a good substrate of protein kinases, but the resultant thiophosphorylated protein kinase substrate is resistant to hydrolysis by phosphatases (Gratecos and Fischer 1974; Gergely et al. 1976; Pires and Perry 1977). For example, ATPγS is a good substrate of smooth muscle MLCK enabling Ca^{2+}, calmodulin-dependent thiophosphorylation of smooth muscle myosin, whereas thiophosphorylated myosin is quite resistant to phosphatase activity. Thiophosphorylation of gizzard myosin resulted in loss of the Ca^{2+} sensitivity of the actin-activated ATPase activity in a system reconstituted from the purified proteins (Sherry et al. 1978), again suggesting that myosin phosphorylation represents the dominant Ca^{2+}-dependent regulatory mechanism. In support of these observations, incubation of skinned smooth muscle fibers with ATPγS in the presence of Ca^{2+} led to thiophosphorylation of myosin LC_{20} and induced Ca^{2+}-insensitive activation of tension upon subsequent addition of ATP (Hoar et al. 1979; Cassidy et al. 1979).

Nucleotides known to be poor substrates of MLCK (ITP, GTP, UTP and CTP) were found to be poor relative to ATP in inducing Ca^{2+}-sensitive superprecipitation of crude gizzard actomyosin or tension development in skinned smooth muscle fibers (Cassidy and Kerrick 1982). Similarly, the thiophosphate analogs, ITPγS and GTPγS, were very poor relative to ATPγS at irreversibly activating superprecipitation (Cassidy and Kerrick 1982).

3.1.4.3 Use of Ca^{2+}-Independent Myosin Light Chain Kinase

We have recently used a different approach to evaluate the phosphorylation theory. This involved production of a Ca^{2+}-independent form of MLCK which enabled phosphorylation of myosin in the absence of Ca^{2+}. This permitted us to distinguish between the effects of myosin phosphorylation and those of other potential Ca^{2+}-dependent regulatory mechanisms which may be involved in the control of actin-myosin interactions. The initial phase of this program involved preparation, purification, and characterization of Ca^{2+}-independent MLCK. Preparation was achieved by limited proteolysis of the Ca^{2+}-dependent enzyme with α-chymotrypsin (Walsh et al. 1982b): a fragment of M_r 80,000 which was fully active in the absence of Ca^{2+} and calmodulin was generated. This active fragment no longer bound calmodulin. A possible explanation of the mechanism whereby Ca^{2+}-dependence of MLCK is removed by limited proteolysis is illustrated in Fig. 4. The apoenzyme of MLCK is inactive since a region of the polypeptide chain (indicated "inhibitory peptide domain") is folded in such a way as to mask the active site. Enzymatic activity can be expressed in one of two ways. Firstly, binding of Ca^{2+}-calmodulin alters the conformation of MLCK so that the inhibitory peptide moves, unmasking the active site. This is the physiological mechanism of activation of MLCK. Secondly, limited proteolysis removes the inhibitory peptide domain, similarly unmasking the active site.

The Ca^{2+}-independent MLCK was purified by ion-exchange chromatography and used in smooth muscle actomyosin (Walsh et al. 1982b) and skinned fiber preparations (Walsh et al. 1982a; Walsh et al. 1983a) to determine the effects of myosin phosphorylation (achieved in the absence of Ca^{2+}) on the actin-activated myosin Mg^{2+}-ATPase activity on the one hand and tension development on the other hand. In the crude gizzard actomyosin preparation, which contained Ca^{2+}-dependent MLCK, calmodulin,

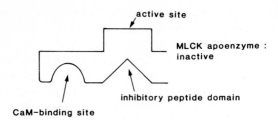

Fig. 4. The mechanism of activation of MLCK by binding of Ca^{2+}-CaM or by limited proteolysis

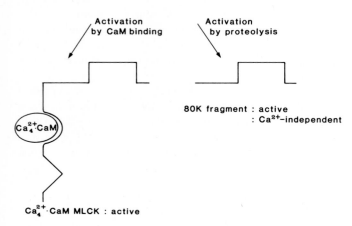

tropomyosin, and many other proteins, the ATPase activity was Ca^{2+}-sensitive: the steady-state ATPase rate was 23.2 nmol P_i released min^{-1} mg^{-1} myosin in the presence of 0.1 mM Ca^{2+}, and 5.6 nmol P_i released min^{-1} mg^{-1} myosin in the presence of 1 mM EGTA. When myosin in this crude actomyosin preparation was phosphorylated by the Ca^{2+}-independent MLCK in the absence of Ca^{2+}, the Ca^{2+}-sensitivity of the ATPase was abolished: the steady-state ATPase rate was then 23.2 nmol P_i released min^{-1} mg^{-1} myosin in the presence of Ca^{2+} and 22.0 nmol P_i released min^{-1} mg^{-1} myosin in the absence of Ca^{2+}. Myosin phosphorylation was, therefore, sufficient to induce actin-activation of the Mg^{2+}-ATPase activity in a crude gizzard actomyosin.

The Ca^{2+}-independent MLCK was also used in a series of experiments designed to observe the effects of myosin phosphorylation on tension development in skinned chicken gizzard fibers. Incubation of the fiber bundles with Ca^{2+}-independent MLCK and ATP in the absence of Ca^{2+} elicited tension development which was reversed when the kinase was washed out (Fig. 5a). Reversible phosphorylation of LC_{20} was demonstrated during a similar contraction-relaxation cycle in the absence of Ca^{2+} induced by the Ca^{2+}-independent MLCK (Fig. 6): tension development was accompanied by specific phosphorylation of LC_{20}; relaxation was accompanied by dephosphorylation of the myosin. Figure 5b shows that addition of Ca^{2+} following almost-maximal contraction induced by Ca^{2+}-independent MLCK in the absence of Ca^{2+} resulted in no further tension development. In other words, maximum tension development could be accounted for solely by myosin phosphorylation.

As discussed earlier, ATPγS has been used successfully to thiophosphorylate myosin irreversibly. In the skinned fiber system, myosin thiophosphorylation and activation of tension could occur only in the presence of Ca^{2+}, reflecting the Ca^{2+}-dependence of the endogenous MLCK. With the aid of the Ca^{2+}-independent MLCK, which also utilizes

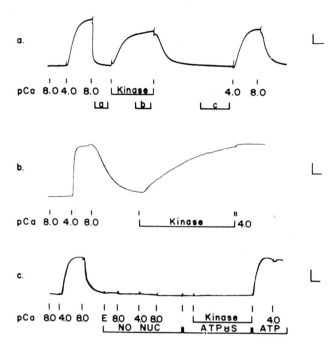

Fig. 5 a-c. The effect of Ca^{2+}-independent MLCK on tension development in skinned chicken gizzard fibers. Small bundles of skinned smooth muscle fibers were prepared from chicken gizzard by gentle homogenization with a Tekmar SDT tissuemizer in a skinning solution as described previously (Cassidy et al. 1980; Hoar et al. 1979; Kerrick and Krasner 1975). The ends of the fiber bundle preparations (1-2 mm in length) were inserted into small stainless steel clamps of a tension transducer similar to that of Hellam and Podolsky (1969). The preparations were then immersed in a relaxing solution containing 1% Triton X-100 for 20 min to solubilize any remaining sarcolemma. This facilitated the diffusion of proteins into the fiber bundle. The test solutions contained 70 mM K^+/Na^+, 2 mM $MgATP^{2-}$, 1 mM Mg^{2+}, 7 mM EGTA, 10^{-8} or 10^{-4} M free Ca^{2+}, propionate as the major anion, and an ATP regenerating system consisting of 15 units of creatine phosphokinase $m\ell^{-1}$ and 15 mM creatine phosphate. In solutions in which ATPγS or [γ-^{32}P]ATP (see Fig. 6) was substituted for ATP, no ATP regenerating system was included and $[K^+]$ was increased to 85 mM. Imidazole propionate was used to maintain the pH at 7 and to adjust the ionic strength to 0.15. The symbol L at the right of each record represents grams of tension in the *ordinate* and time in the *abscissa*. The time increment in all cases was 5 min. The tension increments were: **a** 2.2 mg; **b** 20 mg; **c** 12.5 mg. The amount of tension and the rate of tension development and relaxation depend on the cross-sectional area of the fiber bundle when mounted in the transducer. Ca^{2+}-independent MLCK (Kinase) was added to the solutions where shown at a final concentration of **a** 2.5 μM; **b** 1.3 μM; **c** 2.5 μM. The concentration of ATPγS was 2 mM. *a-c* under experiment *a* refer to time periods utilized in the experiment of Fig. 6. *E* EDTA; *NO NUC* no nucleotide included in the bathing solution, with all other components remaining the same. (Walsh et al. 1982a)

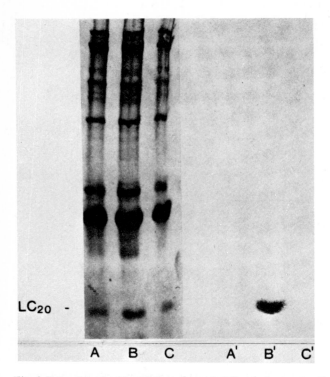

Fig. 6. Reversible phosphorylation of myosin LC_{20} during contraction-relaxation of skinned smooth muscle fibers in the absence of Ca^{2+}. Fiber bundles similar to those mounted in the transducer (Figure 5) were mounted in stainless steel holders and treated as in Fig. 5a. At the times indicated by a, b and c in Fig. 5a, two fiber bundles were removed from these solutions containing $[\gamma-^{32}P]ATP$ (67×10^6 cpm mℓ^{-1} at 15.2 mCi mmol^{-1} of ATP) for SDS-polyacrylamide gel electrophoresis. The 5 $\mu\ell$ samples of A (pCa 8.0), B (pCa 8.0+2.5 μM Ca^{2+}-independent MLCK) and C (pCa 8.0 after wash-out of the kinase) were electrophoresed in a 0.1% SDS, 7.5% acrylamide slab gel containing 5% (v/v) glycerol. The protein bands were bisualized using the silver stain (A, B, and C). Radioactively labeled proteins were visualized by autoradiography of the corresponding gel lanes (A', B', and C'). (Walsh et al. 1982a)

ATPγS as a substrate, myosin thiophosphorylation occurred in the absence of Ca^{2+}. This resulted in contraction of the skinned fibers (Fig. 5c) when ATP was subsequently added (ATPγS is not hydrolyzed by actomyosin and hence will not support contraction).

These and other similar studies we have performed with skinned smooth muscle fibers led us to the following conclusions: Firstly, myosin phosphorylation is the key event in actin-activation of myosin Mg^{2+}-ATPase and initiation of tension development in chicken gizzard skinned fibers. Secondly, it is unnecessary to postulate another Ca^{2+}-dependent regulatory mechanism involved in these processes; if such exists, it can exert only minor modulatory effects which are undetectable in our system. Thirdly, the data do not rule out the possible involvement of a Ca^{2+}-independent mechanism, but such a mechanism would have to be dependent on myosin phosphorylation.

Cande et al. (1983) using isolated smooth muscle (chicken gizzard) cells and cell fragments prepared by glycerination and homogenization, similarly concluded that myosin phosphorylation is the key event in smooth muscle contraction. These cell preparations contracted to $\sim$ one-third their resting length in response to Ca^{2+} and ATP. Ca^{2+}-independent contraction was observed to occur upon addition of ATP following preincubation with Ca^{2+} and ATPγS or Ca^{2+}-independent MLCK.

3.1.4.4 Use of Myosin Phosphatase

Several elegant studies have been performed which demonstrate the reversibility of actin activation of the myosin Mg^{2+}-ATPase and superprecipitation, both of which correlate with the level of myosin phosphorylation. Sellers et al. (1981) used purified MLCK and MLCP to reversibly phosphorylate isolated smooth muscle myosin and measured the actin-activated myosin Mg^{2+}-ATPase activity at each stage of the procedure. The ATPase rate of unphosphorylated myosin (4 nmol P_i released min^{-1} mg^{-1} myosin) rose to 51 nmol min^{-1} mg^{-1} following phosphorylation of the myosin to the extent of 2 mol P_i mol^{-1} myosin. Subsequent dephosphorylation with MLCP decreased the ATPase rate to 5 nmol min^{-1} mg^{-1}. Rephosphorylation with MLCK increased the rate again to 46 nmol min^{-1} mg^{-1}. A similar positive correlation between myosin phosphorylation and actin-activated ATPase activity was observed with heavy meromyosin (HMM) although, as has been shown also by others, the ATPase rate of HMM was much higher ($\sim$ seven fold) than that of intact myosin. Onishi et al. (1982) obtained similar results when examining the effects of reversible myosin phosphorylation on the actin-activated myosin Mg^{2+}-ATPase activity in a system reconstituted from the purified proteins. Similar reversible changes in superprecipitation were observed in the same reconstituted system which also correlated with the level of myosin phosphorylation. This group also showed that ITP, which is not a substrate of MLCK, induced superprecipitation of skeletal acto-gizzard phosphorylated myosin but not of skeletal acto-gizzard unphosphorylated myosin. Furthermore, purified gizzard MLCP reversed the ITP-induced superprecipitation of skeletal acto-gizzard phosphorylated myosin (Onishi et al. 1982). DiSalvo et al. (1983b) similarly observed inhibition of superprecipitation of bovine aortic actomyosin by a spontaneously active phosphatase extracted from this smooth muscle which correlated with inhibition of myosin phosphorylation. All these observations support the phosphorylation theory.

Several recent studies have extended the use of MLCP preparations to skinned smooth muscle fibers. For example, Paul et al. (1982) and Rüegg et al. (1982) showed that a partially purified bovine aortic MLCP accelerated relaxation of skinned hog carotid artery fibers. As discussed in Section 3.1.4.3, skinned chicken gizzard fibers were maximally activated by exposure to Ca^{2+}-independent MLCK in the absence of Ca^{2+}. This tension response was inhibited by the addition of purified gizzard MLCP which induced dephosphorylation of the myosin (Hoar et al. 1983).

3.1.4.5 Use of Calmodulin Antagonists

Several pharmacological agents have been identified which bind to calmodulin in a Ca^{2+}-dependent manner and with high affinity. These include phenothiazine derivatives (Levin and Weiss 1977, 1979) and naphthalenesulfonamide derivatives (Hidaka et al.

1981). These agents appear to bind to hydrophobic sites on CaM which are exposed upon binding Ca^{2+} and thereby block interaction of CaM with its target enzymes. These and other agents have been shown to inhibit MLCK activity and the Ca^{2+}-dependent actin-activated Mg^{2+}-ATPase activity of gizzard actomyosin (Hidaka et al. 1979a; Sheterline 1980; Hidaka et al. 1980). The effects of calmodulin antagonists on tension development in skinned and intact smooth muscle fibers have been widely studied. Tension development in skinned intestinal and arterial muscle fibers (Kerrick et al. 1980; Sparrow et al. 1981) and intact arterial strips (Barron et al. 1980; Kanamori et al. 1981; Asano et al. 1982) and bovine tracheal smooth muscle (Silver and Stull 1983) was inhibited, and relaxation induced in skinned taenia coli strips (Crosby and Diamond 1980) and rabbit aortic strips (Hidaka et al. 1978, 1979b) by a variety of calmodulin antagonists.

These observations are consistent with action of these agents at the level of calmodulin to block its interaction with and activation of MLCK. Caution should be exercised, however, in interpreting results obtained with the use of calmodulin antagonists. Because of their generally hydrophobic nature, such compounds can potentially interact with and perturb numerous cellular sites, notably within membranes, in addition to calmodulin itself. Furthermore, a number of other Ca^{2+}-binding proteins have been shown to interact with calmodulin antagonists in a Ca^{2+}-dependent manner (Marshak et al. 1981; Moore and Dedman 1982). In spite of the nonspecificity of calmodulin antagonists, in many of the studies described in this section changes in myosin phosphorylation were correlated with changes in actin-activated myosin Mg^{2+}-ATPase activity or tension development in response to the antagonist. Such studies, therefore, provide additional and significant evidence in favor of the phosphorylation theory.

3.1.5 The Mechanism of Regulation of Actin-Myosin Interaction by Myosin Phosphorylation

The molecular mechanism whereby myosin phosphorylation leads to or permits actin activation of myosin Mg^{2+}-ATPase is unknown, although recent observations are helping to shed some light on this problem. Filaments of dephosphorylated smooth muscle myosin in 0.15 M KCl were shown to be disassembled by stoichiometric amounts of ATP into a 10S species as shown by ultracentrifugation (Suzuki et al. 1978). At higher ionic strengths, a single species of 6S was found both in the presence and absence of ATP. It was felt initially that this 6S-10S transition indicated dimerization of the myosin, but it was later shown that it was, in fact, due to a substantial change in the radius of gyration, both species of myosin being monomeric (Suzuki et al. 1982a). Electron microscopy of various rotary-shadowed myosins (gizzard, vascular muscle, and thymus) revealed that the 10S form exhibited a looped structure: the tail of the myosin molecule was bent back toward the head region and two hinge regions were apparent, one located approximately one-third and the other two-thirds along the length of the tail (Onishi and Wakabayashi 1982; Trybus et al. 1982; Craig et al. 1983). A portion of the myosin tail near the distal hinge appeared to bind to the neck region of the myosin heads. The 6S species of myosin, on the other hand, exhibited an extended, asymmetric conformation similar to that of skeletal muscle myosin. The transition from 6S to 10S was favored by Mg^{2+}-ATP and was also observed upon dephosphorylation of the myosin. Ikebe et al. (1983) observed a correlation between changes in the relative viscosity of smooth muscle myosin and the 6S-10S transition, providing a simple experimental technique for studying the conformational transition and its effects. The Mg^{2+}-ATPase activity of smooth muscle myosin varied under different conditions of ionic strength, $[Mg^{2+}]$, [ATP] and phosphorylation state, and paralleled the viscosity changes, i.e., the myosin Mg^{2+}-ATPase activity reflected the shape of the myosin molecule. Under approximately physiological conditions of ionic strength, $[Mg^{2+}]$ and [ATP], fully phosphorylated myosin existed in the extended 6S form and exhibited high ATPase activity, whereas dephosphorylated myosin existed in the folded 10S conformation with low ATPase activity. These findings suggest a possible basis for the mechanism whereby myosin phosphorylation regulates smooth muscle contraction. It appears unlikely, however, that the 10S-6S transition accurately reflects the in vivo situation during smooth muscle activation, since myosin filaments apparently exist in both contracted and relaxed smooth muscle (Somlyo et al. 1981), although the existence of disassembled myosin in relaxed smooth muscle has been indicated in several studies (see below). It may be rather that the interactions seen in the 10S conformation in isolated myosin and the conformational changes leading to formation of the 6S species reflect intermolecular interactions within myosin filaments in vivo rather than intramolecular interactions within individual myosin molecules.

Further insight into the mechanism whereby LC_{20} regulates actin-myosin interaction in smooth muscle has come from hybridization experiments with "desensitized" scallop myofibrils. In scallop muscles, contraction is regulated by reversible binding of Ca^{2+} to the myosin regulatory light chains, not by reversible phosphorylation of the light chains. Thus, in native scallop myofibrils the actin-activated myosin Mg^{2+}-ATPase is low in the absence of Ca^{2+} and high in the presence of Ca^{2+} ($> 1 \mu M$). The regulatory light chains of scallop myosin can be selectively removed by washing with EDTA at 25° C (Szent-Györgyi et al. 1973; Chantler and Szent-Györgyi 1980) to yield "desensitized scallop myosin, i.e., the actin-activated myosin Mg^{2+}-ATPase activity is high both in the presence and absence of Ca^{2+} · Ca^{2+} sensitivity can be restored by readdition of the isolated regulatory light chain to desensitized scallop myosin. These observations suggested that in the resting scallop muscle the regulatory light chains inhibit actin-myosin interaction and this inhibition is relieved by the binding of Ca^{2+} to these light chains, i.e., the regulatory light chains operate via a mechanism of repression-derepression.

Regulatory light chains isolated from a variety of myosins bind to desensitized scallop myofibrils, an observation which led Kendrick-Jones et al. (1983) to study the effects of hybridization of the isolated LC_{20} of smooth muscle and desensitized scallop myofibrils on the Ca^{2+} sensitivity of the actin-activated myosin Mg^{2+}-ATPase. They found that Ca^{2+} sensitivity was restored when gizzard LC_{20} bound, to the extent of 2 mol mol^{-1} myosin, to desensitized scallop myofibrils. If this hybrid was phosphorylated on LC_{20} with gizzard MLCK, Ca^{2+} sensitivity was lost and could be restored by subsequent treatment with gizzard phosphatase. Similarly, thiophosphorylation of the hybrid induced loss of Ca^{2+} sensitivity. They concluded that the roles of the regulatory light chains in molluscan muscle and mammalian smooth muscle are similar. Therefore, in the relaxed smooth muscle LC_{20} inhibits actin-myosin interaction. This inhibition is relieved by Ca^{2+}-dependent phosphorylation of LC_{20} leading to contraction. In other words, LC_{20} phosphorylation in smooth muscle plays the same role as Ca^{2+} binding to the regulatory light chains in molluscan muscle.

As mentioned earlier, a number of studies have suggested that smooth muscle and non-muscle myosins exist in a labile state of organization, although this is a controversial point. Several in vitro experiments performed with isolated myosin indicate that LC_{20} phosphorylation regulates thick filament formation (Suzuki et al. 1978; Trybus et al. 1982; Kendrick-Jones et al. 1983; Suzuki et al. 1982b; Kendrick-Jones et al. 1981). At physiological ionic strength, pH 7.0, and in the presence of MgATP, nonphosphorylated smooth muscle myosin exists in a disassembled state. Phosphorylation of LC_{20} promotes thick filament formation, and dephosphorylation results in filament disassembly. It appears, therefore, that the folded 10S conformation, which exists under these conditions when the myosin is nonphosphorylated, is incapable of filament formation and that LC_{20} phosphorylation, leading to unfolding of the molecule to the 6S form, permits filament assembly. It is, however, unclear whether or not these findings bear any relevance to the situation in the living muscle, particularly in view of the demonstration by Somlyo et al. (1981) that myosin filaments with nonphosphorylated LC_{20} are present in relaxed smooth muscle.

The data of Sellers et al. (1981) and Ikebe et al. (1982) on the effects of phosphorylation of smooth muscle heavy meromyosin, which is incapable of filament formation, on the actin-activated myosin Mg^{2+}-ATPase of this myosin fragment indicate that the effect of LC_{20} phosphorylation on myosin filament formation is distinct from its effect on the actin-activated ATPase, i.e., myosin filaments are not required for actin activation of myosin Mg^{2+}-ATPase.

3.2 Other Ca^{2+}-Dependent Regulatory Mechanisms

While most investigators agree that myosin phosphorylation is a prerequisite for smooth muscle contraction, controversy still exists. This ranges from the isolated standpoint of Ebashi's group, who maintain that myosin phosphorylation does not perform a regulatory function in this system, to a rather widespread belief that myosin phosphorylation is involved but a second Ca^{2+}-dependent regulatory mechanism also controls smooth muscle actin-myosin interaction, at least under certain circumstances.

Reports showing variable effects of myosin phosphorylation on the actin-activated myosin Mg^{2+}-ATPase have led to speculation that myosin phosphorylation may not be required for contraction. Thus, in some preparations changes in LC_{20} phosphorylation were observed without changes in ATPase activity or superprecipitation (Cole et al. 1982; Persechini et al. 1981; Grand et al. 1982; Ebisawa 1983). On the other hand, under certain conditions the myosin Mg^{2+}-ATPase was activated by actin without myosin phosphorylation (Cole et al. 1982; Mikawa et al. 1977; Merkel et al. 1982). Several of these studies suggested the requirement of a factor in addition to myosin phosphorylation for actin activation of the myosin Mg^{2+}-ATPase. Cole et al. (1983) recently described the preparation of chicken gizzard myosin which, when fully phosphorylated, possessed an actin-activated Mg^{2+}-ATPase of very high specific activity (up to 600 nmol P_i released min^{-1} mg^{-1} myosin), i.e., comparable to that of skeletal muscle actomyosin. Their data indicated that at < 100% phosphorylation of myosin a Ca^{2+}-sensitive system distinct from MLCK regulates the ATPase.

It is possible that such effects result from artifactual changes in the myosin molecule during preparation. In support of this possibility, Seidel (1979) and Bailin and Lopez

(1981, 1982) showed that modification (e.g., oxidation or dinitrophenylation) of a sulf-hydryl group on the M_r 17,000 light chain of smooth muscle myosin could mimic the effects of LC_{20} phosphorylation. Similarly, limited proteolysis of smooth muscle my-osin can yield a form which is activated by actin without phosphorylation (Ebashi et al. 1977; Okamoto and Sekine 1981; Seidel 1980; Rees and Frederiksen 1981).

As mentioned above, Ebashi's group (Ebashi et al. 1977; Hirata et al. 1977) observed no effect of myosin phosphorylation on superprecipitation or actin-activated Mg^{2+}-ATPase activity of their myosin preparations, but instead obtained evidence that actin-myosin interaction was regulated by a Ca^{2+}-dependent, actin-linked system consisting of three proteins: tropomyosin, leiotonin A (M_r 80,000), and leiotonin C (M_r 17,000) (Ebashi 1980; Nonomura and Ebashi 1980). Leiotonin C was described as a Ca^{2+}-binding protein similar to, but distinct from, calmodulin and troponin C. While calmodulin could sub-stitute functionally for leiotonin C in superprecipitation studies (Mikawa et al. 1978), it was concluded, from comparative studies of the Ca^{2+}- and Sr^{2+}-dependencies of superprecipitation of native aortic actomyosin containing leiotonin C or calmodulin, that leiotonin C and not calmodulin was the physiological regulator. As pointed out by Sellers et al. (1981), since neither the purity of these regulatory proteins, their mode of action, nor their stoichiometry have been completely characterized, it remains difficult to assess the physiological importance of this thin filament-linked system in the regula-tion of smooth muscle contraction.

A number of early reports suggested the presence in smooth muscle of proteins resembling the subunits of striated muscle troponin (Ebashi et al. 1977; Carsten 1971; Sparrow and van Bockxmeer 1972; Shibata et al. 1973; Ebashi et al. 1975; Ito and Hotta 1976; Ito et al. 1976; Head et al. 1977). It became apparent subsequently, how-ever, that these proteins were not, in fact, troponin subunits. For example, the troponin C-like protein isolated from several smooth muscles by Head et al. (1977) was later shown to be calmodulin (Grand et al. 1979). It was then generally accepted that smooth muscle lacks troponin until Marston et al. (1980) suggested thin filament-associated regulation in pig aorta and turkey gizzard. Porcine aorta thin filaments contained proteins resembling troponins I and C which formed a Ca^{2+}-dependent complex that was stable in 6 M urea, i.e., similar to skeletal troponin. These thin filaments also contained kinase and phos-phatase activities directed against an endogenous basic protein of M_r 21,000 (Walters and Marston 1981). Phosphorylation of this protein increased the number of high-affinity Ca^{2+}-binding sites on the thin filaments by four to eight fold and decreased the $[Ca^{2+}]$ required for half-maximal activation of skeletal myosin Mg^{2+}-ATPase from 2.7 to 0.3 μM. It was suggested that phosphorylation of the M_r 21,000 protein induces a con-formational change in troponin C which affects its Ca^{2+}-binding properties.

In support of the existence of smooth muscle troponin, Endo and Obinata (1981) isolated troponin subunits T (M_r 33,000), I (M_r 24,000) and C (M_r 18,000) from as-cidian smooth muscle. Together these proteins conferred Ca^{2+} sensitivity on recon-stituted rabbit skeletal actomyosin in the presence of tropomyosin. Unlike the mam-malian skeletal muscle system, however, ascidian troponin markedly accelerated actin-myosin interaction in the presence of Ca^{2+}, its inhibitory action on the actomyosin ATPase in the absence of Ca^{2+} being less remarkable. It remains, however, to be de-monstrated convincingly that vertebrate smooth muscle contains troponin or a troponin-like system.

A recent series of papers involving mechanical measurements and quantitation of myosin phosphorylation levels in many types of intact smooth muscle fibers contracted in response to a variety of stimuli led Murphy to propose a second Ca^{2+}-dependent regulatory system in smooth muscle (Dillon et al. 1981; Aksoy et al. 1982; Butler and Siegman 1982; Silver and Stull 1982; Gerthoffer and Murphy 1983; Aksoy et al. 1983). Whereas these data were consistent with myosin phosphorylation being necessary for cross-bridge cycling leading to shortening or stress development, the stress developed could be maintained in a Ca^{2+}-dependent manner while myosin was dephosphorylated to close to resting levels. Murphy concluded, therefore, that a second Ca^{2+}-dependent regulatory mechanism came into play which enabled nonphosphorylated cross-bridges to maintain active stress. These nonphosphorylated cross-bridges appeared to be non- or slowly cycling and were termed "latch bridges". Recent evidence, using both intact (Aksoy et al. 1983) and skinned (Chatterjee and Murphy 1983) smooth muscle fibers, suggests that this transient change in myosin phosphorylation with generation and maintenance of force is due to alterations in sarcoplasmic $[Ca^{2+}]$ following stimulation. Evidence that stimulation of arterial smooth muscle induces a transient increase in myoplasmic $[Ca^{2+}]$ has been provided by light emission from aequorin-loaded cells (Morgan and Morgan 1982). Thus there is a rapid rise in sarcoplasmic free $[Ca^{2+}]$ in response to stimulation. This activates the MLCK which phosphorylates myosin allowing rapid cross-bridge cycling. Murphy proposes that the $[Ca^{2+}]$ then declines to a value intermediate between the peak $[Ca^{2+}]$ and resting $[Ca^{2+}]$. At this intermediate $[Ca^{2+}]$ the MLCK is inactivated and MLCP dephosphorylates myosin. However, stress is maintained, apparently as a result of this second Ca^{2+}-dependent regulatory mechanism. This mechanism must have a higher Ca^{2+} sensitivity than MLCK since at the intermediate $[Ca^{2+}]$ the latter is inoperative. This mechanism is supported by the observation of Silver and Stull (1982) that phosphorylase phosphorylation, which is also Ca^{2+}-dependent, increased to a maximum upon stimulation of intact bovine tracheal strips with carbachol, remained elevated for up to 10 min, and then declined while isometric tension was maintained.

A possible physiological role for the latch state would be energy conservation brought about by a reduction in the cross-bridge cycling rate during tonic contractions. The second Ca^{2+}-dependent regulatory mechanism appears to function only following the initial activation and $[Ca^{2+}]$ transient leading to myosin phosphorylation and stress development. The nature of the Ca^{2+}-binding site regulating latch bridge formation is unknown. A potential candidate is LC_{20} itself which binds Ca^{2+} with high affinity. Indeed, several workers have suggested that direct binding of Ca^{2+} to myosin is involved, in addition to myosin phosphorylation, in actin activation of smooth muscle myosin Mg^{2+}-ATPase (Rees and Frederiksen 1981; Chacko et al. 1977; Litten et al. 1977). The possibility that LC_{20} phosphorylation could affect the Ca^{2+}-binding properties of myosin leading to the latch state is one worthy of investigation. Other candidates potentially responsible for the latch state include the troponin-like system of Marston and leiotonin, both referred to above. The final stage in Murphy's hypothesis would be relaxation occurring as a result of a further reduction in sarcoplasmic free $[Ca^{2+}]$ below the threshold for latch bridge formation, i.e., the second Ca^{2+}-dependent regulatory mechanism is turned off. The biochemical basis of the latch state will be the focus of considerable attention in the immediate future.

4 Future Prospects

It is apparent from the discussion to this point that several aspects of the mechanism of regulation of smooth muscle contraction involve ongoing research. In addition to these, a number of recent interesting findings have opened up new avenues of research in this field. The remainder of this chapter will consider these in turn.

4.1 Cyclic AMP-Dependent Phosphorylation of Myosin Light Chain Kinase

The contractile state of many smooth muscles is controlled by cAMP which is synthesized within the cell in response, for example, to binding of β-adrenergic hormones to their specific receptors within the plasma membrane (Bär 1974; Namm and Leader 1976). The possibility that β-adrenergic hormone-induced relaxation may occur via phosphorylation of one of the proteins of the contractile apparatus was originally suggested by Sands et al. (1976). Subsequently, Adelstein et al. (1978) showed that MLCK is an in vitro substrate of the catalytic subunit of cAMP-dependent protein kinase. In the absence of bound CaM the kinase was phosphorylated stoichiometrically at two sites (A and B), whereas only site B was phosphorylated in the $Ca_4^{2+} \cdot CaM \cdot MLCK$ complex (Conti and Adelstein 1981). While site B phosphorylation had no observable effect on MLCK, site A phosphorylation decreased the affinity of MLCK for CaM $\sim$20-fold. This suggested a possible physiological role of cAMP-dependent phosphorylation of MLCK: In response to β-adrenergic hormones, elevated cAMP levels activate the cAMP-dependent protein kinase which phosphorylates MLCK, thereby reducing its affinity for CaM. The kinase may consequently become inactivated so that myosin phosphorylation ceases and phosphorylated myosin is dephosphorylated by the action of MLCP. There are, however, two major problems with this mechanism which remain to be resolved. First, the predominant form of MLCK in the contracting smooth muscle is $Ca_4^{2+} \cdot CaM \cdot MLCK$, which is phosphorylated only at site B and without effect. Second, even if sufficient MLCK apoenzyme is present to be phosphorylated at both sites A and B, the effect of the phosphorylation is relatively minor: the K_d CaM is shifted from 1.2 nM to 25 nM (Conti and Adelstein 1981). The total CaM concentration in gizzard smooth muscle has been estimated at $\sim$ 31 μM (Grant and Perry 1979). In order for the MLCK to be inactivated by phosphorylation, the vast majority of this CaM would have to be unavailable, e.g., by being bound to other CaM-binding proteins. This appears not to be the case: gizzard smooth muscle contains only two CaM-binding proteins in significant amount, namely MLCK and caldesmon (see Sect. 4.3) which we have estimated to be present at concentrations of 4 μM and 20 μM, respectively (Ngai PK and Walsh MP, unpublished results, 1983). The caldesmon concentration of 20 μM refers to the monomer concentration; the native protein is a dimer, each subunit binding one calmodulin molecule (see Sect. 4.3).

Evidence in favor of the proposed role of cAMP-dependent phosphorylation of MLCK in relaxation of smooth muscle has been forthcoming using native actomyosin of aorta (Silver and diSalvo 1979; Silver et al. 1981) and hog carotid artery (Mrwa et al. 1979) and chemically skinned smooth muscle fibers (Mrwa et al. 1979; Kerrick and Hoar 1981; Rüegg et al. 1981). However, an equally significant body of evidence exists against such a role (see, for example, Miller et al. 1983; Sparrow et al. 1983).

4.2 Phosphorylation of Smooth Muscle Myosin by the Ca^{2+}- and Phospholipid-Dependent Protein Kinase

A novel protein kinase which requires Ca^{2+} and a phospholipid (specifically phosphatidylserine) for activity was discovered in rat brain (Inoue et al. 1977) and subsequently shown to be widely distributed in mammalian tissues, including smooth muscle (Kuo et al. 1980; Minakuchi et al. 1981). This Ca^{2+}, phospholipid-dependent protein kinase, abbreviated kinase C, is believed to be involved in the regulation of various physiological phenomena elicited by a variety of extracellular messengers, including some peptide hormones, growth factors, muscarinic cholinergic and α-adrenergic agents. Kinase C can be activated either by an increase in cytosolic $[Ca^{2+}]$ or by unsaturated diacylglycerol without any increase in $[Ca^{2+}]$. A phospholipid, presumably contributed by the inner part of the membrane lipid bilayer, is required for kinase C activity in each case. Unsaturated diacylglycerol may serve as a second messenger, analogous to Ca^{2+} and cAMP. It is produced transiently from membrane-bound phosphatidylinositol via the action of phospholipase C. Interestingly, tumor-promoting phorbol esters can substitute for unsaturated diacylglycerol in the activation of kinase C (Castagna et al. 1982; Niedel et al. 1983). This protein kinase has a relatively broad substrate specificity. Its possible involvement in the regulation of actin-myosin interactions in smooth muscle was originally suggested by the observation of Endo et al. (1982) that the isolated LC_{20} of gizzard myosin serves as a substrate for kinase C. They also demonstrated that intact gizzard myosin could be phosphorylated by kinase C, albeit at a slower rate. The stoichiometry of LC_{20} phosphorylation by kinase C was not measured in this study, making it difficult to assess the possible physiological significance of this observation. However, more recently Naka et al. (1983) demonstrated kinase C-mediated phosphorylation of myosin LC_{20} in intact platelets activated by the tumor-promoting phorbol ester, 12-0-tetradecanoylphorbol-13-acetate (TPA). The site of phosphorylation in LC_{20} in response to TPA (which presumably acts via kinase C) was distinct from the site phosphorylated by Ca^{2+}, calmodulin-dependent MLCK. It will be interesting to know the effects of kinase C-mediated phosphorylation of LC_{20} on the actin-activated myosin Mg^{2+}-ATPase activity of smooth muscle and how it relates to the MLCK-mediated phosphorylation.

4.3 Caldesmon

Sobue et al. (1981a) isolated a major calmodulin-binding protein of subunit M_r 150,000 from chicken gizzard. The native protein was shown to be a dimer and was named caldesmon. Caldesmon, which they estimated to be present to the extent of 240 mg/100 g gizzard, interacted with calmodulin in a Ca^{2+}-dependent manner and with F-actin in a Ca^{2+}-independent manner. Calmodulin was found to block the interaction of caldesmon (CaD) with F-actin in the presence but not absence of Ca^{2+}, suggesting that Ca^{2+} acts as a flip-flop switch for the interconversion of the following two complexes:

$$\text{Actin} + \text{CaM-CaD} \underset{+Ca^{2+}}{\overset{-Ca^{2+}}{\rightleftharpoons}} \text{CaD-Actin} + \text{CaM}$$

Equilibrium between the two complexes occurred at ~ 1 μM Ca^{2+}, indicating the potential physiological significance of this flip-flop mechanism during normal $[Ca^{2+}]$ transients. Using ^{3}H-calmodulin, Sobue et al. (1981b) observed 1:1 binding of CaM to each CaD subunit with apparent $K_d = 1.7$ μM. Caldesmon has been shown, with the aid of antibodies, to be found in aorta, uterus, and platelets and, therefore, may be widespread in smooth muscle and nonmuscle tissues (Kakiuchi et al. 1983).

The interaction of caldesmon with F-actin raised the possibility that this protein may play a role in regulating actin-myosin interaction. In support of this idea, Sobue et al. (1982a), using desensitized chicken gizzard actomyosin, concluded from super-precipitation experiments that myosin phosphorylation was required for actin-myosin interaction, but that caldesmon could then come into play to regulate the actin-myosin interaction. Specifically, caldesmon, in the absence of Ca^{2+}, was found to reverse the superprecipitation induced by MLCK-dependent phosphorylation of myosin without any dephosphorylation of the myosin. This was explained by the flip-flop mechanism described above. Thus, at high $[Ca^{2+}]$, CaD associated with CaM rather than actin filaments allowing actin-myosin interaction to occur as the myosin became phosphorylated. When the free $[Ca^{2+}]$ decreased, CaD dissociated from CaM and bound to actin, thereby blocking actin-myosin interaction, even though the myosin remained phosphorylated since phosphatase activity was absent from the system.

These observations were made exclusively with in vitro systems, hence their significance in regulation of actin-myosin interaction in the intact living system is uncertain. In particular, it is not clear why such a system, which would perform the same function as myosin light chain phosphatase, should exist. Perhaps the role of caldesmon in vivo is unrelated to regulation of actin-myosin interaction but rather involves regulation of gel-sol transformations of actin filaments which are involved in such processes as cell shape changes, locomotion, and phagocytosis. The calmodulin-caldesmon system was observed to confer Ca^{2+}-sensitivity on the filamin-induced actin gel-sol transformation (Sobue et al. 1982b). In the presence of Ca^{2+}, caldesmon bound to calmodulin allowing filamin (a high M_r actin-binding protein) to cross-link actin filaments to form a gel. In the absence of Ca^{2+}, the CaM-CaD complex dissociated, allowing caldesmon to interact with actin, thereby blocking its association with filamin and causing solation.

Acknowledgements. The author is very grateful to Dr. David J. Hartshorne, Department of Nutrition and Food Science, University of Arizona, for introducing him to the fascinating area of control of contractility in smooth muscle, and to numerous colleagues who very kindly provided reprints and preprints of their work which greatly aided in the writing of this chapter. The author's research is supported by grants from the Medical Research Council of Canada and the Alberta Heritage Foundation for Medical Research.

References

Adachi K, Carruthers CA, Walsh MP (1983) Identification of the native form of chicken gizzard myosin light chain kinase with the aid of monoclonal antibodies. Biochem Biophys Res Commun 115: 855-863

Adelstein RS, Klee CB (1981) Purification and characterization of smooth muscle myosin light chain kinase. J Biol Chem 256: 7501-7509

Adelstein RS, Conti MA, Hathaway DR, Klee CB (1978) Phosphorylation of smooth muscle myosin light chain kinase by the catalytic subunit of adenosine 3':5'-monophosphate-dependent protein kinase. J Biol Chem 253: 8347-8350

Aksoy MO, Murphy RA, Kamm KE (1982) Role of Ca^{2+} and myosin light chain phosphorylation in regulation of smooth muscle. Am J Physiol 242 (Cell Physiol 11): C109-C116

Aksoy MO, Mras S, Kamm KE, Murphy RA (1983) Ca^{2+}, cAMP, and changes in myosin phosphorylation during contraction of smooth muscle. Am J Physiol 245 (Cell Physiol 14): C255-C270

Anderson JM, Cormier MJ (1978) Calcium-dependent regulator of NAD kinase in higher plants. Biochem Biophys Res Commun 84: 595-602

Asano M, Suzuki Y, Hidaka H (1982) Effects of various calmodulin antagonists on contraction of rabbit aortic strips. J Pharmacol Exp Ther 220: 191-196

Bär HP (1974) Cyclic nucleotides and smooth muscle. Adv Cyclic Nucleotide Res 4: 195-237

Bailin G, Lopez F (1981) Dinitrophenylation of chicken gizzard myosin: reactivity of the 17,000-dalton light chain. Biochim Biophys Acta 668: 45-56

Bailin G, Lopez F (1982) Thiolysis of dinitrophenylated sulfhydryl groups in gizzard myosin. Restoration of regulatory properties in a reconstituted actomyosin. J Biol Chem 257: 264-270

Barron JT, Bárány M, Bárány K (1979) Phosphorylation of the 20,000-dalton light chain of myosin of intact arterial smooth muscle in rest and in contraction. J Biol Chem 254: 4954-4956

Barron JT, Bárány M, Bárány K, Storti RV (1980) Reversible phosphorylation and dephosphorylation of the 20,000-dalton light chain of myosin during the contraction-relaxation-contraction cycle of arterial smooth muscle. J Biol Chem 255: 6238-6244

Bhalla RC, Sharma RV, Gupta RC (1982) Isolation of two myosin light-chain kinases from bovine carotid artery and their regulation by phosphorylation mediated by cyclic AMP-dependent protein kinase. Biochem J 203: 583-592

Blumenthal DK, Stull JT (1980) Activation of skeletal muscle myosin light chain kinase by calcium (2+) and calmodulin. Biochemistry 19: 5608-5614

Brostrom VO, Huang YC, Breckenridge BM, Wolff DJ (1975) Identification of a calcium-binding protein as a calcium-dependent regulator of brain adenylate cyclase. Proc Natl Acad Sci USA 72: 64-68

Butler TM, Siegman MJ (1982) Chemical energetics of contraction in mammalian smooth muscle. Fed Proc 41: 204-208

Cande WZ, Tooth PJ, Kendrick-Jones J (1983) Regulation of contraction and thick filament assembly-disassembly in glycerinated vertebrate smooth muscle cells. J Cell Biol 97: 1062-1071

Carsten ME (1971) Uterine smooth muscle: troponin. Arch Biochem Biophys 147: 353-357

Cassidy PS, Kerrick WGL (1982) Superprecipitation of gizzard actomyosin, and tension in gizzard muscle skinned fibers in the presence of nucleotides other than ATP. Biochim Biophys Acta 705: 63-69

Cassidy PS, Hoar PE, Kerrick WGL (1979) Irreversible thiophosphorylation and activation of tension in functionally skinned rabbit ileum strips by [^{35}S]ATPγS. J Biol Chem 254: 11148-11153

Cassidy P, Hoar PE, Kerrick WGL (1980) Inhibition of Ca^{2+}-activated tension and myosin light chain phosphorylation in skinned smooth muscle strips by the phenothiazines. Pflügers Arch Eur J Physiol 387: 115-120

Castagna M, Takai Y, Kaibuchi K, Sano K, Kikkawa U, Nishizuka Y (1982) Direct activation of calcium-activated, phospholipid-dependent protein kinase by tumor-promoting phorbol esters. J Biol Chem 257: 7847-7851

Cavadore JC, Molla A, Harricane MC, Gabrion J, Benyamin Y, Demaille JG (1982) Subcellular localization of myosin light chain kinase in skeletal, cardiac and smooth muscles. Proc Natl Acad Sci USA 79: 3475-3479

Chacko S, Conti MA, Adelstein RS (1977) Effect of phosphorylation of smooth muscle myosin on actin activation and Ca^{2+} regulation. Proc Natl Acad Sci USA 74: 129-133

Chantler PD, Szent-Györgyi AG (1980) Regulatory light-chains and scallop myosin: full dissociation, reversibility, and co-operative effects. J Mol Biol 138: 473-492

Chatterjee M, Murphy RA (1983) Calcium-dependent stress maintenance without myosin phosphorylation in skinned smooth muscle. Science (Wash DC) 221: 464-466

Cheung WY (1970) Cyclic 3',5'-nucleotide phosphodiesterase. Demonstration of an activator. Biochem Biophys Res Commun 38: 533-538

Cheung WY (1980) Calmodulin plays a pivotal role in cellular regulation. Science (Wash DC) 207: 19-27

Cheung WY, Bradham LS, Lynch TJ, Lin YM, Tallant EA (1975) Protein activator of cyclic 3':5'-nucleotide phosphodiesterase of bovine or rat brain also activates its adenylate cyclase. Biochem Biophys Res Commun 66: 1055-1062

Cohen P, Burchell A, Foulkes JG, Cohen PTW, Vanaman TC, Nairn AC (1978) Identification of the Ca^{2+}-dependent modulator protein as the fourth subunit of rabbit skeletal muscle phosphorylase kinase. FEBS Lett 92: 287-293

Cole HA, Grand RJA, Perry SV (1982) Non-correlation of phosphorylation of the P-light chain and the actin activation of the ATPase of chicken gizzard myosin. Biochem J 206: 319-328

Cole HA, Patchell VB, Perry SV (1983) Phosphorylation of chicken gizzard myosin and the Ca^{2+}-sensitivity of the actin-activated Mg^{2+}-ATPase. FEBS Lett 158: 17-20

Collins JH (1976) Homology of myosin DTNB light chain with alkali light chains, troponin C and parvalbumin. Nature (Lond) 259: 699-700

Conti MA, Adelstein RS (1981) The relationship between calmodulin binding and phosphorylation of smooth muscle myosin kinase by the catalytic subunit of 3':5' cAMP-dependent protein kinase. J Biol Chem 256: 3178-3181

Cooke P (1982) A reversible change in the functional organization of thin filaments in smooth muscle fibers. Eur J Cell Biol 27: 55-61

Craig R, Smith R, Kendrick-Jones J (1983) Light-chain phosphorylation controls the conformation of vertebrate non-muscle and smooth muscle myosin molecules. Nature (Lond) 302: 436-439

Crosby ND, Diamond J (1980) Effects of phenothiazines on calcium induced contractions of chemically skinned smooth muscle. Proc West Pharmacol Soc 23: 335-338

Dabrowska R, Aromatorio D, Sherry JMF, Hartshorne DJ (1977) Composition of the myosin light chain kinase from chicken gizzard. Biochem Biophys Res Commun 78: 1263-1272

Dabrowska R, Hinkins S, Walsh MP, Hartshorne DJ (1982) The binding of smooth muscle myosin light chain kinase to actin. Biochem Biophys Res Commun 107: 1524-1531

deLanerolle P, Stull JT (1980) Myosin phosphorylation during contraction and relaxation of tracheal smooth muscle. J Biol Chem 255: 9993-10000

deLanerolle P, Adelstein RS, Feramisco JR, Burridge K (1981) Characterization of antibodies to smooth muscle myosin kinase and their use in localizing myosin kinase in nonmuscle cells. Proc Natl Acad Sci USA 78: 4738-4742

Dillon PF, Aksoy MO, Driska SP, Murphy RA (1981) Myosin phosphorylation and the cross-bridge cycle in arterial smooth muscle. Science (Wash DC) 211: 495-497

DiSalvo J, Gifford D, Jiang MJ (1983a) Properties and function of phosphatases from vascular smooth muscle. Fed Proc 42: 67-71

DiSalvo J, Gifford D, Bialojan C, Rüegg JC (1983b) An aortic spontaneously active phosphatase dephosphorylates myosin and inhibits actin-mysoin interaction. Biochem Biophys Res Commun 111: 906-911

Ebashi S (1980) Regulation of muscle contraction. Proc R Soc Lond B Biol Sci 207: 259-286

Ebashi S, Toyo-oka T, Nonomura Y (1975) Gizzard troponin. J Biochem (Tokyo) 78: 859-861

Ebashi S, Mikawa T, Hirata M, Toyo-oka T, Nonomura Y (1977) Regulatory proteins of smooth muscle. In: Casteels R, Godfraind T, Rüegg JC (eds) Excitation-contraction coupling in smooth muscle. Elsevier/North Holland Biomedical, Amsterdam, pp 325-334

Ebisawa K (1983) Ca^{2+} regulation not associated with phosphorylation of myosin light chain in aortic intima smooth muscle. J Biochem (Tokyo) 93: 935-937

Endo T, Obinata T (1981) Troponin and its components from ascidian smooth muscle. J Biochem (Tokyo) 89: 1599-1608

Endo M, Kitazawa T, Yagi S, Iino M, Kakuta Y (1977) Some properties of chemically skinned smooth muscle fibers. In: Casteels R, Godfraind T, Rüegg JC (eds) Excitation-contraction coupling in smooth muscle. Elsevier/North Holland, Amsterdam, pp 199-210

Endo T, Naka M, Hidaka H (1982) Ca^{2+}-phospholipid-dependent phosphorylation of smooth muscle myosin. Biochem Biophys Res Commun 105: 942-948

Fay FS, Cooke PH (1973) Reversible disaggregation of myofilaments in vertebrate smooth muscle. J Cell Biol 56: 399-411

Ferenczi MA, Homsher E, Trentham DR, Weeds AG (1978) Preparation and characterization of frog muscle myosin subfragment 1 and actin. Biochem J 171: 155-163

Filo RS, Bohr DF, Rüegg JC (1965) Glycerinated skeletal and smooth muscle: calcium and magnesium dependence. Science (Wash DC) 147: 1581-1583

Frearson N, Focant BWW, Perry SV (1976) Phosphorylation of a light chain component of myosin from smooth muscle. FEBS Lett 63: 27-32

Gallis B, Edelman AM, Casnellie JE, Krebs EG (1983) Epidermal growth factor stimulates tyrosine phosphorylation of the myosin regulatory light chain from smooth muscle. J Biol Chem 258: 13089-13093

Gergely P, Vereb G, Bot G (1976) Thiophosphate-activated phosphorylase kinase as a probe in the regulation of phosphorylase phosphatase. Biochim Biophys Acta 429: 809-816

Gerthoffer WT, Murphy RA (1983) Myosin phosphorylation and regulation of the cross-bridge cycle in tracheal smooth muscle. Am J Physiol 244 (Cell Physiol 13): C182-C187

Gopinath RM, Vincenzi FF (1977) Phosphodiesterase protein activator mimics red blood cell cytoplasmic activator of (Ca^{2+} + Mg^{2+})ATPase. Biochem Biophys Res Commun 77: 1203-1209

Grand RJA, Perry SV (1979) Calmodulin-binding proteins from brain and other tissues. Biochem J 183: 285-295

Grand RJA, Perry SV, Weeks RA (1979) Troponin C-like proteins (calmodulins) from mammalian smooth muscle and other tissues. Biochem J 177: 521-529

Grand RJA, Patchell VB, Darby MK, Perry SV (1982) A factor inhibiting the Mg^{2+}-stimulated ATPase of gizzard actomyosin. J Muscle Res Cell Motil 3: 479-480

Gratecos D, Fischer EH (1974) Adenosine 5'-0(3-thiotriphosphate) in the control of phosphorylase activity. Biochem Biophys Res Commun 58: 960-967

Greaser ML, Gergely J (1971) Reconstitution of troponin activity from three protein components. J Biol Chem 246: 4226-4233

Greaser ML, Gergely J (1973) Purification and properties of the components from troponin. J Biol Chem 248: 2125-2133

Guerriero V Jr, Rowley DR, Means AR (1981) Production and characterization of an antibody to myosin light chain kinase and intracellular localization of the enzyme. Cell 27: 449-458

Hartshorne DJ, Mrwa U (1980) The phosphorylation of gizzard myosin and isolated light chains by myosin light chain kinase. In: Siegel FL, Carafoli E, Kretsinger RH, McLennan DH, Wasserman RH (eds) Calcium-binding proteins: structure and function. Elsevier/North Holland, Amsterdam, pp 255-262

Hartshorne DJ, Mrwa U (1982) Regulation of smooth muscle actomyosin. Blood Vessels 19: 1-18

Head JF, Weeks RA, Perry SV (1977) Affinity-chromatographic isolation and some properties of troponin C from different muscle types. Biochem J 161: 465-471

Hellam DC, Podolsky RJ (1969) Force measurements in skinned muscle fibers. J Physiol (Lond) 200: 807-819

Hidaka H, Asano M, Iwadare S, Matsumoto I, Totsuka T, Aoki M (1978) A novel vascular relaxing agent, N-(6-aminohexyl)-5-chloro-1-naphthalenesulfonamide which affects vascular smooth muscle actomyosin. J Pharmacol Exp Ther 207: 8-15

Hidaka H, Naka M, Yamaki T (1979a) Effect of novel specific myosin light chain kinase inhibitors on Ca^{2+}-activated Mg^{2+}-ATPase of chicken gizzard actomyosin. Biochem Biophys Res Commun 90: 694-699

Hidaka H, Yamaki T, Totsuka T, Asano M (1979b) Selective inhibitors of Ca^{2+}-binding modulator of phosphodiesterase produce vascular relaxation and inhibit actin-myosin interaction. Mol Pharmacol 15: 49-59

Hidaka H, Yamaki T, Naka M, Tanaka T, Hayashi H, Kobayashi R (1980) Calcium-regulated modulator protein interacting agents inhibit smooth muscle calcium-stimulated protein kinase and ATPase. Mol Pharmacol 17: 66-72

Hidaka H, Asano M, Tanaka T (1981) Activity-structure relationship of calmodulin antagonists. Naphthalenesulfonamide derivatives. Mol Pharmacol 20: 571-578

Higashi K, Fukunaga K, Matsui K, Maeyama M, Miyamoto E (1983) Purification and characterization of myosin light-chain kinase from porcine myometrium and its phosphorylation and modulation by cyclic AMP-dependent protein kinase. Biochim Biophys Acta 747: 232-240

Hirata M, Mikawa T, Nonomura Y, Ebashi S (1977) Ca^{2+} regulation in vascular smooth muscle. J Biochem (Tokyo) 82: 1793-1796

Hoar PE, Kerrick WGL, Cassidy PS (1979) Chicken gizzard: Relation between calcium-activated phosphorylation and contraction. Science (Wash DC) 204: 503-506

Hoar PE, Walsh MP, Pato MD, Adelstein RS, Hartshorne DJ, Kerrick WGL (1983) The functional effect of Ca^{2+}-insensitive myosin light chain kinase, phosphatase, and catalytic subunit of the cAMP-dependent protein kinase on tension in skinned gizzard smooth muscle fibers. Biophys J 41: 152a

Huxley HE, Hanson J (1954) Changes in the cross-striations of muscle during contraction and stretch and their structural interpretation. Nature (Lond) 173: 973-976

Huxley AF, Niedergerke R (1954) Structural changes in muscle during contraction. Nature (Lond) 173: 971-973

Ikebe M, Ogihara S, Tonomura Y (1982) Non-linear dependence of actin activated Mg^{2+}ATPase activity on the extent of phosphorylation of gizzard myosin and H meromyosin. J Biochem (Tokyo) 91: 1809-1812

Ikebe M, Hinkins S, Hartshorne DJ (1983) Correlation of enzymatic properties and conformation of smooth muscle myosin. Biochemistry 22: 4580-4587

Inoue M, Kishimoto A, Takai Y, Nishizuka Y (1977) Studies on a cyclic nucleotide-independent protein kinase and its proenzyme in mammalian tissues. II Proenzyme and its activation by calcium-dependent protease from rat brain. J Biol Chem 252: 7610-7616

Ito N, Hotta K (1976) Regulatory protein of bovine tracheal smooth muscle. J Biochem (Tokyo) 80: 401-403

Ito N, Takagi T, Hotta K (1976) Regulatory protein of vascular smooth muscle. J Biochem (Tokyo) 80: 899-901

Janis RA, Moats-Staats BM, Gualtieri RT (1980) Protein phosphorylation during spontaneous contraction of smooth muscle. Biochem Biophys Res Commun 96: 265-270

Jakes R, Northrop F, Kendrick-Jones J (1976) Calcium-binding regions of myosin "regulatory" light chains. FEBS Lett 70: 229-234

Jarrett HW, Penniston JT (1977) Partial purification of the Ca^{2+}-Mg^{2+} ATPase activator from human erythrocytes: its similarity to the activator of 3':5'-cyclic nucleotide phosphodiesterase. Biochem Biophys Res Commun 77: 1210-1216

Kakiuchi S, Yamazaki R, Nakajima H (1970) Properties of a heat-stable phosphodiesterase activating factor isolated from brain extracts. Studies on cyclic 3',5'-nucleotide phosphodiesterase. II. Proc Jpn Acad 46: 587-592

Kakiuchi R, Inui M, Morimoto K, Kanda K, Sobue K, Kakiuchi S (1983) Caldesmon, a calmodulin-binding, F-actin interacting protein, is present in aorta, uterus and platelets. FEBS Lett 154: 351-356

Kanamori M, Naka M, Asano M, Hidaka H (1981) Effects of N-(6-aminohexyl)-5-chloro-1-naphthalenesulfonamide and other calmodulin antagonists (calmodulin interacting agents) on calcium-induced contraction of rabbit aortic strips. J Pharmacol Exp Ther 217: 494-499

Kelly RE, Rice RV (1969) Ultrastructural studies on the contractile mechanisms of smooth muscle. J Cell Biol 42: 683-694

Kemp BE, Pearson RB, House C (1982) Phosphorylation of a synthetic heptadecapeptide by smooth muscle myosin light chain kinase. J Biol Chem 257: 13349-13353

Kemp BE, Pearson RB, House C (1983) Role of basic residues in the phosphorylation of synthetic peptides by myosin light chain kinase. Proc Natl Acad Sci USA 80: 7471-7475

Kendrick-Jones J, Scholey JM (1981) Myosin-linked regulatory systems. J Muscle Res Cell Motil 2: 347-372

Kendrick-Jones J, Tooth PJ, Taylor KA, Scholey JM (1981) Regulation of myosin-filament assembly by light-chain phosphorylation. Cold Spring Harbor Symp Quant Biol 46:929-938

Kendrick-Jones J, Cande WZ, Tooth PJ, Smith RC, Scholey JM (1983) Studies on the effect of phosphorylation of the 20,000 M_r light chain of vertebrate smooth muscle myosin. J Mol Biol 165: 139-162

Kerrick WGL, Hoar PE (1981) Inhibition of smooth muscle tension by cyclic AMP-dependent protein kinase. Nature (Lond) 292: 253-255

Kerrick WGL, Krasner B (1975) Disruption of the sarcolemma of mammalian skeletal muscle fibers by homogenization. J Appl Physiol 39: 1052-1055

Kerrick WGL, Hoar PE, Cassidy PS (1980) Calcium-activated tension: The role of myosin light chain phosphorylation. Fed Proc 39: 1558-1563

Klee CB, Vanaman TC (1982) Calmodulin. Adv Protein Chem 35: 213-321

Kuo JF, Andersson RGG, Wise BC et al. (1980) Calcium-dependent protein kinase: widespread occurrence in various tissues and phyla of the animal kingdom and comparison of effects of phospholipid, calmodulin and trifluoperazine. Proc Natl Acad Sci USA 77: 7039-7043

LaPorte DC, Wierman BM, Storm DR (1980) Calcium-induced exposure of a hydrophobic surface on calmodulin. Biochemistry 19: 3814-3819

Levin RM, Weiss B (1977) Binding of trifluoperazine to the calcium-dependent activator of cyclic nucleotide phosphodiesterase. Mol Pharmacol 13: 690-697

Levin RM, Weiss B (1979) Selective binding of antipsychotics and other psychoactive agents to the calcium-dependent activator of cyclic nucleotide phosphodiesterase. J Pharmacol Exp Ther 208: 454-459

LePeuch CJ, Haiech H, Demaille JG (1979) Concerted regulation of cardiac sarcoplasmic reticulum calcium transport by cyclic adenosine monophosphate-dependent and calcium-calmodulin-dependent phosphorylations. Biochemistry 18: 5150-5157

Litten RA III, Solaro RJ, Ford GD (1977) Properties of the calcium-sensitive components of bovine arterial actomyosin. Arch Biochem Biophys 182: 24-32

Marcum JM, Dedman JR, Brinkley BR, Means AR (1978) Control of microtubule assembly-disassembly by calcium-dependent regulator protein. Proc Natl Acad Sci USA 75: 3771-3775

Marshak DR, Watterson DM, vanEldik LJ (1981) Calcium-dependent interaction of S-100b, troponin C, and calmodulin with an immobilized phenothiazine. Proc Natl Acad Sci USA 78: 6793-6797

Marston SB, Trevett RM, Walters M (1980) Calcium ion-regulated thin filaments from vascular smooth muscle. Biochem J 185: 355-365

Matsuda G, Suzuyama Y, Maita T, Umegane T (1977) The L-2 light chain of chicken skeletal muscle myosin. FEBS Lett 84: 53-56

Merkel L, Mrwa U, Bachle-Stulz C (1982) Chicken gizzard ATPase can be activated by a protein factor without concomitant change in phosphorylation of the myosin. J Muscle Res Cell Motil 3: 477-478

Mikawa T, Nonomura Y, Ebashi S (1977) Does phosphorylation of myosin light chain have direct relation to regulation in smooth muscle? J Biochem (Tokyo) 82: 1789-1791

Mikawa T, Nonomura Y, Hirata M, Ebashi S, Kakiuchi S (1978) Involvement of an acidic protein in regulation of smooth muscle contraction by the tropomyosin-leiotonin system. J Biochem (Tokyo) 84: 1633-1636

Miller JR, Silver PJ, Stull JT (1983) The role of myosin light chain kinase phosphorylation in β-adrenergic relaxation of tracheal smooth muscle. Mol Pharmacol 24: 235-242

Minakuchi R, Takai Y, Yu B, Nishizuka Y (1981) Widespread occurrence of calcium-activated, phospholipid-dependent protein kinase in mammalian tissues. J Biochem (Tokyo) 89: 1651-1654

Moore PB, Dedman JR (1982) Calcium-dependent protein binding to phenothiazine columns. J Biol Chem 257: 9663-9667

Morgan JP, Morgan KG (1982) Vascular smooth muscle: the first recorded Ca^{2+} transients. Pflügers Arch Eur J Physiol 395: 75-77

Morgan M, Perry SV, Ottaway J (1976) Myosin light chain phosphatase. Biochem J 157: 687-697

Mrwa U, Hartshorne DJ (1980) Phosphorylation of smooth muscle myosin and myosin light chains. Fed Proc 39: 1564-1568

Mrwa U, Troschka M, Rüegg JC (1979) Cyclic AMP-dependent inhibition of smooth muscle actomyosin. FEBS Lett 107: 371-374

Murphy RA (1979) Filament organization and contractile function in vertebrate smooth muscle. Annu Rev Physiol 41: 737-748

Naka M, Nishikawa M, Adelstein RS, Hidaka H (1983) Phorbol ester-induced activation of human platelets is associated with protein kinase C phosphorylation of myosin light chains. Nature (Lond) 306: 490-492

Namm DL, Leader JP (1976) Occurrence and function of cyclic nucleotides in blood vessels. Blood Vessels 13: 24-47

Ngai PK, Carruthers CA, Walsh MP (1984) Isolation of the native form of chicken gizzard myosin light chain kinase. Biochem J 218: 863-870

Niedel JE, Kuhn LJ, Vandenbark GR (1983) Phorbol diester receptor copurifies with protein kinase C. Proc Natl Acad Sci USA 80: 36-40

Noiman ES (1980) Phosphorylation of smooth muscle myosin light chains by cAMP-dependent protein kinase. J Biol Chem 255: 11067-11070

Nonomura Y, Ebashi S (1980) Calcium regulatory mechanism in vertebrate smooth muscle. Biomed Res 1: 1-14

Ohashi M, Nonomura Y (1979) Disappearance of smooth muscle thick filaments during K^+ contraction. Cell Struct Funct 4: 325-329

Okamoto Y, Sekine T (1981) N-terminal region of gizzard myosin heavy chain is critical for the ATPase activity. J Biochem (Tokyo) 90: 833-842

Onishi H, Wakabayashi T (1982) Electron microscopic studies of myosin molecules from chicken gizzard muscle I: the formation of the intramolecular loop in the myosin tail. J Biochem (Tokyo) 92: 871-879

Onishi H, Umeda J, Uchiwa H, Watanabe S (1982) Purification of gizzard myosin light-chain phosphatase, and reversible changes in the ATPase and superprecipitation activities of actomyosin in the presence of purified preparations of myosin light-chain phosphatase and kinase. J Biochem (Tokyo) 91: 265-271

Pato MD, Adelstein RS (1980) Dephosphorylation of the 20,000-dalton light chain of myosin by two different phosphatases from smooth muscle. J Biol Chem 255: 6535-6538

Pato MD, Adelstein RS (1983a) Purification and characterization of a multisubunit phosphatase from turkey gizzard smooth muscle. J Biol Chem 258: 7047-7054

Pato MD, Adelstein RS (1983b) Characterization of a Mg^{2+}-dependent phosphatase from turkey gizzard smooth muscle. J Biol Chem 258: 7055-7058

Paul RJ, diSalvo J, Rüegg JC (1982) Myosin phosphatase fractions enhance relaxation in chemically skinned vascular smooth muscle. Biophys J 37: 186a

Perrie WT, Smillie LB, Perry SV (1973) A phosphorylated light-chain component of myosin from skeletal muscle. Biochem J 135: 151-164

Persechini A, Hartshorne DJ (1981) Phosphorylation of smooth muscle myosin: evidence for cooperativity between the myosin heads. Science (Wash DC) 213: 1383-1385

Persechini A, Hartshorne DJ (1982) Cooperative behavior of smooth muscle myosin. Fed Proc 41: 2868-2872

Persechini A, Hartshorne DJ (1983) Ordered phosphorylation of the two 20,000 molecular weight light chains of smooth muscle myosin. Biochemistry 22: 470-476

Persechini A, Mrwa U, Hartshorne DJ (1981) Effect of phosphorylation on the actin-activated ATPase activity of myosin. Biochem Biophys Res Commun 98: 800-805

Piascik MT, Babich M, Rush ME (1983) Calmodulin stimulation and calcium regulation of smooth muscle adenylate cyclase activity. J Biol Chem 258: 10913-10918

Pires EMV, Perry SV (1977) Purification and properties of myosin light-chain kinase from fast skeletal muscle. Biochem J 167: 137-146

Rees DD, Frederiksen DW (1981) Calcium regulation of porcine aortic myosin. J Biol Chem 256: 357-364

Rüegg JC, Sparrow MP, Mrwa U (1981) Cyclic AMP-mediated relaxation of chemically skinned fibers of smooth muscle. Pflügers Arch Eur J Physiol 390: 198-201

Rüegg JC, diSalvo J, Paul RJ (1982) Soluble relaxation factor from vascular smooth muscle: a myosin light chain phosphatase. Biochem Biophys Res Commun 106: 1126-1133

Sands H, Penberthy W, Meyer TA, Jorgensen R (1976) Cyclic AMP-stimulated phosphorylation of bovine tracheal smooth muscle contractile and noncontractile proteins. Biochim Biophys Acta 445: 791-801

Seidel JC (1979) Activation by actin of ATPase activity of chemically modified gizzard myosin without phosphorylation. Biochem Biophys Res Commun 89: 958-964

Seidel JC (1980) Fragmentation of gizzard myosin by α-chymotrypsin and papain, the effects on ATPase activity, and the interaction with actin. J Biol Chem 255: 4355-4361

Sellers JR, Adelstein RS (1982) Cooperativity and the reversible phosphorylation of smooth muscle heavy meromyosin. Biophys J 37: 262a

Sellers JR, Pato MD, Adelstein RS (1981) Reversible phosphorylation of smooth muscle myosin, heavy meromyosin, and platelet myosin. J Biol Chem 256: 13137-13142

Sherry JMF, Gorecka A, Aksoy MO, Dabrowska R, Hartshorne DJ (1978) Roles of calcium and phosphorylation in the regulation of the activity of gizzard myosin. Biochemistry 17: 4411-4418

Sheterline P (1980) Trifluoperazine can distinguish between myosin light chain kinase-linked and troponin C-linked control of actomyosin interaction by Ca^{2+}. Biochem Biophys Res Commun 93: 194-200

Shibata N, Yamagami T, Yoneda S, Akagami H, Takeuchi K, Tanaka K, Okamura Y (1973) Identification of myosin A, actin and native tropomyosin constitution of arterial contractile protein (myosin B) and their characteristics. Jpn Circ J 37: 229-232

Shoenberg CF (1973) The influence of temperature on the thick filaments of vertebrate smooth muscle. Philos Trans R Soc Lond B Biol Sci 265: 197-202

Silver PJ, diSalvo J (1979) Adenosine 3':5'-monophosphate-mediated inhibition of myosin light chain phosphorylation in bovine aortic actomyosin. J Biol Chem 254: 9951-9954

Silver PJ, Stull JT (1982) Regulation of myosin light chain and phosphorylase phosphorylation in tracheal smooth muscle. J Biol Chem 257: 6145-6150

Silver PJ, Stull JT (1983) Effects of the calmodulin antagonist, fluphenazine, on phosphorylation of myosin and phosphorylase in intact smooth muscle. Mol Pharmacol 23: 665-670

Silver PJ, Holroyde MJ, Solaro RJ, diSalvo J (1981) Ca^{2+}, calmodulin and cyclic AMP-dependent modulation of actin-myosin interactions in aorta. Biochim Biophys Acta 674: 65-70

Singh TJ, Akatsuka A, Huang K-P (1983) Phosphorylation of smooth muscle myosin light chain by five different kinases. FEBS Lett 159: 217-220

Small JV, Sobieszek A (1980) The contractile apparatus of smooth muscle. Int Rev Cytol 64: 241-306

Sobieszek A (1977) Vertebrate smooth muscle myosin. Enzymatic and structural properties. In: Stephens NL (ed) The biochemistry of smooth muscle. University Park Press, Baltimore, Maryland, pp 413-443

Sobue K, Muramoto Y, Fujita M, Kakiuchi S (1981a) Purification of a calmodulin-binding protein from chicken gizzard that interacts with F-actin. Proc Natl Acad Sci USA 78:5652-5655

Sobue K, Muramoto Y, Fujita M, Kakiuchi S (1981b) Calmodulin-binding protein from chicken gizzard that interacts with F-actin. Biochem Int 2: 469-476

Sobue K, Morimoto K, Inui M, Kanda K, Kakiuchi S (1982a) Control of actin-myosin interaction of gizzard smooth muscle by calmodulin- and caldesmon-linked flip-flop mechanism. Biomed Res 3: 188-196

Sobue K, Morimoto K, Kanda K, Maruyama K, Kakiuchi S (1982b) Reconstitution of Ca^{2+}-sensitive gelation of actin filaments with filamin, caldesmon, and calmodulin. FEBS Lett 138: 289-292

Somlyo AV (1980) Ultrastructure of vascular smooth muscle. In: Bohr DF, Somlyo AP, Sparks HV (eds) The cardiovascular system II: vascular smooth muscle. American Physiological Society, Bethesda, MD, pp 33-67 (Handbook of physiology, Section 2)

Somlyo AP, Somlyo AV, Shuman H (1979) Electron probe analysis of vascular smooth muscle: composition of mitochondria, nuclei and cytoplasm. J Cell Biol 81: 316-335

Somlyo AV, Butler TM, Bond M, Somlyo AP (1981) Myosin filaments have nonphosphorylated light chains in relaxed smooth muscle. Nature (Lond) 294: 567-569

Somlyo AP, Somlyo AV, Shuman H, Endo M (1982) Calcium and monovalent ions in smooth muscle. Fed Proc 41: 2883-2890

Sparrow MP, van Bockxmeer FM (1972) Arterial tropomyosin and a relaxing protein fraction from vascular smooth muscle. Comparison with skeletal tropomyosin and troponin. J Biochem (Tokyo) 1075-1080

Sparrow MP, Mrwa U, Hofmann F, Rüegg JC (1981) Calmodulin is essential for smooth muscle contraction. FEBS Lett 125: 141-145

Sparrow MP, Pfitzer G, Gagelmann M, Rüegg JC (1983) Effect of calmodulin, Ca^{2+} and cAMP protein kinase on skinned tracheal smooth muscle. Am J Physiol 246: 308-314

Stewart AA, Ingebritsen TS, Manalan A, Klee CB, Cohen P (1982) Discovery of a Ca^{2+}- and calmodulin-dependent protein phosphatase: probable identity with calcineurin (CaM-BP$_{80}$). FEBS Lett 137: 80-84

Suzuki H, Onishi H, Takahashi K, Watanabe S (1978) Structure and function of chicken gizzard myosin. J Biochem (Tokyo) 84: 1529-1542

Suzuki H, Kamata T, Onishi H, Watanabe S (1982a) Adenosine triphosphate-induced reversible change in the conformation of chicken gizzard myosin and heavy meromyosin. J Biochem (Tokyo) 91: 1699-1705

Suzuki H, Takahashi K, Onishi H, Watanabe S (1982b) Reversible changes in the state of phosphorylation of gizzard myosin, in that of gizzard myosin assembly, in the ATPase activity of gizzard myosin, in that of actomyosin and in the superprecipitation activity. J Biochem (Tokyo) 91: 1687-1698

Szent-Györgyi AG, Szentkiralyi EM, Kendrick-Jones J (1973) The light chains of scallop myosin as regulatory subunits. J Mol Biol 74: 179-203

Tanaka T, Hidaka H (1980) Hydrophobic regions function in calmodulin-enzyme(s) interactions. J Biol Chem 255: 11078-11080

Trybus KM, Huiatt TW, Lowey S (1982) A bent monomeric conformation of myosin from smooth muscle. Proc Natl Acad Sci USA 79: 6151-6155

Uchiwa H, Kato T, Onishi H, Isobe T, Okuyama T, Watanabe S (1982) Purification of chicken gizzard myosin light-chain kinase, and its calcium and strontium sensitivities as compared with those of superprecipitation and ATPase activities of actomyosin. J Biochem (Tokyo) 91: 273-282

Vallet B, Molla A, Demaille JG (1981) Cyclic adenosine 3',5'-monophosphate-dependent regulation of purified bovine aortic calcium/calmodulin-dependent myosin light chain kinase. Biochim Biophys Acta 674: 256-264

Walsh MP, Hartshorne DJ (1982) Actomyosin of smooth muscle. In: Cheung WY (ed) Calcium and cell function, vol III. Academic Press, pp 223-269

Walsh MP, Hartshorne DJ (1983) Calmodulin. In: Stephens NL (ed) Biochemistry of smooth muscle, vol II. CRC Press, pp 1-84

Walsh M, Stevens FC (1978) Chemical modification studies on the Ca^{2+}-dependent protein modulator: the role of methionine residues in the activation of cyclic nucleotide phosphodiesterase. Biochemistry 17: 3924-3930

Walsh MP, Persechini A, Hinkins S, Hartshorne DJ (1981) Is smooth muscle myosin a substrate for the cAMP-dependent protein kinase? FEBS Lett 126: 107-110

Walsh MP, Bridenbaugh R. Hartshorne DJ, Kerrick WGL (1982a) Phosphorylation-dependent activated tension in skinned gizzard muscle fibers in the absence of Ca^{2+}. J Biol Chem 257: 5987-5900

Walsh MP, Dabrowska R, Hinkins S, Hartshorne DJ (1982b) Calcium-independent myosin light chain kinase of smooth muscle. Preparation by limited chymotryptic digestion of the Ca^{2+}-dependent enzyme, purification and characterization. Biochemistry 21: 1919-1925

Walsh MP, Hinkins S, Flink IL, Hartshorne DJ (1982b) Bovine stomach myosin light chain kinase: purification, characterization, and comparison with the turkey gizzard enzyme. Biochemistry 21: 6890-6896

Walsh MP, Bridenbaugh R, Kerrick WGL, Hartshorne DJ (1983a) Gizzard Ca^{2+}-independent myosin light chain kinase: evidence in favor of the phosphorylation theory. Fed Proc 42: 45-50

Walsh MP, Hinkins S, Dabrowska R, Hartshorne DJ (1983b) Smooth muscle myosin light chain kinase. Methods Enzymol 99: 279-288

Walters M, Marston SB (1981) Phosphorylation of the calcium-ion regulated thin filaments from vascular smooth muscle. A new regulatory mechanism? Biochem J 197: 127-139

Wang JH, Desai R (1977) Modulator binding protein. Bovine brain protein exhibiting the Ca^{2+}-dependent association with the protein modulator of cyclic nucleotide phosphodiesterase. J Biol Chem 252: 4175-4184

Werth DK, Haeberle JR, Hathaway DR (1982) Purification of a myosin phosphatase from bovine aortic smooth muscle. J Biol Chem 257: 7306-7309

Zetterquist O, Ragnarsson V (1982) The structural requirements of substrates of cyclic AMP-dependent protein kinase. FEBS Lett 139: 287-290

Pharmacology of Calcium Antagonists

T. GODFRAIND[1]

CONTENTS

1 Introduction

A calcium antagonist may be defined as a drug able to interact with the function of calcium during cellular activation by physiological or pathological processes. This group of drugs has also received other names, such as calcium channels blockers, calcium entry blockers, or calcium overload inhibitors.

The concept of calcium antagonism has been developed on the basis of two different and parallel experimental approaches, one dealing mainly with studies on smooth muscle (see Godfraind 1981), the other on cardiac muscle (see Fleckenstein 1982). The first calcium antagonists were synthetized at the end of the 1950's by pharmaceutical companies that were completely unaware of the mechanism of action of their own compounds. Cinnarizine, the first drug to be recognized as a calcium antagonist (Godfraind and Polster 1968) was initially described as an anti-histaminic and a motion-sickness drug; on the other hand, verapamil was introduced as a β-blocker.

In this chapter, I would like to discuss the criteria allowing an organic compound to be considered as being a calcium antagonist, then to examine the various drug families so far identified, and finally to explore their potential interest for medicine.

[1] Laboratoire de Pharmacodynamie Générale et de Pharmacologie, Université Catholique de Louvain, UCL 7350, Avenue Emmanuel Mounier, 73, B-1200 Bruxelles, Belgium

Many observations show basic similarities in the biochemical mechanisms responsible for calcium handling in various tissues, and indicate that these mechanisms are unevenly distributed between tissues. A comprehensive analysis of the action of calcium antagonists therefore requires their study to be done in various tissues. Such a task would be beyond the scope of this chapter. Therefore, because calcium antagonists were first characterized in smooth muscle (Godfraind and Polster 1968), and because vascular smooth muscle appears to be the main target for the therapeutic effect of most of the clinically used calcium antagonists, their properties will be examined in this tissue.

2 Characterization of Calcium Antagonists

Calcium antagonists are considered as drugs interfering with the function of calcium. Therefore, any organic compound that will mimic calcium withdrawal in any biological system could belong to this group. Although any tissue could be used for this characterization, smooth muscle appears to be one of the most convenient. The simplest experimental procedure with which to assess the ability of a pharmacological compound to antagonize Ca^{2+} function is to preincubate a smooth muscle in Ca^{2+}-free physiological solution and then depolarize it with a KCl-rich solution and gradually increase the Ca^{2+} concentration in the bathing solution. This will result in an increase in tension dependent on Ca^{2+} concentration. When this procedure is repeated in the presence of a calcium antagonist, the contractile responses are depressed in a concentration-dependent manner. When concentration-effect curves are established on the basis of experimental data, the antagonism so far observed may appear as a competitive or a noncompetitive one. This phenomenological appearance of the antagonism not only depends upon the drug used, but also upon the tissue examined. Figure 1 illustrates such Ca^{2+} concentration-effect curves obtained in rabbit mesenteric arteries and in rat aorta using cinnarizine as an antagonist. The preparations are very sensitive to increasing Ca^{2+} concentration and 20 μM of the cation can trigger the contraction. The response to the cation is attenuated by cinnarizine concentration as low as 10^{-9} M. In the case of some dihydropyridines lower concentrations are active (Godfraind 1983a).

The observation of a contractile response to increasing extracellular calcium concentration in a muscle depolarized in Ca-free solution is a strong indication that in such conditions there is an increase in membrane permeability for calcium. The depression of this calcium effect by a drug may be interpreted as a reduction of membrane permeability for calcium. However, as already pointed out by Godfraind and Kaba (1969), other interpretations are possible. The drug could indeed interfere with the contractile machinery, causing a reduction of the contractile response. Such an interpretation has recently been reconsidered by Spedding (1983) for cinnarizine. This author has used taenia preparations from the guinea-pig caecum which were treated with Triton X-100 and glycerol to disrupt the plasma membrane. The preparations contracted in response to low concentration of calcium (10-40 μM) and the contractions were dependent upon exogenous ATP. The author has tested several calcium antagonists at the concentration

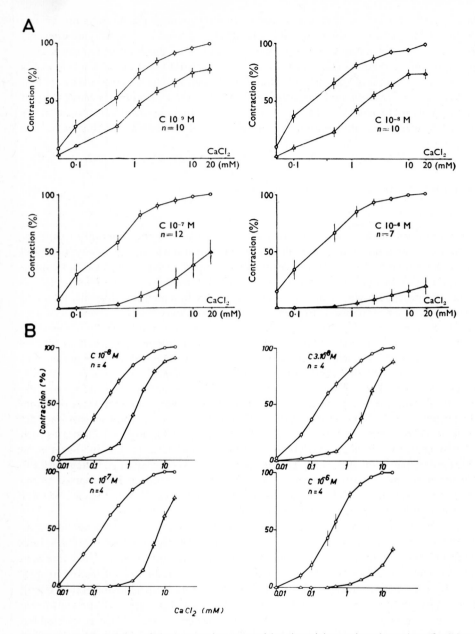

Fig. 1. A, B. Illustration of the antagonism exerted by cinnarizine against the action of calcium in depolarized smooth muscle. **A** Antagonism in isolated rabbit mesenteric arteries. (Godfraind and Kaba 1969). **B** Antagonism observed in isolated rat aorta (unpublished results). Note the appearance of a noncompetitive antagonism in **A** and of a competitive one in **B**

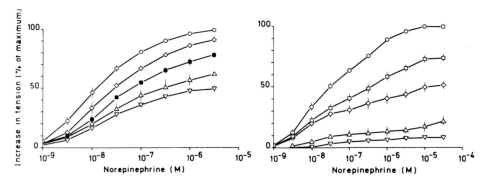

Fig. 2 The inhibition by flunarizine of norepinephrine-induced contractions of rat aorta *(left)* and rat superior mesenteric artery *(right)*. Each graph shows concentration-effect curves obtained in the absence (○) and in the presence of various concentrations of flunarizine [M^{-8} M (◊), 10^{-7} M (■) and 3×10^{-7} M (△) and 3×10^{-6} M (▲)]. (After Godfraind and Dieu 1981). Note the higher sensitivity of mesenteric vessels to flunarizine and the difference in the magnitude of the part of the contraction resistant to the antagonist, this part being much less important in mesenteric artery. This is the case not only for flunarizine but also for nifedipine. (See Godfraind 1983)

of 100 μM. Nifedipine, verapamil, and diltiazem did not inhibit Ca^{2+}-induced activation of the contractile proteins. In contrast, fendiline, cinnarizine, flunarizine, pimozide and trifluoperazine inhibited calcium-induced contraction. The effects of cinnarizine were reversible. These observations are to some extent at variance with those of Ishizuka and Endo (1983), who have reported that diltiazem inhibits the contraction of skinned skeletal muscle fibers but in agreement with those of Kreye et al. (1983), who studied the action of various calcium antagonists on skinned renal rabbit arterial muscle. Verapamil, nifedipine and felodipine were ineffective, whereas trifluoperazine, W-7 and fendiline relaxed Ca-calmodulin-induced contraction. The significance of the observations with skinned fibers must be interpreted with caution, mainly because the concentrations required to observe an effect are very high when compared to the concentrations active on calcium-evoked contraction of depolarized intact muscle. Studies of agonist-evoked contraction may help to obtain a first indication on the site of action: membrane versus contractile proteins. In rat aorta, drugs from the diphenylpiperazines and from the dihydropyridines group reduce the response to various agonists including norepinephrine, prostaglandins and serotonin. One of the characteristics of this antagonism is the existence of a large fraction of the response that is resistant to the blocker (Fig. 2). This resistant contraction represents most of the response that can be elicited in Ca-free medium, the latter being completely resistant to the drugs of the two groups. This residual response is thought to be due to the release of intracellularly bound Ca^{2+} (Godfraind and Kaba 1969). Several treatments are able to greatly attenuate or to abolish this response to agonists in Ca-free solution, one is the duration of the incubation in the presence of a Ca^{2+} chelator such as EGTA, another is the action of the agonist that can totally release the intracellularly stored calcium (Broekaert and Godfraind 1973; Deth and van Breemen 1977; Karaki et al. 1982; Godfraind and Miller 1983). This agonist-evoked response, thought to be due to intracellular Ca release, can

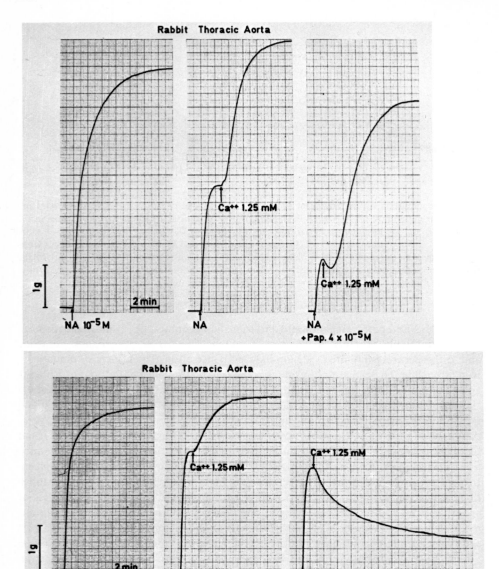

Fig. 3. Rabbit thoracic aorta, dissociation of contraction evoked by internal release of calcium and by calcium entry through receptor-operated channel. *Lower panel: from left to right* response to noradrenaline in physiological solution, response to noradrenaline in Ca-free solution, restablishment of the maximum response after readmission of the normal calcium concentration, the same experiment in the presence of lanthanum showing that only the contraction evoked by calcium readmission is abolished. *Upper panel* a similar experimental sequence with papaverine instead of lanthanum showing that the antispasmodic agent specifically reduces the contraction evoked by noradrenaline in the absence of extracellular calcium. (Experiment from Broeckaert and Godfraind 1979)

also be attenuated by drugs such as papaverine (Fig. 3) or diltiazem (Saida and van Breemen 1983). Figure 3 illustrates experiments performed with papaverine in rabbit isolated vessels. In a Ca^{2+}-free solution, 10^{-5} M norepinephrine evokes a fast non-sustained contraction; after the readmission of Ca^{2+}, a slow contraction develops. The action of papaverine and cinnarizine (Fig. 4) on the two components of the contractile response was studied using concentration of inhibitors that reduced by 50% the contraction evoked by norepinephrine in physiological solution. Cinnarizine selectively reduces the extracellular Ca^{2+}-dependent response, whereas papaverine inhibits the initial fast response more than the contraction elicited when calcium is readmitted to the bathing solution at the peak of noradrenaline response. Such experiments show that it is possible to dissociate the calcium pools involved in the control of contraction, and suggest that the action on membrane is more important than any other hypothetical effect.

However, because drugs may have more than one action site in a biological system, the evidence given by contraction studies may not always appear convincing, and more direct estimates of calcium fluxes are required. Several attempts have been made to estimate the changes in cytoplasmic Ca^{2+} occurring during excitation of smooth muscle. Direct measurements of ^{45}Ca movements are not feasible in smooth muscle, because they are not indicative of changes that could occur within the cell. Indeed, a large amount of calcium is bound on extracellular sites, in order to allow the identification of the biologically active calcium fraction, several attempts have been made; the most successful are based on the use of lanthanum. Lanthanum replaces Ca^{2+} on superficial binding sites and does not penetrate the cell. Because lanthanum blocks transmembrane fluxes of calcium, it has been proposed that the calcium content of a muscle washed in a lanthanum solution could provide an estimate of cellular calcium. The ^{45}Ca fluxes across the smooth muscle cell membrane may be estimated in the rat aorta by measuring the ^{45}Ca turnover in the La^{3+}-resistant Ca fraction. This fraction corresponds to the amount of Ca that is not displaced when the tissue is soaked in 50 mM La^{3+} solution (Godfraind 1976). Noradrenaline increases the rate of uptake of ^{45}Ca into this calcium fraction. There is no net gain in tissue calcium and calcium efflux increases in a similar manner (Fig. 5). The increase in the rate of ^{45}Ca uptake is dose-dependent. In the presence of the α-antagonist phentolamine, the dose-response curves for noradrenaline on ^{45}Ca uptake are displaced to the right in a manner suggesting a competitive antagonism, the pA_2 for phentolamine being equal to 7.8. Because of the resemblance of this pA_2 value to that obtained for the antagonistic effect of phentolamine on the contractile responses to norepinephrine, it appears that the activation of α-adrenoceptors is responsible for both the increased rate of Ca entry and the contraction. These observations suggest that α-adrenoceptor activation does open Ca pathways in the membrane, the actual accepted terminology being to consider that the Ca pathways could consist of calcium channels closely associated to the receptor and coined ROC's (receptor-operated-channels) (Bolton 1979). High K solutions also allow an increased influx of ^{45}Ca, that presents some differences to the agonist-dependent one, namely, as discussed below, its sensitivity to calcium antagonists. Those channels controlled by membrane potential have been named POC's (potential-operated-channels). In addition to ROC's and POC's, Bevan et al. (1982) has proposed the existence in some vessels, mainly in cerebral arteries, of channels sensitive to stretch (SOC's, stretch-operated

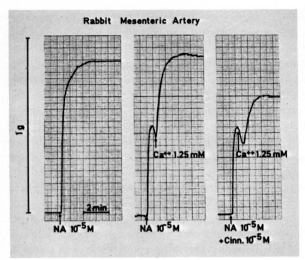

Fig. 4. Rabbit mesenteric artery, experimental protocol similar to Fig. 3 showing that cinnarizine, which is inactive in aorta, inhibits specifically the response observed after readmission of calcium to the perfusing medium. (For ref. see Fig. 3)

channels). The calcium electrochemical gradient is oriented inward, so that there is a powerful driving force that pushes calcium within the cell. The low intracytoplasmic calcium concentration (below 10^{-7} M) may be attributed to the low permeability of the membrane of the resting cell to the cation. The slow turnover of intracellular cal-

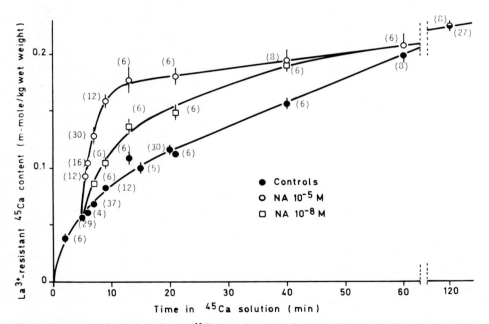

Fig. 5. The action of noradrenaline on ^{45}Ca entry into vascular smooth muscle cell, as a function of the concentration of the agonist. Note that the agonist abruptly increases the rate of calcium entry, but does not alter the maximum uptake measured after 120 min, an indication that noradrenaline enhances calcium turnover. (Godfraind 1976)

cium at rest probably occurs through leak channels. The constancy of cellular Ca^{2+} content over long periods of time requires that calcium efflux exactly balance calcium influx, which can be achieved by pumps located in the plasma membrane. At present two calcium pumps have been identified in the plasma membrane: a Ca pump that actively extrudes Ca^{2+} at the expense of ATP hydrolysis, and a Na-Ca exchange process making use of the Na electrochemical gradient (Godfraind-De Becker and Godfraind 1980; Morel and Godfraind 1984). In addition, intracellular buffering processes may regulate fast changes in the cytoplasmic calcium. Cytoplasmic proteins such as parvalbumins have calcium-binding properties. Mitochondria are able to take up large amounts of calcium and may serve as calcium sinks. Endoplasmic (or sarcoplasmic) reticulum appears to be more efficient than mitochondria for normal regulation of calcium levels in the sarcoplasm. This is illustrated by experiments showing that ER vesicles have a higher affinity for calcium than mitochondria, although their capacity is lower (Verbeke et al. 1977). The membranal Ca-pump allows the extrusion of calcium from the cell against the driving force of the Ca electrochemical gradient. It operates at the expense of ATP-hydrolysis, that permits the phosphorylation of the pump. The rate of ATP hydrolysis by the pump is a function of the calcium concentration. However, calcium does not operate by itself, but after formation of a complex with calmodulin identified by Cheung as a ubiquitous molecule. The interactions of calcium and this protein have been characterized by Kretsinger (for ref. see Tomlinson et al. 1984). This calcium-calmodulin complex (Ca-CAM) is the form under which calcium is operating in the cell.

The other calcium extrusion mechanism is driven by the force given by the inward sodium electrochemical gradient. It is proposed that Na and Ca have an affinity for the same membrane carrier (the Na-Ca antiporter) that binds one Ca^{2+} and three Na^+. Its function, which appears to be electrogenic, is enhanced by phosphorylation. Because of this electrogenecity, the function of the Na-Ca antiporter is controlled by membrane potential. As a result, calcium is extruded into the resting polarized cell, but this system allows calcium entry into the depolarized cell (Morel and Godfraind, 1984).

The two calcium extrusion mechanisms are unevenly distributed among the various tissues. For example, in heart, the Na-Ca exchange has a much higher capacity than the Ca-ATP pump, the reverse situation being found in smooth muscle (Fig. 6).

3 Differentiation Between Calcium Antagonists

As already pointed out in the preceding section, studies on smooth muscle allow characterization of a drug as having calcium-antagonistic properties. Table 1 reports some of the chemical compounds interfering with calcium in order to reduce or to facilitate its biological action. They may be listed under the general heading of "calcium modulators". The most extensively studied of these drugs are calcium entry blockers, because of their wide clinical use.

The simplest experimental procedure to assess the relation between calcium entry and the development of contraction is, as discussed above, to preincubate a smooth muscle in Ca-free physiological solution, to depolarize it with Ca-free KCl-rich solution

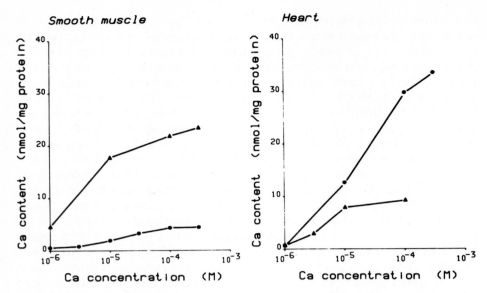

Fig. 6. A comparison of the ATP-dependent and of the Na-driven ^{45}Ca uptake by microsomal fractions prepared from heart and smooth muscle. (Morel and Godfraind 1984) Note that, in heart, the Na-driven calcium transport is more important than the ATP-dependent one; the reciprocal observation is done in smooth muscle

and gradually increase the calcium concentration in the bathing solution. This will result in an increase in tension dependent on Ca^{2+} concentration (Godfraind and Kaba 1969).

When an agonist is used to stimulate the tissue, the calcium dependency of the response may also be examined. It must be pointed out that the response of a vessel to vasoconstrictors may depend upon the interaction of the agonist with its receptors, but that it may be modulated by factors extrinsic to the muscle cell. Some of those factors may depend upon the endothelium. The essential role of endothelium as a mediator of relaxant responses induced in isolated vascular tissues by agonists such as acetylcholine, substance P, and the calcium ionophore A23187 is well accepted (Furchgott and Zawadzki 1980). Bradykinin produces a relaxation dependent on the endothelium in arteries of some species but not of others (Cherry et al. 1982). Other mediators of vascular relaxation such as ATP and ADP are partially dependent on the endothelium in some species, whilst AMP, adenosine, vasoactive intestinal peptide, and isoprenaline exert a direct relaxant effect on the smooth muscle (see Vanhoutte and Rimele 1983). We have examined the role of the endothelial tissue as a modulator of contractile effects of agonists on vascular smooth muscle. We first examined the influence of endothelium on the contractile response evoked by the adrenoceptor agonists noradrenaline and clonidine in isolated rat aorta (Egleme et al. 1984). Selective removal of endothelial cells was confirmed in depolarized preparations by demonstrating the presence or the lack of a relaxation response to acetylcholine (10^{-6} M). Concentration-effect curves obtained with noradrenaline and clonidine in isolated rat aorta with and

Table 1. Calcium modulators

---

1   *Calcium antagonists*

1.1   Calcium entry blockers
1.1.1   Diphenylpiperazines:
        cinnarizine, flunarizine, lidoflazine
1.1.2   Verapamil derivatives:
        verapamil, gallopamil, tiapamil
1.1.3   Dihydropyridines:
        nifedipine, nimodipine, nisoldipine, nitrendipine, etc.
        PY 108-068
1.1.4   Benzothiazepines:
        diltiazem, KB-944
1.1.5   Others:
        bencyclane, bepridil, fendiline, perhexiline, prenylamine
        yohimbine derivatives
        papaverine
        amrinone

1.2   Intracellular calcium antagonists

1.2.1   Trimethoxybenzoates
1.2.2   Methylenedioxyindenes

1.3   Calmodulin antagonists

1.3.1   Phenothiazines:
        trifluoperazine, chlorpromazine
1.3.2   Naphtalene derivatives:
        W-7
1.3.3   Local anesthetics
        dibucaine
1.3.4   Dopamine antagonists:
        pimozide, haloperidol
1.3.5   Calmidazolium (R24571)

2   *Calcium agonists*

2.1   Dihydropyridines:
        BAY K 8644, CGP 28392

---

without endothelium are illustrated in Fig. 7. Noradrenaline concentration-effect curves obtained in the absence of endothelium were shifted to the left as compared to the concentration-effect curves obtained in the presence of endothelium. A similar shift was observed with other agonists, such as serotonin, $PGF_{2\alpha}$, and phenylephrine. On the other hand, removal of endothelium evoked a dramatic effect on the contractile responsiveness of rat aorta to clonidine. The maximum response increased by a factor of 9 and the EC_{50} value decreased. A similar observation has just been made with other α_2 agonists. In recent experiments, we have examined the possibility that endothelial factors could affect membrane potential. Indirect evidence suggests that they could

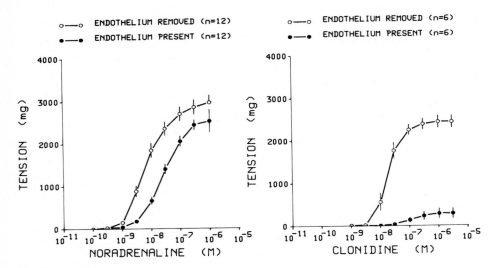

Fig. 7. Role of the presence of endothelium in the contractile response of rat aorta to the α-agonists noradrenaline and clonidine. (After Egleme et al. 1984). Note the important enhancement of the response to clonidine after removal of endothelium, the response to noradrenaline is increased less

slightly hyperpolarize the smooth muscle cell, but this action could be considered as unspecific (unpublished experiments).

In some tissues, depolarization exerts presynaptic influences, this is namely the case in rabbit aorta, where depolarization-evoked catecholamine release could contribute to the contraction, as shown by its sensitivity to α blockers. On the other hand, in rat aorta, there is no adrenergic innervation (Godfraind 1983a). Recently, Rimele and Vanhoutte (1983) have reported that the contraction of the canine saphenous vein in response to KCl, but not to noradrenaline, is enhanced by indomethacine, indicating that calcium entry could activate phospholipase leading to vasoactive products of the arachidonic acid cascade.

The above observations show a possible complexity in the factors controlling or modulating the contractile responses in vessels. This emphazises the need for direct analysis of calcium fluxes and of calcium channels.

The description of agonists and K-depolarization-evoked Ca entry in terms of passage through calcium channels is not only a question of terminology but is also supported by experimental studies. Nelson et al. (1984) have recently reported that they could incorporate calcium channels into planar lipid bilayers containing phosphatidylethanolamine and phosphatidylserine. Within minutes following the addition of brain membrane preparation in the presence of $CaCl_2$, $BaCl_2$ or $SrCl_2$ stepwise current fluctuations were observed by membrane depolarization. These channels were selective for divalent cations over monovalent ions. Lanthanum and cadmium produced a concentration-dependent reduction of the apparent single-channel conductance. The nature of the divalent cation carrying current through the channel affected not only the single-channel conductance, but also the channel open times, with mean open time being shortest for barium.

Another major means with which to analyze the properties of the calcium channels is the study of calcium entry blockers. The interaction of calcium entry blockers with calcium channels may be examined in binding studies and in functional studies. Because most of the drugs so far used are highly lipophilic, binding studies are generally performed on microsomal preparations containing various cellular membranes, including the plasma membrane.

In preparations from striated muscle, it has been shown that tritiated dihydropyridines bind to calcium channels located in the T tubules. Such binding studies have led to the hypothesis that those channels could be constituted by three subunits interacting differently with calcium entry blockers (Glossman et al. 1982). In studies on brain, cardiac, and striated muscle membranes, it has been possible to identify the existence of specific binding sites for so-called calcium antagonists and to characterize the interaction within one single chemical family of calcium entry blockers and between the various families (see Table 1). In addition, such studies have allowed the discovery of dihydropyridines that have an affinity for the binding sites on the channels but that, instead of blocking the open channels, facilitate their opening (Schramm et al. 1983a, b).

Studies in smooth muscle have in addition demonstrated a correlation of binding with the biological activity of the compounds. This was possible by measuring the competition of nitrendipine binding by calcium entry blockers of the various families and their potency to depress the contraction of intestinal smooth muscle (Bolger et al. 1983). We are now conducting a similar study in rat aorta (Godfraind et al. 1984b). In this tissue, we have characterized the interaction of nimodipine with calcium channels by measuring its action on contraction, on ^{45}Ca fluxes and on H^3-nitrendipine binding. In addition, we have observed that the interaction of nimodipine with calcium channels was stereospecific.

A convenient method to assess the ability of a compound to block calcium entry is to examine how drugs modify smooth muscle contractile responses to calcium in a depolarized medium. As Figs. 1 and 8 illustrate, the contractile response is attenuated dose-dependently. Analysis of calcium concentration-response curves sometimes shows the appearance of a competitive antagonism, depending mainly upon the preparation utilized. Figure 8 also shows that the antagonism exerted by flunarizine increases with the duration of depolarization and reaches a steady state after few minutes. This time-course of the contraction in the presence of a calcium entry blocker is not specific for flunarizine, but has been observed with most of the so-called Ca-antagonists, although the kinetics appeared to be different with the molecules studied (Godfraind and Miller 1983). This observation could not be explained by an inappropriate preincubation with flunarizine, before the test depolarization, indeed, successive contractions showed the same pattern. ^{45}Ca fluxes measurements have been performed at peak Ca entry (after 2 min) and when the inhibition was at steady state (after 35 min). The inhibition of calcium entry was increased to the same extent as the contraction. IC_{50} and I_{50} values were within the same range. This indicates that this increase in inhibition may be considered as a use-dependent process (Godfraind and Dieu 1981).

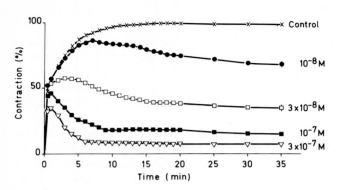

Fig. 8. Contractile response evoked by K-depolarization in the absence (0) and in the presence of various concentrations of flunarizine. Note the typical pattern of the contractile response in the presence of the calcium entry blocker, that is characterized by an initial peak contraction followed by a reduction of the tension to reach a secondary plateau. As discussed in the text, measurements of inhibition of calcium influx also show an accentuation of the inhibition characterized by a similar time-course, the concentration producing 50% inhibition of calcium entry is similar to the one producing 50% of inhibition of the contraction. (See Godfraind and Dieu 1981)

This suggests that the interaction of the calcium entry blocker with calcium channels is increased by depolarization. Recently, Bolton et al. (1983) have reported a similar observation in the taenia coeci stimulated electrically. They have observed that the reduction of the electrical activity was increased by the length of the stimulus. Contractions evoked by agonists do not show a similar increase in inhibition. This may be interpreted in the light of the hypothesis considering the existence of potential and receptor-operated channels (Bolton 1979; Cauvin et al. 1983). One other difference between depolarization- and agonist-evoked contractions is that the former are incompletely blocked by calcium entry blockers, while the latter may be completely suppressed. This has generally been interpreted as due to the release by agonists of intracellular calcium in an amount sufficient to support the contraction (Godfraind and Kaba 1969). In more recent studies on rat aorta, we have observed that the noradrenaline blockade achieved by a maximum effective concentration of several calcium entry blockers has the following features (Godfraind and Dieu 1981; Godfraind and Miller 1983; Godfraind et al. 1984): it does not attain the reduction observed in Ca-free solution and there is an incomplete blockade of noradrenaline-dependent Ca entry. A similar observation has been made with other agonists such as serotonin, and $PGF_{2\alpha}$ (Godfraind et al. 1984, and unpublished data). Nevertheless, dose-effect curves for inhibition of agonist contraction and Ca entry are superimposed, an indication that the inhibition of the two processes is related to an interaction of the blocker with similar sites. In addition, as observed with nimodipine (Godfraind et al. 1984), IC_{50}, I_{50}, and K_i values are closely similar. This indicates that blockade of calcium entry is responsible for inhibition of contraction and is related to the binding sites of calcium entry blockers.

The difference in the blockade of [45]Ca entry dependent on agonists and dependent on membrane depolarization shows that the two modes of stimulation operate Ca

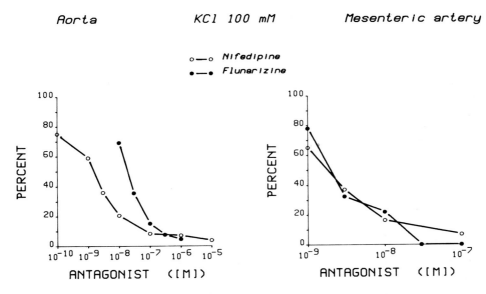

Fig. 9. Depression by flunarizine and nifedipine of the contractile response of rat aorta and mesenteric artery stimulated by K-depolarizing solution. Note difference in sensitivity for flunarizine between the two arteries. (After Godfraind 1983; Godfraind and Dieu 1981)

channels differently and support the concept of Ca channel specificity. However, at the molecular level, it is not yet clear how POC's and ROC's differ.

Calcium entry blockers show differences in their ability to block Ca channels in every tissue so far studied. Comparison of the concentrations required to reduce by 50% the contraction of myocardium and of smooth muscle shows that the former has a very low sensitivity to some drugs. This may be quantified by comparing I_{50} values. As shown by Van Nueten (1982), who compared rat caudal artery and myocardium, the ratio is equal to 3 for verapamil and to 210 to flunarizine. Among smooth muscles, there is also a variation in sensitivity to Ca entry blockers, as shown by the insensitivity of the myogenic contraction of rat portal vein to flunarizine and its high sensitivity to D-600 and nifedipine (Van Nueten 1982).

A comparison between flunarizine and nifedipine may help at illustrating some observations of specificity among arteries. Figure 9 illustrates concentration-inhibition curves of the contraction of rat aorta and rat mesenteric artery evoked by K-depolarization. It is obvious that nifedipine is more potent than flunarizine in aorta but that the two drugs are equipotent in mesenteric artery. Because IC_{50} values of nifedipine are identical in the two vascular preparations, this indicates that flunarizine blockade potency is higher in mesenteric artery than in aorta.

Such changes in sensitivity to calcium entry blockers within arteries were initially observed with cinnarizine (Godfraind et al. 1968). The earlier observation has been confirmed in a study comparing the actions of cinnarizine and papaverine in rabbit aorta and mesenteric artery. Papaverine I_{50} values were similar in the two tissues, whereas cinnarizine was more potent in depressing calcium-evoked contractions in depolarized mesenteric artery than in depolarized aorta. In addition, the noradrenaline-evoked contraction in aorta was resistant to cinnarizine, but the contraction in mesenteric artery was sensitive (Broekaert and Godfraind 1973). Dieu and Godfraind (1981) have studied the inhibition of noradrenaline-evoked change in perfusion pressure in the perfused mesenteric bed of the rat. In this preparation, flunarizine was more potent than cinnarizine, although both drugs were equipotent as inhibitors of noradrenaline contraction in rat aorta.

This tissue selectivity is not only found for diphenylpiperazines but also for dihydropyridines. Kazda and Towart (1982) and Towart (1981) have reported that nimodipine has a preferential action in cerebral arteries. Nimodipine 2×10^{-6} inhibits the contraction evoked by serotonin in rabbit basilar artery, whereas the response of the saphenous artery is not much affected. Such changes in blocking activity were not only observed by measuring contraction but also by measuring calcium fluxes, a strong indication that the interaction of calcium entry blockers with the calcium channel was responsible for tissue selectivity.

It must nevertheless be pointed out that the degree of inhibition of smooth muscle contractile response to an agonist for a given concentration of a calcium entry blocker depends upon at least two factors: the affinity of ROC's for the calcium entry blocker and the relative contribution of extracellular and intracellular calcium pools in the initiation and in the maintenance of the contraction. To understand the general hemodynamic effect of a calcium entry blocker, it is necessary to consider not only the actions discussed here for arteries, but also those observed in veins, where some calcium channels are resistant to calcium entry blockers, even to verapamil (Winquist and Baskin 1983).

Up to now, I have only considered specific calcium entry blockade, assuming that in view of the identity of I_{50} and IC_{50} values, the action of the blocker was located only at the level of calcium channels, However, some drugs may have an additional intracellular effect. This has been documented by Saida and van Breemen (1983) with diltiazem, which depresses responses due to noradrenaline-evoked intracellular calcium release. This effect was observed at concentrations higher than 10^{-6} M when the action on calcium entry was observed below 10^{-6} M. Similarly, the demonstration by Spedding (1983a, b) that cinnarizine and flunarizine could interact directly with the contractile proteins was made using concentrations in the micromolar range when the action on calcium entry may be observed three orders of magnitude lower.

In addition to diltiazem, other chemical compounds may be considered as "intracellular" calcium antagonists because they are believed to inhibit the release of calcium from intracellular sites. These agents include trimethoxybenzoates and methylenedioxyindenes (Rahwan et al. 1977, 1981). This group could also include sodium nitroprusside, ryanodine, dantrolene, and diazoxide. However, as recently discussed by Weishaar et al. (1983), most of these drugs have complex effects including some interaction with calcium channels that could be responsible, at least partly, for their relaxant action in muscles.

Other calcium antagonists belonging to the group of calmodulin antagonists also inhibit the contraction of vascular smooth muscle evoked by K depolarization. Their action on calcium fluxes is generally complex, because they inhibit calcium entry, calcium efflux and calcium release (Karaki et al. 1982). Lugnier et al. (1984) have reported that flunarizine could act as a calmodulin inhibitor. The concentrations used by these authors for biochemical measurements were two or three orders of magnitude higher than those active on Ca channels. Nevertheless, this observation is in agreement with that of Spedding (1983a, b) and it fits within the observation that smooth muscle contraction is activated by the complex calcium-calmodulin (Ebashi et al. 1982; Silver and Stull 1983). As recently shown by Mazzei et al. (1984), who have analyzed the action of calmidazolium, a potent inhibitor of calmodulin, complex kinetics can be observed. The authors have reported that calmidazolium inhibits phospholipid-sensitive Ca-dependent protein kinase with an IC_{50} of 5.3 μM. It inhibits the calcium-calmodulin stimulated enzymes with IC_{50} value of 1.6 μM for myosin light chain kinase and of 0.1 μM for phosphodiesterase. Analysis of these inhibitions shows complex kinetics, suggesting that the agent interacts with the cofactors, the enzyme and/or the cofactor-enzyme complexes. At saturation concentrations of the cofactors, enzyme inhibition was non-competitive with respect to the cofactors. Inhibition of myosine light chain kinase by calmidazolium was completely overcome by phosphatidylserine, indicating a strong hydrophobic reaction between calmidazolium and the phospholipid in the presence of calmodulin. In addition, it was observed that calmidazolium inhibits phosphorylation of various endogenous proteins in brain stimulated specifically by Ca/phosphatidylserine or Ca-calmodulin. Similar observations have been reported with other calmodulin antagonists such as trifluoperazine and W-7. The complexity of molecular action of so-called calmodulin antagonists emphasizes the importance of taking into account the level of the active concentration and not only the existence of an effect if the mode of action in vivo has to be considered. This is illustrated by Fig. 10 for flunarizine. Low concentration inhibit noradrenaline-evoked calcium influx in rat aorta and leave un-altered calcium efflux. The latter is slightly reduced at much higher concentration, an effect that could be related to an inhibition of calcium pump that is activated by Ca-calmodulin (Morel and Godfraind 1984). Such high concentrations inhibit the Ca-Mg-ATPase in vitro. Nevertheless, it leaves unaltered the contraction evoked in calcium-free medium, indicating that the intracellular action of this drug, if any, does not affect in vivo the Ca-calmodulin activation of myosin light chain kinase.

New possibilities for the analysis of the mode of action of calcium entry blockers are presented by the use of dihydropyridine derivatives that, instead of blocking, open the calcium channels. One of the most interesting is BAY K 8644. It differs from nifedipine in only two positions, the 3-ester position in the dihydropyridine ring and the ortho position in the aromatic ring. In vivo, it has positive inotropic and vasoconstrictive activities. In vitro, it contracts partially depolarized aortic strips and increases calcium entry into the smooth muscle cell (Schramm et al. 1983a and b). We have ex-ined the action of BAY K 8644 on the electrical and mechanical activity of the isolated taenia coli of the guinea-pig. We have observed that BAY K 8644 reduced the electrical threshold for opening Ca^{2+} channels by an interaction with sites sensitive to nifedipine (Godfraind 1984, Godfraind et al. 1984). It is to be hoped that the development of this and similar compounds will increase our understanding of the regulation of calcium channels.

It would be unfair to end this section without mentioning that some suggestions have been made that calcium entry blockers could interfere with calcium efflux. Church and Zsoter (1980) have suggested that the action of nifedipine could be explained by

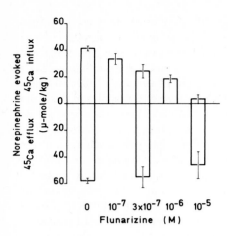

Fig. 10. Noradrenaline-evoked calcium influx and calcium efflux in isolated rat aorta treated with the various concentrations of flunarizine indicated in the *lower part* of the figure. Note that 10^{-5} M blocked nearly all calcium influx but also slightly reduced calcium efflux. (After Godfraind and Dieu, 1981)

an action on calcium efflux rate. However, it seems that the magnitude of this effect was not sufficient to account for the action of the dihydropyridine on muscle contraction. This question needs to be re-examined in view of the recent report by David-Dufilho et al. (1984) that nifedipine 2×10^{-8} M stimulated calcium transport by cardiac sarcolemmal vesicles from hypertensive but not from normotensive rats.

4 Clinical Interest of Calcium Antagonists

The clinical use of calcium antagonists now seems well established. Because most of the compounds utilized belong to the subgroup of calcium entry blockers, it seems of interest to raise the questions of the relation between calcium entry blockade and therapeutic effect.

Although Table 2 is far from being exhaustive, it reports the main indications of calcium antagonists and illustrates not only their diversity, but also their selectivity. This means that drugs utilized to treat cardiac disorders or hypertension appear to have low interest for CNS vascular disorders, the reverse being also true. Experimental observations already discussed above may help toward understanding this selective utilization.

Table 2. Main indications of calcium antagonists (calcium entry blockers)

Cardiac arrhythmias:

    Verapamil

Vasospastic angina:

    A. Prinzmetal's angina
    B. Effort-induced angina with ST-segment elevation
    C. Variable effort angina

Stable (effort induces) and unstable angina:

Verpamil, nifedipine (and other dihydropyridines under study), diltiazem, lidoflazine

Hypertension:

Verapamil, nifedipine (and other dihydropyridines under study), diltiazem

Hypertrophic cardiamyopathy:

Verapamil, lidoflazine, nifedipine

Preservation of myocardium against ischemic damage:

Under study with verapamil, nifedipine, diltiazem, lidoflazine, myoflazine

Peripheral vascular disorders (intermittent claudicatio, . . .):

Cinnarizine, flunarizine

Cerebral vasospasm, cerebral ischemia, postapoplectic states:

Flunarizine, nimodipine

Migraine:

Flunarizine, nimodipine

Epilepsy:

Flunarizine

4.1 Arrhythmia

Because of its potent activity on the slow Na-Ca channel, verapamil appears effective to treat several froms of cardiac arrhythmias. It has been shown to be active in atrial fibrillation (Schamroth 1971), in paroxysmal supraventricular tachycardia. These actions are probably due to prolongation of atrio-ventricular nodal refractoriness by a direct action on the node proximal to the bundle of His (Fleckenstein 1982).

4.2 Angina pectoris

The most spectacular therapeutic effect of calcium antagonists have been obtained in the treatment of Prinzmetal's angina (Maseri 1979). It is now acknowledged that coronary spasm has a prominent pathogenic role not only in classical Prinzmetal's angina but also in unstable rest angina (Braunwald 1981). Furthermore, there is substantial evidence that calcium antagonists are also of value in the treatment of classic, stable, effort-induced angina (Krebs 1982). The imbalance between myocardial oxygen supply and demand, responsible for ischemia in stable angina, can be ameliorated by augmenting the supply, by reducing the demand or both. Studies in vivo, namely by the cold pressor test and in vitro on isolated human coronary arteries have clearly demonstrated that they contract in response to various stimuli. This contraction can be attenuated by nifedipine at a concentration equivalent to that found in plasma of treated patients (Godfraind and Miller 1983). The beneficial effect in stable angina can also be attributed to some decrease in preload and afterload resulting from depression of venomotility and to decrease in tone of resistant vessels.

4.3 Hypertension

The therapeutic effect of calcium antagonists in the treatment of systemic hypertension has been established with verapamil and nifedipine (see Agabati-Rosei et al. 1982). It has been reported that hypotensive action is observable only in hypertensive patients and that normotensive patients receiving a similar posology show no change in blood pressure. Such clinical observations have been interpreted on the basis of the demonstration from animal studies of the existence of abnormalities in the properties of hypertensive vessels (Mulvany and Nyborg 1980). The question is open whether or not calcium antagonists are acting because of their vasodilating action or because of a reversal of a pathogenic change. In vitro experiments show that calcium antagonists of the dihydropyridine type restore relaxation properties in vessels from hypertensive animals (Godfraind 1983b). There is no doubt that further studies will help to clarify this important question and solve the question of sustained antihypertensive therapy.

In spite of the decrease in blood pressure, it has been observed that nifedipine increases urinary volume and sodium excretion, an action that will undoubtedly contribute to the antihypertensive action (Krebs 1982).

4.4 Other Cardiovascular Indications

Some other indications may still be considered as being experimental, as is the case for pulmonary hypertension and hypertrophic cardiomyopathy. An interesting concept is the preservation of myocardium against ischemic damage, an action that could reduce the extension of the primary necrotic area of infarcted myocardium. There are encouraging results, but more has to be done in this field.

4.5 Peripheral Vascular Diseases

One field of possible application of calcium antagonists is intermittent claudicatio (Rudofsky and Nobbe 1982). Careful studies conducted with flunarizine have shown that the condition of patients was improved, but mainly in the case of cigarette smokers. There are some encouraging reports in other peripheral vascular diseases.

4.6 Central Vernous System Disturbances

There are now several reports on the beneficial effect of flunarizine in the treatment of common and classical migraine (Caers 1982). The pharmacological target for this action may be cerebral arteries or neurons, but there is no conclusive proposition at present available.

Clinical trials are now in progress to examine the protective action of specific calcium entry blockers in ischemic brain damage. Actual data in patients with aneurysmal sub-arachnoid hemorrhage indicate that nimodipine, a calcium antagonist with a predilec-tive action on cerebral circulation, could improve the clinical situation of patients at risk (Allen et al. 1983).

Clinical reports have demonstrated the beneficial action of cinnarizine and flunarizine in the treatment of vertigo, and in sleep disorders in aged patients (Vanhoutte 1982). Recent preliminary data suggest a possible action in some forms of epilepsia. Such agents therefore appear active in disseminated neural disturbances.

5 Conclusions

The actual established clinical uses, as well as the prospective ones, indicate that calcium antagonists are active in pathological states where various classes of drug are presently utilized, namely β-blockers, anti-arrhythmics, anti-epileptics, anti-migraine drugs and even sedatives. However, all calcium antagonists do not share a similar pharmacological profile. This can be explained by considering variations in tissues sensitivities, established at the pharmacodynamic level.

References

Agabati-Rosei E, Alicandri C, Beschi M, Castellano M, Fariello R, Romanelli G (1982) In: Godfraind T, Albertini A, Paoletti R (eds) Calcium modulators. Elsevier/North Holland Biomedical Press, Amsterdam New York Oxford, pp 257-269

Allen GS, Ahn HS, Preziosi TJ, et al. (1983) Cerebral arterial spasm — A controlled trial of nimo-dipine in patients with subarachnoid hemorrhage. N Engl J Med 308: 619-624

Bevan JA, Bevan RD, Hwa JJ, Owen MP, Tayo FM, Winquist RJ (1982) In: Godfraind T, Albertini A, Paoletti R (eds) Calcium modulators. Elsevier/North Holland Biomedical Press, Amsterdam New York Oxford, pp 125-132

Bolger GT, Genco P, Klockowski R, Luchowski E, Siegel H, Janis RA, Triggle AM, Triggle DJ (1983) Characterization of binding of the Ca^{2+} channel antagonist, [^{3}H]-nitrendipine, to guinea-pig ileal smooth muscle. J Pharmacol Exp Ther 225: 291-309

Bolton TB (1979) Mechanisms of action of transmitters and other substances on smooth muscle. Physiol Rev 59: 606-718

Bolton TB, Kitamura K, Morel N (1983) Use-dependent effects of calcium entry blocking drugs on the electrical and mechanical activities of guinea-pig taenia coli. Br J Pharmacol 78: 174P

Braunwald E (1981) Coronary artery spasm as a cause of myocardial ischemia. J Lab Clin Med 97: 299-312

Broekaert A, Godfraind T (1973) The actions of ouabain on isolated arteries. Arch Int Pharmacodyn Ther 203: 393-395

Broekaert A, Godfraind T (1979) A comparison of the inhibitory effect of cinnarizine and papaverine on the noradrenaline- and calcium-evoked contraction of isolated rabbit aorta and mesenteric arteries. Eur J Pharmacol 53: 281-288

Caers LI (1982) Flunarizine, a calcium entry blocker, in the treatment of common and classical migraine. In: Godfraind T, Albertini A, Paoletti R (eds) Calcium modulators. Elsevier/North Holland Biomedical Press, Amsterdam New York Oxford pp 223-231

Cauvin C, Loutzenhiser R, van Breemen C (1983) Mechanisms of calcium antagonist-induced vaso-dilation. Annu Rev Pharmacol Toxicol 23: 373-396

Cherry PD, Furchgott RF, Zawadzski JV, Jothianandan D (1982) Role of endothelial cells in re-laxation of isolated arteries by bradykinin. Proc Natl Acad Sci USA 79: 2106-2110

Church J, Zsoter TT (1980) Calcium antagonistic drugs. Mechanism of action. Can J Physiol Pharmac 58: 254-264

David-Dufilho M, Devynck M-A, Kazda S, Meyer P (1984) Stimulation by nifedipine of calcium transport by cardiac sarcolemmal vesicles from spontaneously hypertensive rats. Eur J Pharma-col 97: 121-127

Deth R, van Breemen C (1977) Agonist induced $^{45}Ca^{2+}$ release from smooth muscle cells of the rabbit aorta. J Membr Biol 30: 363-380

Dieu D, Godfraind T (1981) A comparison of flunarizine and cinnarizine. Br J Pharmacol 72: 583P

Ebashi S, Nakamura S, Nakasoni H, Kohama K, Nonomura Y (1982) Differences and similarities of contractile mechanism in muscle. In: Godfraind T, Albertini A, Paoletti R (eds) Calcium mod-ulators. Elsevier/North Holland Biomedical Press, Amsterdam New York Oxford, pp 39-49

Egleme C, Godfraind T, Miller RC (1984) Enhanced responsiveness of rat isolated aorta to clonidine after removal of the endothelial cell. Br J Pharmacol 81: 16-18

Fleckenstein A (1982) Basic membrane actions of calcium antagonists with special reference to verapamil. In: Godfraind T, Albertini A, Paoletti E (eds) Calcium modulators. Elsevier/North Holland Biomedical Press, Amsterdam New York Oxford, pp 297-310

Furchgott RF, Zawadzki JV (1980) The obligatory role of endothelial cells in the relaxation of arterial smooth muscle by acetylcholine. Nature 288: 373-376

Glossman H, Ferry DR, Lubbecke F, Meeves R, Hoffmann F (1982) Calcium channels: direct identi-fication with radioligand-binding studies. Trends Pharmacol Sci 3: 431-437

Godfraind-De Becker A, Godfraind T (1980) Calcium-transport system: a comparative study in dif-ferent cells. Int Rev Cytol 67: 141-170

Godfraind T (1976) Calcium exchange in vascular smooth muscle, action of noradrenaline and lanthanum. J Physiol (Lond) 260: 21-35

Godfraind T (1981) Mechanisms of action of calcium entry blockers. Fed Proc 40: 2866-2871

Godfraind T (1983a) Actions of nifedipine on calcium fluxes and contraction in isolated rat arteries. J Pharmacol Exp Ther 224: 443-450

Godfraind T (1983b) Mechanism of calcium channel blockade in physiological and pathological sit-uations. Naunyn-Schmiedeberg's Arch Pharmacol 324: R7

Godfraind T (1984) Relaxation of aorta from spontaneously hypertensive rats. Arch Int Pharma-codyn Ther. In press

Godfraind T, Dieu D (1981) The inhibition by flunarizine of the norepinephrine-evoked contrac-tion and calcium influx in rat aorta and mesenteric arteries. J Pharmacol Exp Ther 217: 510-515

Godfraind T, Kaba A (1969) Blockade or reversal of contraction induced by calcium and adrenaline in depolarized arterial smooth muscle. Br J Pharmacol 36: 549-560

Godfraind T, Miller RC (1983) Specificity of action of Ca^{++} entry blockers. A comparison of their actions in rat arteries and in human coronary arteries. Circ Res 52 (Suppl I) 1-81-1-91

Godfraind T, Polster P (1968) Etude comparative de medicaments inhibant la reponse contractile de vaisseaux isoles d'origine humaine ou animale. Therapie 23: 1209-1220

Godfraind T, Hardy J-P, Morel N (1984a) The action of the spasmogenic dihydropyridine BAY K 8644 on the mechanical and the electrical activity of guinea-pig taenia coli. Arch Int Pharma-codyn Ther. In press

Godfraind T, Wibo M, Egleme C, Wauquaire J (1984b) The interaction of nimodipine with calcium channels in rat isolated aorta and in human neuroblastoma cells. In: Proc First Int Nimotop Congress. In press

Ishizuka T, Endo M (1983) Effects of diltiazem on skinned skeletal muscle fibers of the African clawed toad. Circ Res 52: 110-114

Karaki H, Weiss GB (1979) Alterations in high and low affinity binding of ^{45}Ca in rabbit aortic smooth muscle by norepinephrine and potassium after exposure to lanthanum and low temperature. J Pharmacol Exp Ther 211: 92-96

Karaki H, Murakami K, Nakagawa H, Ozaki H, Urakawa N (1982) Effects of calmodulin antagonists on tension and cellular calcium content in depolarized vascular and intestinal smooth muscles. Br J Pharmacol 77: 661-666

Kazda S, Towart R (1982) Nimodipine: a new calcium antagonistic drug with a preferential cerebrovascular action. Acta Neuroch 63: 259-265

Krebs R (1982) The therapeutic profile of nifedipine. In: Godfraind T, Albertini A, Paoletti R (eds) Calcium modulators. Elsevier/North Holland Biomedical Press, Amsterdam New York Oxford, pp 311-326

Kreye VAW, Rüegg JC, Hofmann F (1983) Effect of calcium-antagonist and calmodulin-antagonist drugs on calmodulin-dependent contractions of chemically skinned vascular smooth muscle from rabbit renal arteries. Naunyn-Schmiedeberg's Arch Pharmacol 323: 85-89

Lugnier C, Follenius A, Gerard D, Stoclet JC (1984) Bepridil and flunarizine as calmodulin inhibitors. Eur J Pharmacol 98: 157-158

Maseri A (1979) Variant angina and coronary vasospasm. Clues to a broader understanding of angina pectoris. Cardiovasc Res 4: 647-667

Mazzei G, Schatzman RC, Scott Turner R, Vogler WR, Kuo JF (1984) Phospholipid-sensitive Ca^{2+}-dependent protein kinase inhibition by R-24571, a calmodulin antagonist. Biochem Pharmacol 33: 125-130

Morel N, Godfraind T (1984) Na-Ca exchange in smooth-muscle microsomal fractions. Biochem J 218: 421-427

Mulvany MJ, Nyborg N (1980) An increased calcium sensitivity of mesenteric resistance vessels in young and adult spontaneously hypertensive rats. Br J Pharmacol 71: 585-596

Nelson MT, French RJ, Krueger BK (1984) Voltage-dependent calcium channels from brain incorporated into planar lipid bilayers. Nature 308: 77-80

Rahwan RG, Faust MM, Witiak DT (1977) Pharmacological evaluation of new calcium antagonists: 2-Substituted 3-dimethylamino-5,6-methylenedioxyindenes. J Pharmacol Exp Ther 201: 126-137

Rahwan RG, Witiak DT, Muir WW (1981) Calcium antagonists. Annu Rev Med Chem 16: 257-268

Rimele TJ, Vanhoutte PM (1983) Effects of inhibitors of arachidonic acid metabolism and calcium entry on responses to acetylcholine, potassium and norepinephrine in the isolated canine saphenous vein. J Pharmacol Exp Ther 225: 720-728

Rudofsky G, Nobbe F (1982) Treatment of patients with intermittent claudication. In: Godfraind T, Albertini A, Paoletti R (eds) Calcium modulators. Elsevier/North Holland Biomedical Press, Amsterdam New York Oxford, pp 287-294

Saida K, van Breemen C (1983) Mechanism of Ca^{2+}-antagonist-induced vasodilation. Intracellular actions. Circ Res 52: 137-142

Schamroth L (1971) Immediate effects of intravenous verapamil on atrial fibrillation. Cardiovasc Res 5: 419-424

Schramm M, Thomas G, Towart R, Franckowiak G (1983a) Novel dihydropyridines with positive inotropic action through activation of Ca^{2+} channels. Nature 303: 535-537

Schramm M, Thomas G, Towart R, Franckowiak G (1983b) Activation of calcium channels by novel 1,4-dihydropyridines. A new mechanism for positive inotropics or smooth muscle stimulants. Arzneim Forsch/Drug Res 33: 1268-1272

Silver PJ, Stull JT (1983) Effects of the calmodulin antagonist, fluphenazine, on phosphorylation of myosin and phosphorylase in intact smooth muscle. Mol Pharmacol 23: 665-670

Spedding M (1983a) Direct inhibitory effects of some "calcium-antagonists" and trifluoperazine on the contractile proteins in smooth muscle. Br J Pharmacol 79: 225-231

Spedding M (1983b) Functional interactions of calcium-antagonists in K^+-depolarized smooth muscle. Br J Pharmacol 80: 485-488

Tomlinson S, MacNeil S, Walker SW, Ollis CA, Merritt JE, Brown BL (1984) Calmodulin and cell function. Clin Sci 66: 497-508

Towart R (1981) The selective inhibition of serotonin-induced contractions of rabbit cerebral vascular smooth muscle by calcium antagonistic dihydropyridines: an investigation of the mechanism of action of nimodipine. Circ Res 48: 650-657

Vanhoutte PM (1982) Cinnarizine, flunarizine, lidoflazine. In: Godfraind T, Albertini A, Paoletti R (eds) Calcium modulators. Elsevier/North Holland Biomedical Press, Amsterdam New York Oxford, pp 351-362

Vanhoutte PM, Rimele TJ (1983) Role of the endothelium in the control of vascular smooth muscle function. J Physiol (Paris) 78: 681-686

Van Nueten JM (1982) Selectivity of calcium entry blockers. In: Godfraind T, Albertini A, Paoletti R (eds) Calcium modulators. Elsevier/North Holland Biomedical Press, Amsterdam New York Oxford, pp 199-208

Verbeke N, Morel N, Godfraind T (1977) Microsomal and mitochondrial calcium pumps in smooth muscle. In: Casteels R, Godfraind T, Rüegg JC (eds) Excitation-contraction coupling in smooth muscle. Elsevier/North Holland Biomedical Press, Amsterdam New York Oxford, pp 219-224

Weishaar RE, Quade M, Schenden JA, Harvey RK (1983) The methylenedioxyindenes, a novel class of "intracellular calcium antagonists": Effects on contractility and on processes involved in regulating intracellular calcium homeostasis. J Pharmacol Exp Ther 227: 767-778

Winquist RJ, Baskin EP (1983) Calcium channels resistant to organic calcium entry blockers in a rabbit vein. Am J Physiol 245: H1024-H1030

Biochemistry of the Ca²⁺- and Calmodulin-Dependent Regulation of the Cytoskeleton

S. KAKIUCHI[1]

CONTENTS

1 Introduction

Following the discovery of calmodulin (Kakiuchi et al. 1969, 1970; Cheung 1970), it was proposed that the contractile device of smooth muscle and the cytoskeleton of nonmuscle tissues may be regulated by Ca^{2+}/calmodulin. Recently, we have obtained considerable amounts of evidence to support the view that the regulatory actions of Ca^{2+}/calmodulin are mediated by a number of specific calmodulin-binding proteins which also bind to F-actin filaments or to tubulin (Kakiuchi and Sobue 1983). Caldesmon (Sobue et al. 1981a, b) from smooth muscle and tau factor (Sobue et al. 1981c) from brain microtubules are calmodulin-binding proteins which also bind to F-actin or tubulin, respectively. The Ca^{2+}-dependent binding of calmodulin to these proteins obviated the interaction between these proteins and F-actin or tubulin. Therefore, the binding of these proteins to calmodulin and cytoskeletal proteins alternates, depending on the concentration of Ca^{2+} (Fig. 1). As the binding of the calmodulin-binding proteins to the target cytoskeletal proteins modulates the function of the latter proteins, calmodulin regulates the function of cytoskeletal proteins via these calmodulin-binding proteins. We named this type of regulatory action of calmodulin flip-flop regulation (Kakiuchi and Sobue 1981, see also Kakiuchi and Sobue 1983). This article is a brief description of these calmodulin-binding proteins.

[1] Department of Neurochemistry and Neuropharmacology, Institute of Higher Nervous Activity, Osaka University Medical School, 4-3-57, Nakanoshima, Kita-ku Osaka 530, Japan

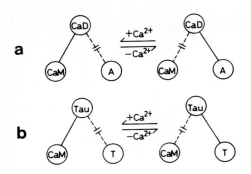

Fig. 1 a,b. Calmodulin-binding proteins which bind to F-actin (**a**) or tubulin (**b**) in a flip-flop fashion. *CaM* calmodulin; *CaD* caldesmon; *A* F-actin; *T* tubulin

2 Caldesmon

Caldesmon was first purified from chicken gizzard smooth muscle and characterized as a Ca^{2+}-dependent calmodulin-binding protein which also binds to F-actin, in a flip-flop fashion (Sobue et al. 1981a, b). We named this protein caldesmon, a word derived from calmodulin and the Greek word *desmos* which means binding. Upon gel electrophoresis in the presence of SDS, caldesmon sample from gizzard resolved into two protein bands of equal staining intensities on the gel with M_r's of 150,000 and 147,000 (Sobue et al. 1982a). Caldesmon is the major calmodulin-binding protein in gizzard tissue because its concentration in this tissue was 240 mg/100 g (8 μM) (Sobue et al. 1981a). Since 2 mol of calmodulin bound to 1 mol (as a heterodimer) of caldesmon, the amount of caldesmon found in 100 g of gizzard can take up about 27 mg of calmodulin, which is about 70% of the total amount of calmodulin in this tissue.

Caldesmon is also present in bovine aorta, uterus, and in human platelets (Kakiuchi et al. 1983). Its presence was demonstrated on the SDS gels by means of protein staining, immune blotting using the antibody raised against chicken gizzard caldesmon, and of gel overlay method using ^{125}I-calmodulin. Subsequently, with use of the indirect immunofluorescence technique, the caldesmon antibody was demonstrated in smooth muscle cells of the intestinal wall, and of blood vessels, and in the apical portion of the absorptive epithelial cells of rat's small intestine (Ishimura et al. 1984). More recently, caldesmon was also demonstrated to be present in various lines of cultured fibroblast cells and kidney cells (Owada et al. 1984). The anti-caldesmon (raised against gizzard caldesmon) immunofluorescence in these cells coincided with the cellular stress fibers which are known to be composed of the actomyosin-containing microfilaments. All these caldesmon samples from a variety of cell types possess the same general properties that they bind to calmodulin in the presence of Ca^{2+} and also bind to F-actin and react with the antibody made to gizzard caldesmon. In spite of these similarities, they differ in their molecular weights determined on the SDS gels. Like gizzard caldesmon, platelet caldesmon was composed of two polypeptide bands of M_r 150,000 and 147,000, but caldesmon in the aorta and uterus smooth muscles gave a single band of 150,000 (Kakiuchi et al. 1983). The estimated molecular weight of cultured cell caldesmon was 77,000 (Owada et al. 1984). This 77,000 type caldesmon was then purified from bovine adrenal medulla recently (unpublished result). It seems that caldesmon of smooth mus-

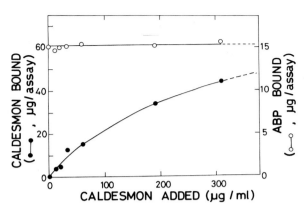

Fig. 2. Binding of caldesmon and ABP to F-actin (Maruyama et al. manuscript submitted). Increasing concentrations of caldesmon as indicated in the figure were added to fixed concentrations of ABP (100 μg ml^{-1}) and F-actin (1.8 mg ml^{-1}), and the preparation was incubated for 30 min at 30°C. The mixtures were centrifuged for 60 min at 105,000 g and the pellets analyzed by SDS-poly-acrylamide gel electrophoresis for F-actin-associated proteins. The concentrations of proteins were determined by densitometric scanning

cle type are composed of M_r 150,000 (or 150,000/147,000) polypeptides, whereas caldesmons from nonmuscle cells have an estimated molecular weight of 77,000 on the SDS gel.

Caldesmon was shown to act as an inhibitor of the interactions between F-actin and actin-binding protein (ABP; also called filamin) and between F-actin and myosin. There-fore, the ABP-induced gelation (crosslink formation) of F-actin filaments in smooth muscle was inhibited by the binding of caldesmon to F-actin in the absence of Ca^{2+}, and this inhibition was abolished by the calmodulin-caldesmon interaction in the pres-ence of Ca^{2+} (Sobue et al. 1982a). The molar ratio of F-actin (as monomer): caldes-mon (as dimer): ABP (as dimer) that led to a complete inhibition of gelation was 240:4:1 (Maruyama et al., manuscript submitted). Thus, the caldesmon-calmodulin system can confer the Ca^{2+}-sensitivity upon the gel-sol transformation of actin fila-ments in the presence of ABP. The inhibition of actin gelation by caldesmon is not due to the competition of caldesmon and ABP for the same binding site on F-actin. This can be seen in Fig. 2 in which increasing concentrations of caldesmon were add-ed to a fixed concentration of ABP-F-actin mixture, and caldesmon and ABP associ-ated with F-actin were separated from their free forms by centrifugation. The result showed that the amount of ABP associated with F-actin was not affected by addition of increasing concentrations of caldesmon, and was the same as without the addition of caldesmon (Maruyama et al., manuscript submitted). In addition, the amount of F-actin recovered in the pellets remained constant, regardless of the amount of cal-desmon added, indicating that caldesmon did not sever the actin filaments.

Contraction of smooth muscle is controlled by micromolar concentrations of Ca^{2+}, and calmodulin is considered to be a mediator of this action of Ca^{2+}. A generally ac-cepted view is that phosphorylation of two of the M_r 20,000 light chains of myosin catalyzed by a specific Ca^{2+}- and calmodulin-dependent myosin light chain kinase is responsible for activation of the myosin-actin interaction (for reviews, see Adelstein et al. 1982; Walsh et al. 1983; Stull et al. 1983). On the other hand, Ebashi et al. (1982) postulated the presence of a control via thin filaments. The controversy be-tween the two theories has not been solved. Recently we observed that the activity of the actin-myosin interaction of gizzard smooth muscle, as determined by superprecipitation of actomyosin ATPase activity, is controlled by the caldesmon-linked flip-flop mechanism, i.e., in the absence of Ca^{2+}, binding of caldesmon to F-actin filaments inhibits the actin-myosin interaction and Ca^{2+}-dependent binding of calmodulin to caldesmon relieves this inhibition (Morimoto et al. 1982). In our experimental conditions, phosphorylation of myosin light chains catalyzed by Ca^{2+}- and calmodulin-dependent myosin light chain kinase and the presence of tropomyosin were prerequisites for the actin-myosin interaction. With the phosphorylated myosin, caldesmon, in combination with calmodulin, conferred the Ca^{2+} sensitivity upon the actomyosin activity. Therefore, our results are in favor of the dual regulation on the smooth muscle contraction through the myosin phosphorylation and the caldesmon-linked flip-flop control of thin filaments. In both mechanisms, calmodulin is implicated as a mediator of the Ca^{2+} signal.

3 135 K Protein

A Ca^{2+}-dependent calmodulin-binding protein distinct from caldesmon on the SDS gel was purified from chicken gizzard smooth muscle (Sobue et al. 1982b). This protein migrated slightly faster than caldesmon on electrophoresis in the presence of SDS and its molecular weight was estimated to be 135,000 on the gel. Like caldesmon, this protein was able to bind to F-actin, in the flip-flop fashion. The purified protein has myosin light chain kinase activity and, in light of its substrate specificity and specific activity, it is probably the enzyme itself. Dabrowska et al. (1982) also reported that this enzyme binds to actin filaments. However, in their conditions, the binding was not influenced by the presence of calmodulin and/or Ca^{2+}. The reason for this discrepancy is unclear at the moment. The indirect evidence for the interaction of myosin light chain kinase with actin filaments has been obtained by studies on the immunofluorescence localization of this enzyme in the cells. Thus, de Lanerolle et al. (1981), Guerriero et al. (1981), and Cavadore et al. (1982) found that anti-myosin light chain kinase immunofluorescence coincides with actin bundles in muscle cells and nonmuscle cells. Therefore, in addition to its well-known role in regulating the myosin function by catalyzing phosphorylation of myosin light chains, this enzyme may directly interact with actin filaments. Re-evaluation of its role in the regulation of smooth muscle contraction is needed. The concentration of 135 K protein in chicken gizzard is 3 μM (Dabrowska et al. 1982) to 4.7 μM (Sobue et al. 1982b).

4 Spectrin in Mammalian Erythrocytes

The biconcave shape of the red cell as well as its elasticity is attributed to its submembranous structure. This structure, which appears as a fibrous network in electron microscopy and is called cytoskeleton conventionally, is composed of several protein species that include spectrin, actin, and band 4.1. These proteins account for about 76, 5, and 5%, respectively, of the total cytoskeletal protein mass (for review, see Cohen 1983). Spectrin consists of two types of polypeptide, $M_r=240,000$ (α subunit) and $M_r=220,000$ (β subunit), and a native form of spectrin is composed of two of such heterodimers presumably formed by head-to-head associations of two dimers. As the major component of the erythrocyte cytoskeleton, spectrin tetramers are associated with the inner surface of the plasma membrane through ankyrin and band 4.1, spectrin tetramers being bound to ankyrin at the head portions of β subunit and to band 4.1 at the tail portions (reviewed by Cohen 1983). Spectrin tetramers also bind to actin oligomers at their tail ends.

We found recently that spectrin purified from human erythrocytes bound to calmodulin in the presence of Ca^{2+} (Sobue et al. 1980; 1981d). The calmodulin-binding ability of spectrin is attributable to its α subunit because the isolated α subunit showed Ca^{2+}-dependent binding of calmodulin (Kakiuchi et al. 1982a). Kd value of spectrin for calmodulin was 2.8 μM (Sobue et al. 1981d), this being comparable to 2.5 μM of the concentration of calmodulin in human erythrocytes (Kakiuchi et al. 1982b). Also the concentration of calmodulin in erythrocytes (2.5 μM=140,000 molecules per cell) was close to that of spectrin in these cells (about 200,000/cell for α subunit; Steck 1974). Taken together, the formation of spectrin-calmodulin complexes in the cells may be well balanced against their dissociation to free forms. The functional significance of the formation of spectrin-calmodulin complexes is unclear and waits further studies.

5 Spectrin-Related Proteins in Nonerythroid Cells

Spectrin was thought to be a specific protein in erythrocytes because attempts to detect spectrin in different cell types by immunological means were not successful (reviewed by Branton et al. 1981). However, previous work from this laboratory demonstrated the presence of a Ca^{2+}-dependent calmodulin-binding protein which is asso-

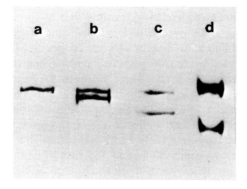

Fig. 3 a-d. Comparison of calspectin and spectrin on SDS-polyacrylamide gel electrophoresis (Kakiuchi et al. 1982a). a M_r 240,000 calmodulin-binding protein isolated in an earlier experiment (Kakiuchi et al. 1981); b calspectin; c spectrin of human erythrocytes; d chicken gizzard smooth muscle ABP *(upper band)* and myosin heavy chain *(lower band)*. The acrylamide concentration was 5%

ciated with particulate fractions of homogenates of brain (Teshima and Kakiuchi 1978) and of other tissues (Sobue et al. 1979). The calmodulin-binding protein subsequently purified from the particulate fraction of a bovine homogenate had an electrophoretic mobility in the presence of SDS identical to α subunit of spectrin (Fig. 3) (Kakiuchi et al. 1981). In combination with their earlier finding that spectrin is a Ca^{2+} dependent calmodulin-binding protein (Sobue et al. 1980, 1981d), they proposed the presence of a spectrin-related calmodulin-binding protein in the submembranous structure (cytoskeleton) of nerve cells (Kakiuchi et al. 1981). This proposal was substantiated when a doublet form of protein (α subunit = M_r 240,000; β subunit = M_r 235,000) with "spectrin-like" properties (i.e., formation of tetramers ($\alpha_2\beta_2$), rod-like appearance of tetramers in electron microscopy, and capability of binding to F-actin to form a viscous gel) was purified from bovine brain by the same group (Kakiuchi et al. 1982a, c). They called this protein calspectin (=calmodulin-binding spectrin-like protein). The calmodulin-binding polypeptide purified earlier (Kakiuchi et al. 1981) was then identified as α subunit of calspectin (Fig. 3). This protein is a major constituent of brain [>1 mg g^{-1} of brain (wet weight)], of which about 70% was associated with the particulate fraction (Kakiuchi et al. 1982c). In another experiment, the protein accounted for about 3% of the membrane protein (Kakiuchi et al. 1981). Therefore, the concentration of this protein is almost comparable to that of spectrin in erythrocyte (1.4 mg spectrin g^{-1} of packed erythrocytes).

Quite independently, the same protein was detected in brain by Levine and Willard (1981), Shimo-oka and Watanabe (1981), and Davies and Klee (1981) as an axonally transported protein, an actin-binding protein, and as a calmodulin-binding protein, respectively. Obviously, in 1981, this protein was viewed from different angles by each of the research groups. Levine and Willard gave the name fodrin to this protein as it was also detected in the cortical cytoplasma of many cells including neurons. It was in the next year that independent lines of research (Glenney et al. 1982a; Kakiuchi et al. 1982a, c; Palfrey et al. 1982; Bennett et al. 1982) reached the same conclusion that this protein is a spectrin-related protein. A wide distribution of this type of protein in vertebrate tissues and cells was also recognized (for reviews, Lazarides and Nelson 1982; Baines 1983; Kakiuchi and Sobue 1983).

On rotary shadowing electron microscopy, calspectin appeared as extended flexible rods of about 200 nm in length, presumably formed by head-to-head association of pairs of heterodimers (Kakiuchi et al. 1982a; Glenney et al. 1982a; Bennett et al. 1982). This molecular shape is quite similar to that of spectrin, although the image of calspectin is less tortuous than that of spectrin. Using rotary-shadowing electron microscopy coupled with ferritin-labeling of calmodulin, Tsukita et al. (1983) mapped the calmodulin-binding sites on calspectin. They found that calmodulin molecules (indicated by ferritin particles) are attached to the head parts of calspectin dimers at a position 10-20 nm from the top of the head, and the number of the calmodulin-binding sites is one per each dimer or two per each tetramer. This result, combined with our earlier finding that the isolated α subunit of calspectin is a calmodulin-binding polypeptide (Kakiuchi et al. 1981), clearly shows that the calmodulin binding of calspectin is attributed solely to its α subunit. This is illustrated schematically in Fig. 4. Earlier observations indicating the inability of the β subunit polypeptide of calspectin to bind to calmodulin (Kakiuchi et al. 1982a; Glenney et al. 1982b) could not be inter-

Fig. 4. Schematic representation of calspectin molecule (tetramer): Its binding
sites for calmodulin and actin oligomer

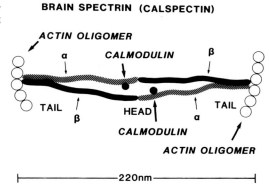

preted unambiguously, because in those experiments, calmodulin binding was examined
by the gel overlay method on SDS gels and the possibility was not excluded that the
binding ability of the β subunit may have been destroyed during the SDS treatment.
For comparison of other properties of spectrin and calspectin, see Kakiuchi and Sobue
(1983).

The physiological significance of interaction of calspectin with calmodulin is unclear. Most recently, we made two interesting observations in this respect. One is that
there is a flip-flop type regulation for calspectin. In previous studies using purified calspectin, we were unable to demonstrate the flip-flop type binding for calspectin, i.e.,
its binding to F-actin was not affected by addition of calmodulin plus Ca^{2+}. However,
we found that, in the presence of supernatants of tissue homogenates, binding of calspectin to calmodulin and F-actin did occur in a flip-flop fashion (Sobue et al. 1983).
This factor (DE-200 factor), which was partially purified from a brain homogenate by
DEAE-cellulose column chromatography, is nondialyzable, heat-labile, and devoid of
calspectin kinase activity. More work is needed for the further information. The other
one is the discovery of protein kinase(s) in brain that phosphorylates calspectin in the
presence of Ca^{2+} and calmodulin. Brain synaptosomal fraction is particularly rich in
calspectin (Kakiuchi et al. 1982c), and we found that this endogenous calspectin is
phosphorylated by protein kinase activity associated with synaptosomes (Sobue et al.
1982c). In a subsequent study, this protein kinase activity was solubilized from a synaptic membrane fraction and partially purified (Sobue et al. 1982d). The enzyme thus
obtained had an estimated M_r of 800,000 and required Ca^{2+} and calmodulin for full
activity.

6 Tau Factor from Brain Microtubules

Microtubules have been shown to be involved in mitosis, cell motility, and maintenance
of cell shape. Microtubules are in a state of dynamic equilibrium with tubulin, and the
control mechanism that governs the assembly-disassembly of microtubules in the cell

has been the subject of a thorough investigation. Weisenberg (1972) was the first to succeed in microtubule assembly in vitro by chelating Ca^{2+} using EGTA and to suggest that the assembly-disassembly of microtubules is regulated by concentrations of Ca^{2+} in the cell. Subsequently, calmodulin has been proposed for the mediator of this effect of Ca^{2+} (Means and Dedman 1980).

Microtubules, purified by cycles of assembly and disassembly in vitro, are composed of tubulin and several microtubule-associated proteins (abbreviated to MAP's). When brain tubulin is freed of MAP's, the tubulin no longer assembles into microtubules under standard polymerization conditions (for review, Timasheff and Grisham 1980). MAP's consist of several proteins. MAP1 and MAP2 are high molecular weight species and tau (τ) factor is a family of four closely related lower M_r (55,000-62,000) proteins (Cleveland et al. 1977). We found that τ factor is a Ca^{2+}-dependent calmodulin binding protein (Sobue et al. 1981c). Subsequently, we were able to reconstitute a Ca^{2+}-sensitive tubulin polymerization system using purified tubulin, τ factor and calmodulin (Kakiuchi and Sobue 1981). As shown in Fig. 1b schematically, the system is controlled by micromolar concentrations of Ca^{2+} in the flip-flop fashion. In another line of research, Kumagai et al. (1982) reported that tubulin is a Ca^{2+}-dependent calmodulin-binding protein and association of calmodulin with tubulin led to a protein complex incapable of assembly into microtubules.

7 Conclusions

Table 1 summarizes calmodulin-binding proteins that interact with cytoskeletal proteins in a flip-flop fashion. The binding of these proteins to target cytoskeletal proteins modulates their functions, and Ca^{2+}-dependent binding to calmodulin eliminates the interaction between the calmodulin-binding proteins and target cytoskeletal proteins. In this respect, these are key proteins through which regulatory actions of calmodulin are

Table 1. Cytoskeleton-related calmodulin-binding proteins (cytocalbins)

Calmodulin-binding protein	Target cytoskeletal protein	Tissue	Note
Caldesmon	F-actin	Gizzard, aorta, uterus, platelets, cultured cells (fibroblast cells, kidney cells)	
135 K protein	F-actin	Gizzard	Myosin-light chain kinase
Calspectin (fodrin)	F-actin	Brain, adrenal, pituitary	Spectrin-related protein
Tau factor	Tubulin	Brain (microtubules)	A member of MAP's

transmitted to the contractile device and cytoskeleton of smooth muscles and non-muscle tissues. We have proposed the name "cytocalbin" for these *cyto*skeleton-related *cal*modulin-*bin*ding proteins.

Acknowledgement. We thank T. Nagasaka for excellent typing servies and M. Ohara for comments on the manuscript.

References

Adelstein RS, Sellers JR, Conti MA, Pato MD, de Lanerolle P (1982) Regulation of smooth muscle contractile proteins by calmodulin and cyclic AMP. Fed Proc 41:2873-2878

Baines AJ (1983) The spread of spectrin. Nature (Lond) 301:377-378

Bennett V, Davis J, Fowler WE (1982) Brain spectrin, a membrane-associated protein related in structure and function to erythrocyte spectrin. Nature (Lond) 299:126-131

Branton D, Cohen CM, Tyler J (1981) Interaction of cytoskeletal proteins on the human erythrocyte membrane. Cell 24:24-32

Cavadore JC, Molla A, Harricane MC, Gabrion J, Benyamin Y, Demaille JG (1982) Subcellular localization of myosin light chain kinase in skeletal, cardiac and smooth muscles. Proc Natl Sci USA 79:3475-3479

Cheung WY (1970) Cyclic 3',5'-nucleotide phosphodiesterase: demonstration of an activator. Biochem Biophys Res Commun 38:533-538

Cleveland DW, Hwo SY, Kirschner MW (1977) Purification of tau, a microtubule-associated protein that induces assembly of microtubules from purified tubulin. J Mol Biol 116:207-225

Cohen CM (1983) The molecular organization of the red cell membrane skeleton. Semin Hematol 20:141-158

Dabrowska R, Hinkins S, Walsh MP, Hartshorne DJ (1982) The binding of smooth muscle myosin light chain kinase to actin. Biochem Biophy Res Commun 107:1524-1531

Davies PJA, Klee CB (1981) Calmodulin-binding proteins: a high molecular weight calmodulin-binding protein from bovine brain. Biochem Int 3:203-212

Ebashi S, Nanomura Y, Nakamura S, Nakasone H, Kohama K (1982) Regulatory mechanism in smooth muscle: actin-linked regulation. Fed. Proc 41: 2863-2867

Glenney JR, Glenney P, Osborn M, Weber K (1982a) An F-actin and calmodulin-binding protein from isolated intestinal brush borders has a morphology related to spectrin. Cell 28:843-854

Glenney JR, Glenney P, Weber K (1982b) Erythroid spectrin, brain fodrin and intestinal brush-border proteins (TW-260/240) are related molecules containing a common calmodulin-binding subunit bound to a variant cell type-specific subunit. Proc Natl Acad Sci USA 79:4002-4005

Guerriero Jr V, Rowley DR, Means AR (1981) Production and characterization of an antibody to myosim light chain kinase and intracellular localization of the enzyme. Cell 27:449-458

Ishimura K, Ban T, Matsuda H, Fujita H, Sobue K, Kakiuchi S (1984) Immunocytochemical demonstration of caldesmon (a calmodulin-binding F-actin-interacting protein) in smooth muscle fibers and absorptive epithelial cells of the rat small intestine. Cell and Tissue Res 235:207-209

Kakiuchi S, Sobue K (1981) Ca^{2+} and calmodulin-dependent flip-flop mechanism in the microtubule assembly-disassembly. FEBS Lett 132:141-143

Kakiuchi S, Sobue K (1983) Control of cytoskeleton by calmodulin and calmodulin-binding proteins. Trends Biochem Sci 8:59-62

Kakiuchi S, Yamazaki R, Nakajima H (1969) Studies on brain cyclic 3',5'-nucleotide phosphodiesterase: its purification and properties (in Japanese). Bull Jpn Neurochem Soc 8:17-20

Kakiuchi S, Yamazaki R, Nakajima H (1970) Properties of a heat-stable phosphodiesterase activating factor isolated from brain extract: studies on cyclic 3',5'-nucleotide phosphodiesterase. II. Proc Jpn Acad 46:587-592

Kakiuchi S, Sobue K, Fujita M (1981) Purification of a 240,000 dalton calmodulin-binding protein from a microsomal fraction of brain. FEBS Lett 132:144-148

Kakiuchi S, Sobue K, Kanda K et al. (1982a) Correlative biochemical and morphological studies of brain calspectin: a spectrin-like calmodulin-binding protein. Biomed Res 3:400-410

Kakiuchi S, Yasuda S, Yamazaki R, Teshima Y, Kanda K, Kakiuchi R, Sobue K (1982b) Quantitative determinations of calmodulin in the supernatant and particulate fractions of mammalian tissues. J Biochem (Tokyo) 92:1041-1048

Kakiuchi S, Sobue K, Morimoto K, Kanda K (1982c) A spectrin-like calmodulin-binding protein (calspectin) of brain. Biochem Int 5:755-762

Kakiuchi R, Inui M, Morimoto K, Kanda K, Sobue K, Kakiuchi S (1983) Caldesmon, a calmodulin-binding, F actin-interacting protein, is present in aorta, uterus and platelets. FEBS Lett 154:351-356

Kumagai H, Nishida E, Sakai H (1982) The interaction between calmodulin and microtubule proteins. (IV) Quantitative analysis of the binding between calmodulin and tubulin dimer. J Biochem (Tokyo) 91:1329-1336

de Lanerolle P, Adelstein RS, Feramisco JR, Burridge K (1981) Characterization of antibodies to smooth muscle myosin kinase and their use in localizing myosin kinase in nonmuscle cells. Proc Natl Acad Sci USA 78:4738-4742

Lazarides E, Nelson WJ (1982) Expression of spectrin in nonerythroid cells. Cell 31:505-508

Levine J, Willard M (1981) Fodrin: axonally transported polypeptides associated with the internal periphery of many cells. J Cell Biol 90:631-643

Means AR, Dedman JR (1980) Calmodulin: an intracellular calcium receptor. Nature (Lond) 285:73-77

Morimoto K, Kambayashi J, Kosaki G, Kanda K, Sobue K, Kakiuchi S (1982) Calmodulin is the sole Ca^{2+}-sensitizing factor in platelet myosin B. Biomed Res 3:83-90

Owada MK, Hakura A, Iida K, Yahara I, Sobue K, Kakiuchi S (to be published 1984) Occurrence of caldesmon (a calmodulin-binding protein) in cells: comparison of normal and transformed cells. Proc Natl Acad Sci USA

Palfray HC, Schiebler W, Greengard P (1982) A major calmodulin-binding protein common to various vertebrate tissues. Proc Natl Acad Sci USA 79:3780-3784

Shimo-oka T, Watanabe Y (1981) Stimulation of actomyosin Mg^{2+}-ATPase activity by a brain microtubule-associated protein fraction. High-molecular weight actin-binding protein is the simulating factor. J Biochem (Tokyo) 90:1297-1307

Sobue K, Muramoto Y, Yamazaki R, Kakiuchi S (1979) Distribution in rat tissues of modulator-binding protein of particulate nature: studies with [^{3}H]modulator protein. FEBS Lett 105:105-109

Sobue K, Fujita M, Muramoto Y, Kakiuchi S (1980) Spectrin as a major modulator binding protein of erythrocyte cytoskeleton. Biochem Int 1:561-566

Sobue K, Muramoto Y, Fujita M, Kakiuchi S (1981a) Calmodulin-binding protein from chicken gizzard that interacts with F-actin. Biochem Int 2:469-476

Sobue K, Muramoto Y, Fujita M, Kakiuchi S (1981b) Purification of a calmodulin-binding protein from chicken gizzard that interacts with F-actin. Proc Natl Acad Sci USA 78:5652-5655

Sobue K, Fujita M, Muramoto U, Kakiuchi S (1981c) The calmodulin-binding protein in microtubules is TAU factor. FEBS Lett 132:137-140

Sobue K, Muramoto Y, Fujita M, Kakiuchi S (1981d) Calmodulin-binding protein of erythrocyte cytoskeleton. Biochem Biophys Res Commun 100:1063-1070

Sobue K, Morimoto K, Kanda K, Maruyama K, Kakiuchi S (1982a) Reconstitution of Ca^{2+}-sensitive gelation of actin filaments with filamin, caldesmon and calmodulin. FEBS Lett 138:289-292

Sobue K, Morimoto K, Kanda K, Fukunaga K, Miyamoto E, Kakiuchi S (1982b) Interaction of 135000-M_r calmodulin-binding protein (myosin kinase) and F-actin: another Ca^{2+}- and calmodulin-dependent flip-flop switch. Biochem Int 5:503-510

Sobue K, Kanda K, Yamagami K, Kakiuchi S (1982c) Ca^{2+}- and calmodulin-dependent phosphorylation of calspectin (spectrin-like calmodulin-binding protein; fodrin) by protein kinase system in synaptosomal cytosol and membranes. Biomed Res 3:561-570

Sobue K, Kanda K, Kakiuchi S (1982d) Solubilization and partial purification of protein kinase systems from brain membranes that phosphorylate calspectin, a spectrin-like calmodulin-binding protein (fodrin). FEBS Lett 150:185-190

Sobue K, Kanda K, Adachi J, Kakiuchi S (1983) Calmodulin-binding protein that interact with actin filaments in a Ca^{2+}-dependent flip-flop manner: survery in brain and secretory tissues. Proc Natl Acad Sci USA 80:6868-6871

Steck TL (1974) The organization of proteins in the human red blood cell membrane. J Cell Biol 62:1-19

Stull JT, Silver PJ, Miller JR, Blumenthal DK, Botterman BR, Klug GA (1983) Phosphorylation of myosin light chain in skeletal and smooth muscles. Fed Proc 42:21-26

Teshima Y, Kakiuchi S (1978) Membrane-bound forms of Ca^{2+}-dependent protein modulator: Ca^{2+}-dependent and independent bindings of modulator protein to the particulate fraction from brain. J Cyclic Nucleotide Res 4:219-231

Timasheff SN, Grisham LM (1980) In vitro assembly of cytoplasmic microtubules. Annu Rev Biochem 49:565-591

Tsukita S, Tsukita S, Ishikawa H, Kurokawa M, Morimoto K, Sobue K, Kakiuchi S (1983) Binding sites of calmodulin and actin on the brain spectrin, calspectin. J Cell Biol 97:574-578

Walsh MP, Bridenbaugh R, Glenn W, Kerrick L, Hartshorne DJ (1983) Gizzard Ca^{2+}-independent myosin light chain kinase: evidence in favor of the phosphorylation theory. Fed Proc 42:45-50

Weisenberg RC (1972) Microtubule formation in vitro in solutions containing low calcium concentrations. Science (Wash DC) 177:1104-1105

Calcium-Regulated Protein Phosphorylation in Mammalian Brain

S. I. WALAAS and A. C. NAIRN [1]

CONTENTS

1 Introduction

Calcium (Ca^{2+}) has been recognized as an important intracellular messenger in mammalian tissues, regulating, for example, skeletal muscle contraction (Huxley 1964) and excitation-secretion coupling in secretory glands and peripheral nerves (Douglas 1968, Katz and Miledi 1969). In the central nervous system (CNS), Ca^{2+} has been found to be involved in a large variety of nerve cell functions, for example in neurotransmitter synthesis and release, neuronal activity-regulated glycogenolysis and mitochondrial respiration, exocytotic and endocytotic phenomena, generation of Ca^{2+} spikes, and in the control of ion channels, such as Ca^{2+}-dependent K^+-channels (see, e.g., Baker 1972; Rubin 1972; Blaustein et al. 1981; Reichardt and Kelly 1983).

[1] Laboratory of Molecular and Cellular Neuroscience, The Rockefeller University, 1230 York Avenue, New York, NY 10021, USA

The mechanisms by which Ca^{2+} achieves its intracellular effects appear to be diverse. In many instances, the actions of Ca^{2+} are mediated by certain Ca^{2+}-binding proteins such as calmodulin and troponin C (for reviews, see Cheung 1980; Kakiuchi et al. 1982, Klee and Vanaman 1982). Strong evidence suggests that many of the effects of calmodulin, in turn, may be mediated through the regulation of protein phosphorylation, which is believed to be one of the major biochemical mechanisms by which extracellular stimuli regulate intracellular functions (Greengard 1978; Cohen 1982; Nestler and Greengard 1983, 1984). Such protein phosphorylation systems consist in their simplest form of three components. These are a protein kinase, which catalyzes the incorporation of phosphate from a phosphate donor (usually adenosine triphosphate (ATP), a protein substrate, which binds phosphate on either serine, threonine, or tyrosine residues, and a phosphoprotein phosphatase, which removes the phosphate from the substrate. Changes in the state of phosphorylation of a specific substrate bring about changes in the physical properties of the protein, and in turn lead to changes in the functional properties of the particular protein.

Protein phosphorylation systems have been found to have the necessary specificity and flexibility to regulate a great number of major physiological functions in mammalian tissues (Greengard 1978; Cohen 1982; Nestler and Greengard 1984). Several signal molecules are known to activate specific protein kinases, the best-studied being the ubiquitous "second messenger" cyclic adenosine monophosphate (cyclic AMP) (reviews: see Nathanson and Kebabian 1982). Recent work has shown that Ca^{2+} can also regulate specific protein phosphorylation systems in various tissues (review: Schulman 1982). In nervous tissue, a role for Ca^{2+}-dependent protein phosphorylation was suggested by studies using purified nerve terminals from cerebral cortex (Krueger et al. 1976, 1977). These so-called synaptosomes were first preincubated with $[^{32}P]0_4$ to label the intracellular ATP pools. Increasing intra-terminal Ca^{2+}-concentration, by depolarization with veratridine or high K^+-concentrations, led to an increase in the ^{32}P-content of several proteins. The result of such an experiment is shown in Fig. 1. The most prominent phosphorylated proteins consisted of two polypeptides which had apparent molecular weights, on sodium dodecyl sulphate-polyacrylamide gel electrophoresis (SDS-PAGE), of M_r 86,000/80,000. These two phosphoproteins, named synapsin Ia and Ib, respectively, were purified and characterized, and will be described in detail in Section 3. The effects of veratridine and K^+ were totally dependent on the presence of Ca^{2+} in the external medium (Fig. 1), and were rapid, reversible, and concomitant with Ca^{2+} influx into the nerve terminals. In addition, the effects were independent of cyclic nucleotides (Krueger et al. 1977).

Further studies on broken cell preparations confirmed and extended these findings. The addition of Ca^{2+} to synaptosomal lysates was shown to stimulate the phosphorylation of several proteins (DeLorenzo 1976, Schulman and Greengard 1978a). Subsequent analysis showed that Ca^{2+}-regulated protein phosphorylation in both calmodulin-depleted particulate and cytosol fractions from brain was mediated by the ubiquitous Ca^{2+}-binding protein calmodulin (Schulman and Greengard, 1978 a, b; Yamauchi and Fujisawa 1979; O'Callaghan et al. 1980). Ca^{2+}/calmodulin-dependent protein phosphorylation appears to be particularly important in brain. The levels of Ca^{2+}/calmodulin-dependent protein kinase activity are very high in this tissue (see Sect. 2), and a large number of brain-specific endogenous substrates have been discovered

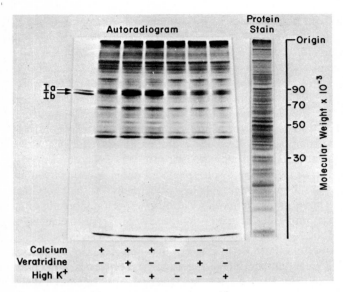

Fig. 1. Effect of depolarization-induced Ca^{2+}-influx on the phosphorylation of endogenous proteins in synaptosomes from rat cerebral cortex. The crude synaptosomal fraction was preincubated with $[^{32}P]O_4$ for 30 min and aliquots were then incubated for 30 s in the absence or presence of 0.1 mM veratridine or 60 mM KCl (High K+). 1 mM $CaCl_2$ was present where indicated. The incubation was terminated by the addition of SDS, and the samples were analyzed by SDS-polyacrylamide gel electrophoresis, protein staining and autoradiography. *Left lane* indicates the position of standard synapsin I (Ia and Ib) which was included as a marker. (Wu et al. 1982)

(see Sect. 3) (Schulman and Greengard 1978a, b, Walaas et al. 1983a, b; Wrenn et al. 1980).

More recently, a second class of Ca^{2+}-dependent protein kinase, the Ca^{2+}/phospholipid-dependent protein kinase (also known as Protein kinase C), has been described (Takai et al. 1979a). Studies on peripheral, non-neuronal tissues have indicated that this protein kinase may mediate certain effects of those Ca^{2+}-mobilizing hormones and neurotransmitters which increase phosphatidylinositol turnover (see, e.g., Michell 1975, 1983; Nishizuka 1983) and that it may be involved in regulation of exocytotic release, tumor promotion, and a number of other functions (Takai et al. 1981, Nishizuka 1983). The Ca^{2+}/phospholipid-dependent protein kinase is present in brain in very high concentration (Kuo et al. 1980; Walaas et al. 1983b), where it catalyzes the phosphorylation of a distinct set of proteins which have a wide distribution throughout the CNS (Walaas et al. 1983a, b; Wrenn et al. 1980).

The purpose of this chapter is to review particular aspects of Ca^{2+}-dependent protein phosphorylation in the mammalian CNS. We will discuss the Ca^{2+}/calmodulin-dependent and Ca^{2+}/phospholipid-dependent protein kinases involved, as well as several of the specific substrates so far identified for these kinases. A brief discussion of Ca^{2+}-regulated dephosphorylation of phosphoproteins is also included. No complete coverage of the literature will be attempted; rather, we concentrate on specific, well-studied systems which exemplify the properties of the various brain Ca^{2+}-regulated protein phosphorylation systems.

2 Ca^{2+}-Dependent Protein Kinases in Brain

Ca^{2+}-dependent activation of protein kinases was originally discovered for the enzymes phosphorylase kinase (Brostrom et al. 1971; Cohen 1973) and myosin light chain kinase (Pires and Perry 1977). Although the exact molecular mechanism for this activation was not known, Kretsinger (1977) suggested that phosphorylase kinase might contain an "EF" hand, the protein structure he had predicted from X-ray crystallography studies of the Ca^{2+}-binding protein parvalbumin, which bound Ca^{2+} with a dissociation constant of $10^{-6}-10^{-5}$ M. The discovery that the Ca^{2+}-dependent protein phosphorylation system in brain required for activity the ubiquitous multifunctional Ca^{2+}-binding protein, calmodulin (for review, see Cheung 1980), led to the suggestion that Ca^{2+} and calmodulin-dependent protein phosphorylation would be widespread in various tissues (Schulman and Greengard 1978b). Such a role was supported by evidence that Ca^{2+}-dependent activation of myosin light chain kinase from smooth (Dabrowska et al. 1977, 1978) and skeletal muscle (Perry et al. 1978; Yagi et al. 1978) and phosphorylase kinase (Cohen et al. 1978) were also dependent on calmodulin. In each of these systems, calmodulin, which is evolutionarily related to parvalbumin, was found to fulfill the role of a Ca^{2+}-binding "target" protein, as suggested by Kretsinger.

It was originally believed that most Ca^{2+}-dependent protein phosphorylation was controlled by calmodulin (Schulman and Greengard 1978b; Yamauchi and Fujisawa 1979). More recently, however, a Ca^{2+} and phospholipid-dependent protein kinase has been found in high concentrations in brain as well as in other tissues (Nishizuka 1980, 1983). This enzyme does not require calmodulin for activity, but is activated by Ca^{2+} and certain phospholipids (Takai et al. 1979a; Kaibuchi et al. 1981). Several distinct Ca^{2+}/calmodulin-dependent protein kinases from mammalian brain have been characterized. In contrast, only one Ca^{2+}/phospholipid-dependent protein kinase has been found.

2.1 Ca^{2+}/Calmodulin-Dependent Protein Kinases

2.1.1 Ca^{2+}/Calmodulin-Dependent Synapsin I Kinases

Following the observation that synaptosomal membranes contain a Ca^{2+}/calmodulin-dependent protein kinase (Schulman and Greengard 1978 a, b), Sieghart and coworkers (Sieghart et al. 1979) showed that two of the major substrates for Ca^{2+}-dependent protein phosphorylation in synaptosomes were identical to the proteins synapsin Ia and Ib. Synapsin Ia and Ib (collectively known as synapsin I, see Sect. 3 for discussion) were previously known to be two of the major substrates for cyclic AMP-dependent protein kinase in the nervous system, and had been purified and extensively characterized (Ueda and Greengard 1977). Further studies (Huttner and Greengard 1979; Kennedy and Greengard 1981) showed that synapsin I contained one site (Site I) which was phosphorylated by cyclic AMP-dependent protein kinase and by a Ca^{2+}/calmodulin-dependent protein kinase, and two additional sites (collectively termed Site II) which were phosphorylated by a Ca^{2+}/calmodulin-dependent protein kinase. In these studies *Staphylococcus aureus* V8 protease was used to analyze the different phosphorylation sites. Digestion

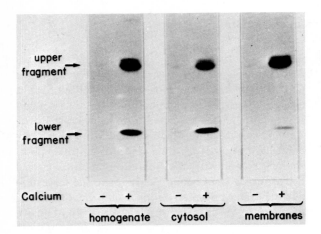

Fig. 2. Differential phosphorylation of Site I and Site II in synapsin I by rat brain Ca^{2+}/calmodulin-dependent protein kinases. Purified synapsin I was incubated with crude brain homogenate, cytosol or membranes plus Mg-[γ-^{32}P]ATP in the absence or presence of 0.3 mM free Ca^{2+}, following which the reactions were stopped by SDS and the samples analysed by SDS-PAGE. Phosphorylated synapsin I was located by autoradiography, cut out from the gel, and subjected to incomplete digestion with *Staphylococcus aureus* V8 protease. The resulting phosphopeptide fragments were then separated by SDS-PAGE and analyzed by autoradiography. *Lower fragment* indicates the phosphopeptide containing Site I; *upper fragment* indicates the phosphopeptide containing Site II. (Kennedy and Greengard 1981)

of synapsin I with this protease produced a fragment which contained Site I and a fragment which contained Site II (Huttner and Greengard 1979; Huttner et al. 1981). Analysis of the subcellular distribution of Ca^{2+}/calmodulin-dependent protein kinase activity using this technique (Fig. 2) showed that the kinase activity which phosphorylated Site I (lower fragment) was present only in the cytosolic fraction while the kinase activity which phosphorylated Site II (upper fragment) was distributed in both cytosolic and membrane fractions. These and other studies have established that two distinct Ca^{2+}/calmodulin-dependent protein kinases are responsible for this site specificity (Kennedy and Greengard 1981; Kennedy et al. 1983a; Nairn and Greengard 1983; Lai et al. 1983; McGuinness et al. 1984). Ca^{2+}/calmodulin-dependent protein kinase I (Ca^{2+}/calmodulin kinase I) specifically phosphorylates Site I of synapsin I and Ca^{2+}/calmodulin-dependent protein kinase II (Ca^{2+}/calmodulin kinase II) specifically phosphorylates Site II. These two enzymes, which have been purified and characterized and have been shown to be distinct from myosin light chain kinase and phosphorylase kinase, probably represent two of the major Ca^{2+}/calmodulin-dependent protein kinases in brain.

2.1.1.1 Ca^{2+}/Calmodulin Kinase I

Ca^{2+}/calmodulin kinase I was purified (Nairn and Greengard 1983, 1984) more than 1000-fold from bovine brain using synapsin I as substrate, by DEAE-cellulose and hydroxylapatite chromatography, gel filtration and Sepharose-calmodulin affinity chromatography. Ca^{2+}/calmodulin kinase I had a K_m for synapsin I (Site I) of approximately 2-3 μM, which is similar to that of cyclic AMP-dependent protein kinase, and K_m of approximately 50 μM for ATP. Under optimal assay conditions, the purified kinase had a specific activity of up to 3 $\mu mol\ min^{-1}\ mg^{-1}$ using synapsin I as substrate. In addition to synapsin I, Ca^{2+}/calmodulin kinase I phosphorylated smooth muscle myosin light chain. Skeletal muscle myosin light chain, glycogen synthase, tubulin and microtubule-associated-protein 2 (MAP-2) were not phosphorylated to any significant extent by the enzyme.

The molecular weight of the native enzyme was estimated to be 48,000 from gel filtration studies. The purified enzyme preparation contained two major proteins, in approximately equal amounts, with molecular weights estimated to be 37,000 and 39,000 using SDS-PAGE. Both proteins bound calmodulin using the ^{125}I-calmodulin overlay technique and copurified with the kinase activity through a number of chromatographic procedures. In addition, both proteins were phosphorylated with low stoichiometry in a Ca^{2+}/calmodulin-dependent manner. These and other results suggest that the enzyme activity is associated with the two polypeptides of 37,000 and 39,000 molecular weight.

Ca^{2+}/calmodulin kinase I, although found in highest concentration in the brain, appears to have a widespread tissue distribution (Nairn and Greengard 1984). Ca^{2+}/calmodulin kinase I activity was found in the cytosol from all rat tissues examined, including (in decreasing order of activity), forebrain, pancreas, spleen, lung, adrenal gland, heart, skeletal muscle, liver, and kidney. Furthermore, Ca^{2+}/calmodulin kinase I activity was partially purified from bovine heart and found to have properties similar to those of the enzyme prepared from brain. Ca^{2+}/calmodulin kinase I also appears to be widely distributed in the CNS, where it is probably present in all parts of the neuron (A. C. Nairn and P. Greengard, unpublished results). Since synapsin I is present only in the nerve terminal, the widespread distribution of Ca^{2+}/calmodulin I suggests the existence of additional physiological substrates for this enzyme in both neuronal and non-neuronal mammalian tissues.

2.1.1.2 Ca^{2+}/Calmodulin Kinase II

Ca^{2+}/calmodulin kinase II has recently been purified from rat forebrain by two different research groups using similar procedures (Bennett et al. 1983; Lai et al. 1983; McGuinness et al. 1984). The enzyme was purified more than 200-fold using DEAE-cellulose chromatography, ammonium sulfate precipitation, gel filtration and Sepharose-calmodulin affinity chromatography. The purified enzyme exhibited a broad substrate specificity (Lai et al. 1983; Bennett et al. 1983) with synapsin I (2-4 $\mu mol\ min^{-1}\ mg^{-1}$ being the best substrate tested. Ca^{2+}/calmodulin kinase II also phosphorylated MAP-2 and smooth muscle myosin light chain, but not phosphorylase b or tubulin. Bennett et al. (1983) reported that the enzyme phosphorylated casein and phosvitin, but not glycogen synthase. In contrast, Lai et al. (1983) reported significant phosphorylation of glycogen synthase but not of casein or phosvitin. Several other Ca^{2+}/calmodulin-dependent protein kinases have been prepared from rat forebrain which have physico-chemical properties very similar to those of Ca^{2+}/calmodulin kinase II (Fukunaga et al. 1982; Sobue et al. 1982; Goldenring et al. 1983; Schulman et al. 1983; Yamauchi and Fujisawa 1983a). There were, however, several differences in substrate specificity among these different enzymes preparations. MAP-2, smooth muscle myosin light chain and casein but not phosphorylase b were good substrates for all the kinase preparations. Synapsin I was tested as a substrate for only one of these kinase preparations but apparently was not phosphorylated (Goldenring et al. 1983). Tubulin was a relatively good substrate for one preparation (Goldenring et al. 1983), and tryptophan and tyrosine hydroxylase were potential substrates for another (Yamauchi and Fujisawa 1983a). It is likely that the different enzyme preparations used in these studies all represent Ca^{2+}/calmodulin kinase II, and that the reported differences in substrate specificity probably reflect variations in assay procedure or in the substrate preparations used, rather than in the properties of the kinase.

The purified Ca^{2+}/calmodulin kinase II (Bennett et al. 1983; Lai et al. 1983; McGuinness et al. 1984) had a molecular weight of 600,000-650,000 as measured using gel filtration and contained a major M_r 50,000 polypeptide and less prominent polypeptides of M_r 60,000 and 58,000. The M_r 60,000 and 50,000 polypeptides, although immunologically related (Kelly et al. 1984) were shown to be distinct polypeptides. The M_r 60,000 and 58,000 polypeptides were, however, very similar, suggesting that the M_r 58,000 polypeptide may have been generated from the M_r 60,000 polypeptide by proteolysis (Kennedy et al. 1983a). Several results suggest that enzyme activity is associated with all three subunits. Each protein was autophosphorylated in a Ca^{2+}/calmodulin-dependent manner, and each protein was shown to bind calmodulin using a ^{125}I-calmodulin-overlay technique; preliminary results also suggested that each subunit bound ATP, as analyzed with the photoaffinity label 8-azido ATP (Y. Lai, T.L. McGuiness, and P. Greengard, personal communication).

It is not known whether the individual polypeptides of Ca^{2+}/calmodulin kinase II prepared from forebrain are subunits of a single enzyme or constitute different isozymes. Support for the possibility that the forebrain enzyme itself may be an isozyme of a multifunctional Ca^{2+}/calmodulin-dependent protein kinase has, however, recently been obtained. Ca^{2+}/calmodulin kinase II purified from rat cerebellum, although having properties almost identical to the enzyme prepared from forebrain, contained predominantly 60,000 and 58,000 molecular weight polypeptides (T.L. McGuiness, Y. Lai and P. Greengard, personal communication) (see Sect. 3 for further discussion). A number of Ca^{2+}/calmodulin-dependent protein kinases which have similar properties to those of Ca^{2+}/calmodulin kinase II have also been prepared from tissues other than brain. These properties include: similar substrate specificities, native molecular weights of 500,000-800,000, and autophosphorylatable subunits with molecular weights ranging between 60,000 and 50,000. These kinases were prepared from electric organ of *Torpedo californica* (Palfrey et al. 1983), from turkey erythrocytes (Palfrey et al., 1984) and from bovine heart (Palfrey 1984); and glycogen synthase kinase purified from liver (Ahmed et al. 1982; Payne et al. 1983) and skeletal muscle (Woodgett et al. 1982). Glycogen synthase kinase from skeletal muscle, for example, which has a substrate specificity identical to that of Ca^{2+}/calmodulin kinase II (McGuinness et al. 1983; Schworer and Soderling 1983), has a native molecular weight of 800,000 and contains autophosphorylated subunits of M_r 59,000, 58,000, and 54,000 (Woodgett et al. 1982; McGuinness et al. 1983). In addition, monoclonal antibodies prepared against Ca^{2+}/calmodulin kinase II have been shown to cross-react with each subunit of glycogen synthase kinase (McGuinness et al. 1983). The similar physicochemical and immunological properties and broad substrate specificities of these different Ca^{2+}/calmodulin-dependent protein kinases strongly suggest that they are isozymes of a multifunctional Ca^{2+}/calmodulin-dependent protein kinase.

Ca^{2+}/calmodulin kinase II is present in very high concentration in the brain, comprising perhaps as much as 0.3% of total brain protein (Bennett et al. 1983; McGuinness et al. 1984). The enzyme, which is present only in neurons (Sieghart et al. 1980), is found throughout the cell, being particularly enriched in dendrites (McGuinness et al. 1984) and post-synaptic densities (Kennedy et al. 1983b; Kelly et al. 1984). In post-synaptic densities the "major PSD protein" (Cotman and Kelly 1980) has been shown to be identical to the 50,000 molecular weight subunit of Ca^{2+}/calmodulin kinase II. Ca^{2+}/calmodulin kinase II is found in both soluble and particulate fractions of brain homogenates (Kennedy et al. 1983a), and it has been suggested that autophosphorylation may regulate the intracellular distribution of the enzyme (Bennett et al. 1983). In a regional survey of Ca^{2+}/calmodulin kinase II activity in rat brain, the enzyme was found to have a widespread though variable distribution (Walaas et al. 1983a, b). High activity was found in cortical regions, particularly in the hippocampus and in most subcortical forebrain regions, while relatively low activity was found in the cerebellum, diencephalon, mesencephalon, pons/medulla, and spinal cord. It is an interesting possibility that different isozymes of Ca^{2+}/calmodulin kinase II may be responsible for the activity found in particular brain regions (see Sect. 3 for further discussion). The widespread distribution in brain and the broad specificity of Ca^{2+}/calmodulin kinase II suggests that the enzyme mediates or modulates a variety of Ca^{2+}-regulated mechanisms in the nervous system.

2.1.2 Myosin Light Chain Kinase and Phosphorylase Kinase

Myosin light chain kinase and phosphorylase kinase, two Ca^{2+}/calmodulin-dependent protein kinases which have been purified from several non-neuronal tissues and characterized extensively, have been identified in brain. Myosin light chain kinase has been purified from forebrain (Dabrowska and Hartshorne 1978; Hathaway et al. 1981) and found to have properties similar to those of the smooth muscle enzyme. The purified kinase has a molecular weight of 130,000 and it specifically phosphorylates smooth muscle myosin light chain (Hathaway et al. 1981). A number of other studies have reported partial purifications of myosin light chain kinase from brain (Yamauchi and Fujisawa 1980; Miyamoto et al. 1981; Kennedy and Greengard 1981). However, both Ca^{2+}/calmodulin kinase I and II (see above) and several additional protein kinases (Matsumara et al. 1982) phosphorylate myosin light chain. It is not clear, therefore, whether these partial preparations represent myosin light chain kinase.

Although myosin has been isolated from brain, its precise function in this tissue is unknown (Trifaro 1978). Phosphorylation of smooth muscle or nonmuscle myosin light chain is believed to be a prerequisite for interaction between myosin and actin (Adelstein 1980; Mrwa and Hartshorne 1980), and myosin light chain kinase probably plays the same role in brain as it does in smooth muscle and nonmuscle cells.

Phosphorylase kinase has been identified in, but not purified from, brain (Osawa 1973; Taira et al. 1982). The brain enzyme, which appears to be similar to phosphorylase kinase from skeletal muscle, is activated both by Ca^{2+} and by phosphorylation by cyclic AMP-dependent protein kinase (Taira et al. 1982). The brain kinase, however, is only partly cross-reactive with antibody prepared against the enzyme from skeletal muscle, suggesting that it may be a distinct phosphorylase kinase isozyme (Taira et al. 1982). Low concentrations of glycogen and glycogen metabolizing enzymes are found throughout the brain and are present in both glia and neurons (Knull and Khandelwal 1982). It is likely that Ca^{2+} activation of phosphorylase kinase regulates glycogen breakdown, a process which has been shown to be enhanced by electrical stimulation (King et al. 1967).

2.2 Ca^{2+}/Phospholipid-Dependent Protein Kinase

The Ca^{2+}/phospholipid-dependent protein kinase (also known as Protein Kinase C) was first purified from cerebellum as a cyclic nucleotide protein-independent kinase which could be activated by a Ca^{2+}-dependent protease also found in brain (Takai et al. 1977; Inoue et al. 1977). The enzyme was subsequently found to be activated by the addition of membrane phospholipid, of which phosphatidylserine and phosphatidylinositol were the best tested, and Ca^{2+}, the K_a for which varied with the particular membrane source used (Takai et al. 1977a, b). This variability was found to depend on low concentrations of diacylglycerol, which (under optimal conditions) was shown to decrease the apparent K_a value for Ca^{2+} from approximately $7 \times 10^{-5} M$ to $5 \times 10^{-6} M$ (Kishimoto et al. 1980).

Based on a series of studies in platelets, a general model has been proposed (Nishizuka 1983) by which the enzyme may be activated by an increase in the concentration of diacylglycerol, which in turn is produced by receptor-stimulated phosphatidylinositol turnover (for review see Michell 1975, 1983). According to this scheme the activation of the kinase by diacylglycerol, although dependent on a low concentration of Ca^{2+}, is not dependent on an increase in intracellular Ca^{2+} concentration. The exact mechanisms, however, by which Ca^{2+}, phospholipid, and diacylglycerol activate the enzyme is not known. Ca^{2+} does not interact directly with the kinase in the absence of phospholipid, and there does not appear to be a "calmodulin-like" calcium-binding domain associated with the kinase monomer (Nishizuka 1983). In the presence of high Ca^{2+} (10^{-5} -$10^{-4} M$), or in the presence of low Ca^{2+} ($10^{-6} M$) plus diacylglycerol, cytosolic kinase associates with plasma membrane phospholipid and is activated (Nishizuka 1983). Interestingly, tumor-promoting phorbol esters may exert some of their actions by substituting for diacylglycerol and causing activation of the Ca^{2+}/phospholipid-dependent protein kinase (Castagna et al. 1982). These phorbol esters, when added to intact cells, also cause the translocation of the kinase from cytosol to the plasma membrane (Kraft et al. 1982; Kraft and Anderson 1983).

The Ca^{2+}/phospholipid-dependent protein kinase has been purified from brain (Kikkawa et al. 1982; Albert et al. 1984) and other tissues (Wise et al. 1982; Schatzman et al. 1983a) and the properties of the enzyme from different sources appear to be similar. The enzyme from brain is a monomer of M_r 82,000 estimated from SDS-polyacrylamide electrophoresis and M_r 87,000 estimated from gel filtration (Kikkawa et al. 1982). The enzyme has a broad substrate specificity which is distinctly different from those of both cyclic nucleotide-dependent and Ca^{2+}/calmodulin-dependent protein kinases. Both myelin basic protein (Kikkawa et al. 1982; Turner et al. 1982), MAP-2 (Walaas et al. 1983a, b), an M_r 87,000 protein (Wu et al. 1982), the B-50 protein (Aloyo et al. 1983) and several other unidentified proteins (Walaas et al. 1983a, b), have been identified as possible physiological substrates for the enzyme in brain (see Sect. 3). In addition, histone H1 (Kikkawa et al. 1982; Wise et al. 1982), ribosomal protein S6 (Le Peuch et al. 1983), vinculin (Werth et al. 1983), initiation factor eIF-2 (Schatzman et al. 1983b), troponin T (Katoh et al. 1982), and smooth muscle myosin light chain (Endo et al. 1982; Nishikawa et al. 1983) have been shown to be good substrates for the enzyme.

Ca^{2+}/phospholipid-dependent protein kinase has a broad species, tissue, and cellular distribution (Kuo et al. 1980; Minakuchi et al. 1981). The enzyme is highly concentrated and widely distributed in brain (Walaas et al. 1983b). In a regional survey of kinase activity in rat brain, the highest activity was found in cortical regions and in the cerebellum, intermediate activity was found in subcortical telecephalic regions, and the lowest activity was found in the diencephalon, mesencephalon, rhombencephalon (except the cerebellum) and spinal cord (Walaas et al. 1983b).

The subcellular distribution of the kinase in brain is of particular interest. In studies of the distribution of kinase activity in brain (Walaas et al. 1983a, b) the enzyme appeared to be predominantly soluble, but other workers have suggested that a significant amount of the enzyme in brain is particulate (Katoh and Kuo 1982; Kikkawa et al. 1982). The particulate kinase differs from the soluble enzyme in that it will not phosphorylate exogenously added substrates without the addition of a detergent, for example, Triton X-100. The nature of the association of the particulate kinase with the membrane is not known, however.

Ca^{2+}/phospholipid-dependent protein kinase is activated by K^+ − or veratridine-induced Ca^{2+} influx in intact synaptosomes (Wu et al. 1982). In addition, depolarization of brain slices or intact synaptosomes, and activation of, for example, muscarinic receptors, stimulate polyphosphatidylinositol turnover (for review see Hawthorne and Pickard 1979). Physiological activity in intact CNS preparations therefore appears to activate Ca^{2+}/phospholipid-dependent protein kinase. It is possible, at least in brain, that the soluble fraction of the kinase might be activated mainly by Ca^{2+} influx into the cell, whereas the particulate fraction might be activated mainly by an increae in diacylglycerol concentration brought about by phosphatidylinositol turnover.

It is likely that Ca^{2+}/phospholipid-dependent protein phosphorylation is involved in many functions in the brain. The brain contains high levels of the enzyme as well as a large number of specific substrate proteins. It will be of particular interest to investigate the intracellular localization of the different substrate proteins, as well as that of the kinase itself. Such a study of the possible association or reassociation of the proteins with various cellular components, for example the plasma membrane, synaptic and coated vesicles or cytoskeleton would add greatly to our understanding of the action of the Ca^{2+}/phospholipid-dependent protein kinase in vivo.

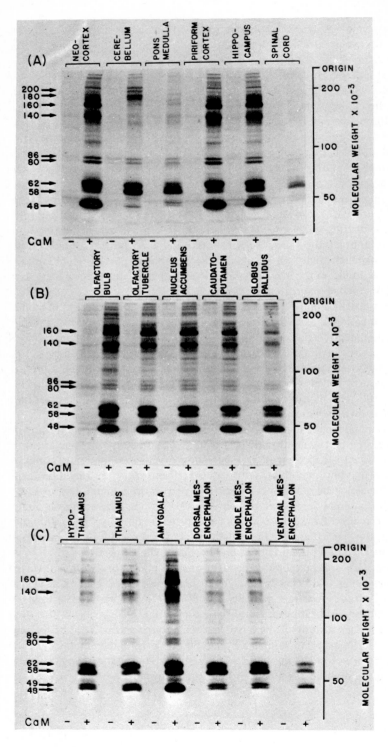

Fig. 3, A-C

3 Ca^{2+}-Dependent Phosphorylation of Brain Proteins

A complete molecular understanding of those Ca^{2+}-regulated physiological effects in brain which are mediated or modulated by protein phosphorylation obviously requires the identification and characterization of the brain proteins which are phosphorylated by Ca^{2+}-dependent protein kinases. There are two distinct approaches to such studies. The first is to characterize the possible Ca^{2+}-regulated phosphorylation of known proteins which either have established functions (such as enzymes, receptors, ion channels), or which are well characterized but whose exact function is unclear. This approach has been used successfully in the case of proteins such as the catecholamine-synthesizing enzyme tyrosine hydroxylase (Yamauchi et al. 1981), MAP-2 (Vallee 1980; Yamauchi and Fujisawa 1982) and myelin basic protein (Sulakhe et al. 1980; Turner et al. 1982), all of which are discussed below. This approach is limited, how-ever, since many of the proteins involved in Ca^{2+}-regulated functions in the CNS have yet to be identified.

The second approach is to search for previously unknown proteins by virtue of their ability to be phosphorylated in response to Ca^{2+}. This approach is potentially open-ended in the sense that, depending on the type of preparation used, it should ultimately be possible to detect most or all of the substrates for Ca^{2+}-dependent pro-tein kinases present in the tissues. These protein substrates may then be characterized with respect to their biochemical, physiological and anatomical properties, and a hy-pothesis about their functional importance might be formulated and experimentally tested (Nestler and Greengard 1983, 1984). Examples of investigations of Ca^{2+}-de-pendent protein phosphorylation in brain where this open-ended approach has been taken include studies with intact synaptosomes (Krueger et al. 1977; Huttner and Greengard 1979; Wu et al. 1982), synaptosomal lysates, synaptic membranes, and syn-aptosol (Schulman and Greengard 1978a, b; Hershkowitz 1978; Sorensen and Mahler 1983; O'Callaghan et al. 1980), total cytosol (Yamauchi and Fujisawa 1979, Walaas et al. 1983b), synaptic vesicles (DeLorenzo and Freedman 1977; Burke and DeLorenzo 1982a; Moskowitz et al. 1983), and synaptic junctions and postsynaptic densites (Grab et al. 1981; Kennedy et al. 1983b; Mahler et al. 1982). As an example of such studies, some aspects of the complexity of brain Ca^{2+}/calmodulin-regulated membrane pro-tein phosphorylation are demonstrated in Fig. 3. In this study (Walaas et al. 1983a) rat brain was micro-dissected into about 20 distinct regions, and calmodulin-depleted membranes from these samples were incubated in a Ca^{2+}-containing medium with [γ-^{32}P]ATP for 10 s in the absence of presence of calmodulin. The phosphorylated pro-teins were then separated by SDS-PAGE and visualized by autoradiography. The ex-periment shows that addition of calmodulin markedly stimulated the phosphorylation of a great number of membrane proteins. Similar results were obtained in studies on the

◀ **Fig. 3 A-C.** Regional distribution of Ca^{2+}/calmodulin-dependent endogenous phosphorylation of membrane proteins in rat CNS. Rat brain and spinal cord were microdissected and membranes pre-pared and incubated with Mg-[γ-^{32}P]ATP and Ca^{2+} (0.5 mM) in the absence or presence of calmodu-lin (CaM, 10 μg ml^{-1}). The reaction was terminated with SDS and the phosphorproteins analyzed by SDS-PAGE and autoradiography. *Numbers* indicate the apparent molecular weight of major phos-phoproteins. (Walaas et al. 1983a)

cytosol, where, in addition to calmodulin-dependent protein phosphorylation, a high level of Ca^{2+}/phospholipid-dependent phosphorylation was found (Walaas et al. 1983b).

From these and other studies, the substrate proteins were divided into three general categories according to their distribution: some are present throughout the CNS in approximately equal amounts and may be present in all neurons (for example in Fig. 3 the M_r 86,000/80,000 doublet representing synapsin I, see discussion below); others are widely but more unevenly distributed and may not be equally concentrated in all cells (for example the major M_r 62,000, 58,000 and 48,000 phosphopeptides which represent subunits of isozymes of the Ca^{2+}/calmodulin kinase II, see Sect. 2 and below); and finally, some phosphoproteins appear to be restricted to one region only (for example in Fig. 3 the M_r 180,000 and 200,000 proteins seen in the cerebellum only), indicating that they may be present in only one or a few cell types (Walaas et al. 1983a, b). These results therefore show that the Ca^{2+}-dependent protein kinases in brain phosphorylate a wide variety of substrate proteins. We will now describe several of the well-characterized substrate proteins.

3.1 Autophosphorylation of Ca^{2+}/Calmodulin Kinase II

The most prominent phosphorylated substrates observed when homogenates, membranes or cytosol fractions prepared from forebrain are phosphorylated in the presence of Ca^{2+} plus calmodulin are phosphopeptides with apparent molecular weights of 62,000, 58,000, and 48,000 (Fig. 3) (e.g., Schulman and Greengard 1978a; DeLorenzo 1976; Yamauchi and Fujisawa 1979; Walaas et al 1983a, b). Recent studies have indicated that these are identical to the autophosphorylatable subunits of M_r 60,000, 58,000 and 50,000 of Ca^{2+}/calmodulin kinase II (Kennedy et al. 1983a; Bennett et al. 1983; McGuinness et al. 1983) (the minor differences in molecular weight are probably caused by different SDS-PAGE systems used). The results shown in Fig. 3 indicate, however, that although the M_r 62,000, 58,000 and 48,000 phosphopeptides are found throughout the brain, the concentration of each varies in different brain regions. For example, the M_r 50,000 subunit is enriched in the forebrain, while the M_r 58,000 and 60,000 subunits are the major components in the cerebellum, brain stem and spinal cord (Walaas et al. 1983a, b). Peptide mapping studies have shown that these M_r 62,000, 58,000 and 48,000 phosphopeptides found in every brain region shown in Fig. 3 are identical to the respective subunits of Ca^{2+}/calmodulin kinase II purified from rat forebrain (S. I. Walaas and T.L. McGuinness, unpublished). Ca^{2+}/calmodulin kinase II from forebrain appears to be an isozyme of a multifunctional Ca^{2+}/calmodulin-dependent protein kinase (see Sect. 2) (McGuinness et al. 1983). The differential distribution of the various autophosphorylated subunits therefore suggests a variable isozyme composition of the Ca^{2+}/calmodulin kinase II in different brain regions.

3.2 Ca^{2+}/Calmodulin-Dependent Phosphorylation of Synapsin I

Synapsin I (previously called Protein I), consists of two polypeptides of M_r 86,000 and 80,000 designated synapsin Ia and Ib. It is found only in nervous tissue, where it is enriched in virtually all nerve terminals and appears to be associated with the external surface of synaptic vesicles (DeCamilli et al. 1983a, b). It represents a major neuron-specific protein, making up about 0.5% of total neuronal protein and approximately 6-10% of synaptic vesicle protein (Huttner et al. 1983). It is unusually basic, having an isoelectric point of 10.2-10.3, and appears to consist of a proline-rich "tail-region" and a globular "head-region" (Ueda and Greengard 1977), both of which are phosphorylated by Ca^{2+}-dependent protein kinases. Ca^{2+}/calmodulin kinase I phosphorylates Site I which is present in the head region (Site I is also phosphorylated by cyclic AMP-dependent protein kinase), and Ca^{2+}/calmodulin kinase II phosphorylates Site II which is present in the tail region (see Sect. 2 for further discussion).

Several studies have demonstrated that both Sites I and II in synapsin I can be phosphorylated by Ca^{2+}-dependent mechanisms in various intact tissue preparations. Examples include intact synaptosomes (Huttner and Greengard 1979) and brain slices (Forn and Greengard 1978) which were depolarized by chemical stimuli, and the superior cervical ganglion (Nestler and Greengard 1982) and the posterior pituitary (Tsou and Greengard 1982), where the protein was phosphorylated both by chemical depolarization and by electrically induced impulse conduction. In these studies the incorporation of ^{32}P into both sites I and II was found to be rapid, and reversible within seconds following termination of the stimulus (see, e.g., Forn and Greengard 1978; Nestler and Greengard 1982; Tsou and Greengard 1982).

The functional importance of Ca^{2+}-regulated phosphorylation of the two domains of synapsin I is not completely understood. The specific localization of the protein on the cytoplasmic, external surface of the synaptic vesicles suggests an involvement in vesicle function, possibly in the release of neurotransmitters (Greengard 1981). Synapsin I is bound through its tail region to a specific high affinity saturable site on the synaptic vesicles (Huttner et al. 1983; Schiebler et al. 1983; Ueda 1981). Phosphorylation of this tail region (Site II) by Ca^{2+}/calmodulin kinase II has been shown to influence this binding (Huttner et al. 1983). It is not known if phosphorylation of the head region (Site I) of the molecule by Ca^{2+}/calmodulin kinase I or by cyclic AMP-dependent protein kinase influences the binding of synapsin I to synaptic vesicles. Phosphorylation of this part of the protein might, however, be a possible final common pathway through which those neurotransmitters which work by increasing cyclic AMP, and those stimuli which lead to increases in Ca^{2+}, could affect synaptic vesicle function by, for example, facilitating neurotransmitter release at the synapse (Greengard 1981; Nestler and Greengard 1983).

3.3 Ca^{2+}-Dependent Phosphorylation of Microtubule Proteins

Microtubules are present throughout the nerve cell, both in association with postsynaptic membranes, dendrites, somata, axons and in presynaptic terminals (see e.g., Cotman and Kelly 1980; Gordon-Weeks et al. 1982). Microtubules consist of tubulin and microtubule-associated proteins (MAP's), both of which are phosphorylated in a Ca^{2+}-dependent manner.

3.3.1 Tubulin

Tubulin, the main protein component of microtubules, consists of α-tubulin (M_r 57,000) and β-tubulin (M_r 54,000) (Perry and Wilson 1982). Activation of endogenous Ca^{2+}/calmodulin-dependent protein kinase activity in both cytosol fractions, synaptosomes and synaptic vesicles have been reported to induce phosphorylation of two polypeptides which had apparent molecular weight similar to α-tubulin and β-tubulin (Burke and DeLorenzo 1981, 1982a, b). A brain Ca^{2+}/calmodulin-dependent protein kinase which phosphorylates tubulin has also been purified and characterized (Goldenring et al. 1983). This protein kinase appears, however, to be similar to Ca^{2+}/calmodulin kinase II (See Sect. 2), an enzyme for which tubulin has been found to be a poor substrate (Bennett et al. 1983; Kennedy et al. 1983a; McGuinness et al 1983). In addition, no Ca^{2+}/calmodulin-dependent phosphorylation of prominent M_r 57,000/54,000 proteins was seen in membranes or cytosol fractions prepared from several rat brain regions in which tubulin is probably present (Walaas et al. 1983a, b). The significance of tubulin as a brain phosphoprotein therefore remains to be determined.

3.3.2 Microtubule-Associated Protein 2 (MAP-2)

MAP-2, a high molecular weight protein of 280,000-300,000, is enriched in neurons, where it is present mostly in dendrites (Matus et al. 1981; Miller et al. 1983). MAP-2 was originally found to be a substrate for cyclic AMP-dependent protein kinase (Sloboda et al. 1975; Vallee 1980), and more recent studies have shown that the protein also is phosphorylated by both Ca^{2+}/calmodulin-dependent and Ca^{2+}/phospholipid-dependent protein kinases in brain cytosol (Yamauchi and Fujisawa 1982; Walaas et al. 1983b). The protein is phosphorylated on more than one site, the number of which has been reported to vary between 1-2 per molecule (Sloboda et al. 1975) and up to 20-22 sites per molecule (Theurkauf and Vallee 1983). (These discrepancies appear to be dependent on the method of preparation of the protein, which may result in variable states of phosphorylation of the molecule prior to the in vitro phosphorylation). MAP-2 has been shown to be a good substrate for Ca^{2+}/calmodulin-dependent protein kinase II (Sect. 2) (Bennett et al. 1983; Goldenring et al. 1983; McGuinness et al. 1983). Phosphorylation of purified MAP-2 with Ca^{2+}/calmodulin kinase II led to ^{32}P-incorporation into at least five sites, which appeared to be distinct from those phosphorylated by cyclic AMP-dependent protein kinase (Yamauchi and Fujisawa 1982). Most of the ^{32}P incorporated into MAP-2 by this kinase was found to be bound to serine residues (Goldenring et al. 1983).

It is clear that MAP-2 is a major phosphate acceptor protein in brain tissue (Walaas et al. 1983a, b). However, methods capable of studying changes in the state of phosphorylation of the protein in intact tissue have not been reported, and the extracellular signals which might regulate MAP-2 phosphorylation are not known. Phosphorylation of MAP-2 by Ca^{2+}/calmodulin kinase II has been reported to induce disassembly of microtubules (Yamauchi and Fujisawa 1983b). Ca^{2+}-regulated phosphorylation of MAP-2 may therefore represent an important mechanisms for the control of the assembly-disassembly cycle of microtubules.

3.4 Ca^{2+}/Phospholipid-Dependent Phosphorylation of an M_r 87,000 Protein

Studies of the intact synaptosome preparation prelabelled with ^{32}P have recently led to the discovery of a Ca^{2+}/phospholipid-dependent protein phosphorylation system in nerve terminals (Wu et al. 1982). Depolarization-induced calcium influx into synaptosomes resulted in the phosphorylation of an M_r 87,000 protein. Further studies using crude cytosol or membrane preparations showed that the M_r 87,000 protein could be phosphorylated by the addition of Ca^{2+} plus phosphatidylserine but not Ca^{2+} plus calmodulin (Fig. 4) (Wu et al. 1982; Walaas et al. 1983a, b). These results, therefore suggests that the M_r 87,000 protein is a physiological substrate for Ca^{2+}/phospholipid-dependent protein kinase in brain.

The M_r 87,000 phosphoprotein appears to be widely distributed throughout the CNS (Walaas et al. 1983a, b). Subcellular fractionation of rat cerebral cortex or cerebellum has shown that it is enriched in the crude synaptosomal fraction and in the cytosol (Wu et al. 1982; S. I. Walaas, unpublished). A fraction of the M_r 87,000 protein is also reproducibly found in particulate preparations, indicating that the protein may be partly associated with membranes (Walaas et al. 1983a, b). Biochemical studies have shown that the protein is acidic (isoelectric point of approximately 4.5 in the presence of urea) (Wu et al. 1982; S. I. Walaas, unpublished). It contains two phosphorylation sites, both of which can be phosphorylated by endogenous and exogenous Ca^{2+}/phospholipid-dependent protein kinase, but not by Ca^{2+}/calmodulin-dependent protein kinase (Fig. 4) (Albert et al. 1984). Unexpectedly, addition of calmodulin to the phosphorylation mixture reproducibly inhibits the Ca^{2+}/phospholipid-catalyzed phosphorylation of the protein, an effect that may be of physiological significance (Wu et al. 1982; Albert et al. 1984).

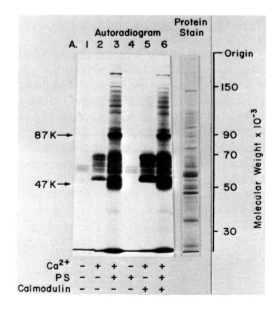

Fig. 4. Demonstration of Ca^{2+}/phospholipid-dependent phosphorylation of an 87 kDa protein in brain nerve terminals. Brain synaptosomes were lyzed by osmotic shock, the synaptosol isolated by centrifugation, and incubated with Mg-[γ-^{32}P]ATP in the absence or presence of 0.5 mM free Ca^{2+}, 50 μg ml^{-1} phosphatidylserine, or 10 μg ml^{-1} calmodulin as indicated. The reaction was terminated with SDS, and the phosphoproteins were analyzed by SDS-PAGE and autoradiography. 87K, M_r 87,000 substrate for Ca^{2+}/phospholipid-dependent protein kinase; 47K, M_r 47,000 substrate for Ca^{2+}/phospholipid-dependent protein kinase. (After Wu et al. 1982)

The physiological function of the M_r 87,000 protein is not known. Extensive studies have indicated that activation of the Ca^{2+}/phospholipid-dependent protein kinase in blood platelets may be an important molecular mechanism for the regulation of the release of serotonin from platelet granules (Takai et al. 1981; Nishizuka 1983). The Ca^{2+}/phospholipid-dependent protein kinase has also been suggested to be involved in catecholamine secretion from chromaffin cells (Knight and Baker 1983) and in insulin release from pancreatic islets (Tanigawa et al. 1982). By analogy, Ca^{2+}/phospholipid-dependent phosphorylation of the M_r 87,000 protein in nerve terminals may be involved in Ca^{2+}-dependent mediation or modulation of the release of CNS neurotransmitters.

3.5 Ca^{2+}/Phospholipid-Dependent Phosphorylation of the B-50 Protein

B-50 is the name given to a brain membrane protein of M_r 48,000 which is phosphorylated by endogenous membrane protein kinase in a reaction which originally was found to be inhibited by high concentrations of ACTH (Zwiers et al. 1976). This phosphoprotein and the protein kinase catalyzing its phosphorylation have been partially purified and characterized. Recent evidence suggests that the protein kinase is identical to the Ca^{2+}/phospholipid-dependent protein kinase (Aloyo et al. 1983). The B-50 protein appears to be a distinct, specific phosphoprotein substrate for Ca^{2+}/phospholipid-dependent protein kinase, since it is not phosphorylated by either cyclic AMP-dependent or Ca^{2+}/calmodulin-dependent protein kinase. Biochemical studies indicate that it is an acidic protein (isoelectric point 4.5) with a high content of glutamate, aspartate and alanine (Zwiers et al. 1980). The B-50 protein has been localized specifically to nervous tissue by both biochemical and immunochemical methods. Immunohistochemical studies suggest that it is enriched in brain regions with a high density of synapses, where it appears to be associated with presynaptic plasma membranes, possibly as an integral membrane protein (Sorensen et al. 1981; Kristjansson et al. 1982; Oestreicher et al. 1981).

ACTH and related peptides, and an affinity-purified B-50 protein antibody, have been found to inhibit the phosphorylation of the B-50 protein in both membranes and in partially purified preparations (Zwiers et al. 1976; Oestreicher et al. 1983). Furthermore, the partially purified B-50/protein kinase complex was found to have phosphatidylinositol 4-phosphate kinase activity. Recent studies have shown that in addition to inhibiting the phosphorylation of B-50, both ACTH and the anti-B-50 antibody increased the formation of phosphatidylinositol 4,5-diphosphate in both membranes and extracts (Jolles et al. 1980; Oestreicher et al. 1983). Therefore, the Ca^{2+}-dependent phosphorylation of B-50 may be involved in the mediation or modulation of the phosphatidylinositol-linked type of signal transmission (Michell 1975, 1983; Hawthorne and Pickard 1979; Berridge 1981) in the brain.

3.6 Ca^{2+}-Dependent Phosphorylation of Myelin Basic Protein

Myelin basic protein is a major component of myelin sheets, which in the CNS are generated by oligodendrocytes and are wrapped around axons to insulate and increase conduction efficiency. Several studies have shown that myelin basic protein is phosphorylated on multiple sites in vivo (Martenson et al. 1983; Agrawal et al. 1982). Despite this, the protein kinase(s) responsible for this phosphorylation has not been conclusively identified. Myelin contains small amounts of cyclic AMP-dependent protein kinase (Miyamoto 1976; Turner et al. 1982). More recently, endogenous myelin basic protein phosphorylation was found to be stimulated by Ca^{2+} (Sulakhe et al. 1980; Walaas et al. 1983a), and further work has implicated calmodulin in this reaction (Endo and Hidaka 1980). Myelin basic protein may also be a substrate for Ca^{2+}/calmodulin kinase II in vitro (Goldenring et al. 1983).

More recently, however, brain myelin was reported to contain a highly active Ca^{2+}/phospholipid-dependent protein kinase, which phosphorylated myelin basic protein in intact membranes, in detergent extracts and in a partially purified preparation of myelin basic protein (Turner et al. 1982). These studies found no evidence for a Ca^{2+}/calmodulin-dependent phosphorylation of myelin basic protein.

The functional significance of myelin basic protein phosphorylation is unknown. The state of phosphorylation of the protein is increased when myelinated axons in the optic nerve are depolarized with high K$^+$ concentrations, with veratridine or through impulse conduction (Murray and Steck 1983). Ca^{2+}-dependent phosphorylation may therefore be involved in a potential interplay between the state of phosphorylation of myelin basic protein and impulse conduction in adjacent axons.

3.7 Ca^{2+}-Dependent Phosphorylation of Tyrosine Hydroxylase

Tyrosine hydroxylase exemplifies a protein with known function, which has been studied in detail as a substrate for Ca^{2+}-dependent phosphorylation. Tyrosine hydroxylase is the enzyme catalyzing the first and rate-limiting step in the synthesis of the catecholamine neurotransmitters dopamine, noradrenaline and adrenaline. The enzyme is specifically localized in catecholaminergic neurons and chromaffin cells (see, e.g., Roth 1979). It is a highly regulated enzyme, responding to various signal molecules which rapidly change the catalytic efficiency of the enzyme (review: Roth 1979). One such mechanism involves cyclic AMP-dependent phosphorylation of the enzyme, which leads to increased activity (Morgenroth et al. 1975; Joh et al. 1978). More recent studies have demonstrated two possible mechanism by which Ca^{2+}-dependent phosphorylation may regulate the enzyme. In one of these (Yamauchi et al. 1981), the enzyme was reported to be phosphorylated by Ca^{2+}/calmodulin kinase II. This phosphorylation did not change the enzyme properties per se, but allowed an "activator protein" to activate the enzyme. This activator protein, which has been purified and characterized, does not appear to be specifically localized in catecholamine-producing cells, but is found as an abundant protein throughout brain and several peripheral organs (Yamauchi et al. 1981). Since Ca^{2+}/calmodulin kinase II is also distributed throughout the brain while tyrosine hydroxylase is restricted to catecholaminergic neurons, the specificity of this activation mechanism may be questioned.

A second example of a Ca^{2+}-dependent phosphorylation of tyrosine hydroxylase has been studied in intact chromaffin cells from bovine adrenal medulla, where the Ca^{2+}/calmodulin-dependent mechanism apparently is absent (Haycock et al. 1982). Application of cyclic AMP analogs to these cells will activate and at the same time phosphorylate tyrosine hydroxylase at a single site (Meligeni et al. 1982). In contrast, application of acetylcholine, the natural secretagogue, which also activates the enzyme, stimulates in a Ca^{2+}-dependent manner the phosphorylation of two or more sites in the protein, one of which is clearly different from that phosphorylated by cyclic AMP-dependent protein kinase (Haycock et al. 1982). Furthermore, this "multisite" phosphorylation of tyrosine hydroxylase can be mimicked in vitro by activation of Ca^{2+}/phospholipid-dependent protein kinase. Ca^{2+}/phospholipid-dependent protein kinase may therefore be involved in the mediation of acetylcholine-induced activation of tyrosine hydroxylase in adrenal chromaffin cells (Haycock et al. 1982).

Recent evidence suggests that tryptophan hydroxylase, the enzyme which synthesizes serotonin (see, e.g., Cooper et al. 1982), is also regulated by Ca^{2+}/calmodulin-dependent phosphorylation (Kuhn et al. 1980; Yamauchi et al. 1981). It is therefore possible that Ca^{2+}-dependent phosphorylation of transmitter-synthesizing enzymes may represent a common regulatory mechanism.

4 Ca^{2+}-Dependent Dephosphorylation of Brain Proteins

Although Ca^{2+}-dependent protein phosphorylation is believed to be regulated predominantly by activation of protein kinases, Ca^{2+} also regulated the state of phosphorylation of proteins by activation of phosphoprotein phosphatases. In brain, and other tissues, Ca^{2+} increases the rate of dephosphorylation of the α-subunit of pyruvate dehydrogenase, probably as a result of interaction of Ca^{2+} with the substrate (Randle 1981). Interestingly, the state of phosphorylation of pyruvate dehydrogenase in slices prepared from the hippocampal cortex of rat brain is regulated by repetitive electrical stimulation (Browning et al. 1979; Magilen et al. 1981). The mechanism and significance of this regulation are not well understood, however (Magilen et al. 1981).

Calcineurin, a Ca^{2+}/calmodulin-dependent phosphoprotein phosphatase, has been purified from brain and other tissues (Stewart et al. 1982; Yang et al. 1982). The protein, originally discovered as an inhibitor of Ca^{2+}/calmodulin-dependent cyclic nucleotide phosphodiesterase (Wang and Desai 1977; Klee and Krinks 1978), is found in high concentration in brain (Wallace et al. 1980). The enzyme readily dephosphorylates a number of substrates in vitro (Stewart et al. 1982), including several neuron-specific phosphoproteins (King et al. 1984) that are phosphorylated by cyclic AMP-dependent, cyclic GMP-dependent and cyclic nucleotide-independent protein kinases. These include DARPP-32, a substrate for cyclic AMP-dependent protein kinase found in dopamine-innervated brain regions (Walaas et al. 1983c), and G-substrate, a substrate for cyclic GMP-dependent protein kinase found in Purkinje cells of the cerebellum (Detre et al. 1984). It will be of interest to study if these, or other brain phosphoproteins, are substrates for calcineurin in vivo.

5 Conclusion

The overriding importance of reversible protein phosphorylation in cellular regulation is now generally acknowledged (Greengard 1978; Cohen 1982), and some of the mechanisms by which, for example, cyclic nucleotides influence these systems have been the subject of intensive study (e.g., Nathanson and Kebabian 1982). In contrast, our knowledge on the importance of Ca^{2+}-regulated protein phosphorylation is incomplete. The data reviewed in this chapter demonstrate the existence of a rich variety of Ca^{2+}-dependent protein kinases, protein phosphatases and specific protein substrates for these enzymes in the mammalian CNS. In addition, numerous other brain-specific protein phosphorylation systems undoubtedly exist which have yet to be identified. The Ca^{2+}-dependent phosphorylation systems so far described in brain show marked differences in biochemical properties and have distinct regional and cellular localizations. While many of the molecular characteristics of both Ca^{2+}/calmodulin-dependent protein kinases and the Ca^{2+}/phospholipid-dependent protein kinase have been described, the role of the substrates for these enzymes have not been well defined. Given the abundance of Ca^{2+}-regulated protein phosphorylation systems in the brain, it seems reasonable to expect that such mechanisms will turn out to be of major functional importance in the mammalian nervous system.

References

Adelstein RS (1980) Phosphorylation of muscle contractile proteins. Fed Proc 39:1544-1546

Agrawal HC, O'Connell K, Randle CL, Agrawal D (1982) Phosphorylation in vivo of four basic proteins of rat brain myelin. Biochem J 201:39-47

Ahmed Z, DePaoli-Roach AA, Roach PJ (1982) Purification and characterization of a rabbit liver calmodulin-dependent protein kinase able to phosphorylate glycogen synthase. J Biol Chem 257:8348-8355

Albert KA, Wu W C-S, Nairn AC, Greengard P (1984) Inhibition by calmodulin of calcium phospholipid-dependent protein phosphorylation. Proc Natl Acad Sci USA 81: 3622-3625

Aloyo VJ, Zwiers H, Gispen WH (1983) Phosphorylation of B-50 protein by calcium-activated phospholipid-dependent protein kinase and B-50 kinase. J Neurochem 41:649-653

Baker PF (1972) Transport and metabolism of calcium ions in nerve. Prog Biophys Mol Biol 24:177-223

Bennett MK, Erondu NE, Kennedy MB (1983) Purification and characterization of a calmodulin-dependent protein kinase that is highly concentrated in brain. J Biol Chem 258:12735-12744

Berridge MJ (1981) Phosphatidylinositol hydrolysis and calcium signaling. In: Dumont JE, Greengard P, Robinson GA (eds) Adv Cyclic Nucleotide Res 14:289-299

Blaustein MP, Nachshen DA, Drapeau P (1981) Excitation-secretion coupling: the role of calcium. In: Stjärne L, Hedqvist P, Lagercrantz H, Wennmalm A (eds) Chemical Neurotransmission 75 years. Academic, London, pp 125-138

Brostrom CO, Hunkeler FL, Krebs EG (1971) The regulation of skeletal muscle phosphorylase kinase by Ca^{2+}. J Biol Chem 246:1961-1967

Browning M, Dunwiddie T, Bennett W, Gispen W, Lynch G (1979) Synaptic phosphoproteins: specific changes after repetitive stimulation of the hippocampal slice. Science (Wash DC) 203:60-62

Burke BE, DeLorenzo RJ (1981) Ca^{2+}- and calmodulin-stimulated endogenous phosphorylation of neurotubulin. Proc Natl Acad Sci USA 78: 991-995

Burke BE, DeLorenzo RJ (1982a) Ca^{2+} and calmodulin-dependent phosphorylation of endogenous synaptic vesicle tubulin by a vesicle-bound calmodulin kinase system. J Neurochem 38:1205-1218

Burke BE, DeLorenzo RJ (1982b) Ca^{2+}- and calmodulin-regulated endogenous tubulin kinase activity in presynaptic nerve terminal preparations. Brain Res 236:393-415

Castagna M, Takai Y, Kaibuchi K, Sano K, Kikkawa U, Nishizuka Y (1982) Direct activation of calcium-activated phospholipid-dependent protein kinase by tumor-promoting phorbol esters. J Biol Chem 257:7847-7851

Cheung WU (1980) Calmodulin plays a pivotal role in cellular regulation. Science (Wash DC) 207:19-27

Cohen P (1973) The subunit structure of rabbit skeletal muscle phosphorylase kinase and the molecular basis of its activation reactions. Eur J Biochem 34:1-4

Cohen P (1982) The role of protein phosphorylation in neural and hormonal control of cellular activity. Nature (Lond) 296:617-620

Cohen P, Burchell A, Foulkes JG, Cohen PTW, Vanaman TC, Nairn AC (1978) Identification of the Ca^{2+}-dependent modulator protein as the fourth subunit of rabbit skeletal muscle phosphorylase kinase. FEBS Lett 92:287-293

Cooper JR, Bloom FE, Roth RH (1982) The biochemical basis of neuropharmacology. Oxford University Press, New York

Cotman CW, Kelly PT (1980) Macromolecular architecture of CNS synapses. In: Cotman CW, Poste G, Nicolson GL (eds) The cell surface and neuronal function. Elsevier North-Holland, Amsterdam, pp 505-533 (Cell Surface Reviews, vol 6)

Dabrowska R, Hartshorne DJ (1978) A Ca^{2+}- and modulator-dependent myosin light chain kinase from non-muscle cells. Biochem Biophys Res Commun 85:1352-1359

Dabrowska R, Aramatorio D, Sherry JMF, Hartshorne DJ (1977) Composition of the myosin light chain kinase from chicken gizzard. Biochem Biophys Res Commun 78:1263-1271

Dabrowska R, Sherry JMF, Aramatorio DK, Hartshorne DJ (1978) Modulator protein as a component of the myosin light chain kinase from chicken gizzard. Biochemistry 17:253-258

DeCamilli P, Cameron R, Greengard P (1983a) Synapsin I (protein I), a nerve terminal-specific phosphoprotein. I. Its general distribution in synapses of the central and peripheral nervous system demonstrated by immunofluorescence in frozen and plastic sections. J Cell Biol 96:1337-1354

DeCamilli P, Harris SM, Huttner WB, Greengard P (1983b) Synapsin I (protein I), a nerve terminal-specific phosphoprotein. II. Its specific association with synaptic vesicles demonstrated by immonocytochemistry in agarose-embedded synaptosomes. J Cell Biol 96: 1355-1373

DeLorenzo RJ (1976) Calcium-dependent phosphorylation of specific synaptosomal fraction proteins: possible role of phosphoproteins in mediating neurotransmitter release. Biochem Biophys Res Commun 71:590-597

De Lorenzo RJ, Freedman SD (1977) Calcium-dependent phosphorylation of synaptic vesicle proteins and its possible role in mediating neurotransmitter release and vesicle function. Biochem Biophys Res Commun 77:1036-1043

Detre JA, Nairn AC, Aswad DW, Greengard P (1984) Localization of G-substrate, a specific substrate for cyclic GMP-dependent protein kinase in mammalian brain. J Neurosci 4: 2843-2849

Douglas WW (1968) Stimulus-secretion coupling: the concept and clues from chromaffin and other cells. Br J Pharmacol 34:451-474

Endo T, Hidaka H (1980) Ca^{2+}-calmodulin dependent phosphorylation of myelin isolated from rabbit brain. Biochem Biophys Res Commun 97:553-558

Endo T, Naka M, Hidaka H (1982) Ca^{2+}-phospholipid dependent phosphorylation of smooth muscle myosin. Biochem Biophys Res Commun 105:942-948

Forn J, Greengard P (1978) Depolarizing agents and cyclic nucleotides regulate the phosphorylation of specific neuronal proteins in rat cerebral cortex slices. Proc Natl Acad Sci USA 75:5195-5199

Fukunaga K, Yamamoto H, Matsui K, Higashi K, Miyamoto E (1982) Purification and characterization of a Ca^{2+}- and calmodulin-dependent protein kinase from rat brain. J Neurochem 39:1607-1617

Goldenring JR, Gonzales B, McGuire JS Jr, DeLorenzo RJ (1983) Purification and characterization of a calmodulin-dependent kinase from rat brain cytosol able to phosphorylate tubulin and microtubule-associated proteins. J Biol Chem 258:12632-12640

Gordon-Weeks PR, Burgoyne RD, Gray EG (1982) Presynaptic microtubules: organization and assembly/disassembly. Neuroscience 7:739-749

Grab DJ, Carlin RK, Siekevitz P (1981) Function of calmodulin in postsynaptic densites II. Presence of a calmodulin-activatable protein kinase activity. J Cell Biol 89:440-448

Greengard P (1978) Phosphorylated proteins as physiological effectors. Science (Wash DC) 199:146-152

Greengard P (1981) Intracellular signals in the brain. In: The Harvey lectures, series 75. Academic, New York, pp 277-331

Hathaway DR, Adelstein RS, Klee CB (1981) Interaction of calmodulin with myosin light chain kinase and cAMP-dependent protein kinase in bovine brain. J Biol Chem 256:8183-8189

Hawthorne JN, Pickard MR (1979) Phospholipids in synaptic function. J Neurochem 32:5-14

Haycock JW, Bennett WF, George RJ, Waymire JC (1982) Multiple site phosphorylation of tyrosine hydroxylase. J Biol Chem 257:13699-13703

Hershkowitz M (1978) Influence of calcium on phosphorylation of a synaptosomal protein. Biochim Biophys Acta 542:274-283

Huttner WB, Greengard P (1979) Multiple phosphorylation sites in protein I and their differential regulation by cyclic AMP and calcium. Proc Natl Acad Sci USA 76:5402-5406

Huttner WB, DeGennarro LJ, Greengard P (1981) Differential phosphorylation of multiple sites in purified protein I by cyclic AMP-dependent and calcium-dependent protein kinases. J Biol Chem 256:1482-1488

Huttner WB, Schiebler W, Greengard P, DeCamilli P (1983) Synapsin I (protein I), a nerve terminal-specific phosphoprotein. III. Its association with synaptic vesicles studied in a highly-purified synaptic vesicle preparation. J Cell Biol 96:1374-1388

Huxley AF (1964) Muscle. Annu Rev Physiol 26:131-152

Inoue M, Kishimoto A, Takai Y, Nishizuka Y (1977) Studies on a cyclic nucleotide-independent protein kinase and its proenzyme in mammalian tissues. J Biol Chem 252:7610-7616

Joh TH, Park DH, Reis DJ (1978) Direct phosphorylation of brain tyrosine hydroxylase by cyclic AMP-dependent protein kinase: mechanism of enzyme activation. Proc Natl Acad Sci USA 75:4744-4748

Jolles J, Zwiers H, van Dongen CJ, Schotman P, Wirtz KWA, Gispen WH (1980) Modulation of brain polyphosphoinositide metabolism by ACTH-sensitive protein phosphorylation. Nature (Lond) 286:623-625

Kaibuchi K, Takai Y, Nishizuka Y (1981) Cooperative roles of various membrane phospholipids in the activation of calcium-activated, phospholipid-dependent protein kinase. J Biol Chem 256:7146-7149

Kakiuchi S, Hidaka H, Means AR (eds) (1982) Calmodulin and intracellular Ca^{++} receptors. Plenum, New York

Katoh N, Kuo JF (1982) Subcellular distribution of phospholipid-sensitive calcium-dependent protein kinase in guinea pig heart, spleen, and cerebral cortex, and inhibition of the enzyme by Triton X-100. Biochem Biophys Res Commun 106:590-595

Katoh N, Wise BC, Kuo JF (1982) Phosphorylation of cardiac troponin subunit (troponin I) and tropomyosin-binding subunit (troponin T) by cardiac phospholipid-sensitive Ca^{2+}-dependent protein kinase. Biochem J 209:189-195

Katz B, Miledi R (1969) Tetrodotoxin-resistant electrical activity in presynaptic terminals. J Physiol (Lond) 203:459-487

Kelly PT, McGuiness TL, Greengard P (1984) Evidence that the major postsynaptic densitiy protein is a component of a Ca^{2+}/calmodulin-dependent protein kinase. Proc Natl Acad Sci USA 81: 945-949

Kennedy MB, Greengard P (1981) Two calcium/calmodulin-dependent protein kinases, which are highly concentrated in brain, phosphorylate protein I at distinct sites. Proc Natl Acad Sci USA 78:1293-1297

Kennedy MB, McGuinness T, Greengard P (1983a) A calcium/calmodulin-dependent protein kinase from mammalian brain that phosphorylates synapsin I: partial purification and characterization. J Neurosci 3:818-831

Kennedy MB, Bennett MK, Erondu NE (1983b) Biochemical and immunochemical evidence that the "major postsynaptic density protein" is a subunit of a calmodulin-dependent protein kinase. Proc Natl Acad Sci USA 80:7357-7361

Kikkawa U, Takai Y, Minakuchi R, Inohara S, Nishizuka Y (1982) Calcium-activated, phospholipid-dependent protein kinase from rat brain. Subcellular distribution, purification and properties. J Biol Chem 257:13341-13348

King LJ, Lowry OH, Passonneau JV, Venson V (1967) Effects of convulsants on energy reserves in the cerebral cortex. J Neurochem 14:599-611

King MM, Huang CY, Chock BP, Nairn AC, Hemmings HC, Chan K-F J, Greengard P (1984) Identification of four phosphoproteins from mammalian brain as substrates for calcineurin. J Biol Chem 259: 8080-8083

Kishimoto A, Takai Y, Mori T, Kikkawa U, Nishizuka Y (1980) Activation of calcium and phospholipid-dependent protein kinase by diacylglycerol, its possible relation to phosphatidylinositol turnover. J Biol Chem 255: 2273-2278

Klee CB, Krinks MH (1978) Purification of cyclic 3' ,5'-nucleotide phosphodiesterase inhibitory protein by affinity chromatography on activator protein coupled to Sepharose. Biochemistry 17:120-126

Klee CB, Vanaman TC (1982) Calmodulin. Adv Protein Chem 35:213-321

Knight DE, Baker PF (1983) The phorbol ester TPA increases the affinity of exocytosis for calcium in 'leaky' adrenal medullary cells. FEBS Lett 160:98-100

Knull HR and Khandelwal RL (1982) Glycogen metabolizing enzymes in brain. Neurochem Res 7:1307-1317

Kraft AA, Anderson WB (1983) Phorbol esters increase the amount of Ca^{2+}, phospholipid-dependent protein kinase associated with plasma membrane. Nature (Lond) 301:621-623

Kraft AS, Anderson WB, Cooper HL and Sando JJ (1982) Decreases in cytosolic calcium/phospholipid-dependent protein kinase activity following phorbol ester treatment of E 24 thyoma cells. J Biol Chem 257:13193-13196

Kretsinger RH (1977) Evolution of the informational role of calcium in eukaryotes. In: Wasserman RH (ed) Calcium-binding proteins and calcium function. Elsevier/North-Holland, New York, p. 63

Kristjansson GI, Zwiers H, Oestreicher AB, Gispen WH (1982) Evidence that the synaptic phosphoprotein B-50 is localized exclusively in nerve tissue. J Neurochem 39:371-378

Krueger BK, Forn J, Greengard P (1976) Calcium-dependent protein phosphorylation in rat brain synaptosomes. Abst of the 6th meeting of the Society of Neuroscience, p 1007

Krueger BK, Forn J, Greengard P (1977) Depolarization-induced phosphorylation of specific proteins, mediated by calcium ion influx in rat brain synaptosomes. J Biol Chem 252:2764-2773

Kuhn DM, O'Callaghan JP, Juskevich J, Lovenberg W (1980) Activation of brain tryptophan hydroxylase by ATP-Mg^{2+}: dependence on calmodulin. Proc Natl Acad Sci USA 77:4688-4691

Kuo JF, Andersson RGG, Wise BC et al. (1980) Calcium-dependent protein kinase: wide-spread occurrence in various tissues and phyla of the animal kingdom and comparison of effects of phospholipid, calmodulin, and trifluoperazine. Proc Natl Acad Sci USA 77:7039-7043

Lai Y, McGuinness TL, Greengard P (1983) Purification and characterization of brain Ca^{2+}/calmodulin-dependent protein kinase II that phosphorylates synapsin I. Soc Neurosci Abst 9:1029

LePeuch CJ, Ballester R, Rosen OM (1983) Purified rat brain calcium-and phospholipid-dependent protein kinase phosphorylates ribosomal protein S6. Proc Natl Acad Sci USA 80:6858-6862

Magilen G, Gordon A, Au A, Diamond I (1981) Identification of a mitochondrial phosphoprotein in brain synaptic membrane preparations. J Neurochem 36:1861-1864

Mahler HR, Kleine LP, Ratner N, Sorensen RG (1982) Identification and topography of synaptic phosphoproteins. In: Gispen WH, Routtenberg A (eds) Brain phosphoproteins, characterization and function. Elsevier Biomedical, Amsterdam, pp 27-48 (Progr Brain Res, vol 56)

Martenson RE, Law MJ, Deibler GE (1983) Identification of multiple in vivo phosphorylation sites in rabbit myelin basic protein. J Biol Chem 258:930-937

Matsumura S, Murakami N, Yasuda S, Kumon A (1982) Site-specific phosphorylation of brain myosin light chains by calcium-dependent and calcium-independent myosin kinases. Biochem Biophys Res Commun 109:683-688

Matus A, Bernhardt R, Hugh-Jones T (1981) High molecular weight microtubule-associated proteins are preferentially associated with dendritic microtubules in brain. Proc Natl Acad Sci USA 78:3010-3014

McGuinness TL, Lai Y, Greengard P, Woodgett JR, Cohen P (1983) A multifunctional calmodulin-dependent protein kinase. Similarities between skeletal muscle glycogen synthase kinase and a brain synapsin I kinase. FEBS Lett 163:329-334

McGuiness TL, Lai Y, Ouimet CC, Greengard P (1984) Calcium/calmodulin-dependent protein phosphorylation in the nervous system. In: Rubin RP, Putney JW, Weiss E (eds) Calcium in biological systems. Plenum, New York, pp 291-305

Meligeni JA, Haycock JW, Bennett WF, Waymire JC (1982) Phosphorylation and activation of tyrosine hydroxylase mediate the cAMP-induced increase in catecholamine biosynthesis in adrenal chromaffin cells. J Biol Chem 257:12632-12640

Michell RH (1975) Inositol phospholipids and cell surface receptor function. Biochem Biophys Acta 415:81-147

Michell RH (1983) Polyphosphoinositide breakdown as the initiating reaction in receptor-stimulated inositol phospholipid metabolism. Life Sci 32:2083-2085

Miller P, Walter U, Theurkauf WE, Vallee RB, DeCamilli P (1982) Frozen tissue sections as an experimental system to reveal specific binding sites for the regulatory subunit of type II cAMP-dependent protein kinase in neurons. Proc Natl Acad Sci USA 79:5562-5566

Minakuchi R, Takai Y, Yu B, Nishizuka Y (1981) Widespread occurence of calcium-activated phospholipid-dependent protein kinase in mammalian tissues. J Biochem (Tokyo) 89:1651-1654

Miyamoto E (1976) Phosphorylation of endogenous proteins in myelin of rat brain. J Neurochem 26:573-577

Miyamoto E, Fukunaga K, Matsui K, Iwasa Y (1981) Occurence of two types of Ca^{2+}-dependent protein kinases in the cytosol fraction of the brain. J Neurochem 37:1324-1330

Morgenroth VH III, Hegstrand LH, Roth RH, Greengard P (1975) Evidence for involvement of protein kinase in the activation by adenosine 3':5'-monophosphate of brain tyrosine 3-monoxygenase. J Biol Chem 250:1946-1948

Moskowitz N, Glassman A, Ores C, Schook W, Puszkin S (1983) Phosphorylation of brain synaptic and coated vesicle proteins by endogenous Ca^{2+}/calmodulin- and cAMP-dependent protein kinases. J Neurochem 40:711-718

Mrwa U, Harthshorne DJ (1980) Phosphorylation of smooth muscle myosin and myosin light chain. Fed Proc 39:1564-1568

Murray N, Steck AJ (1983) Depolarizing agents regulate the phosphorylation of myelin basic protein in rat optic nerves. J Neurochem 41:543-548

Nairn AC, Greengard P (1983) Purification and characterization of a brain Ca^{2+}/calmodulin-dependent protein kinase I that phosphorylates synapsin I. Soc Neurosci Abst 9:1029

Nairn AC, Greengard P (1984) Widespread tissue distribution of Ca^{2+}/calmodulin-dependent protein kinase I that phosphorylates synapsin I. Fed Proc 43:1467

Nathanson JA, Kebabian JW (eds) (1982) Cyclic nucleotides I. Springer, Berlin Heidelberg New York (Handbook of Experimental Pharmacology, vol 58/I)

Nestler E, Greengard P (1982) Nerve impulses increase the phosphorylation state of protein I in rabbit superior cervical ganglion. Nature (Lond) 296:452-454

Nestler E, Greengard P (1983) Protein phosphorylation in the brain. Nature (Lond) 305:583-588

Nestler E, Greengard P (1984) Protein phosphorylation and neuronal function. Wiley, New York

Nishikawa M, Hidaka H and Adelstein RS (1983) Phosphorylation of smooth muscle heavy meromyosin by calcium-activated, phospholipid-dependent protein kinase. The effect on actin-activated MgATPase activity. J Biol Chem 258:14069-14072

Nishizuka Y (1980) Three multifunctional protein kinase systems in transmembrane control. Mol Biol Biochem Biophys 32:113-135

Nishizuka Y (1983) Phospholipid degradation and signal translation for protein phosphorylation. Trends Biochem Sci 8:13-16

O'Callaghan JP, Dunn LA, Lovenberg W (1980) Calcium-regulated phosphorylation in synaptosomal cytosol: dependence on calmodulin. Proc Natl Acad Sci USA 77:5812-5816

Oestreicher AB, Zwiers H, Schotman P, Gispen WH (1981) Immunohistochemical localization of a phosphoprotein (B-50) isolated from rat brain synaptosomal plasma membranes. Brain Res Bull 6:145-153

Oestreicher AB, van Dongen CJ, Zwiers H, Gispen WH (1983) Affinity-purified anti-B-50 antibody: interference with the function of the phosphoprotein B-50 in synaptic plasma membranes. J Neurochem 41:331-340

Ozawa E (1973) Activation of phosphorylase kinase from brain by small amounts of calcium ion. J Neurochem 20:1487-1488

Palfrey HC (1984) Two novel calmodulin-dependent protein kinases from mammalian myocardium. Fed Proc 43:1466

Palfrey HC, Rothlein JE, Greengard P (1983) Calmodulin-dependent protein kinase and associated substrates in *Torpedo* electric organ. J Biol Chem 258:9496-9503

Palfrey HC, Lai Y, Greengard P (1984) Calmodulin-dependent protein kinase from avian erythrocytes. Brewer, G (ed) Proc 6th Int Conf Red Cell Struct and Metab. Liss, New York, pp 291-301

Payne ME, Schworer CM, Soderling TR (1983) Purification and characterization of rabbit liver calmodulin-dependent glycogen synthase kinase. J Biol Chem 258:2376-2382

Perry SV, Cole HA, Frearson N, Moir AJG, Nairn AC, Solaro RJ (1978) Phosphorylation of the myofibrillar proteins. Proc FEBS 12th Meet 54:147-159

Perry GW, Wilson DL (1982) On the identification of α- and β-tubulin subunits. J Neurochem 38-1155-1159

Pires EMV, Perry SV (1977) Purification and properties of myosin light chain kinase from fast skeletal muscle. Biochem J 167:137-146

Randle P (1981) Phosphorylation-dephosphorylation cycles and the regulation of fuel selection in mammals. Curr Top Cell Regul 18:107-129

Reichardt LF, Kelly RB (1983) A molecular description of nerve terminal function. Annu Rev Biochem 52:871-926

Roth RH (1979) Tyrosine hydroxylase. In: Horn AS, Korf J, Westerink BHC (eds) The neurobiology of dopamine. Academic, London, pp 101-122

Rubin RP (1972) The role of calcium in the release of neurotransmitter substances and hormones. Pharmacol Rev 22:389-428

Schatzmann RC, Rayner RL, Fritz RB, Kuo JF (1983a) Purification to homogeneity, characterization, and monoclonal antibodies of phospholipid-sensitve Ca^{2+}-dependent protein kinase from spleen. Biochem J 209:435-443

Schatzmann RC, Grifo JA, Merrick WC, Kuo JF (1983b) Phospholipid-sensitive Ca^{2+}-dependent protein kinase phosphorylates the β subunit of eukaryotic initiation factor 2 (eIF-2). FEBS Lett 159:167-170

Schiebler W, Rothlein J, Jahn R, Doucet JP, Greengard P (1983) Synapsin I (protein I) binds specifically and with high affinity to highly purified synaptic vesicles from rat brain. Soc Neurosci Abst 9:882

Schulman H (1982) Calcium-dependent protein phosphorylation. In: Nathanson JA, Kebabian JW (eds) Cyclic nucleotides I. Springer, Berlin Heidelberg New York, pp 425-478 (Handbook of Experimental Pharmacology, vol 58/I)

Schulman H, Greengard P (1978a) Stimulation of brain membrane protein phosphorylation by calcium and an endogenous heat-stable protein. Nature (Lond) 271:478-479

Schulman H, Greengard P (1978b) Ca^{2+}-dependent protein phosphorylation system in membrane from various tissues, and its activation by "calcium-dependent regulator". Proc Natl Acad Sci USA 75:5432-5436

Schulman H, Kuret JA and Spitzer KH (1983) Calcium and calmodulin-dependent phosphorylation of microtubules. Fed Proc 42:2250

Schworer CM, Soderling TR (1983) Substrate specificity of liver calmodulin-dependent glycogen synthase kinase. Biochem Biophys Res Commun 116:412-416

Sieghart W, Forn J, Greengard P (1979) Ca^{2+} and cyclic AMP regulate phosphorylation of the same two membrane-associated proteins specific to nerve tissue. Proc Natl Acad Sci USA 76:2475-2479

Sieghart W, Schulman H, Greengard P (1980) Neuronal localization of Ca^{2+}-dependent protein phosphorylation in brain. J Neurochem 34:548-553

Sloboda RD, Rudolph SA, Rosenbaum JL, Greengard P (1975) Cyclic AMP-dependent endogenous phosphorylation of a microtubule-associated protein. Proc Natl Acad Sci USA 72:177-181

Sobue K, Kanda K, Kakiuchi S (1982) Solubilization and partial purification of protein kinase systems from brain membranes that phosphorylate calspectin. FEBS Lett 150:185-190

Sorensen RG, Mahler HR (1983) Calcium-stimulated protein phosphorylation in synaptic membranes. J Neurochem 40:1349-1365

Sorensen RG, Kleine LP, Mahler HR (1981) Presynaptic localization of phosphoprotein B-50. Brain Res Bull 7:57-61

Stewart AA, Ingebritsen TS, Manalan A, Klee CB, Cohen P (1982) Discovery of a Ca^{2+}- and calmodulin protein phosphatase. Probable identity with calcineurin (CaM-BP80). FEBS Lett 137:80-84

Sulakhe PV, Petrali EH, Thiessen BJ, Davis ER (1980) Calcium ion-stimulated phosphorylation of myelin proteins. Biochem J 186:469-473

Taira T, Kii R, Sakai K et al. (1982) Comparison of glycogen phosphorylase kinase of various rat tissues. J Biochem (Tokyo) 91:883-888

Takai Y, Kishimoto A, Inoue M, Nishizuka Y (1977) Studies on a cyclic nucleotide-independent protein kinase and its proenzyme in mammalian tissues. J Biol Chem 252:7603-7609

Takai Y, Kishimoto A,, Iwasa Y, Kawahara Y, Mori T, Nishizuka Y (1979a) Calcium-independent activation of a multifunctional protein kinase by membrane phospholipids. J Biol Chem 254:3692-3695

Takai Y, Kishimoto A, Kikkawa U, Mori T, Nishizuka Y (1979b) Unsaturated diacylglycerol as a possible messenger for the activation of calcium-activated, phospholipid-dependent protein kinase system. Biochem Biophys Res Commun 91:1218-1224

Takai Y, Kishimoto A, Kawahara Y et al. (1981) Calcium and phosphatidylinositol turnover as signalling for transmembrane control of protein phosphorylation. In: Dumont JE, Greengard P, Robinson GA (eds) Adv Cyclic Nucleotide Res 14:301-313

Tanigawa K, Kuzuya H, Imura H, Taniguchi H, Baba S, Takai Y, Nishizuka Y (1982) Calcium-activated, phospholipid-dependent protein kinase in rat pancreas islets of Langerhans. FEBS Lett 138:183-186

Theurkauf WE, Vallee RB (1983) Extensive cAMP-dependent and cAMP-independent phosphorylation of microtubule-associated protein 2. J Biol Chem 258:7883-7886

Trifaro JM (1978) Contractile proteins in tissues originating in the neural crest. Neuroscience 3:1-24

Tsou K, Greengard P (1982) Regulation of phosphorylation of proteins I, IIIa, and IIIb in rat neurohypophysis in vitro by electrical stimulation and by neuroactive agents. Proc Natl Acad Sci USA 79:6075-6079

Turner RS, Chou C-H J, Kibler RF, Kuo JF (1982) Basic protein in brain myelin is phosphorylated by endogenous phospholipid-sensitive Ca^{2+}-dependent protein kinase. J Neurochem 39:1397-1404

Ueda T (1981) Attachment of the synapse-specific phosphoprotein Protein I to the synaptic membrane: a possible role of the collagenase-sensitive region of Protein I. J Neurochem 36:297-300

Ueda T, Greengard P (1977) Adenosine 3',5'-monophosphate-regulated phosphoprotein system of neuronal membranes. I. Solubilisation, purification and some properties of an endogenous phosphoprotein. J Biol Chem 252:5755-5763

Vallee R (1980) Structure and phosphorylation of microtubule-associated protein 2 (MAP 2). Proc Natl Acad Sci USA 77:3206-3210

Walaas SI, Nairn AC, Greengard P (1983a) Regional distribution of calcium and cyclic adenosine 3':5'-monophosphate-regulated protein phosphorylation system in mammalian brain. I. Particulate systems. J Neurosci 3:291-301

Walaas SI, Nairn AC, Greengard P (1983b) Regional distribution of calcium-and cyclic adenosine 3':5'-monophosphate-regulated protein phosphorylation systems in mammalian brain. II. Soluble systems. J Neurosci 3:302-311

Walaas SI, Aswad DW, Greengard P (1983c) A dopamine- and cyclic AMP-regulated phosphoprotein enriched in dopamine-innervated brain regions. Nature (Lond) 301:69-72

Wallace RW, Tallant EA, Cheung WY (1980) High levels of a heat-labile calmodulin-binding protein (CaM-BP 80) in bovine neostriatum. Biochemistry 19:1831-1837

Wang JH, Desai R (1977) Modulator binding protein. Bovine brain protein exhibiting the Ca^{2+}-dependent association with the protein modulator of cyclic nucleotide phosphodiesterase. J Biol Chem 252:4175-4184

Werth DK, Niedel JE, Pastan I (1983) Vinculin, a cytoskeletal substrate of protein kinase C. J Biol Chem 258:11423-11426

Wise BC, Raynor RL, Kuo JF (1982) Phospholipid-sensitive Ca^{2+}-dependent protein kinase from heart. J Biol Chem 257:8481-8488

Woodgett JR, Tonks NK, Cohen P (1982) Identification of calmodulin-dependent glycogen synthase kinase in rabbit skeletal muscle, distinct from phosphorylase kinase. FEBS Lett 148:5-11

Wrenn RW, Katoh N, Wise BC, Kuo JF (1980) Stimulation by phosphatidylserine and calmodulin of calcium-dependent phosphorylation of endogenous proteins from cerebral cortex. J Biol Chem 255:12042-12046

Wu W C-S, Walaas SI, Nairn AC, Greengard P (1982) Calcium/phospholipid regulates phosphorylation of a M_r "87K" substrate protein in brain synaptosomes. Proc Natl Acad Sci USA 79:5249-5253

Yagi K, Yazawa M, Kakiuchi W, Ohshima M, Uenishi K (1978) Identification of an activator protein for myosin light chain kinase as the Ca^{2+}-dependent modulator protein. J Biol Chem 253:1338-1340

Yamauchi T, Fujisawa H (1979) Most of the Ca^{2+}-dependent endogenous phosphorylation of rat brain cytosol proteins requires Ca^{2+}-dependent regulator protein. Biochem Biophys Res Commun 90:1172-1178

Yamauchi T, Fujisawa H (1980) Evidence for three distinct forms of calmodulin-dependent protein kinases from rat brain. FEBS Lett 116:141-144

Yamauchi T, Fujisawa H (1982) Phosphorylation of microtubule-associated protein 2 by calmodulin-dependent protein kinase (kinase II) which occurs only in the brain tissues. Biochem Biophys Res Commun 109:975-981

Yamauchi T, Fijuisawa H (1983a) Purification and characterization of the brain calmodulin-dependent protein kinase (kinase II), which is involved in the activation of tryptophan 5-monooxygenase. Eur J Biochem 131:15-21

Yamauchi T, Fujisawa H (1983b) Disassembly of microtubules by the action of calmodulin-dependent protein kinase (kinase II) which occurs only in the brain tissues. Biochem Biophys Res Commun 110:287-291

Yamauchi T, Nakata H, Fujisawa H (1981) A new activator protein that activates tryptophan 5-monooxygenase and tyrosine 3-monooxygenase in the presence of Ca^{2+}, calmodulin-dependent protein kinase. J Biol Chem 256:5404-5409

Yang S, Tallant EA, Cheung WY (1982) Calcineurin is a calmodulin-dependent protein phosphatase. Biochem Biophys Res Commung 106:1419-1425

Zwiers H, Veldhuis HD, Schotman P, Gispen WH (1976) ACTH, cyclic nucleotides, and brain protein phosphorylation in vitro. Neurochem Res 1:669-677

Zwiers H, Schotman P, Gispen WH (1980) Purification and some characteristics of an ACTH-sensitive protein kinase and its substrate in rat brain membranes. J Neurochem 34:1689-1699

Calcium and Calmodulin Control of Neurotransmitter Synthesis and Release

R. J. DeLorenzo[1]

CONTENTS

1 Introduction

Calcium ions play a major role in the activity and function of nervous tissue (Rubin 1972; Rasmussen and Goodman 1977). One of the most widely recognized roles of Ca^{2+} in synaptic function is its action in synaptic modulation and neurotransmission. Earlier studies demonstrated that the relase of neurotransmitter substances by verte-brate neuromuscular junctions was dependent on the Ca^{2+} ion concentration in the media (Rubin 1972; DelCastillo and Stark 1952). Studies at the synaptic level showed that the effects of Ca^{2+} on neurotransmission were not secondary to effects of Ca^{2+} on the presynaptic action potential, but were directly dependent on the entry of Ca^{2+} into the nerve terminal (Katz and Miledi 1969, 1970; Miledi and Slater 1966; Miledi

[1] Department of Neurology, Yale University School of Medicine, New Haven, CT 06510 (203) 785-4085, USA

1973). The role of Ca^{2+} in stimulus-secretion coupling has also been demonstrated in a variety of tissues (Douglas 1968).

The function of Ca^{2+} as a second messenger in synaptic function is well established. However, a question in neuroscience research at the present time is what is the molecular mechanism mediating the effects of Ca^{2+} on synaptic activity, especially its effects on neurotransmitter release and turnover. Research in my laboratory over the last 10 years has been directed at providing a molecular approach to studying the biochemistry of the Ca^{2+}-signal in neurotransmitter release and synaptic modulation. In vitro and in vivo preparations were developed and employed to study the effects of Ca^{2+} on neurotransmitter release (DeLorenzo 1980a; DeLorenzo et al. 1979), synaptic protein phosphorylation (DeLorenzo 1980a, 1982, 1983), and synaptic vesicle and synaptic membrane interactions (DeLorenzo 1980a, 1982). These studies provided an experimental framework to demonstrate that calmodulin (CaM), a major Ca^{2+}-receptor protein in brain, modulates of the biochemical effects of Ca^{2+} on synaptic preparations. From this evidence the calmodulin hypothesis of neuronal transmission was developed (DeLorenzo 1981a). This hypothesis states that as Ca^{2+} enters the presynaptic nerve terminal, it binds to calmodulin and activates several Ca^{2+}-calmodulin-regulated processes that modulate synaptic activity. In this paper, the evidence from my laboratory and others for the role of calcium and calmodulin in neurotransmitter release and synthesis will be presented.

2 Synaptic Calmodulin

Calmodulin is a heat-stable, Ca^{2+}-binding protein that is structurally very similar to troponin C from skeletal muscle (Cheung 1980). Calmodulin is present in many tissues and animal and plant species. Calmodulin is also found in high concentrations in brain (Cheung 1980; Klee et al. 1980). However, to implicate this Ca^{2+}-regulator protein in synaptic function, it is necessary to demonstrate that calmodulin is present at the synapse.

A vesicle-bound heat-stable protein was isolated from highly enriched preparations of synaptic vesicles from rat cortex that had the same molecular weight as calmodulin (DeLorenzo 1980a, 1981b). This vesicle-bound protein could be removed from the vesicles in the presence of EGTA and was found to bind Ca^{2+} at micromolar concentrations (DeLorenzo 1981a). When compared to calmodulin isolated from whole rat brain, the vesicle-Ca^{2+}-binding protein was found the be identical to calmodulin in molecular weight, amino acid composition, isoelectric point, and in its ability to stimulate vesicle protein kinase, adenylate cyclase and phosphodiesterase activity (DeLorenzo 1980a, 1981b). Vesicle calmodulin represented 0.92% of the total protein in synaptic vesicle fraction.

A heat-stable, Ca^{2+}-binding protein was also isolated from nerve terminal synaptoplasm prepared by standard procedures (DeLorenzo 1980a, 1981a). This synaptic protein was found to be identical with whole brain calmodulin in molecular weight, isoelectric point, amino acid composition and ability to activate adenylate cyclase and phosphodiesterase activity. Synaptic CaM comprised 0.71% of the total protein in the synaptoplasm preparation. Because the concentration of CaM in whole brain fractions is approximately 1% of the total brain protein, the high percentage of CaM in synaptoplasm and synaptic vesicle fractions strongly indicates the presence of this Ca^{2+}-receptor protein in the presynaptic nerve terminal. Calmodulin has also been isolated and characterized from post-synaptic density preparations (Grab et al. 1979). The ability to isolate CaM from highly enriched preparations of pre- and post-synaptic fractions strongly indicates that CaM is a trans-synaptic protein. Therefore, CaM is very well suited to mediate the effects of Ca^{2+} on both the pre- and post-synaptic sides of the synapse.

3 Calmodulin and Ca^{2+}-Dependent Neurotransmitter Release

Since the evidence indicates that CaM is present at the synapse, this Ca^{2+}-receptor protein is a possible pre-synaptic protein for modulating the effects of Ca^{2+} on neurotransmitter release and turnover. As Ca^{2+} enters the presynaptic ending, it can bind to this high-affinity receptor and initiate several biochemical processes involved in synaptic function. Initial evidence for a role of calmodulin involvement in neurotransmitter release was developed employing preparations of intact synaptosomes and isolated synaptic vesicles.

3.1 Synaptic Vesicle Studies

A more physiological procedure for isolating synaptic vesicles has been developed (DeLorenzo 1981b; DeLorenzo and Freedman 1977a). Vesicles from this isolation procedure were shown to be much more responsive to Ca^{2+} than vesicles prepared under the standard hypotonic isolation methods (DeLorenzo and Freedman 1978). Ca^{2+} in the presence of ATP and Mg^{2+} simultaneously initiated the release of vesicle neurotransmitter substances, vesicle protein phosphorylation, and vesicle and membrane interactions (DeLorenzo 1980a, 1981a, b, 1982). The Ca^{2+}-responsive synaptic preparation was then studied to determine if calmodulin mediated the effects of Ca^{2+} on vesicle neurotransmitter release.

The vesicles prepared under more physiological procedures that simulated the intracellular environment also contained calmodulin (DeLorenzo 1981a). The calmodulin in the vesicle preparation was tightly bound to the vesicle surface and could be selectively removed by washing the vesicles with the Ca^{2+}-chelating agent, EGTA. Thus, it was possible to obtain preparations of calmodulin-containing (plain vesicles) and calmodulin-depleted (treated vesicles) vesicles. The calmodulin-depleted vesicle fractions were then studies for neurotransmitter release (Table 1).

Ca^{2+} in the presence of calmodulin stimulates the release of norepinephrine and acetylcholine (DeLorenzo et al. 1979; DeLorenzo 1982) from calmodulin-depleted vesicles (Table 1). Ca^{2+} or calmodulin alone, however, had no significant effect on neurotransmitter release (Table 1). Trifluoperazine, a phenothiazine that inactivates

Table 1. Effects of calmodulin and Ca^{2+}-calmodulin kinase inhibitors on Ca^{2+}-calmodulin-stimulated protein phosphorylation and neurotransmitter release in isolated synaptic vesicles

Condition	Neurotransmitter release (%) Acetylcholine	Norepinephrine	Protein DPH-M Phosphorylation (%)
Control	34	38	21
Ca^{2+}	41	44	25
Calmodulin	36	39	22
Ca^{2+} + calmodulin	100	100	100
Ca^{2+} + calmodulin			
+ trifluoperazine	62	68	55
+ phenytoin	69	72	49
+ diazepam	61	73	47

Calmodulin-depleted synaptic vesicles were isolated and studied for neurotransmitter release and protein DPH-M phosphorylation as described previously (DeLorenzo et al. 1979). The data give the means of 10 determinations and are expressed as percent of the maximally stimulated condition (100%), The largest ± S.E.M. was 5.7. The effects of trifluoperazine (15 μM), phenytoin (80 μM), and diazepam (75 μM) were found to be statistically significant in comparison to maximally stimulated values
P < 0.001

calmodulin (Weiss et al. 1980), also inhibited Ca^{2+}-calmodulin stimulated vesicle neurotransmitter release (Table 1). The calmodulin kinase inhibitors, phenytoin and diazepam (DeLorenzo 1980b, 1983; DeLorenzo et al. 1981), were found to inhibit Ca^{2+}-calmodulin-stimulated vesicle neurotransmitter release (Table 1). The Ca^{2+}-calmodulin stimulation of release was also shown to be dependent on Mg^{2+} and ATP (DeLorenzo 1981a) and vesicles prepared under hypotonic conditions (DeLorenzo and Freedman 1978) did not show significant Ca^{2+}-calmodulin-stimulated release of neurotransmitter substances.

3.2 Intact Synaptosome Studies

Although vesicle preparations offer several advantages for studying the effects of Ca^{2+} on vesicular transmitter release, it is important to correlate the results from the isolated vesicle fractions with data obtained from neurotransmitter release studies of intact nerve terminal preparations. Isolated intact nerve terminals (synaptosomes) are excellent preparations for studying the effects of Ca^{2+} and membrane depolarization on neurotransmitter release (Blaustein et al. 1972).

In my laboratory we have employed the intact synaptosome system to study the role of CaM in neurotransmitter release as summarized in Table 2. The disadvantage of the synaptosome system for studying the effects of CaM on neurotransmitter release is that it is not yet possible to remove CaM from the synaptosome without destroying the via-

Table 2. Effects of calmodulin and Ca^{2+}-calmodulin kinase inhibitors on Ca^{2+} uptake, neurotransmitter release and protein phosphorylation in intact nerve terminal preparations [a]

| Condition | Ca^{2+} Uptake, % | Neurotransmitter release, % | | Protein phosphorylation, % | |
		Acetylcholine	Norepinephrine	Whole synaptosomes	Synaptic vesicles
Control	–	45	52	58	39
Ca^{2+}	40	51	56	61	44
Ca^{2+}, K	100	100	100	100	100
Ca^{2+}, K					
+ trifluoperazine	61	63	68	69	61
+ phenytoin	65	68	70	68	67
+ diazepam	66	64	59	72	67
Ca^{2+}, A23187	95	94	98	91	96
Ca^{2+}, A23187					
+ trifluoperazine	68	69	73	72	76
+ phenytoin	73	74	76	73	70
+ diazepam	77	78	75	75	72

*[a] Intact synaptosomes were incubated under various conditions after preincubation with ^{32}P followed by quantitation of Ca^{2+} uptake, neurotransmitter release and protein DPH-M phosphorylation as described (DeLorenzo 1981a, 1982). Concentrations of trifluoperazine, phenytoin, and diazepam were 15, 80, and 20 μM, respectively. The data give the means of eight determinations and are expressed as percentage of the maximally stimulated condition (100%). The largest ±SEM was 6.3. The effects of changes produced by all three drugs in comparision to the maximally stimulated condition were statistically significant $P < 0.001$

bility of the preparation. However, various pharmacological inhibitors of CaM (e.g., trifluoperazine) and Ca^{2+}-CaM protein kinase activity (e.g., phenytoin and diazepam), which inhibit Ca^{2+}-CaM release in vesicle preparations, have been used to probe the possible involvement of CaM in neurotransmitter release from intact nerve terminals.

Conditions that induce Ca^{2+} entry by the depolarization of the synaptosome membrane (e.g., high K^+ or veratridine) or by producing Ca^{2+} channels (e.g., Ca^{2+} ionophore A23187) caused significant synaptosomal release of norepinephrine and acetylcholine (Table 2). This increased release of neurotransmitter substances produced by both elevated K^+ and A23187 was significantly inhibited by trifluoperazine (Table 2). These results suggest that inhibition of CaM by trifluoperazine blocks the release process. However, it is not possible from these experiments to determine if trifluoperazine is inhibiting release by blocking Ca^{2+} uptake or by inhibiting a specific Ca^{2+}-regulated process within the nerve terminal.

To test these possibilities, the effects of trifluoperazine on Ca^{2+} uptake (Table 2) was investigated. Trifluoperazine inhibits the depolarization-dependent uptake of Ca^{2+} into intact synaptosomes induced by both elevated K^+ and veratridine (DeLorenzo 1981a, 1982). However, the Ca^{2+} uptake produced by A23187 was not inhibited by trifluoperazine. This data indicates that trifluoperazine inhibits release in two ways: (1) by inhibiting depolarization-dependent Ca^{2+} uptake, and (2) by blocking a Ca^{2+}-regulated process that modulates release even when Ca^{2+} is entering the nerve terminal in the presence of A23187.

The anticonvulsant, phenytoin, and the benzodiazepine, diazepam, also block norepinephrine and acetylcholine release from intact synaptosomes produced by A23187 under conditions where they do not block Ca^{2+} uptake, (Table 2). These inhibitors of CaM kinase activity (DeLorenzo et al. 1981) also blocked the Ca^{2+}-CaM-dependent release of neurotransmitter substances from isolated vesicles, further suggesting that CaM is involved in neurotransmission. Studies from both isolated vesicles and intact synaptosome preparations indicate that CaM may act as a Ca^{2+} receptor mediating some of the effects of Ca^{2+} on neurotransmission.

4 A Molecular Approach to the Actions of Calcium and Calmodulin on Neurosecretion

The evidence presented previously suggests that Ca^{2+}-CaM-regulated synaptic biochemical processess may regulate the effect of Ca^{2+} on synaptic activity. Therefore, it is important to determine which CaM-regulated enzyme systems are involved in specific aspects of synaptic function. Evidence from several laboratories has suggested that Ca^{2+}-calmodulin-regulated protein phosphorylation may play a role in regulating the release of neurotransmitter substances. Studies from Puszkin's laboratory have indicated that Ca^{2+}-calmodulin may regulate vesicle-membrane interactions through alteration of membrane lipid environments. Although the precise biochemical mechanisms mediating the effects of Ca^{2+} on release are not known, research is now being focused on several promising areas.

5 Ca^{2+}-Stimulated Protein Phosphorylation

Calcium-stimulated protein phosphorylation in brain was initially described in whole brain homogenates and highly enriched preparations of synaptosomes (DeLorenzo 1976, 1977; DeLorenzo and Freedman 1977a, b). These results demonstrated that Ca^{2+} stimulated the endogenous phosphorylation of many brain proteins, but particularly proteins in the 10,000-20,000, 50,000-54,000, 60,000-64,000, and 150,000-300,000 molecular weight ranges. Two protein bands with molecular weights of 52,000-54,000 and 60,000-64,000 (proteins DPH-M and DPH-L, respectively) were of particular interest because they were most dramatically stimulated by Ca^{2+} and inhib-

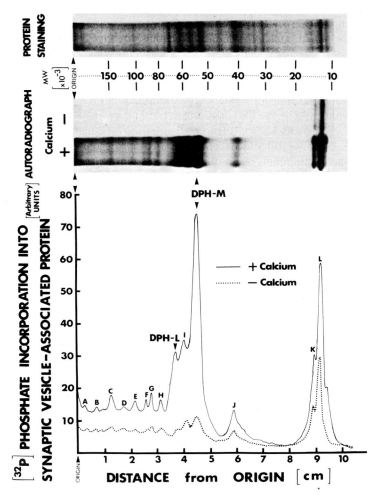

Fig. 1. Effects of calcium on the endogenous phosphorylation of synaptic vesicle-associated proteins. Synaptic vesicles were obtained and incubated under standard conditions in the presence of $MgCl_2$ (4 mM) and $(\gamma\text{-}^{32}P)ATP$ (25 μM) plus or minus $CaCl_2$ (10 μM). Following the isolation of the vesicles after the reaction by centrifugation, the vesicle pellets were subjected to SDS-polyacrylamide gel electrophoresis, protein staining, autoradiography, and quantitation, as described previously. The results shown are representative of 12 individual experiments. Each arbitrary unit approximately 38.6 cpm. Molecular weight determinations were performed as described previously. (DeLorenzo 1981b)

ited by phenytoin, an anti-convulsant that blocks several Ca^{2+}-dependent processes, including neurotransmitter release (DeLorenzo 1983). The phosphorylation of synaptosomal proteins DPH-L and DPH-M was also enriched in synaptic vesicle fractions (Fig. 1) prepared from intact synaptosomes (DeLorenzo and Freedman 1977a). These results demonstrated that the Ca^{2+}-stimulated phosphorylation observed in synaptosome fractions was occurring within the synaptosomes and not in some other membrane contaminations in the preparations.

A hypothesis was developed from these findings, suggesting that Ca^{2+}-dependent protein phosphorylation (a new phosphorylation system distinct from cyclic AMP kinase) may regulate the effects of Ca^{2+} on synaptic function and neurotransmitter release. The Ca^{2+}-dependent pattern of endogenous protein phosphorylation has been observed by other researchers in several isolated brain fractions (Ehrlich 1978; Carlin et al. 1980) and in preparations of other tissues such as the adrenal medulla (Amy and Kirshner 1981) and the electric organ of Torpedo (Michaelson and Advissar 1979).

Depolarization-dependent Ca^{2+} uptake in intact synaptosomes was shown to stimulate the phosphorylation of an 80,000-mol. wt. protein (protein I) in intact synaptosomes (Krueger et al. 1977). The levels of phosphorylation of proteins with identical molecular weights to protein DPH-L and DPH-M seen in isolated synaptosomes were shown to be stimulated in intact synaptosomes by depolarizing conditions that stimulated Ca^{2+} entry and simultaneously initiated neurotransmitter release from intact synaptosomes (DeLorenzo 1981a). The phosphorylation of these proteins in intact preparations was also shown to be occurring in the synaptic vesicle, synaptic membrane, and post-synaptic density preparations from these intact fractions (DeLorenzo 1980a). These results provided the first evidence that the depolarization-dependent phosphorylation of specific proteins in intact synaptosome preparations was actually occurring within the synaptosomes and not associated with other contaminants in the synaptosome preparations. The level of phosphorylation of mol. wt. 50,000-60,000 proteins also correlate with norepinephrine release in intact adrenal medulla cells (Amy and Kirshner 1981).

6 Calmodulin Mediation of the Effects of Ca^{2+} on Protein Kinase Activation

Ca^{2+}-stimulated endogenous membrane protein phosphorylation was shown to be dependent on CaM in crude preparations of brain membrane and in several other tissues (Schulman and Greengard 1978a, b). Calmodulin was subsequently demonstrated to mediate the effect of Ca^{2+} on the phosphorylation of specific synaptic vesicle proteins (DeLorenzo et al. 1979). Calmodulin was also shown to modulate the effects of Ca^{2+} on the endogenous phosphorylation of highly enriched synaptic membrane (DeLorenzo 1980a), synaptic junctional complex (DeLorenzo 1981a, 1981b), and postsynaptic density (Grab et al. 1980) preparations. Furthermore, depolarization-dependent Ca^{2+} uptake by intact synaptosomes was shown to simultaneously stimulate the release of neurotransmitter substances and the phosphorylation of specific (especially proteins DPH-L and DPH-M) synaptic vesicle, synaptic membrane, synaptic junctional and post-synaptic density proteins isolated from the intact synaptosomes following depolarization-dependent Ca^{2+} uptake (DeLorenzo 1980a).

In summary, evidence from several laboratories has confirmed with both intact and broken synaptosome fractions the initial observations in isolated synaptosome fractions that Ca^{2+} stimulates the phosphorylation of synaptic proteins. Phosphorylation of DPH-L and DPH-M and possibly other mol. wt. 50,000-60,000 proteins appears to be consistently observed and correlates with neurotransmitter release; however, protein I (Krueger et al. 1977), low molecular weight proteins (Ehrlich 1978), and several high molecular weight proteins (DeLorenzo 1976) have also been described as serving as endogenous substrates for endogenous Ca^{2+}-CaM kinase activity.

7 Possible Role of Protein Phosphorylation in Neurosecretion

To determine the role of protein phosphorylation in neurotransmitter release, studies were conducted on isolated synaptic vesicle fractions. This system has the advantage of allowing the easy manipulation of the vesicle environment while simultaneously studying the effects of Ca^{2+} and calmodulin on protein phosphorylation and neurotransmitter release. By employing this system, it was shown that protein phosphorylation and neurotransmitter release (norepinpephrine and acetylcholine) had the same requirements for Mg^{2+}, ATP, and Ca^{2+} (DeLorenzo 1980a, 1981b). In addition, various incubation conditions such as pH and buffer solutions that produced maximal Ca^{2+}-stimulated neurotransmitter release also produced maximal levels of protein phosphorylation. Both vesicle neurotransmitter release and phosphorylation were clearly shown to be dependent on the presence of calmodulin. Trifluoperazine, also inhibited both protein phosphorylation and neurotransmitter release in the synaptic vesicles (Table 1). Thus, there is strong evidence that both protein phosphorylation and vesicular neurotransmitter release are calmodulin-dependent in this vesicle system.

Phenytoin and diazepam have been shown to directly inhibit the Ca^{2+}-calmodulin kinase system (DeLorenzo and Glaser 1976; DeLorenzo et al. 1977; DeLorenzo 1983) without inactivating calmodulin, as in the case with trifluoperazine. These drugs can thus be used to further explore the relationship between release and calmodulin kinase activation. Both phenytoin and diazepam have also been shown in this laboratory to inhibit Ca^{2+}-CaM stimulated vesicle neurotransmitter release (DeLorenzo 1982), which suggests that direct inactivation of the calmodulin kinase system by these drugs also inhibits neurotransmitter release (Table 2). In addition, vesicles prepared by standard procedures that do not provide rapid isolation or more physiological media show essentially normal morphology, but lose their ability to demonstrate endogenous Ca^{2+}- and calmodulin-activated protein phosphorylation and neurotransmitter release, further indicating that the labile Ca^{2+}-stimulated protein kinase may be responsible for vesicle release (DeLorenzo 1980a).

Although the isolated vesicle preparation offers several advantages for studying protein phosphorylation and neurotransmitter release, it is an isolated system and must be compared to intact synaptosome preparations. To demonstrate a relationship between release and phosphorylation in the intact synaptosome system, the drugs trifluoperazine, phenytoin, and diazepam were used to inhibit calmodulin and the Ca^{2+}-kinase system, respectively. As described above for neurotransmitter release, these drugs were tested with K^+-induced depolarization and with Ca^{2+} uptake produced by the Ca^{2+} ionophore A23187. The results (Table 2) demonstrate that both trifluoperazine and phenytoin inhibit protein phosphorylation and neurotransmitter release in intact synaptosomes independent of their effects on depolarization-dependent Ca^{2+} influx in the presence of A23187. Thus, inhibition of calmodulin by trifluoperazine and direct inhibition of the Ca^{2+}-calmodulin protein kinase by phenytoin and diazepam produce the same effects on neurotransmitter release and protein phosphorylation in intact synaptosomes. These results provide pharmacological evidence, further indicating that activation of Ca^{2+}-calmodulin kinase systems may be required to initiate or maintain neurotransmitter release in intact synaptosome preparations.

8 Ca^{2+}-Calmodulin Tubulin and Microtubule-Associated Protein Kinase System in Synaptic Modulation

Previous investigations have demonstrated the existence of Ca^{2+}-calmodulin-dependent endogenous tubulin kinase activity, which is distinct from Mg^{2+} and cAMP-dependent kinases, associated with synaptosome (Burke and DeLorenzo 1981), synaptic vesicle (Burke and DeLorenzo 1982b), and synaptic cytosol preparations (Burke and DeLorenzo 1982a) (Fig. 2). Since tubulin is a major "functional" and "structural" protein in nerve cells, Ca^{2+}-calmodulin-regulated phosphorylation of cytoskeletal proteins, especially at the synapse, may be of importance in the control of some aspects of neuronal function. Thus, purification of a tubulin kinase may provide insight into the role of Ca^{2+} and calmodulin in nerve function. Initial efforts to purify this tubulin kinase were unsuccessful due to the instability of the enzyme activity post-mortem (Goldenring et al. 1982). However, we recently succeeded in stabilizing this kinase ac-

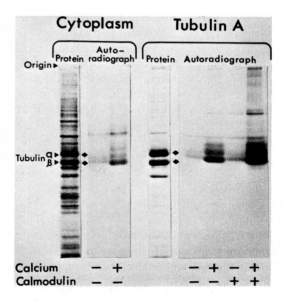

Fig. 2. Protein patterns and autoradiographs illustrating endogenous calcium-calmodulin stimulated phosphorylation in brain cytoplasm and microtubule preparations (*tubulin A*). Cytoplasm and microtubule fractions were prepared and phosphorylated as described previously. (DeLorenzo et al. 1982)

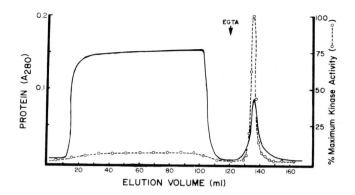

Fig. 3. Calmodulin-affinity chromatography. Phosphocellulose-purified kinase was chromatographed on calmodulin-affinity resin and activity was eluted with 2 mM EGTA (o - - - o). The elution of protein was monitored at 280 nm (——). (Goldenring et al. 1983)

tivity and separating a crude fraction of brain cytosolic tubulin kinase activity from both calmodulin and endogenous substrate tubulin (Goldenring et al. 1983). This initial stabilization and separation of this kinase system away from endogenous calmodulin and tubulin has now permitted further purification and characterization of this brain enzyme (Fig. 3).

Work from this laboratory has succeeded in purifying to apparent homogeneity a brain cytosolic calmodulin-dependent protein kinase which phosphorylates tubulin and microtubule-associated proteins as major substrates. The highly purified enzyme preparation contained two calmodulin-binding autophosphorylating subunits, designated rho and sigma, with molecular weights of 52,000 and 63,000, respectively, and similar isoelectric points between 6.7 and 7.2 (Fig. 4). The rho and sigma subunits of the kinase demonstrated significant structural homology based on two-dimensional tryptic peptide-mapping studies. The isolation and characterization of this calmodulin-binding enzyme complex provides additional insight into the mechanism of Ca^{2+}-dependent regulation of phosphorylation in the brain. This enzyme is clearly distinguishable by several criteria from other brain calmodulin-dependent kinases which have been reported previously.

Tubulin and cytoskeletal proteins play a structural and dynamic role in many aspects of neuronal function (Olmsted and Borisy 1973) and are found in high concentrations in nerve terminal preparations. MAP-2, the heat-stable microtubule-associated protein, is involved in the assembly of microtubules (Kim et al. 1979), and its phosphorylation may regulate microtubule formation (Jameson et al. 1980). Ca^{2+}-calmodulin regulate the stability of microtubules (Marcum et al. 1978) and some microtubule-associated proteins bind calmodulin (Sobue et al. 1981). We have reported that Ca^{2+}- calmodulin-dependent phosphorylation of tubulin- and microtubule-associated proteins may mediate the effects of Ca^{2+} on microtubule polymerization (DeLorenzo et al. 1982). Recent work in this laboratory indicates that the kinase purified here can be isolated complexed with tubulin (Fig. 2) and that the calmodulin-binding microtubule-associated protein is homologous with the subunits of the enzyme. In addition, while endogenous Mg^{2+}-dependent kinase phosphorylates tubulin on only serines, the

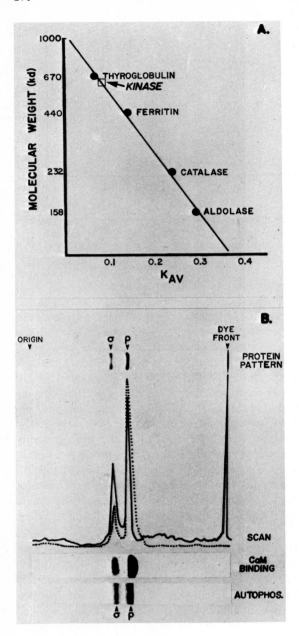

Fig. 4 A, B. Chromatography on Sephacryl S-300 Superfine. **A** Fractogel-purified enzyme chromatographed on Sephacryl S-300 with and apparent molecular weight of approximately 600,000. **B** the kinase fraction obtained from chromatography on Sephacryl displayed only two silver staining protein components of 52,000 mol. wt. (ρ) and 63,000 (δ), respectively, when resolved on one-dimensional SDS-PAGE. Both the ρ and δ subunits of the kinase bound calmodulin (CaM) in denaturing gels and displayed characteristic autophosphorylation. The densitometric scan demonstrates that the protein staining (. . . .) and calmodulin binding (——) coincide (Goldenring et al. 1983)

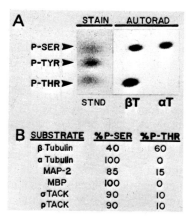

Fig. 5

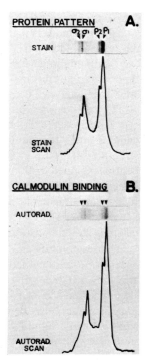

Fig. 6.

Fig. 5 A, B. Phosphoamino acid analysis of phosphorylated substrates. **A** α- (αT) and β-tubulin (βT) were phosphorylated by purified kinase under standard conditions and resolved on two-dimensional gels. The phosphorylated spots were then subjected to phosphoamino acid analysis by standard procedures. Ninhydrin staining of phosphoamino acid standards show that the procedure separated in onedimensional phosphoserine, phosphothreonine, and phosphotyrosine. The autoradiographs demonstrate that while α-tubulin was phosphorylated predominantly on serine residues, β-tubulin was phosphorylated on both threonine and serine residues. **B** amino acid phosphorylation of α- and β-tubulin, MAP-2, myelin basic protein (MBP) by purified kinase, and the autophosphorylation of the ρ and δ subunits of the enzyme were quantitated by densitometric scanning of autoradiographic images or scintillation counting of isolated phosphoamino acid spots. The data are expressed as percent of serine and threonine phosphorylation. No tyrosine phosphorylation was observed in any of the substrates tested. (Goldenring et al. 1983)

Fig. 6 A, B. Electrophoresis of the enzyme on high resolution one-dimensional SDS-PAGE. Purified kinase (10 μg) was resolved on high resolution one-dimensional SDS-PAGE gels and processed for calmodulin binding. **A** high resolution SDS-PAGE separated the ρ subunit of the enzyme into major (ρ_1) and minor (ρ_2) protein staining bands. The δ subunit of the enzyme was also separated into two components (δ_1 and δ_2). The scan of the dried gel is shown for reference. **B** the autoradiograph of the calmodulin-binding corresponding to the protein pattern in **A** demonstrates that all four of the tubulin-associated calmodulin kinase subunits bind calmodulin in denaturing gels. The scan of the autoradiograph shows that the calmodulin binding coincided with the protein staining. ^{125}I-Calmodulin binding was completely inhibited in the presence of 2 mM EGTA or an excess of nonradioactive calmodulin. (Goldenring 1983)

Ca^{2+}-calmodulin-dependent kinase phosphorylates beta-tubulin on both threonine and serine residues (Fig. 5). Preliminary studies indicate that phosphothreonine phosphorylation occurs at a single site on beta-tubulin, further suggesting a unique role for Ca^{2+}-calmodulin-dependent phosphorylation in the regulation of tubulin function. All these data indicate that the Ca^{2+}-calmodulin-dependent tubulin and MAP-2 kinase isolated here may mediate the effects of Ca^{2+} on tubulin-microtubule dynamics.

The extended resolution of the purified kinase on 20-cm one-dimensional SDS-PAGE gels separated the rho and sigma subunits into doublets. All four of these subunits demonstrated calmodulin binding (Fig. 6). In order to investigate further the novel structure of the four calmodulin-binding subunits, all four of the kinase subunits were peptide mapped with two-dimensional tryptic fingerprints (Goldenring et al. 1983). These maps demonstrated that all four peptides contain significant homology among their major fragments. Thus, the native enzyme system which chromatographs with an apparent molecular weight of 600,000 appears to consist of a complex of homologous calmodulin-binding autophosphorylating subunits.

The homologies among the enzyme subunits underscore the presence of a series of specific properties of this kinase system which comprise the "signatures" to identify this kinase (Goldenring et al. 1983). (1) The enzyme contains two calmodulin-binding doublets, rho and sigma, of mol. wt. 52,000 and 63,000, respectively. (2) Both the rho and sigma subunits demonstrate isoelectric points between 6.7 and 7.2 (3) Both the rho and sigma subunits demonstrate autophosphorylation. (4) The rho and sigma subunits show significant homologies as assessed by tryptic digest fingerprings. (5) Phosphorylation of the enzyme subunits in the absence of substrate elicits the production of lower mobility phosphorylated species. (6) The kinase phosphorylates beta-tubulin on both threonine and serine residues. These "signatures" may now permit the study of similar enzymes associated with kinase systems in cytosol, membrane, and the trans-synaptic junctional complex.

The observations presented here indicate that this kinase may mediate important interactions between Ca^{2+} and the cytoskeleton through phosphorylation of tubulin and microtubule-associated proteins. Elucidation of the exact role of calmodulin-dependent phosphorylation in regulating tubulin dynamics and the possible homology of a number of calmodulin-dependent kinase systems awaits further investigation. Although the kinase phosphorylates tubulin and microtubule-associated proteins, the results do not exclude the possibility that the enzyme may have a broader substrate range in neuronal tissue. The ability to identify this enzyme system by its characteristic properties should provide the basis for investigating the functional significance of this kinase in cellular activity. Tubulin and cytoskeletal elements play a major role in maintaining the structural and functional integrity of the synapse. Thus, regulation of this system by Ca^{2+} and calmodulin through specific protein kinase systems may play a role in regulating some aspects of synaptic function.

9 Identification of the Major Components of Protein Bands DPH-L and DPH-M as Alpha and Beta Tubulin and the Subunits of a Calmodulin Kinase System

The identification and characterization of a specific calmodulin kinase system in synaptosome fractions provides the biochemical means to identify this kinase in intact synaptosome and other preparations employed to study the effects of Ca^{2+} on synaptic modulation. Employing two-dimensional gel electrophosresis and peptide mapping studies, experiments in this laboratory have identified the autophosphorylation of the rho and sigma subunits of the calmodulin kinase system as a major component of the Ca^{2+}-CaM stimulated endogenous phosphorylation of protein bands DPH-L and DPH-M. As previously reported, Ca^{2+}-CaM stimulated phosphorylation of alpha and beta tubulin also comprises a significant fraction of these bands in brain cytosol synaptic and vesicle fractions (Burke and DeLorenzo 1981; 1982a, b).

The level of phosphorylation of proteins DPH-L and DPH-M have been correlated with depolarization-dependent Ca^{2+} uptake and neurotransmitter release in intact synaptosomes (DeLorenzo 1980a, 1981a, 1982). Thus, the Ca^{2+}-CaM stimulated phosphorylation of tubulin and this kinase system may play an important role in regulating some of the effects of calcium at the synapse.

10 Phosphorylation of Synapsin I and Other Proteins in Synaptic Function

The role of protein phosphorylation as a major regulator in mediating the effects of numerous second messangers and hormones in brain has been well established especially by the extensive studies of Greengard's group (Greengard 1976, 1978, 1981). The evidence indicates that numerous kinases and protein substrates are involved in regulating various aspects of neuronal activity. A discussion of this diverse field is beyond the scope of this paper. However, the role of synapsin I phosphorylation in synaptic regulation is important, since synapsin I phosphorylation occurs in the presynaptic nerve terminal in response to depolarization dependent Ca^{2+} uptake (Krueger et al. 1977). Synapsin I (Protein I) is a doublet of closely related proteins found in nervous tissue and associated with synaptic vesicle and membrane fractions (Greengard 1981). This protein was discovered and extensively studied by Greengard's group, and represents the major model system for demonstrating the role of cyclic AMP and Ca^{2+}-stimulated protein kinase systems in mediating neuronal function in brain. Although the functional significance of synapsin I has not been determined, the presence of this protein at the synapse certtainly indicates that this major and well-characterized kinase system (Kennedy et al. 1983; Greengard 1981; Bennett et al. 1983) may play an important role in mediating some of the effects of Ca^{2+} on synaptic function.

11 Ca^{2+}-Dependent Phospholipase A$_2$ and Synaptic Modulation

Phospholipase A$_2$ (PLA$_2$) is a ubiquitously distributed enzyme that catalyzes the hydrolysis of phosphoglycerides into lysophosphoglyceride and fatty acid. This enzyme is significantly regulated by Ca^{2+} (Vogt 1978). Evidence has accumulated that endogenous brain PLA$_2$ may play a major role in membrane interactions and function (Bazan 1971; Goracci et al. 1978). PLA$_2$ has recently been shown to be present in synaptic plasma membrane and synaptic vesicle fractions (Moskowitz et al. 1982). Puszkin's group has suggested that endogenous synaptic vesicle PLA$_2$ activity is regulated by calmodulin and induces interaction of brain synaptic vesicles (Moskowitz et al. 1982). Thus, Ca^{2+} regulation of this enzyme system and potentially other enzymes involved in lipid metabolism may also provide an important insight into synaptic modulation.

12 Ca^{2+}-Calmodulin-Stimulated Membrane Interactions

Calmodulin was found to mediate Ca^{2+}-dependent synaptic vesicles and synaptic membrane interactions (DeLorenzo 1980a, 1981a). Membrane interactions and fusion were initiated in a Ca^{2+}-calmodulin-dependent fashion in these systems, under conditions that stimultaneously stimulated protein phosphorylation (Table 3). The ability of Ca^{2+} and calmodulin to stimulate vesicle membrane interactions was confirmed by Moskowitz et al. (1982), and it was indicated that calmodulin regulates Ca^{2+}-dependent phospholipase A$_2$ in the vesicle fraction. These results suggest that both calmodulin kinase and phospholipase A$_2$ activity may modulate membrane interactions in a Ca^{2+}-calmodulin-dependent manner.

Table 3. Effects of calmodulin and Ca^{2+} on synaptic vesicle and synaptic membrane interactions

Condition	Synaptic vesicle and synaptic membrane interactions (Number vesicles/μ membrane)		
Control	2.58	±	0.21
Ca^{2+}	5.33*[a]	±	0.36
Calmodulin	2.66	±	0.19
Ca^{2+} + calmodulin	12.36*†[a,b]	±	0.91
+ trifluoperazine	6.33*[a]	±	0.55

Calmodulin-depleted synaptic vesicles were incubated under various conditions with synaptic membrane, prepared for electronmicroscopy, and quantitated for vesicle-membrane interactions as described previously (DeLorenzo 1980a). The data give the mean values of 500 determinations
*[a]P < 0.001 in comparision to control condition
†[b]P < 0.001 in comparision to Ca^{2+} condition

13 Ca^{2+} and Calmodulin Regulation of Tryptophan Hydroxylase Activity and Neurotransmitter Turnover

Calcium has been shown to regulate the turnover of numerous neurotransmitter substances. Independent studies from several groups indicates that Ca^{2+} activates tryptophan hydroxylase in vitro and that this effect was regulated by a Ca^{2+} kinase system (Lysz and Sze 1978; Hamon et al. 1978; Kuhn et al. 1978; Kuhn and Lovenberg 1982; Yamauchi and Fujisawa 1979a, b). Tryptophan hydroxylase catalyzes the initial rate-limiting step in the biosynthesis of the neurotransmitter seritonin. Thus, regulation of this major synaptic pathway by Ca^{2+} provided initial evidence for a role of Ca^{2+} in the synthesis and turnover of neurotransmitter substances, and suggested that the activity of this enzyme system is modulated by phosphorylation. The results from these studies indicated that the regulation of tryptophan hydroxylase was not mediated by a cyclic AMP kinase system.

The regulation of tryptophan hydroxylase was subsequently shown to be dependent on Ca^{2+} and calmodulin (Kuhn and Lovernberg 1982; Aymauchi and Fujisawa 1979a, b). Yamauchi and Fujisawa (1979a, b) were able to separate tryptophan hydroxylase from endogenous calmodulin by differential ammonium sulfate precipitation. The CaM-depleted enzyme fraction could then be studied for Ca^{2+} and calmodulin activation. The calmodulin-depleted tryptophan hydroxylase could not be activated by Ca^{2+} alone, but required the presence of CaM. Kuhn and Lovenberg (1982) also documented the dependence of this Ca^{2+} effect on calmodulin by demonstrating that trifluoperazine and other phenothiazine CaM inhibitors antagonized the effects of Ca^{2+} on tryptophan hydroxylase. Recent studies (Yamauchi and Fujisawa 1983) have resulted in the purification of a calmodulin-dependent kinase system that phosphorylates tryptophan 5-mono-oxygenase and regulates the activity of this enzyme system. Thus, it is clear from these studies that calmodulin-dependent phosphorylation of this major enzyme system regulates the rate-limiting step in the synthesis of seritonin.

14 Conclusion

The results reviewed in this presentation indicate that calmodulin regulates several of the effects of Ca^{2+} on synaptic biochemical processes. Although the full functional significance of the role of calmodulin in synaptic function is not known, this data indicates that calmodulin plays a role in modulating synaptic function and possibly neurosecretion. The role of calmodulin in regulating neurotransmitter turnover, cytoskeletal dynamics, synaptic vesicle function, and synaptic protein phosphorylation is an important area for further investigation.

References

Amy CM, Kirshner N (1981) Phosphorylation of adrenal medulla cell proteins in conjunction with stimulation of catecholamine secreation. J Neurochem 3:847-854

Bazan NG Jr (1971) Phospholipase A_1 and A_2 in brain subcellular fractions. Acta Physiol Lat Am 21:101-106

Bennett MK, Erondu NE, Kennedy MB (1983) Purification and characterization of a calmodulin-dependent protein kinase that is highly concentrated in brain. J Biol Chem 258:12735-12744

Blaustein MP, Johnson EM, Needleman P (1972) Calcium-dependent norepinephrine release from presynaptic nerve endings in vitro. Proc Natl Acad Sci USA 69:2237-2240

Burke BE, DeLorenzo RJ (1981) Calcium- and calmodulin-dependent phosphorylation of neuro-tubulin. Proc Natl Acad Aci USA 78:991-995

Burke BE, DeLorenzo RJ (1982a) Ca^{2+} and calmodulin-regulated endogenous tubulin kinase activity in presynaptic nerve terminal preparations. Brain Res 236:393-415

Burke BE, DeLorenzo RJ (1982b) Ca^{2+}- and calmodulin-dependent phosphorylation of endogenous synaptic vesicle tubulin by a vesicle-bound calmodulin kinase system. J Neurochem 38:1205-1218

Carlin RK, Grab DJ, Cohen RS, Siekevitz P (1980) Isolation and characterization of postsynaptic densites from various brain regions: enrichment of different types of postsynaptic densites. J Cell Biol 86:831-843

Cheung WY (1980) Calmodulin plays a pivotol role in cellular regulation. Science (Wash DC) 207:19-27

Del Castillo J, Stark L (1952) The effects of calcium ions on the motor end-plate potentials. J Physiol (Lond) 124:553-559

DeLorenzo RJ (1976) Calcium-dependent phosphorylation of specific synaptosomal fraction proteins: possible role of phosphorylation in mediating neurotransmitter release. Biochem Biophys Res Commun 71:590-597

DeLorenzo RJ (1977) Antagonistic action of diphenylhydantoin and calcium on the level of phosphorylation of particular rat and human brain proteins. Brain Res 134:125-138

DeLorenzo RJ (1980a) Role of calmodulin in neurotransmitter release and synaptic function. Ann NY Acad Sci 356:92-109

DeLorenzo RJ (1980b) Phenytoin: calcium and calmodulin dependent protein phosphorylation and neurotransmitter release. Adv Neurol 27:399-444

DeLorenzo RJ (1981a) The calmodulin hypothesis of neurotransmission. Cell Calcium 2:365-385

DeLorenzo RJ (1981b) Calcium, calmodulin and synaptic function: modulation of neurotransmitter release, nerve terminal protein phosphorylation and synaptic vesicle morphology by calcium and calmodulin. In: Tapi R and Cotman CW (eds) Regulatory mechanisms of synaptic transmission. Plenum, New York, pp 205-240

DeLorenzo RJ (1982) Calmodulin in neurotransmitter release and synaptic function. Fed Proc Fed Am Soc Exp Biol 41:2265-2272

DeLorenzo RJ (1983) Calcium-calmodulin in protein phosphorylation in neuronal excitability and anticonvulsant drug action. Adv. Neurol. 34:325-338

DeLorenzo RJ, Freedman SD (1977a) Calcium-dependent phosphorylation of synaptic vesicle proteins and its possible role in mediating neurotransmitter release and vesicle function. Biochem Biophys Res Commun 77:1036-1043

DeLorenzo RJ, Freedman SD (1977b) Possible role of calcium-dependent protein phosphorylation in mediating neurotransmitter release and anticonvulsant action. Epilepsia 18:357-365

DeLorenzo RJ, Glaser GH (1976) Effects of diphenylhydantoin on the endogenous phosphorylation of brain protein. Brain Res 105:381-386

DeLorenzo RJ, Emple GP, Glaser GH (1977) Regulation of the level of endogenous phosphorylation of specific brain proteins by diphenylhydantoin. J Neurochem 28:21-30

DeLorenzo RJ, Freedman SD, Yohe WB, Maurer SC (1979) Stimulation of Ca^{2+}-dependent neurotransmitter release and presynaptic nerve terminal protein phosphorylation by calmodulin and a calmodulin-like protein isolated from synaptic vesicles. Proc Natl Acad Sci USA 76:1838-1842

DeLorenzo RJ, Burdette S, Holderness J (1981) Benzodiazepine inhibition of the calcium-calmodulin protein kinase systems in brain membranes. Science (Wash DC) 212:1157-1159

DeLorenzo RJ, Gonzales B, Goldenring JR, Bowling A, Jacobson R (1982) Calcium-calmodulin tubulin kinase system and its role in mediating the calcium signal in brain. Prog Brain Res 56:255-286

Douglas WW (1968) Stimulus-secretion coupling: the concept and clues from chromaffin and other cells. Br J Pharmacol 34:451-474

Ehrlich YH (1978) Phosphoproteins as specifiers for mediators and modulators in neuronal function. In: Ehrlich YH, Volarka J, Davis LO, Brunngraber EG (eds) Modulators, mediators and specifiers in brain function. Plenum, New York, pp 75-101

Goldenring JR, Gonzalez B, DeLorenzo RJ (1982) Isolation of brain calcium-calmodulin tubulin kinase containing calmodulin-binding proteins. Biochem Biophys Res Commun 108:421-428

Goldenring JR, Gonzalez B, McGuire JS, DeLorenzo RJ (1983) Purification and characterization of a calmodulin-dependent kinase from rat brain cytosol able to phosphorylate tubulin and microtubule-associated proteins. J Biol Chem 258:12632-12640

Goracci G, Porcellati G, Woelk H (1978) Subcellular localization and distribution of phospholipases A in liver and brain tissue. In: Galli et al. (eds) Advances in prostaglandin and thromboxane research. Raven, New York, pp 55-67

Grab DJ, Carlin RK, Siekevitz P (1980) The presence and functions of calmodulin in the postsynaptic density. Ann NY Acad Sci 356:55-72

Grab DJ, Berzins K, Cohen RS, Siekevitz P (1979) Presence of calmodulin in postsynaptic densities isolated from canine cerebral cortex. J Biol Chem 254:8690-8696

Greengard P (1976) Possible role for cyclic nucleotides and phosphorylated membrane proteins in postsynaptic actions of neurotransmitters. Nature (Lond) 260:101-8

Greengard P (1978) Phosphorylated proteins as physiological effectors. Science (Wash DC) 199:146-152

Greengard P (1981) Intracellular signals in the brain. Harvey Lect 75:277-331

Hamon M, Bourgoin S, Hery F, Simonnet G (1978) Activation of tryptophan hydroxylase by adenosine triphosphate, magnesium and calcium. Mol Pharmacol 14:99-110

Jameson L, Frey T, Zeeberg B, Daeldorf F, Caplow M (1980) Inhibition of microtubule assemb y by phosphorylation of microtubule-associated proteins. Biochemistry 19:2472-2479

Katz B, Miledi R (1969) Spontaneous and evoked activity of motor nerve endings in calcium ringer. J Physiol (Lond) 203:689-706

Katz B, Miledi R (1970) Further study of the role of calcium in synaptic transmission. J Physiol (Lond) 207:789-801

Kennedy MB, McGuinness T, Greengard P (1983) A calcium/calmodulin-dependent protein kinase from mammalian brain that phosphorylates synaptsin I: partial purification and characterization. J Neurosci 3:818-831

Kim H, Binder LI, Rosenbaum JL (1979) The periodic association of MAP 2 with brain microtubules in vitro. J Cell Biol 80:266-276

Klee CB, Crouch TH, Richman PG (1980) Calmodulin. Annu Rev Biochem 49:489-515

Krueger B, Forn J, Greengard P (1977) Depolarization-induced phosphorylation of specific proteins, mediated by calcium influx, in rat brain synaptosomes. J Biol Chem 252:2764-2773

Kuhn DM, Lovernberg W (1982) Role of calmodulin in the activation of tryptophan hydroxylase. Fed Proc 41:2258-2264

Kuhn DM, Vogel RL, Lovenberg W (1978) Calcium-dependent activation of tryptophan hydroxylase by ATP and magnesium. Biochem Biophys Res Commun 82:759-766

Lysz TW, Sze PY (1978) Activation of brain tryptophan hydroxylase by a phosphorylating system. J Neurosci Res 3:411-418

Marcum JR, Dedman JR, Brinkley BR, Means AR (1978) Control of microtubule assembly and disassembly by calcium-dependent regulator protein. Proc Natl Acad Sci USA 75:3771-3775

Michaelson DM, Advissar S (1979) Ca^{2+}-dependent phosphorylation of purely cholinergic torpedo synaptosomes. J Biol Chem 254:12542-12546

Miledi R (1973) Transmitter release induced by injection of calcium ions into nerve terminals. Proc R Soc (Lond) B Biol Sci 183:421-425

Miledi R, Slater CR (1966) The action of calcium on neuronal synapses in the squid. J Physiol (Lond) 184:473-478

Moskowitz N, Schook W, Puszkin S (1982) Interaction of brain synaptic vesicles induced by endogenous Ca^{2+}-dependent phospholipase A_2. Science (Wash DC) 216:305-307

Olmsted JB, Borisy GG (1973) Microtubules. Annu Rev Biochem 42:507-531

Rasmussen H, Goodman DBP (1977) Relationship between calcium and cyclic nucleotides in cell activation. Physiol Rev 57:421-509

Rubin RP (1972) The role of calcium in the release of neurotransmitter substances and hormones. Pharmacol Rev 22:389-428

Schulman H, Greengard P (1978a) Stimulation of brain membrane protein phosphorylation by calcium and endogenous heat-stable protein. Nature (Lond) 271:478-479

Schulman H, Greengard P (1978b) Calcium-dependent protein phosphorylation system in membranes from various tissues, and its activation by "calcium-dependent regulator." Proc Natl Acad Sci USA 75:5432-5436

Sobue K, Fujita M, Muramoto Y, Kakiuchi S (1981) The calmodulin-binding protein in microtubules is tau factor. FEBS Lett 132:137-140

Vogt W (1978) Role of phospholipase A_2 in prostaglandin formation. In: Galli C et al. (eds) Advances in prostaglandin and thromboxane research, vol. 3. Raven, New York, pp 89-95

Weiss B, Proxialeck W, Cimino M, Barnette MS, Wallace TL (1980) Pharmacological regulation of calmodulin. Ann NY Acad Sci 356:319-345

Yamauchi T, Fujisawa H (1979a) Regulation of rat brainstem tryptophan 5-monooxygenase-calcium-dependent reversible activation by ATP and magnesium. Arch Biochem Biophys 198:219-226

Yamauchi T, Fujisawa H (1979b) Activation of tryptophan 5-monooxygenase by calcium-dependent regulator protein. Biochem Biophys Res Commun 90:28-35

Yamauchi T, Fujisawa H (1983) Purification and characterization of the brain calmodulin-dependent protein kinase (Kinase II), which is involved in the activation of tryptophan 5-monooxygenase. Eur J Biochem 132:15-21

The Role of Calcium in Prostaglandin and Thromboxane Biosynthesis

L. R. BALLOU and W. Y. CHEUNG[1]

CONTENTS

1 Introduction

Prostaglandins and thromboxanes are biologically potent compounds derived from arachidonic acid. They are synthesized in a wide variety of tissues in response to various stimuli, and are known to modulate numerous physiological processes. These compounds are not stored within the cell, and their effects are restricted to the cell in which they are synthesized or to neighboring cells. Their precise mechanisms of action are not well understood; in general, their actions are believed to be mediated via cyclic AMP, or Ca^{2+}.

Because the prostaglandins exhibit a variety of biological activities, the underlying mechanisms regulating their biosynthesis have attracted much attention. The pathway leading to their production has been well established (Flower 1979; Hinman 1972; Lands 1979; Samuelsson et al. 1978; Samuelsson et al. 1975); the rate-limiting step appears to be the availability of arachidonic acid. Since the intracellular level of free arachidonic acid is low, its release from phospholipid is a key point in the regulation of prostaglandin biosynthesis.

Three pathways have been proposed for the release of arachidonic acid from endogenous sources, all of which involve the action of a Ca^{2+}-dependent phospholipase. One involves the action of a nonspecific phospholipase A_2 which cleaves arachidonic

[1] Departments of Biochemistry, St. Jude Children's Research Hospital and The University of Tennessee Center for the Health Sciences, Memphis, TN 38101, USA

acid directly from several phospholipids. Another involves the sequential action of a phosphatidylinositol-specific phospholipase C and a diglyceride lipase. A third involves a phospholipase A_2 specific for phosphatidic acid. The relative importance of each of the proposed pathways is currently a controversial topic and probably varies from tissue to tissue. It appears, however, that Ca^{2+} is essential for the activity of the phospholipases (Ballou and Cheung 1983; Irvine et al. 1979; Lapetina and Michell 1973).

It is the purpose of this review to discuss the role of Ca^{2+} in prostaglandin metabolism by examining its role in the activation of the enzymes catalyzing the release of arachidonic acid. Since the biosynthesis of prostaglandins and thromboxanes is dependent upon the availability of arachidonic acid, elucidation of the mechanism regulating its release is important to the understanding of cellular physiology.

2 Ca^{2+} Flux and Phospholipid Metabolism

The plasma membrane is generally considered to be composed of a phospholipid bilayer with its associated proteins. Membrane-associated proteins may be embedded into the lipid matrix, or they may be loosely associated with the membrane. Until recently, the membrane was generally considered to be a structural component of the cell playing more or less a passive role in cellular metabolism. It is now becoming increasingly evident that the membrane serves not only as a cellular boundary, regulating the passage of materials into and out of the cell, but is also actively involved in the recognition and transduction of incoming stimuli. An important example is the Ca^{2+}-dependent release of arachidonic acid, a process involving at least two discrete phases. First, the receptor-mediated recognition by the cell membrane of an external signal initiates the degradation of phosphatidylinositol, a process often referred to as the "phosphatidylinositol response" or "phosphatidylinositol turnover". This results in increased calcium gating (Berridge 1980, 1981, 1982; Billah et al. 1980; Hawthorne and Pickard 1979; Lapetina et al. 1978a; Michell 1975, 1977, 1979; Michell and Kirk 1981a, b; Michell et al. 1977; Fain 1982), and the production of phosphatidic acid (Gerrard 1979; Imai et al. 1982, 1983; Lapetina and Cuatrecasas 1979). Second, the increase of available Ca^{2+} activates the phospholipases (Bell et al. 1980; Flower and Blackwell 1976; Hamberg and Samuelsson 1973), which catalyze the release of arachidonic acid. Thus, the degradation of phospholipids may not only influence the transport of Ca^{2+} across the membrane, but also provide substrates for the biosynthesis of prostaglandins, themselves important cellular regulators.

3 The Biosynthesis of Prostaglandins and Thromboxanes

The arachidonic acid cascade (Fig. 1) which ultimately produces prostaglandins, thromboxanes, and leukotrienes has been extensively studied and reviewed (Flower 1979; Hinman 1972; Lands 1979; Samuelsson et al. 1978, 1975). The conversion of arachidonic

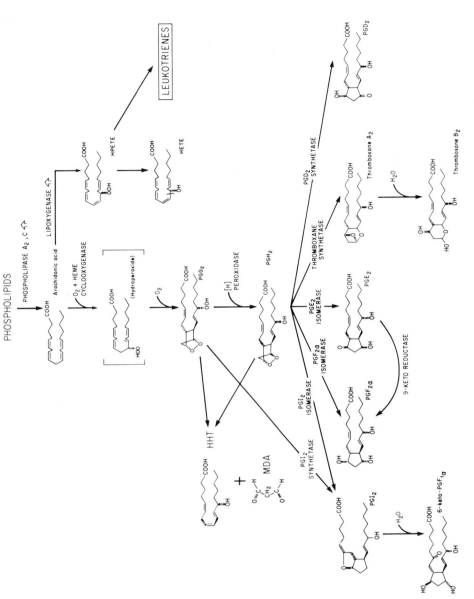

Fig. 1. The biosynthetic pathway of prostaglandins, thromboxanes and leukotrienes. *HETE* 12L-hydroxy-5,8,10,14-eicosatetraenoic acid; *HPETE* 12L-hydroperoxy-5,8,10,14-eicosatetraenoic acid; *MDA* malondialdehyde; *HHT* 17-C-hydroxy acid; *TxA₂*, thromboxane A₂; *TxB₂* thromboxane B₂. ⇧ Indicates stimulation by Ca²⁺

acid into various biologically active products is accomplished by means of two major pathways, the cyclooxygenase pathway and the lipoxygenase pathway. The rate-limiting step in both pathways appears to be the availability of a common substrate, arachidonic acid.

The conversion of arachidonic acid into prostaglandins was first demonstrated by van Dorp et al. (1964) and by Bergström et al. (1964). The initial step in prostaglandin synthesis involves the insertion of molecular oxygen into arachidonic acid to yield an intermediate hydroperoxide (Hamberg and Samuelsson 1967) which is rapidly converted to a cyclic endoperoxide, PGG_2 (Hamberg and Samuelsson 1973). PGG_2 is reduced in the next step to PGH_2 (Hamberg and Samuelsson 1973). Both PGG_2 and PGH_2 are unstable and decompose spontaneously to form a mixture of prostaglandins (Nugteren and Hazelhof 1973). As much as 30% of the total PGG_2 and PGH_2 is converted to a mixture of a hydroxy acid (HHT) and malondialdehyde (MDA) (Hamberg and Samuelsson 1974a). Alternatively, PGH_2 is converted to PGE_2 by PG endoperoxide isomerase, or to PGI_2 (prostacyclin) by prostacyclin synthase. PGI_2 is unstable and is spontaneously hydrolyzed to 6-keto-$PGF_{1\alpha}$. $PGF_{2\alpha}$ may be synthesized directly from PGH_2 by a $PGF_{2\alpha}$ isomerase or nonenzymatically from PGH_2 (Qureshi and Cagen 1982), bypassing the PGE_2 step; PGE_2 is converted to $PGF_{2\alpha}$ by 9-keto-reductase. PGH_2 is reduced to form PGD_2 by PGD synthetase.

Several workers observed that in platelet and lung a majority of the PGG_2 was converted into nonprostaglandin end products (Hamberg and Samuelsson 1974a, b), the thromboxanes. Hamberg and Samuelsson (1974a, b) originally named the end product PHD and found that in stimulated platelets, 30% of the product resulting from the oxidation of arachidonic acid was PHD, with the remainder equally made up of HHT and HETE, derived from the lipoxygenase pathway. Hamberg et al. (1975) later demonstrated that in platelets, PGG_2 was enzymatically converted to PHD, subsequently named thromboxane B_2, through an unstable intermediate thromboxane A_2. The thromboxanes are potent agents for platelet aggregation and smooth muscle contraction.

The biosynthesis of the leukotrienes via the 5-lipoxygenase pathway is less well understood and will not be discussed further, except to point out that, like prostaglandins and thromboxanes, their synthesis is dependent upon the availability of arachidonic acid, in apparent competition with the cyclooxygenase pathway. Unlike cyclooxygenase, lipoxygenase appears to be stimulated by Ca^{2+} (Rubin et al. 1982).

4 Ca^{2+}-Dependent Release of Arachidonic Acid

A major objective of this article is to define the role of Ca^{2+} in the biosynthesis of prostaglandins and thromboxanes. We will focus our attention on that portion of the biosynthetic pathway which involves the Ca^{2+}-dependent release of arachidonic acid. As mentioned earlier, prostaglandins are not stored in the cell and the cellular level of arachidonic acid is low. Since the rate-limiting step of the cyclooxygenase pathway is the availability of arachidonic acid (Ballou and Cheung 1983; Bell et al. 1980; Craven

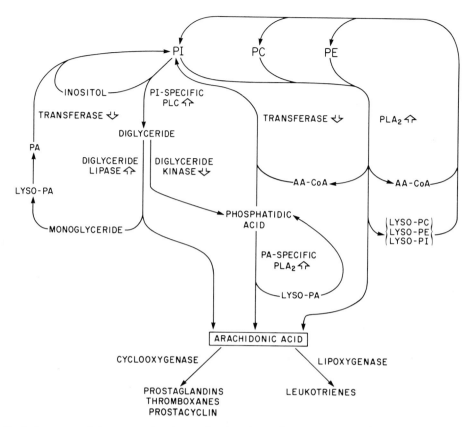

Fig. 2. Summary of the proposed pathways for the release of arachidonic acid. *PLA₂* phospholipase A₂; *PLC* phospholipase C; *PC* phosphatidylcholine; *transferase* CTP-phosphatidate: cytidyl transferase and CDP-1,2-diacylglycerolinositol phosphatidyl transferase; *PI* phosphatidylinositol; *PE* phosphatidylethanolamine; *PA* phosphatidic acid; and AA-CoA, arachidonyl CoA. ⤒ stimulation by Ca^{2+} and ⤓ inhibition by Ca^{2+}

and DeRubertis 1983; Diegel et al. 1980; Jesse and Franson 1979; Lapetina et al. 1978b), the importance of the Ca^{2+}-dependent phospholipases is evident.

The enzymatic pathways proposed for the release of arachidonic acid from phospholipids are summarized in Fig. 2. A traditional pathway involves phospholipase A₂ (Flower and Blackwell 1976; Lapetina et al. 1978b; Silver et al. 1978; Van den Bosch 1980; van Dorp et al. 1964), which catalyzes the direct release of arachidonic acid from the 2-position of several phospholipids in a Ca^{2+}-dependent manner. More recently, a 2-step pathway involving a phospholipase C specific for phosphatidylinositol has been proposed (Bell et al. 1979; Bell and Majerus 1980; Bell et al. 1980; Irvine et al. 1979); the enzyme converts phosphatidylinositol to a diacylglycerol, which is further degraded by a diglyceride lipase (Rittenhouse-Simmons 1980) to glycerol, stearic acid, and arachidonic acid. A third pathway involves phosphorylation of the diacylglycerol (generated by phospholipase C) by a diglyceride kinase to yield phosphatidic acid, which is then acted on by a phosphatidic acid-specific phospholipase (Billah et al. 1981; Broekman et al. 1980; Lapetina et al. 1981) to yield arachidonic acid and lysophosphatidic acid. In those proposed pathways producing phosphatidic acid, Ca^{2+} has been shown to inhibit its re-synthesis to phosphatidylinositol (Broekman et al. 1980; Lapetina et al. 1981).

The effect of ionophores on arachidonic acid metabolism has been examined. In the presence of 5 mM EGTA, the Ca^{2+} ionophore A23187 induced the release of arachidonic acid from platelets (Lapetina et al. 1978a; Pickett et al. 1977; Rittenhouse-Simmons 1981, 1977) suggesting that mobilization of intracellular Ca^{2+} was sufficient to initiate the release of arachidonic acid. Knapp et al. (1977) reported a similar observation in renal medulla.

Rittenhouse-Simmons and Deykin (1978) found that thrombin-induced arachidonic acid release was inhibited by dibutyryl cAMP and by 8-(N,N diethylamine)octyl 3,4,5 trimethoxybenzoic acid, a putative intracellular Ca^{2+} antagonist. The addition of Ca^{2+} did not reverse the inhibition, unless A23187 was also present. Dibutyryl cAMP inhibited arachidonic acid release, probably by reducing intracellular Ca^{2+} levels. In a variety of other cells, however, dibutyryl cAMP appears to stimulate prostaglandin synthesis.

Michell (1975) has associated the elevation of cytoplasmic Ca^{2+} concentration with phosphatidylinositol-turnover in the platelet; thrombin induces this response, which increases intracellular Ca^{2+} levels. Serhan et al. (1981) postulated that phosphatidic acid, an intermediate of phosphatidylinositol-turnover, could act as a Ca^{2+} ionophore. Imai et al. (1982, 1983) later showed that the conversion of diacylglyceride to phosphatidic acid in thrombin-activated platelets was coincident with Ca^{2+} influx, activation of phospholipase A_2 and arachidonic acid release. They argued that the phosphatidic acid acted as an endogenous ionophore rather than as a source of arachidonic acid.

The relative importance of the enzymes involved in arachidonate release has not been established, and will be discussed briefly. The substrate specificity of phospholipase A_2 has been studied extensively, and appears to vary depending on the class of phospholipids and the fatty acid at the sn-2-position (Bills et al. 1976, 1977; Blackwell et al. 1977; Blackwell 1978; Derksen and Cohen 1975; Schoene 1978; Vogt 1978). Bills et al. (1977) found that in human platelets pre-labeled with arachidonic acid, thrombin stimulation resulted in a loss of radioactivity from phosphatidylcholine (PC) and phosphatidylinositol (PI). Others reported that arachidonate was released equally from PC, PI and phosphatidylethanolamine (PE) (Lapetina et al. 1978 b). However, platelets treated with A23187 released arachidonic acid only from PC and PE (Lapetina et al. 1978a). The relative amounts of radioactive arachidonic acid incorporated into lipid vary greatly among lipid classes (Blackwell 1978), as do the specific radioactivities of each lipid class, making it difficult to assess the relative contribution of each lipid. In addition, different cell types have different fatty acid composition in their phospholipids which contribute differently to the arachidonic acid released. For example, PE is hydrolyzed in rabbit (Blackwell 1978) and horse platelets (Lapetina et al. 1978a), but not in human (Bills et al. 1976, 1977; van Dorp et al. 1964).

In other studies, Bills et al. (1976) reported that in stimulated platelets which had been pre-labeled with various unsaturated fatty acids, PC was the only phospholipid hydrolyzed, suggesting the presence of a phospholipase A_2 specific for arachidonyl-phospholipid, a finding confirmed by others (Marcus et al. 1969). Such specificity with respect to the acyl chain is not generally found among phospholipases of the A_2 type; perhaps compartmentalization of substrate, or of PLA_2, or both, may explain why only PC was hydrolyzed.

Walenga et al. (1981) demonstrated in thrombin-stimulated platelets that phospholipase A_2 accounted for virtually all of the arachidonic acid released and converted to prostaglandins. McKean et al. (1981) found that a significant amount of lysophosphatidylcholine accumulated within 1 min following platelet activation; the increase of lysophosphatidylcholine was comparable to the decrease in phosphatidylcholine, indicating the quantitative importance of the phospholipase A_2 pathway in the release of arachidonic acid.

Some argue that phospholipase A_2 plays a minor role in arachidonic acid release because of its apparently low activity in platelet extract (Bell et al. 1979; Bell and Majerus 1980; Bell et al. 1980). A possible explanation for the low phospholipase A_2 activity in platelet extract is the recent observation of an endogenous inhibitor of the enzyme in such preparations (Ballou and Cheung 1983). The inhibitor has since been identified as a mixture of unsaturated fatty acids (Ballou and Cheung 1984), the removal of which results in a marked increase in phospholipase A_2 activity (Ballou and Cheung 1983). Franson et al. (1974) previously reported that polymorphonuclear leukocyte phospholipase A_2 was inhibited by unsaturated fatty acids. Thus, the apparent low phospholipase A_2 activity in vitro may have led some to underestimate its potential role in contribution to the release of arachidonic acid in vivo.

An alternative pathway for the generation of arachidonic acid in platelets involves the sequential action of two enzymes. Rittenhouse-Simmons (1979) demonstrated the existence of a phosphatidylinositol-specific phospholipase C in human platelets. She showed that, upon stimulation, platelet accumulated a 30-fold increase in diglycerides within 5 s. The accumulation was transient; the diglycerides returned to a basal level by 2 min. Platelets pre-labeled with arachidonic acid lost a significant amount of radioactivity from PI, indicating that the diglyceride was derived from it (Mauco et al. 1979). A similar enzyme has been described in other tissues (Hofman and Majerus 1982; Irvine et al. 1979).

Subsequently, Bell et al. (1979) and Rittenhouse-Simmons (1980) found a Ca^{2+}-dependent diglyceride lipase associated with a platelet particulate fraction. Bell et al. (1979) suggested that the activity level of this enzyme was sufficient to account for initial accumulation of diglyceride and the subsequent release of arachidonic acid in activated platelets. The suggestion was based on the apparent V_{max} of the enzyme, not its total measurable activity. Others argue that the diglyceride is enzymatically converted to phosphatidic acid by diglyceride kinase (Billah et al. 1981; Broekman et al. 1980; Lapetina et al. 1981), suggesting that the diglyceride does not serve as a substrate for diglyceride lipase in vivo. Walenga et al. (1981) and others (Billah et al. 1980, 1981; Billah and Lapetina 1982; Broekman et al. 1980, 1981; Lapetina and Cuatrecasas 1973; Lapetina et al. 1978a) showed that in thrombin-stimulated platelets, a vast majority of the diglyceride produced by phospholipase C was phosphorylated to phosphatidic acid. This would deprive the diglyceride lipase of its substrate, making it unlikely to play a major role in the release of arachidonic acid. In an apparent contradiction, however, Broekman et al. (1980) have shown that diglyceride kinase is inhibited by Ca^{2+}.

A phospholipase A_2 specific for phosphatidic acid has been isolated from platelets (Billah et al. 1981; Lapetina et al. 1981). The observation that diglycerides are phosphorylated to phosphatidic acid prior to the release of arachidonate has led some (Billah et al. 1981; Lapetina et al. 1981) to suggest that this phosphatidic acid-specific PLA_2 may play a significant role in the release of arachidonate. However, like the nonspecific PLA_2, its in vitro activity appears insufficient to account for the amount of arachidonate released in vivo.

Evidence supporting each proposed pathway deserves careful evaluation. The apparent high phospholipase C-diglyceride lipase activity in broken cell preparations would favor a major role. However, phospholipases of the A_2 type appear to be membrane-associated, while phospholipase C is soluble; the accessibility of these enzymes to various substrates may be quite different in the extract than in intact cells. Phospholipase A_2 activity has been reported to be enhanced by disruption of the plasma membrane (Blackwell 1978), but unsaturated fatty acids associated with membrane components act as potent inhibitors of phospholipase A_2 activity. While phospholipase C activity can be simulated by unsaturated fatty acids (Hofmann and Majerus 1982; Irvine et al. 1979), the same substances inhibit phospholipase A_2, making it impossible to accurately assess the relative contributions of these enzymes in vitro. Thus, extrapolation of the in vitro activities of phospholipase A_2, phospholipase C, and diglyceride lipase to whole cells may not be warranted.

Perhaps a more realistic concept is one in which a combination of the proposed pathways contributes to arachidonate release. Craven and DeRubertis (1983) have argued that in renal medulla several phospholipases may be involved in arachidonate release, which may also be true in other tissues.

The Ca^{2+}-dependence of the phospholipases has been well established; but the relative sensitivity of these enzymes to Ca^{2+} remains unclear. The concentration of Ca^{2+} giving maximum activity of phospholipase A_2 from various tissues varies greatly. Many estimations are in the millimolar range, and would appear nonphysiological since the cellular level of Ca^{2+} under steady state is approximately 10^{-7} M. In renal medulla, phospholipase C appears to be sensitive to micromolar Ca^{2+}; phospholipase A_2, however, to millimolar Ca^{2+} (Craven and DeRubertis 1983). We have found that phospholipase A_2 from human platelets is activated by micromolar concentrations of Ca^{2+} (unpublished) as is the phosphatidic acid-specific PLA_2 (Lapetina et al. 1981)

5 Does Calmodulin Stimulate Arachidonic Acid Release?

The Ca^{2+}-dependent nature of the phospholipases has led some investigators to examine the effect, if any, of calmodulin (Cheung 1980) on the activity of the phospholipases. Wong and Cheung (1979) reported a slight activation by calmodulin of PLA_2 associated with a membrane fraction of human platelets. However, when the enzyme was later purified in a soluble form, the sensitivity to calmodulin was not detected (Ballou and Cheung 1983), suggesting that the effect of calmodulin was indirect. A highly purified pancreatic phospholipase A_2 was also insensitive to calmodulin (Withnall and Brown 1982). Conversely, Moskowitz et al. (1982, 1983) showed that calmodulin slightly activated PLA_2 in brain synaptic vesicles and snake venom.

Walenga et al. (1981) studied the possible effect of calmodulin with phospholipase A_2 and phospholipase C in human platelets. They noted that trifluoperazine (TFP), a calmodulin antagonist, inhibited the release of arachidonic acid from phosphatidylcholine while stimulating its release from phosphatidylinositol with accumulation of arachidonyl-phosphatidic acid. The differential effects of TFP on the two enzymes appeared in accord with the notion that phospholipase A_2 is calmodulin-activated. However, we have found that TFP has no inhibitory effect on purified platelet PLA_2 activity (Ballou and Cheung 1983).

Craven and DeRubertis (1983) showed that TFP suppressed arachidonic acid release and PGE synthesis in renal medullary tissue. W-7, another calmodulin antagonist, inhibited arachidonic acid release by the particulate forms of phospholipase A_2 and C, but not soluble phospholipase C, implying that both particulate enzymes were calmodulin-activated. It is not known whether soluble and particulate phospholipase C's are different enzymes. Irvine and Dawson (1978) argue that phospholipase C is a soluble enzyme and any activity associated with the membrane is artifactual. Since TFP and W-7 seemed to affect only the particulate enzymes, the possibility remains that the inhibition resulted from nonspecific hydrophobic interactions on the membrane. Another perplexing finding is the increased synthesis of thromboxane A_2 by exogenous calmodulin in human platelets (Wong and Cheung 1979, 1982).

Wong et al. (1980) originally noted that 15-hydroxyprostaglandin dehydrogenase, an enzyme involved in the degradation of all prostaglandins, was inhibited by calmodulin in a calcium-dependent manner. Further studies in our laboratory (Suzuki, Padh and Cheung, unpublished experiment) suggested that the inhibition was independent of Ca^{2+}, and was observed only with high concentrations of calmodulin ($20\,\mu M$ or higher).

From the available evidence, it is not possible to conclude that calmodulin directly stimulates arachidonic acid release by interacting with phospholipases. More work with purified enzymes from various tissues is required before phospholipases or other prostaglandin enzymes can be added to the list of calmodulin-dependent enzymes.

6 Conclusions

Prostaglandins are produced by mammalian cells in response to various stimuli. Their biosynthesis is preceded by the release from membrane phospholipids of arachidonic acid, the sole substrate for prostaglandin synthesis. Several pathways have been proposed for the release of arachidonic acid catalyzed by phospholipases; however, it is difficult to quantitate the relative contribution of each pathway. The activities of the phospholipases are Ca^{2+}-dependent. This implies that the release of arachidonic acid is dependent on the intracellular level of Ca^{2+}. The increase of intracellular Ca^{2+} is probably mediated by increased calcium gating induced by receptor-mediated alteration in membrane permeability. Activation of the phospholipases leads to the release of arachidonic acid, which is converted to prostaglandins and thromboxanes apparently independent of Ca^{2+}. Thus, one major role of Ca^{2+} in the regulation of the biosynthesis of prostaglandins and thromboxanes would appear to be the activation of the enzymes responsible for arachidonic acid release.

The potential role of calmodulin in the Ca^{2+} regulation of prostaglandin metabolism has been examined, but the evidence appears fragmentary and in most cases, inconclusive.

Acknowledgements. We thank Dr. K.U. Malik for reviewing the manuscript. The work in our laboratory has been supported by Cancer Center Support grant CA 21765, by project grants NS 08059 and GM 28178 from the National Institutes of Health and by ALSAC. L.R.B. is a recipient of National Research Service Award CA 09346.

References

Ballou LR, Cheung WY (1983) Marked increase of phospholipase A_2 activity in vitro and demonstration of an endogenous inhibitor. Proc Natl Acad Sci USA 80: 5203-5207

Ballou LR, Cheung WY (1984) Unsaturated fatty acids inhibit human platelet phospholipase A_2 activity. Fed Proc 43: 1463

Bell RL, Kennerly DA, Stanford N, Majerus PW (1979) Diglyceride lipase: a pathway for arachidonate release from human platelets. Proc Natl Acad Sci USA 76: 3238-3241

Bell RL, Majerus PW (1980) Thrombin-induced hydrolysis of phosphatidylinositol in human platelets. J Biol Chem 255: 1790-1792

Bell RL, Stanford N, Kennerly D, Majerus PW (1980) Diglyceride lipase: a pathway for arachidonate release from human platelets. Adv Prostaglandin Thromboxane Res 6: 219-224

Bergström S, Danilesson H, Samuelsson B (1964) The enzymatic formation of prostaglandin E_2 from arachidonic acid. Prostaglandin and related factors. Biochim Biophys Acta 90: 207-210

Berridge MJ (1980) Receptors and calcium signaling. Trends Pharmacol Sci 1: 419-424

Berridge MJ (1981) Phosphatidylinositol hydrolysis — a multifunctional transducing mechanism. Mol Cell Endocr 24: 115-140

Berridge MJ (1982) A novel cellular signaling system based on the integration of phospholipid and calcium metabolism. In: Cheung WY (ed) Calcium and Cell Function, vol III. Academic, New York, pp 1-36

Billah MM, Lapetina EG, Cuatrecasas P (1980) Phospholipase A_2 and phospholipase C activities of platelets: differential substrate specificity, Ca^{2+} requirement, pH dependence and cellular localization. J Biol Chem 255: 10227-10231

Billah MM, Lapetina EG, Cuatrecasas P (1981) Phospholipase A_2 activity specific for phosphatidic acid: a possible mechanism for the production of arachidonic acid in platelets. J Biol Chem 256: 5399-5403

Billah MM, Lapetina EG (1982) Formation of lysophosphatidylinositol in platelets stimulated with thrombin or ionophore A23187. J Biol Chem 257: 5196-5200

Bills TK, Smith JB, Silver MJ (1976) Metabolism of [^{14}C]arachidonic acid by human platelets. Biochim Biophys Acta 424: 303-314

Bills TK, Smith JB, Silver MJ (1977) Selective release of arachidonic acid from the phospholipids of human platelets in response to thrombin. J Clin Invest 60: 1-6

Blackwell GJ, Duncombe WG, Flower RJ, Parsons MF, Vane JR (1977) The distribution and metabolism of arachidonic acid in rabbit platelets during aggregation and its modification by drugs. Br J Pharmacol 59: 353-366

Blackwell GJ (1978) Phospholipase A_2 and platelet aggregation. Adv Prostaglandin Thromboxane Res 3: 137-142

Broekman MJ, Ward JW, Marcus AJ (1980) Phospholipid metabolism in stimulated human platelets: changes in phosphatidylinositol, phosphatidic acid and lysophospholipids. J Clin Invest 66: 275-283

Broekman MJ, Ward JW, Marcus AJ (1981) Fatty acid composition of phosphatidylinositol and phosphatidic acid in stimulated platelets. Persistence of arachidonyl-stearyl structure. J Biol Chem 256: 8271-8274

Cheung WY (1980) Calmodulin plays a pivotal role in cellular regulation. Science (Wash DC) 207: 19-27

Craven PA, DeRubertis FR (1983) Ca^{2+} Calmodulin-dependent release of arachidonic acid for renal medullary prostaglandin synthesis: evidence for involvement of phospholipases A_2 and C. J Biol Chem 258: 4814-4823

Diegel J, Cunningham M, Coburn RF (1980) Calcium dependence of prostaglandin release from the guinea pig taenia coli. Biochim Biophys Acta 619: 482-493

Derksen A, Cohen P (1975) Patterns of fatty acid release from endogenous substrates by human platelet homogenates and membranes. J Biol Chem 250: 9342-9347

Fain JN (1982) Involvement of phosphatidylinositol breakdown in elevation of cytosol Ca^{2+} by hormone and relationship to prostaglandin formation. Horiz Biochem Biophys 6: 237-276

Flower RJ (1979) Prostaglandins and related compounds. In: Vane JR, Ferriera JH (eds) Anti-Inflammatory Drugs. Springer, Berlin Heidelberg New York, pp 374-422 (Handbook of Experimental Pharmacology, vol 50/I)

Flower RJ, Blackwell GJ (1976) The importance of phospholipase-A_2 in prostaglandin biosynthesis. Biochem Pharmacol 25: 285-291

Franson R, Patriarcha P, Elsbach P (1974) Phospholipid metabolism by phagocytic cells. Phospholipase A_2 associated with rabbit polymorphonuclear leukocyte granules. J Lipid Res 15: 380-388

Gerrard JM (1979) Lysophosphatidic acids. Influence on platelet aggregation and intracellular calcium flux. Am J Pathol 96: 423-438

Hamberg M, Samuelsson B (1967) On the mechanism of the biosynthesis of prostaglandins E_1 and $F_{1\alpha}$. J Biol Chem 242: 5336-5342

Hamberg M, Samuelsson B (1973) Detection and isolation of an endoperoxide intermediate in prostaglandin biosynthesis. Proc Natl Acad Sci USA 70: 899-903

Hamberg M, Samuelsson B (1974a) Prostaglandin endoperoxides. Novel transformation of arachidonic acid in human platelets. Proc Natl Acad Sci USA 71: 3400-3404

Hamberg M, Samuelsson B (1974b) Prostaglandin endoperoxides VII. Novel transformations of arachidonic acid in guinea pig lung. Biochem Biophys Res Commun 61: 942-949

Hamberg M, Svensson J, Samuelsson B (1975) Thromboxanes: a new group of biologically active compounds derived from prostaglandin endoperoxides. Proc Natl Acad Sci USA 72: 2994-2998

Hawthorne JN, Pickard MR (1979) Phospholipids in synaptic function. J Neurochem 32: 5-14

Hinman JW (1972) Prostaglandins. Annu Rev Biochem 41: 161-178

Hofmann SL, Majerus PW (1982) Modulation of phosphatidylinositol-specific phospholipase C activity by phospholipid interaction, diglyceride, and calcium ions. J Biol Chem 257: 14359-14364

Imai A, Ishizuka Y, Kawai K, Nozawa Y (1982) Evidence for coupling of phosphatidic acid formation and calcium influx in thrombin-activated human platelets. Biochem Biophys Res Commun 108: 752-759

Imai A, Nakashima S, Nozawa Y (1983) The rapid polyphosphoinositide metabolism may be a triggering event for thrombin-mediated stimulation of human platelets. Biochim Biophys Res Commun 110: 108-115

Irvine RF, Dawson RMC (1978) Is there a membrane-bound, Ca^{2+}-dependent phosphatidylinositol phosphodiesterase in rat brain? Biochem Soc Trans 6: 1020-1021

Irvine RF, Hemington N, Dawson RMC (1979) The calcium-dependent phosphatidylinositol-phosphodiesterase of rat brain. Eur J Biochem 99: 525-530

Irvine RF, Letcher AJ, Dawson RMC (1979) Fatty acid stimulation of membrane phosphatidylinositol hydrolysis by brain phosphatidylinositol phosphodiesterase. Biochem J 178: 497-500

Jesse RL, Franson RC (1979) Modulation of purified phospholipase A_2 activity from human platelets by calcium and indomethacin. Biochim Biophys Acta 575: 467-470

Knapp HR, Olez O, Roberts LJ, Sweetman BJ, Oates JA, Reed PW (1977) Ionophores stimulate prostaglandin and thromboxane biosynthesis. Proc Natl Acad Sci USA 74: 4251-4255

Lands WEM (1979) The biosynthesis and metabolism of prostaglandins. Annu Rev Physiol 41:
    633-652
Lapetina EG, Cuatrecasas P (1979) Stimulation of phosphatidic acid production in platelets pre-
    cedes the formation of arachidonate and parallels the release of serotonin. Biochim Biophys
    Acta 573: 394-402
Lapetina EG, Michell RH (1973) A membrane-bound activity catalyzing phosphatidylinositol
    breakdown to 1,2-diacylglycerol, D-myo-inositol 1,2 cyclic phosphate and D-myo-inositol-1-
    phosphate. Biochem J 131: 433-442
Lapetina EG, Chandrabose KA, Cuatrecasas P (1978a) Ionophore A23187- and thrombin-induced
    platelet aggregation: independence from cyclooxygenase products. Proc Natl Acad Sci USA 75:
    818-822
Lapetina EG, Schmitges CJ, Chandrabose K, Cuatrecasas P (1978b) Regulation of phospholipase ac-
    tivity in platelets. Adv Prostaglandin Thromboxane Res 3: 127-135
Lapetina EG, Billah MM, Cuatrecasas P (1981) The phosphatidylinositol cycle and the regulation
    of arachidonic acid production. Nature (Lond) 292: 367-369
Marcus AJ, Ullman HL, Safier LB (1969) Lipid composition of subcellular particles of human blood
    platelets. J Lipid Res 10: 108-114
Mauco G, Chap H, Douste-Blazy L (1979) Characterization and properties of a phosphatidylinositol
    phosphodiesterase (phospholipase C) from platelet cytosol. FEBS Lett 100: 367-370
McKean ML, Smith JB, Silver MJ (1981) Formation of lysophosphatidylcholine by human platelets
    in response to thrombin. J Biol Chem 256: 1522-1524
Michell RH (1975) Inositol phospholipids and cell surface receptor function. Biochim Biophys Acta
    415: 81-147
Michell RH (1977) The possible involvement of phosphatidylinositol breakdown in the mechanism
    of stimulus-response coupling at receptors which control cell-surface calcium gates. Adv Exp
    Med Biol 83: 447-464
Michell RH (1979) Inositol phospholipids in membrane function. Trends Biochem Sci 4: 128-131
Michell RH, Kirk CJ (1981a) Why is phosphatidylinositol degraded in response to stimulation of
    certain receptors? Trends Pharmacol Sci 2: 86-89
Michell RH, Kirk CJ (1981b) Studies of receptor-stimulated inositol lipid metabolism should focus
    upon measurement of inositol lipid breakdown. Biochem J 198: 247-248
Michell RH, Jones LM, Jafferji SS (1977) A possible role for phosphatidylinositol breakdown in
    muscarinic cholinergic stimulus-response coupling. Biochem Soc Trans 5: 77-81
Moskowitz N, Schook W, Puszkin S (1982) Interaction of brain synaptic vesicles induced by endo-
    genous Ca^{2+}-dependent phospholipase A_2. Science (Wash DC) 216: 305-307
Moskowitz N, Shapiro L, Schook W, Puszkin S (1983) Phospholipase A_2 modulation by calmodulin,
    prostaglandins, and cyclic nucleotides. Biochem Biophys Res Commun 115: 94-99
Nugteren DH, Hazelhof D (1973) Isolation and properties of intermediates in prostaglandin bio-
    synthesis. Biochim Biophys Acta 326: 448-461
Pickett WC, Jesse RL, Cohen P (1977) Initiation of phospholipase A_2 activity in human platelets
    by the calcium ionophore A23187. Biochim Biophys Acta 486: 209-213
Qureshi Z, Cagen LM (1982) Prostaglandin $F_{2\alpha}$ produced by rabbit renal slices is not a metabolite
    of prostaglandin E_2. Biochem Biophys Res Commun 104: 1255-1263
Rittenhouse-Simmons S (1979) Production of diglyceride from phosphatidylinositol in activated
    human platelets. J Clin Invest 63: 580-587
Rittenhouse-Simmons S (1980) Indomethacin-induced accumulation of digylceride in activated
    human platelets. J Biol Chem 255: 2259-2262
Rittenhouse-Simmons S (1981) Differential activation of platelet phospholipase by thrombin and
    ionophore A23187. J Biol Chem 256: 4153-4155
Rittenhouse-Simmons S (1977) The mobilization of arachidonic acid in platelets exposed to
    thrombin or ionophore A23187. J Clin Invest 60: 495-498
Rittenhouse-Simmons S, Deykin D (1978) The activation by Ca^{2+} of platelet phospholipase A_2:
    effects of dibutyryl cyclic adenosine monophosphate and 8-(N-N-diethylamino)octyl-3,4,5-
    trimethoxybenzoate. Biochim Biophys Acta 543: 409-422
Rubin RP, Kelly KL, Halenda SP, Laychock SG (1982) Arachidonic acid metabolism in rat pan-

creatic acinar cells: calcium-mediated stimulation of the lipoxygenase system. Prostaglandins 24: 179-193

Samuelsson B, Goldyne M, Granström E, Hamberg M, Hammarström S, Malmsten C (1978) Prostaglandins and thromboxanes. Annu Rev Biochem 47: 997-1029

Samuelsson B, Grantström E, Green K, Hamberg M, Hammarström S (1975) Prostaglandins. Annu Rev Biochem 44: 669-695

Schoene NW (1978) Properties of platelet phospholipase A_2. Adv Prostaglandin Thromboxane Res 3: 121-126

Serhan C, Anderson P, Goodman E, Dunham E, Dunham P, Weissman G (1981) Phosphatidate and oxidized fatty acids are calcium ionophores. J Biol Chem 256: 2736-2741

Silver MJ, Bills TK, Smith JB (1978) Platelets and prostaglandins: the key role of platelet phospholipase A_2 activity. In: deGaetano G, Gatalini S (eds) Platelets: A Multidisciplinary Approach. Raven, New York, pp 213-225

Van den Bosch H (1980) Intracellular phospholipases A. Biochim Biophys Acta 604: 191-246

van Dorp DA, Beerthius RK, Nugteren DH, Vonkeman H (1964) Enzymatic conversion of all-cis-polyunsaturated fatty acids into prostaglandins. Nature (Lond) 203: 839-843

Vogt W (1978) Role of phospholipase A_2 in prostaglandin formation. Adv Prostaglandin Thromboxane Res 3: 89-95

Walenga RW, Opas EE, Feinstein MB (1981) Differential effects of calmodulin antagonists on phospholipase A_2 and C in thrombin-stimulated platelets. J Biol Chem 256: 12523-12528

Withnall MT, Brown TJ (1982) Pancreatic phospholipase A_2 is not regulated by calmodulin. Biochem Biophys Res Commun 106: 1049-1055

Wong PY-K, Cheung WY (1979) Calmodulin stimulates human platelet phospholipase A_2. Biochem Biophys Res Commun 90: 473-480

Wong PY-K, Cheung WY (1982) Calmodulin stimulates thromboxane synthesis in human platelets: studies with thromboxane synthetase inhibitors. Adv Lipid Res 19: 447-452

Wong PY-K, Lee WH, Chao PH-W, Cheung WY (1980) The role of calmodulin in prostaglandin metabolism. Ann NY Acad Sci 356: 179-189

Calcium Regulation of Insulin Release

W. J. Malaisse, P. Lebrun and A. Herchulez[1]

CONTENTS

1 Introduction

It will soon be 20 years since it was proposed that Ca^{2+} plays a major role in the process of glucose-stimulated insulin release from the pancreatic B-cell (Grodsky and Bennett 1966; Milner and Hales 1967). This view is now confirmed by a large body of evidence. At its simplest, it is presently believed that nutrient (e.g., D-glucose or L-leucine) or non-nutrient (e.g., hypoglycemic sulfonylureas) secretagogs induce a rise in the cytosolic concentration of ionized free Ca^{2+} and, by doing so, activate, either directly or indirectly (e.g., through calmodulin), the machinery mediating the translocation and exocytosis of secretory granules (Malaisse et al. 1978a; Wollheim and Sharp 1981). Although postulated for many years, the rise in cytosolic free Ca^{2+} concentration has only recently been documented (Wollheim et al. 1983). Thus, both nutrient (D-glyceraldehyde and L-glutamine) and non-nutrient secretagogs (e.g., high extracellular K^+) were found to increase cytosolic Ca^{2+}, as measured by the fluorescent Ca^{2+} indicator

[1] Laboratories of Experimental Medicine and Pharmacology, Brussels Free University School of Medicine, Brussels, Belgium

Quin 2, in an insulin-producing cell line (RINm5F). It is not the aim of the present chapter to review the steps which led to the discovery of the role played by Ca^{2+} in insulin release. This has been the subject of several reviews in recent years (Wollheim and Sharp 1981; Hellman et al. 1979; Herchuelz and Malaisse 1982). We intend here to focus attention on (1) the regulation of Ca^{2+} fluxes in the pancreatic B-cell and (2) the effects of Ca^{2+} on target systems in the insulin-producing cell. Although this re-view is restricted to the process of insulin release, it should be realized that the con-cepts under consideration may also apply, within limits, to a number of other hor-monal secretory processes.

2 Regulation of Ca^{2+} Fluxes

Theoretically, insulin secretagogs may increase the cytosolic free Ca^{2+} concentration by at least three mechanisms, i.e., (1) by increasing Ca^{2+} inflow into the islet cells, (2) by decreasing Ca^{2+} outflow from such cells, and (3) by mobilizing Ca^{2+} from binding or sequestration sites or inhibiting its uptake by intracellular organelles. The available evi-dence indeed suggests that these three modalities may participate in the regulation of cytosolic Ca^{2+} concentration. We wish to illustrate this concept, using the process of glucose-induced insulin release as our leading example.

2.1 Stimulation of Ca^{2+} Inflow

Glucose is currently thought to increase Ca^{2+} inflow into the B-cell by gating voltage-sensitive Ca^{2+} channels (Donatsch et al. 1977). This would result from depolarization of the plasma membrane itself attributable to a decrease in K^+ permeability (Henquin 1978a; Atwater 1980; Malaisse and Herchuelz 1982). Both electrophysiological and radioisotopic observations support such a concept.

2.1.1 Electrophysiological Observations

In the absence of any secretagog or at very low glucose concentration, the B-cell dis-plays a stable membrane potential of about -60 mV (Atwater et al. 1978; Meissner et al. 1980). Under the latter condition, the membrane potential depends mainly, but not exclusively, on the membrane K^+ permeability (Atwater et al. 1978; Meissner et al. 1978). When exposed to a concentration of glucose which stimulates insulin release (e.g., 11.1 mM), the B-cell displays bioelectrical activity characterized by the regular alternation of depolarized phases with bursts of spikes and silent polarized periods (Atwater 1980). The view that the pancreatic B-cell may be equipped with voltage-sen-sitive Ca^{2+}-channels is evidenced by the fact that the excess noise of the membrane potential, induced by K^+ (50 mM) depolarization, is inhibited at low extracellular Ca^{2+} concentration and abolished by Co^{2+} or Mn^{2+} (Atwater et al. 1981) which specifically block voltage-sensitive Ca^{2+} channels in nerves (Baker et al. 1973). Under glucose stimulation, the gating of voltage-sensitive Ca^{2+} channels is thought to occur during the ascending phase of the spikes and probably also during the rapid depolarization seen at the start of each burst (Atwater et al. 1982; Meissner and Preissler 1979). Indeed, during the active phase of the burst, the input resistance is decreased (Atwater et al. 1982). Furthermore, a continuous depolarizing current stimulates, while a continuous hyperpolarizing current inhibits the frequency and the amplitude of the spikes (Atwater et al. 1982). Although such currents also alter the amplitude of the underlying potential oscillation (the burst pattern), they fail to affect their time course (Atwater et al. 1982). This is in agreement with the view that the burst pattern also involves cyclic variations in K^+ permeability (Atwater 1980; Meissner and Preissler 1979; Ribalet and Beigelman 1979).

2.1.2 Radioisotopic Observations

The view that glucose may increase Ca^{2+} inflow by gating voltage-sensitive Ca^{2+} channels was also assessed by the measurement of the sensitivity toward verapamil of the inflow of Ca^{2+} induced by the sugar. Organic Ca^{2+}-antagonists like verapamil act preferentially on voltage-sensitive Ca^{2+} channels and have been successfully used in smooth muscles to identify distinct modalities of Ca^{2+} inflow (Bolton 1979; Meisheri et al. 1981). With this approach, it was observed that the increase in Ca^{2+} inflow induced by a rise in the glucose concentration from an intermediate (8.3 mM) to a high value (16.7 mM) displayed the same sensitivity toward verapamil as the increase in Ca^{2+} inflow observed in response to membrane depolarization by high extracellular K^+ (Lebrun et al. 1982a). This confirms that glucose may increase Ca^{2+} inflow into the B-cell by gating voltage-sensitive Ca^{2+} channels. A similar conclusion was reached in the case of hypoglycemic sulfonylureas, at least when used in the absence of glucose (Lebrun et al. 1982b). In these studies it was also disclosed that the inflow of Ca^{2+} stimulated by either an increase in the concentration of glucose from zero to 8.3 mM (Lebrun et al. 1982a) or by arginine (Herchuelz et al. 1984) displayed a lower sensitivity toward verapamil than that evoked by K^+. The latter finding provides evidence for the existence in the B-cell of a second modality of Ca^{2+} transport characterized by a low sensitivity to verapamil and possibly coinciding with voltage-insensitive Ca^{2+} channels. Arginine may stimulate Ca^{2+} inflow by gating voltage-insensitive Ca^{2+} channels, while glucose may act by gating both voltage-sensitive and voltage-insensitive Ca^{2+} channels (Lebrun et al. 1982a; Herchuelz et al. 1984).

2.2 Inhibition of Ca^{2+} Outflow

The view that glucose may decrease Ca^{2+} outflow from the B-cell was initially inferred from the observation that the sugar reduced ^{45}Ca outflow from prelabeled islets (Malaisse et al. 1973). Such a decrease may result from the inhibition by glucose of a process of Na^+-Ca^{2+} counter-transport across the plasma membrane. This process may account for about 70% of the outflow of Ca^{2+} from the B-cell (Herchuelz et al. 1980).

It was proposed, but remains to be proved, that the inhibition of Na^+-Ca^{2+} counter-transport is linked to an increased production of protons in glucose-stimulated islets, with competition between H^+ and Ca^{2+} for a common ionophoretic system mediating the counter-transport process (Anjaneyulu et al. 1980). This could lead to an increase in cytosolic free Ca^{2+} (Malaisse et al. 1973; Kikuchi et al. 1978, Siegel et al. 1980). A Ca^{2+}-responsive ATPase activity has been identified in highly purified plasma membrane fraction from islet cells (Pershadsingh et al. 1980). Although such an ATPase may participate to the regulation of cytosolic free Ca^{2+} (Pershadsingh et al. 1980), no evidence is available to suggest that glucose inhibits such an ATPase. It is also conceivable that the glucose-induced decrease in ^{45}Ca outflow reflects stimulation of Ca^{2+} sequestration in intracellular organelles (see below).

2.3 Intracellular Redistribution of Ca^{2+}

Histochemical studies and radioisotopic measurements in subcellular fractions indicate that, in intact islets, glucose causes the accumulation of Ca^{2+} in various organelles (Herman et al. 1973; Ravazzola et al. 1976; Kohnert et al. 1979). Within 3-5 min exposure to the sugar, an accumulation of Ca^{2+} is also observed at the inner face of the B-cell membrane (Klöppel and Bommer 1979). Whether glucose increases the total Ca content of islet cells remains a matter of debate (Malaisse et al. 1978; Wolters et al. 1982). It was both proposed (Wollheim and Sharp 1981; Wollheim et al. 1978) and denied (Somers et al. 1979b; Henquin 1978b) that the first phase of glucose-induced insulin release depends more on the release of Ca^{2+} from intracellular binding sites than on the availability of extracellular Ca^{2+}. Although still speculative, the concept of a nutrient-induced intracellular redistribution of Ca^{2+} is compatible with the observation that 2-ketoisocaproate and D-glyceraldehyde stimulate ^{45}Ca release from islets prelabeled with ^{45}Ca in the absence of glucose — in which case the cellular ^{45}Ca is apparently more readily releasable — and perifused in the absence of extracellular Ca^{2+} in order to prevent the apparent exchange between influent ^{40}Ca and effluent ^{45}Ca (Lebrun et al. 1982c). Moreover, it was recently proposed that the latter exchange involves a process of Ca^{2+}-stimulated Ca^{2+} release (Scholler et al. 1984a).

2.4 Ca^{2+}-ATPase and Ca^{2+} Sequestration

Ca^{2+}-activated ATPase activity was first identified in subcellular fractions of mouse pancreatic islets by Formby et al. (1976) and further examined in rat islet homogenates by Kasson and Levin (1981) and Owen et al. (1983). These studies were generally conducted at very low Mg^{2+} concentrations, such as those attributable to contamination by the islet material or other reagents (Kasson and Levin 1981; Owen et al. 1983). Although both Ca^{2+} and Mg^{2+} cause a dose-related activation of ATPase activity in these acellular systems, the Ca^{2+}-activated and Mg^{2+}-activated enzymes were thought to represent, at least to some extent, distinct enzymatic entities. The Ca^{2+}-ATPase displays two K_m's for Ca^{2+} close to 0.1 μM and 4-6 μM, respectively.

More recently, Colca et al. (1983) have provided evidence to support the existence of two fundamentally distinct active Ca^{2+} transport systems located, respectively, in the endoplasmic reticulum and plasma membrane of islet cells. The apparent K_m for calcium uptake amounted to 0.05 μM in the plasma membrane preparation and 1.8 μM in the endoplasmic reticulum subcellular fraction. The Ca^{2+}-ATPase activity in the plasma membrane preparation displayed a double K_m (0.2 μM and 0.04 mM, respectively) and was stimulated by calmodulin. The Ca^{2+}-ATPase activity in the endoplasmic reticulum fraction displayed a single affinity for Ca^{2+} (K_m 2-3 μM) and failed to be affected by calmodulin. The two preparations also differed from one another by their pH optimum and sensitivity to oxalate.

The characterization of such Ca^{2+}-ATPases is of obvious interest in considering the modality of either Ca^{2+} extrusion from the islet cell into the extracellular fluid against the prevalent electrochemical gradient or the sequestration of Ca^{2+} in intracellular organelles. In the latter perspective, mitochondria are likely to represent a further major site participating in the regulation of cellular calcium homeostasis. For instance, Prentki et al. (1983) have recently indicated that mitochondria isolated from a rat insulinoma were able to maintain an ambient free Ca^{2+} concentration of about 0.3 and 0.9 μM in the absence and presence of Mg^{2+} (1 mM) respectively. The addition of Na^+ caused a dose-related increase in steady-state extramitochondrial Ca^{2+} concentration.

2.5 Models for the Nutrient-Induced Remodeling of Ca^{2+} Fluxes in Islet Cells

There remains considerable doubt as to the precise mechanism by which glucose induces the changes in Ca^{2+} fluxes so far described in this report. A current theory (Malaisse et al. 1979b) postulates that the coupling of metabolic to ionic events represents a multifactorial process including changes in the generation rate of reducing equivalents (NADH, NADPH), protons (H^+) and high-energy phosphate intermediates (ATP). Thus, distinct coupling factors could regulate distinct cationic movements. The alleged role of a change in cytosolic pH, which could affect the outflow of both K^+ and Ca^{2+} (Henquin 1981; Carpinelli et al. 1980), calls for direct measurements of cytosolic pH in islet cells. The postulated role of reducing equivalents (Hellman et al. 1974; Malaisse et al. 1979a) raises a question as to the target system affected by a change in redox state. The glucose-induced increase in ATP availability is viewed by certain authors (Atwater 1980; Atwater et al. 1980) as allowing facilitated Ca^{2+} extrusion from the islet cells or uptake by intracellular organelles, leading initially to a fall in cytosolic Ca^{2+} concentration and inactivation of the Ca^{2+}-sensitive modality of K^+ extrusion. In a recent mathematical model for stimulus-secretion coupling in the B-cell, it was shown that a facilitated sequestration of Ca^{2+} is not necessarily incompatible with a rapid increase in cytosolic Ca^{2+} concentration and subsequent stimulation of insulin release (Scholler et al. 1984b).

It should be stressed that we have so far purposely restricted the discussion mainly to the effect of nutrient secretagogs upon Ca^{2+} fluxes. The modality by which certain other insulinotropic factors, e.g., adrenergic and cholinergic agents, may affect Ca^{2+} movement in the islet cells will be considered later in this report.

3 Regulatory Roles of Ca^{2+} in Islet Cells

The view that a rise in cytosolic Ca^{2+} concentration triggers insulin release raises the question as to the Ca^{2+}-responsive target system(s) involved in such a functional response. This section deals mainly but not exclusively with such target systems.

3.1 Ca^{2+} as a Trigger for Insulin Release

At least two proposals were initially made to account for the stimulation of insulin release in response to the cytosolic accumulation of Ca^{2+}. First, the ectoplasmic accumulation of Ca^{2+} could cause the collapse of a potential energy barrier between the inner face of the plasma membrane and outer face of secretory granules (Dean 1974). Second, an increase in cytosolic Ca^{2+} concentration could cause contraction of the microfilamentous cell web controlling the access of secretory granules to their exocytotic site at the plasma membrane (Orci et al. 1972). In other words, Ca^{2+} could trigger motile events controled by the B-cell microtubular-microfilamentous system (Van Obberghen et al. 1975). Time-lapse cinematography of monolayer cultures of islet cells provided support for such a view. Thus, Lacy et al. (1975) observed that the capacity of glucose to enhance the saltatory movements of secretory granules along oriented microtubular pathways was abolished in the absence of extracellular Ca^{2+}. Somers et al. (1979a) indicated that the ionophore A23187 markedly stimulated the contractile activity of the cell web in the islet cells. Moreover, in the latter two reports, microtubular poisons were also found to impair the mobility of secretory granules. These biophysical observations should be considered in the light of more recent biochemical investigations indicating the presence of Ca^{2+}-calmodulin-dependent myosin light chain kinase activity in insulin-secreting tissues (Mac Donald and Anjaneyulu 1982; Penn et al. 1982). This topic was recently reviewed, with further emphasis on the possible participation of Ca^{2+}-calmodulin in tubulin phosphorylation and/or polymerization to microtubules (Howell and Tyhurst 1982).

3.2 Calmodulin-Mediated Processes

Sugden et al. (1979) and Valverde et al. (1979) independently reported on the presence of calmodulin in islet cells, with an apparent concentration close to 50 μM. Hutton et al. (1981) were able to isolate and characterize the amino-acid composition and Ca^{2+}-binding capacity of calmodulin extracted from a rat islet cell tumor. Incidentally, the pancreatic B-cells also contain other Ca^{2+}-binding proteins, such as the vitamin D-inducible Ca^{2+}-binding protein (Roth et al. 1982).

There are a number of enzymes whose activity could be regulated by Ca-calmodulin in the islet cells. Three examples only will be cited here, in addition to those already mentioned above (e.g., Ca^{2+}, Mg^{2+}-ATPase, and myosin light chain kinase).

Valverde et al. (1979) first reported on the activation of rat islet adenylate cyclase by Ca-calmodulin. This phenomenon is currently considered as a mechanism by which an increase in cytosolic Ca^{2+} accumulation may lead to stimulation of cyclic AMP synthesis in intact islet cells exposed to suitable nutrient or non-nutrient secretagogues. The view is supported by the finding that the capacity of D-glucose and other nutrients to stimulate cyclic AMP production is abolished in the absence of extracellular Ca^{2+} (Valverde et al. 1983). The stimulation of adenylate cyclase by Ca-calmodulin could account for amplification of the secretory response, since cyclic AMP itself is known to enhance insulin release evoked by a variety of secretagogs. Incidentally, the stimulation of cyclic AMP production, such as that evoked by forskolin, is insufficient per se to cause sustained stimulation of insulin release (Malaisse et al. 1984).

Ca-calmodulin was also reported to activate to a modest extent cyclic AMP phosphodiesterase in rat islet homogenates (Lipson and Oldham 1983; Sharp et al. 1983). The functional significance of such an effect is open to speculation (Sharp et al. 1983).

Calmodulin-activated protein kinase activity was reported by several authors in either insulinoma cells or normal islet tissue (Schubart et al. 1980a, b; Gagliardino et al. 1980; Landt et al. 1982). The major substrate in rat islets seems to be a 53,000-57,000 mol. wt. endogenous protein (Gagliardino et al. 1980; Landt et al. 1982). Further work is required to define the function of the calmodulin-activated protein kinase and the identity of its protein substrate (Gagliardino et al. 1980; Landt et al. 1982).

Although these considerations clearly suggest that the activation of calmodulin-sensitive enzymes may represent a major modality for the functional response to cytosolic Ca^{2+} accumulation in the islet cells, the relative importance of such a process remains to be precisely evaluated. Indeed, exposure of intact islet cells to alleged "specific calmodulin antagonists" provided experimental results difficult to interpret in view of the suspected unspecificity of these drugs (Sharp et al. 1983; Valverde et al. 1981).

3.3 Ca^{2+} and Cyclic AMP

The preceding section indicates that cytosolic Ca^{2+} may play a role in the regulation of cyclic AMP production. The possible effect of cyclic AMP on the regulation of Ca^{2+} fluxes should not be overlooked. On the basis of bioelectrical data, Henquin and Meissner (1983) recently proposed that activation of adenylate cyclase may lead to facilitated Ca^{2+} influx into the B-cell, this representing a hitherto unrecognized effect of cyclic AMP. However, there are reasons to believe that the major effect of cyclic AMP on calcium fluxes consists in an intracellular redistribution of this ion (Brisson et al. 1972; Brisson and Malaisse 1973). For instance, forskolin enhances and clonidine inhibits glucose-induced insulin release without affecting glucose-stimulated ^{45}Ca net uptake by the islets (Malaisse et al. 1984; Leclercq-Meyer et al. 1980). It could be argued that the stimulation of adenylate cyclase activity by forskolin and its inhibition by clonidine, which were both documented in an islet subcellular particulate fraction incubated in the absence of Ca^{2+}, affects the release of insulin by a mechanism independent of any change in Ca^{2+} fluxes, such as the regulation of a cyclic AMP-dependent protein kinase (Montague and Howell 1972). However, several reports suggest the view that cyclic AMP also favors the cytosolic accumulation of Ca^{2+}, e.g., by inhibiting its uptake by certain organelles (Brisson et al. 1972; Sehlin 1976).

3.4 Phospholipid and Protein Kinase C

Changes in the turnover of phospholipids, e.g., in the so-called phosphatidylinositol cycle, may represent a phenomenon tightly linked to the movements of Ca^{2+} in islet cells. This topic, which was recently reviewed (Best and Malaisse 1983d), will only be discussed shortly in the present report. Thus, such phenomena as the stimulation of phospholipid methylation (Saceda et al. 1983), the accelerated generation of metabolites in the arachidonate cascade (Laychock 1982) and the decrease in membrane viscosity associated with B-cell activation (Deleers et al. 1981) will not be further considered.

Instead, we wish to emphasize that certain insulinotropic agents, especially nutrient secretagogues and cholinergic neurotransmitters, do cause the breakdown of phosphoinositides in islet cells (Best and Malaisse 1983a, b, c). This effect is abolished in the absence of Ca^{2+} and presence of EGTA, but persists — to a significant extent — when only $CaCl_2$ is removed from the incubation medium. This "PI" effect could participate to the Ca^{2+}-dependent process of insulin release in several fashions. First, the cleavage of the polar, acidic head-group of phosphoinositides may release a proportion of divalent cations bound in the membrane. Second, it may lead to formation of endogenous ionophores, e.g., phosphaticid acid (Dunlop et al. 1982). Last, it may cause activation of a Ca^{2+}-sensitive, phospholipid-dependent protein kinase (C-kinase) which can be activated by diacylglycerol.

Protein kinase C activity in islet homogenate was reported by Tanigawa et al. (1982) and Hubinont et al. (1984). The latter authors have demonstrated activation of this enzyme by a tumor-promoting phorbol ester, 12-O-tetradecanoylphorbol-13-acetate (TPA). In the presence of TPA and phosphatidylserine, the Ka for Ca^{2+} was close to 5 μM (Hubinont et al. 1984). Since TPA is a potent insulin secretagog but fails to affect K^+ conductance and $^{45}Ca^{2+}$ net uptake in intact islets (Malaisse et al. 1980a), it is tempting to propose that the C-kinase may represent an effector of insulin release, which would be operative even at close-to-basal cytosolic Ca^{2+} concentrations. This is not meant to deny that TPA exerts modest but significant changes in the handling of Ca^{2+} in the islet cells (Malaisse et al. 1980a).

3.5 Miscellaneous

The Ca^{2+}-responsive target systems so far mentioned in this review do not represent an exhaustive list. There are indeed other enzymatic and biophysical processes susceptible to be modulated by the extracellular Ca^{2+} concentration, cytosolic Ca^{2+} concentration, or Ca^{2+} content of intracellular organelles. Two examples will here be mentioned.

First, transglutaminase activity was recently identified in islet homogenates (Bungay et al. 1982; Gomis et al. 1983). This enzyme catalyzes in a Ca^{2+}-dependent fashion the cross-linking of proteins and could, therefore, participate in mechanical events involved in the insulin release process. As judged from the incorporation of $[2,5^{-3}H]$ histamine or $[^{14}C]$methylamine in N, N-dimethylcasein, the threshold concentration for activation of transglutaminase by Ca^{2+} is close to 0.05 μM, with a Ka for Ca^{2+} close to 70-90 μM (Gomis et al. 1983). Considerable difficulties were encountered, however, in our attempt to document the possible role of transglutaminase in the secretory process. Indeed, all drugs so far characterized for their inhibitory action upon transglutaminase activity in islet homogenates were judged rather inadequate for study in intact islets (Gomis et al. 1983; Lebrun et al. 1984) either because of suspected undesirable damaging effect (monodansylcadaverine), side effects (methylamine) and poor cellular penetration (N-p-tosylglycine) or known interference with insulin immunoassay (bacitracin) or cationic fluxes in the islet cells (hypoglycemic sulfonylureas).

Second, a change in Ca^{2+} availability to the B-cell, such as that induced by the removal of $CaCl_2$ from the incubation medium or the addition of excess $MgCl_2$, affects the rate of oxidation of nutrients (e.g., D-glucose or L-glutamine) in islet cells (Malaisse et al. 1978b; Sener et al. 1982). This effect could conceivably be the result of either a direct modulation of enzymes involved in nutrient catabolism or changes in energy expenditure. In the latter respect, the metabolic changes evoked by Ca^{2+} would illustrate a modality for the feedback control of early metabolic events by more distal Ca^{2+}-dependent events in the secretory sequence (Malaisse et al. 1980b).

4 Conclusion

The present review, which should be considered far from exhaustive, emphasizes the concept that changes in Ca^{2+} fluxes in the islet cells play a key role in the process of insulin release. Further progress in this field can be expected, in the not-too-distant future, from the identification and characterization of new molecular determinants involved in either the regulation of calcium movements or the functional response to this divalent cation. For instance, the view that a transient activation of ornithine decarboxylase associated with a sustained increase in polyamine tissue content may serve as messengers to generate a Ca^{2+} signal in the process of cell activation (Koenig et al. 1983) needs now to be investigated in pancreatic islet cells.

References

Anjaneyulu K, Anjaneyulu R, Malaisse WJ (1980) The stimulus-secretion coupling of glucose-induced insulin release. XLIII. Na-Ca counter-transport mediated by pancreatic islet native ionophores. J Inorg Biochem 13:178-188

Atwater I (1980) Control mechanisms for glucose-induced changes in the membrane potential of mouse pancreatic B-cell. Cienc Biol (Luanda) 5:299-314

Atwater I, Ribalet B, Rojas E (1978) Cyclic changes in potential and resistance of the B-cell membrane induced by glucose in islets of Langerhans from mouse. J Physiol (Lond) 278:117-139

Atwater I, Dawson CM, Scott A, Eddlestone G, Rojas E (1980) The nature of the oscillatory behaviour of electrical activity from pancreatic B-cell. Horm Metab Res (suppl) 10:100-107

Atwater I, Dawson CM, Eddlestone GT, Rojas E (1981) Voltage noise measurements across the pancreatic B-cell membrane: calcium channels characteristics. J Physiol (Lond) 314:195-212

Atwater I, Goncalves AA, Rojas E (1982) Electrophysiological measurement of an oscillating potassium permeability during the glucose-stimulated burst activity in mouse pancreatic B-cell. Biomed Res 3:645-648

Baker PF, Meves H, Ridgway EB (1973) Effect of manganese and other agents on the calcium uptake that follows depolarization of squid axons. J Physiol (Lond) 231:511-526

Best L, Malaisse WJ (1983a) Phosphatidylinositol and phosphatidic acid metabolism in rat pancreatic islets in response to neurotransmitter and hormonal stimuli. Biochim Biophys Acta 750:157-163

Best L, Malaisse WJ (1983b) Effects of nutrient secretagogues upon phospholipid metabolism in rat pancreatic islets. Mol Cell Endocr 32:205-214

Best L, Malaisse WJ (1983c) Stimulation of phosphoinositide breakdown in rat pancreatic islets by glucose and carbamylcholine. Biochem Biophys Res Commun 116:9-16

Best L, Malaisse WJ (1983d) Phospholipids and islet function. Diabetologia 25:299-305

Bolton TB (1979) Mechanisms of action of transmitters and other substances on smooth muscle. Physiol Rev 59:606-717

Brisson GR, Malaisse WJ (1973) The stimulus-secretion coupling of glucose-induced insulin release. XI. Effects of theophylline and epinephrine on ^{45}Ca efflux from perifused islets. Metabolism 22:455-465

Brisson GR, Malaisse-Lagae F, Malaisse WJ (1972) The stimulus-secretion coupling of glucose-induced insulin release. VII. A proposed site of action for adenosine-3',5'-cyclic monophosphate. J Clin Invest 51:232-241

Bungay PJ, Griffin M, Potter JM (1982) Evidence for the involvement of transglutaminase in insulin secretion in the rat. Diabetologia 23:159

Carpinelli AR, Sener A, Herchuelz A, Malaisse WJ (1980) Stimulus-secretion coupling of glucose-induced insulin release. Effect of intracellular acidification upon calcium efflux from islet cells. Metabolism 29:540-545

Colca JR, Kotagal N, Lacy PE, McDaniel ML (1983) Comparison of the properties of active Ca^{2+} transport by the islet-cell endoplasmic reticulum and plasma membrane. Biochim Biophys Acta 729:176-184

Dean PM (1974) Surface electrostatic-charge measurement on islet and zymogen granules: effect of calcium ions. Diabetologia 10:427-430

Deleers M, Ruysschaert JM, Malaisse WJ (1981) Glucose induces membrane changes detected by fluorescence polarization in endocrine pancreatic islets. Biochem Biophys Res Commun 98:255-260

Donatsch P, Lowe DA, Richardson BP, Taylor P (1977) The functional significance of sodium channels in pancreatic beta-cells membranes. J Physiol (Lond) 267:357-376

Dunlop M, Larkins RG, Court JM (1982) Endogenous ionophoretic activity in the neonatal rat pancreatic islet. FEBS Lett 144:259-263

Formby B, Capito K, Egeberg J, Hedeskov CJ (1976) Ca-activated ATPase activity in subcellular fractions of mouse pancreatic islets. Am J Physiol 230:441-448

Gagliardino JJ, Harrison DE, Christie MR, Gagliardino EE, Ashcroft SJH (1980) Evidence for the participation of calmodulin in stimulus secretion coupling in the pancreatic B-cell. Biochem J 192:919-927

Gomis R, Sener A, Malaisse-Lagae F, Malaisse WJ (1983) Transglutaminase activity in pancreatic islets. Biochim Biophys Acta 760:384-388

Grodsky GM, Bennett LL (1966) Cation requirements for insulin secretion in the isolated perfused pancreas. Diabetes 15:910-913

Hellmann B, Idahl LÄ, Lernmark Å, Sehlin J, Täljedal IB (1974) Membrane sulphydrylgroups and the pancreatic beta-cell recognition of insulin secretagogues. In: Malaisse WJ, Pirart J (eds) Diabetes, Excerpta Medica. International Congress Series 312, Amsterdam, pp 65-78

Hellman B, Andersson T, Berggren PO, Flatt P, Gylfe E, Kohnert KD (1979) The role of calcium in insulin secretion. In: Dumont J, Nunez J (eds) Hormones and cell regulation. Elsevier/North Holland Biomedical, Amsterdam, pp 69-76

Henquin JC (1978a) D-glucose inhibits potassium efflux from pancreatic islet cells. Nature (Lond) 271:271-273

Henquin JC (1978b) Relative importance of extracellular calcium for the two phases of glucose-stimulated insulin release: studies with theophylline. Endocrinology 102:723-730

Henquin JC (1981) The effect of pH on [86]rubidium efflux from pancreatic islet cells. Mol Cell Endocr 21:119-128

Henquin JC, Meissner HP (1983) Dibutyryl cyclic AMP triggers Ca^{2+} influx and Ca^{2+}-dependent electrical activity in pancreatic B-cells. Biochem Biophys Res Commun 112:614-620

Herchuelz A, Malaisse WJ (1982) Calcium and insulin release. In: Anghileri LJ, Tuffet-Anghileri AM (eds) The role of calcium in biological systems, vol III. CRC, Boca Raton, pp 17-32

Herchuelz A, Sener A, Malaisse WJ (1980) Regulation of calcium fluxes in rat pancreatic islets. IV. Calcium extrusion by sodium-calcium counter-transport. J Membr Biol 57:1-12

Herchuelz A, Lebrun P, Boschero AC, Malaisse WJ (1984) Mechanisms of arginine-stimulated Ca^{2+} influx into pnacreatic B-cell. Am J Physiol 246:E38-E43

Herman L, Sato T, Hales CN (1973) The electron microscopic localization of cations to pancreatic islets of Langerhans and their possible role in insulin secretion. J Ultrastruct Res 42:298-311

Howell SL, Tyhurst M (1982) Microtubules, microfilaments and insulin secretion. Diabetologia 22:301-308

Hubinont CJ, Malaisse WJ (1984) Activation of protein kinase C by a tumor-promoting phorbol ester in pancreatic islets. FEBS Lett 170:247-253

Hutton JC, Penn EJ, Jackson P, Hales CN (1981) Isolation and characterization of calmodulin from an insulin-secreting tumour. Biochem J 193:875-885

Kasson BG, Levin SR (1981) Characterization of pancreatic islet Ca^{2+}-ATPase. Biochim Biophys Acta 662:30-35

Kikuchi M, Wollheim CB, Cuendet GS, Renold AE, Sharp GWG (1978) Studies on the dual effects of glucose on [45]Ca efflux from isolated rat islets. Endocrinology 102:1339-1349

Klöppel G, Bommer G (1979) Ultracytochemical calcium distribution in B-cells in relation to biphasic glucose-stimulated insulin release by the perfused rat pancreas. Diabetes 28:585-592

Koenig H, Goldstone A, Lu CY (1983) Polyamines regulate calcium fluxes in a rapid plasma membrane response. Nature (Lond) 305:530-534

Kohnert KD, Hahn HJ, Gylfe E, Borg H, Hellman B (1979) Calcium and pancreatic B-cell function. 6. Glucose and intracellular [45]Ca distribution. Mol Cell Endocr 16:205-220

Lacy PE, Finke EH, Codilla RC (1975) Cinemicrographic studies on B granule movement in monolayer culture of islet cells. Lab Invest 33:570-576

Landt M, McDaniel ML, Bry CG, Kotagal N, Colca JR, Lacy PE, McDonald JM (1982) Calmodulin-activated protein kinase activity in rat pancreatic islet cell membranes. Arch Biochem Biophys 213:148-154

Laychock SG (1982) Phospholipase A_2 activity in pancreatic islets is calcium-dependent and stimulated by glucose. Cell Calcium 3:43-54

Lebrun P, Malaisse WJ, Herchuelz A (1982a) Evidence for two distinct modalities of Ca^{2+} influx into pancreatic B-cell. Am J Physiol 242:E59-E66

Lebrun P, Malaisse WJ, Herchuelz A (1982b) Modalities of gliclazide-induced Ca^{2+} influx into the pancreatic B-cell. Diabetes 81:1010-1015

Lebrun P, Malaisse WJ, Herchuelz A (1982c) Nutrient-induced intracellular calcium movement in rat pancreatic B-cell. Am J Physiol 243:E196-E205

Lebrun P, Gomis R, Deleers M et al. (1984) Methylamines and islet function. Cationic aspects. J Endocrinol Invest 7:347-355

Leclercq-Meyer V, Herchuelz A, Valverde I, Couturier E, Marchand J, Malaisse WJ (1980) Mode of action of clonidine upon islet function. Dissociated effects upon the time course and magnitude of insulin release. Diabetes 29:193-200

Lipson LG, Oldham SB (1983) The role of calmodulin in insulin secretion: the presence of a calmodulin-stimulatable phosphodiesterase in pancreatic islets of normal and pregnant rats. Life Sci 32:775-780

MacDonald MJ, Anjaneyulu K (1982) Calcium-calmodulin-dependent myosin phosphorylation by pancreatic islets. Diabetes 31:566-570

Malaisse WJ, Herchuelz A (1982) Nutritional regulation of K^+ conductance: an unsettled aspect of pancreatic B-cell physiology. In: Litwack G (ed) Biochemical action of hormones, vol IX. Academic, New York, pp 69-92

Malaisse WJ, Brisson GR, Baird LE (1973) Stimulus-secretion coupling of glucose-induced insulin release. X. Effect of glucose on ^{45}Ca efflux from perifused islets. Am J Physiol 224:389-394

Malaisse WJ, Herchuelz A, Devis G et al. (1978a) Regulation of calcium fluxes and their regulatory roles in pancreatic islets. Ann NY Acad Sci 307:562-582

Malaisse WJ, Hutton JC, Sener A, Levy J, Herchuelz A, Devis G, Somers G (1978b) Calcium antagonists and islet function. VII. Effect of calcium deprivation. J Membr Biol 38:193-208

Malaisse wJ, Sener A, Herchuelz A, Hutton JC (1979a) insulin release: the fuel hypothesis. Metabolism 28:373-386

Malaisse WJ, Hutton JC, Kawazu S, Herchuelz A, Valverde I, Sener A (1979b) The stimulus-secretion coupling of glucose-induced insulin release. XXXV. The links between metabolic and cationic events. Diabetologia 16:331-341

Malaisse WJ, Sener A, Herchuelz A, Carpinelli AR, Poloczek P, Winand J, Castagna M (1980a) Insulinotropic effect of the tumor promoter 12-O-tetradecanoylphorbol-13-acetate in rat pancreatic islets. Cancer Res 40:3827-3831

Malaisse WJ, Sener A, Herchuelz A, Valverde I, Hutton JC, Atwater I, Leclercq-Meyer V (1980b) The interplay between metabolic and cationic events in islet cells: coupling factors and feedback mechanisms. Horm Metab Res (suppl) 10:61-66

Malaisse WJ, Garcia-Morales P, Dufrane SP, Sener A, Valverde I (1984) Forskolin-induced activation of adenylate cyclase, cyclic adenosine monophosphate production and insulin release in rat pancreatic islets. Endocrinology 115:2015-2020

Meisheri KD, Hwang O, Van Breemen C (1981) Evidence for two separate Ca^{2+} pathways in smooth muscle plasmalemma. J Membr Biol 59:19-25

Meissner HP, Preissler M (1979) Glucose-induced changes of the membrane potential of pancreatic B-cells: their significance for the regulation of insulin release. In: Camerini-Davalos RA, Hanover B (eds) Treatment of early diabetes. Plenum, New York, pp 97-107

Meissner HP, Henquin JC, Preissler M (1978) Potassium dependence of the membrane potential of pancreatic B-cells. FEBS Lett 94:87-89

Meissner HP, Preissler M, Henquin JC (1980) Possible ionic mechanisms of the electrical activity induced by glucose and tolbutamide in pancreatic B-cells. In: Waldhäusel WK (ed) Diabetes 1979, Excerpta Medica. International Congress Series 500, Amsterdam, pp 166-171

Milner RDG, Hales CN (1967) The role of calcium and magnesium in insulin secretion from rabbit pancreas studied in vitro. Diabetologia 3:47-49

Montague W, Howell SL (1972) The mode of action of adenosine-3':5'-cyclic monophosphate in mammalian islets of Langerhans. Preparation and properties of islet-cell protein phosphokinase. Biochem J 129:551-560

Orci L, Gabbay KH, Malaisse WJ (1972) Pancreatic beta-cell web: its possible role in insulin secretion. Science (Wash DC) 175:1128-1130

Owen A, Sener A, Malaisse WJ (1983) Stimulus-secretion coupling of glucose-induced insulin release. LI. Divalent cations and ATPase activity in pancreatic islets. Enzyme (Basel) 29:2-14

Penn EJ, Brocklehurst KW, Sopwith AM, Hales CN, Hutton JC (1982) Ca^{2+}-calmodulin-dependent myosin light-chain phosphorylating activity in insulin-secreting tissues. FEBS Lett 139:4-8

Pershadsingh HA, McDaniel ML, Landt M, Bry CG, Lacy PE, McDonald JM (1980) Ca^{2+} activated ATPase and ATP-dependent calmodulin-stimulated Ca^{2+} transport in islet cell plasma membrane. Nature (Lond) 288:492-494

Prentki M, Janjic D, Wollheim CB (1983) The regulation of extramitochondrial steady state free Ca^{2+} concentration by rat insulinoma mitochondria. J Biol Chem 258:7597-7602

Ravazzola M, Malaisse-Lagae F, Amherdt M, Perrelet A, Malaisse WJ, Orci L (1976) Patterns of calcium localization in pancreatic endocrine cells. J Cell Sci 27:107-117

Ribalet B, Beigelman PM (1979) Cyclic variation of K^+ conductance in pancreatic B-cells: Ca^{2+} and voltage dependence. Am J Physiol 237:C137-C146

Roth J, Bonner-Weir S, Norman AW, Orci L (1982) Immunocytochemistry of vitamin D-dependent calcium-binding protein in chick pancreas: exclusive localization in B-cells. Endocrinology 110:2216-2218

Saceda M, Garcia-Morales P, Valverde I, Malaisse WJ (1983) Glucose-induced stimulation of lipid methylation in pancreatic islets. Diabetes 32 (suppl 1):39A

Scholler Y, De Maertelaer V, Malaisse WJ (1984a) Mathematical modelling of stimulus-secretion coupling in the pancreatic B-cell. II. Calcium-stimulated calcium release. Computer Programs in Biomedicine 19: in press

Scholler Y, De Maertelaer V, Malaisse WJ (1984b) Mathematical modelling of stimulus-secretion coupling in the pancreatic B-cell. III. Glucose-induced inhibition of calcium efflux. Biophys J 46:439-446

Schubart UK, Erlichman J, Fleischer N (1980a) The role of calmodulin in the regulation of protein phosphorylation and insulin release in Hamster insulinoma cells. J Biol Chem 255:4120-4124

Schubart UK, Fleischer N, Erlichman J (1980b) Ca^{2+}-dependent protein phosphorylation and insulin release in intact hamster insulinoma cells. J Biol Chem 255:11063-11066

Sehlin J (1976) Calcium uptake by subcellular fractions of pancreatic islets. Effects of nucleotides and theophylline. Biochem J 156:63-69

Sener A, Malaisse-Lagae F, Malaisse WJ (1982) The stimulus-secretion coupling of glucose-induced insulin release. Environmental influences on L-glutamine oxidation in pancreatic islets. Biochem J 202:309-316

Sharp GWG, Wiedenkeller DE, Lipson LG et al. (1983) The multiple roles of calmodulin in the endocrine pancreas and the control of insulin release. In: Mngola EN (ed) Diabetes 1982, Excerpta Medica. International Congress Series 600, Amsterdam, pp 329-336

Siegel EG, Wollheim CB, Renold AE, Sharp GWG (1980) Evidence for the involvement of Na-Ca exchange in glucose-induced insulin release from rat pancreatic islets. J Clin Invest 66:996-1003

Somers G, Blondel B, Orci L, Malaisse WJ (1979a) Motile events in pancreatic endocrine cells. Endocrinology 104:255-264

Somers G, Devis G, Malaisse WJ (1979b) Calcium antagonists and islet function. IX. Is extracellular calcium required for insulin release? Acta Diabetol Lat 16:9-18

Sugden MC, Christie MR, Ashcroft SJH (1979) Presence and possible role of calcium-dependent regulator (calmodulin) in rat islets of Langerhans. FEBS Lett 105:95-100

Tanigawa K, Kuzuya H, Imura H, Taniguchi H, Baba S, Takai Y, Nishizuka Y (1982) Calcium-activated, phospholipid-dependent protein kinase in rat pancreatic islets of Langerhans. FEBS Lett 138:183-186

Valverde I, Vandermeers A, Anjaneyulu R, Malaisse WJ (1979) Calmodulin activation of adenylate cyclase in pancreatic islets. Science (Wash DC) 206:225-227

Valverde I, Sener A, Lebrun P, Herchuelz A, Malaisse WJ (1981) The stimulus-secretion coupling of glucose-induced insulin release. XLVII. The possible role of calmodulin. Endocrinology 108:1305-1312

Valverde I, Garcia-Morales P, Ghiglione M, Malaisse WJ (1983) The stimulus-secretion coupling of glucose-induced insulin release. LIII. Calcium-dependency of the cyclic AMP response to nutrient secretagogues. Horm Metab Res 115:62-68

Van Obberghen E, Somers G, Devis G, Ravazzola M, Malaisse-Lagae F, Orci L, Malaisse WJ (1975) Dynamics of insulin release and microtubular-microfilamentous system. VII. Do microfilaments provide the motive force for the translocation and extrusion of beta granules. Diabetes 24: 892-901

Wollheim CB, Sharp GWG (1981) Regulation of insulin release by calcium. Physiol Rev 61:914-973

Wollheim CB, Kikuchi M, Renold AE, Sharp GWG (1978) The roles of intracellular and extracellular Ca^{++} in glucose-stimulated biphasic insulin release by rat islets. J Clin Invest 62:451-458

Wollheim CB, Tsien RY, Pozzan T (1983) Stimulation of insulin release in accompanied by an increase in cytosolic free Ca^{2+}. Diabetologia 25:205

Wolters GHJ, Wiegman JB, Konijnendijk W (1982) The effect of glucose stimulation on ^{45}calcium uptake of rat pancreatic islets and their total calcium content as measured by a fluorometric micro-method. Diabetologia 22:122-127

Calcium Regulation of Histamine Secretion from Mast Cells

P. T. PEACHELL and F. L. PEARCE [1]

CONTENTS

1 Mast Cells and Immediate Hypersensitivity Reactions

The mast cell is centrally involved in the immunopathology of acute allergic and in-flammatory disease (Pepys and Edwards 1979). These cells are widely distributed in the human body, but are found at the highest concentrations in those areas which come into direct contact with foreign substances, namely in association with blood vessels and nerves in the loose connective tissue of the bronchi, conjunctiva, gut, ear, nose, throat and skin. As such, they are uniquely placed to participate in a variety of allergic disorders and have been incriminated in the aetiology of asthma, rhinitis, con-junctivitis and inflammatory diseases of the gut and skin (Pepys and Edwards 1979).

Mast cells are characterized morphologically by the presence of large numbers of modified lysozomal granules which contain a range of biogenic substances including histamine, neutrophil and eosinophil chemotactic factors, heparin or related pro-teoglycans, and hydrolytic enzymes (Metcalfe et al. 1981). Activation of the mast cell leads to the exocytosis of these granules and the release of their associated preformed mediators, and also evokes the de novo generation of further biologically active prod-ucts of arachidonic acid metabolism, principally leukotrienes, thromboxanes and

[1] Department of Chemistry, University College London, 20 Gordon Street, London WCIH OAJ, Great Britain

prostaglandins (Lewis and Austen 1981; Samuelsson 1983). These spasmogenic and vasoactive mediators then act on distinct effector cells in the bronchi and vasculature to produce the immediate clinical symptoms of allergy. The release of chemotactic factors and hydrolases leads to a subsequent more prolonged response and exacerbation of the inflammatory condition.

The pathophysiological stimulus for the release of chemical mediators from the mast cell is provided by the combination of specific antigen with reaginic (IgE) antibody fixed to the cell surface (for references, see Kazimierczak and Diamant 1978). The consequent dimerization of receptor sites may also be induced experimentally by lectins such as concanavalın A (Ennis et al. 1981; Truneh and Pearce 1981) and by antibodies to IgE or to the IgE-receptor molecule (Ishizaka 1982). Mediator release may additionally be triggered by a variety of pharmacological stımuli including polybasic ligands such as compound 48/80, peptide 40l [the mast-cell-degranulating (MCD) peptide from bee venom] and polylysine, split complement components, calcium ionophores, the polysaccharide dextran, adenosine 5'-triphosphate (ATP), certain neuropeptides and a diversity of drugs and miscellaneous organic molecules (for reviews, see Kazimierczak and Diamant 1978; Lagunoff et al. 1983). While many of these agents exhibit a high degree of tissue and species specificity in their action (Pearce 1982b, 1983), this range of potential agonists has provided the immunopharmacologist with a number of important tools with which to study the release process.

2 Calcium and Histamine Secretion

2.1 Evidence for the Central Role of Calcium in Histamine Secretion

As in other secretory systems, the primary trigger for the release of chemical mediators from the mast cell is believed to be a rise in the concentration of ionized calcium in the cell cytosol. Strong evidence in support of this view has come from experiments in which degranulation and histamine release have been induced by direct introduction of calcium into the cell either by microinjection (Kano et al. 1973), by means of the ionophores A23187 (Foreman et al. 1973), ionomycin (Bennet et al. 1979) and chlortetracycline (Pearce et al. 1983), or by fusion of the cell with liposomes loaded with the cation (Theoharides and Douglas 1978). Moreover, mast cells rendered permeable to divalent cations by pretreatment with ATP (Bennet et al. 1981) or Sendai virus (Gomperts et al. 1983) release histamine when subsequently exposed to calcium buffered in the micromolar range. Similar results have been obtained, albeit with higher concentrations of calcium, following permeabilization with chelating agents (Douglas and Kagayama 1977). In total, these data clearly indicate that a rise in the intracellular concentration of calcium is a necessary and sufficient stimulus for histamine secretion.

2.2 Calcium Pools Involved in Histamine Secretion

2.2.1 Extracellular Calcium

It has long been recognized that optimal, antigen-induced release of histamine from the mast cell requires the presence of calcium ions in the external incubation medium. This has led to the hypothesis that activation of the mast cell transiently increases the permeability of the membrane to calcium ions, that is, opens a receptor-mediated calcium channel or gate in the membrane (Foreman et al. 1976). Influx of the ion down its electrochemical gradient then triggers the release process. The first step in the internalization of the ion by this route appears to involve combination with a specific membrane receptor (Grosman and Diamant 1974; White and Pearce 1981) and, consistently, secretion is inhibited by ions of the lanthanide series which are competitive antagonists of binding of this type (Foreman and Mongar 1973; Pearce and White 1981). The pA_2 value for this interaction varies with the secretagogue, suggesting that different channels and receptors may be involved according to the nature of the inducer (Pearce and White 1981). In further support of this model, Foreman et al. (1977) have shown that stimulation of rat peritoneal mast cells with IgE-directed and other ligands induces an uptake of radioactive calcium from the external environment. There is a strong, positive correlation between the magnitude of the uptake and the degree of induced histamine secretion. Both processes exhibit a similar time-course and dependence on pH. A comparable uptake has also been observed following mast cell activation with the band 2 cationic protein from rabbit neutrophil lysozomes (Ranadive and Dhanani 1980) and with the chemical histamine liberators ATP (Dahlquist 1974) and compound 48/80 (Cochrane and Distel 1982; Ranadive and Lewis 1982; Spataro and Bosmann 1976). It should be noted, however, that major reservations have been expressed about these experiments and it has been suggested that such studies may merely reflect the nonspecific binding of calcium to newly exposed membrane sites revealed by the degranulation process (Grosman and Diamant 1978; Sugiyama 1971). As such, they would represent the consequence rather than the cause of exocytosis. In an attempt to exclude this possibility, Foreman et al. (1977) reported that metabolic inhibitors totally blocked antigen-induced histamine release from the mast cell, but did not strikingly reduce calcium uptake. The latter process was then considered to occur independently of granule extrusion. In contrast, other workers have reported that calcium uptake induced by a variety of pharmacological and immunological stimuli is totally inhibited by metabolic blockers (Cochrane and Distel 1982; Grosman and Diamant 1978; Ranadive and Dhanani 1980). The reasons for these differing results are not clear and the exact interpretation of these experiments remains to be agreed. These discrepancies do not, however, detract from the basic concept of the involvement of receptor-mediated calcium channels in mast cell activation.

2.2.2 Intracellular Calcium

While maximal histamine secretion is dependent on the presence of calcium ions in the external medium, suboptimal responses may be obtained in the absence of the

Table 1. Histamine release from rat peritoneal mast cells in various media

Secretagogue	Histamine release (%) in media containing:			
	No calcium	EDTA (0.1 mM)	Calcium (1 mM)	Calcium (20 mM)
Antigen	22.3 ± 2.0	47.7 ± 1.4	41.4 ± 6.2	13.2 ± 2.0
Anti-IgE	10.7 ± 2.5	20.3 ± 2.8	25.5 ± 4.9	13.0 ± 3.5
Concanavalin A	14.0 ± 2.4	39.9 ± 3.3	31.4 ± 7.5	15.8 ± 3.8
Dextran	2.3 ± 1.0	4.7 ± 1.4	33.2 ± 2.6	12.0 ± 5.0
Chlortetracycline	1.0 ± 0.5	1.5 ± 0.5	72.5 ± 1.0	57.5 ± 2.0
Ionophore A23187	7.6 ± 0.9	35.3 ± 1.9	76.2 ± 1.0	31.6 ± 1.2
Ionomycin	21.0 ± 1.1	40.7 ± 6.6	72.0 ± 3.5	17.9 ± 3.9
Compound 48/80	43.4 ± 7.8	64.2 ± 4.6	68.0 ± 5.4	21.4 ± 4.1
Peptide 401	50.1 ± 3.4	67.7 ± 4.0	69.2 ± 8.7	3.3 ± 0.5
Polylysine	40.3 ± 6.9	51.6 ± 7.7	53.3 ± 7.7	31.0 ± 2.7
Polymyxin	31.1 ± 4.4	38.4 ± 2.4	70.2 ± 2.1	58.5 ± 4.3

Cells were preincubated (5 min, 37°C) in the media shown and then challenged. Secretion was allowed to proceed for a further 10 min. All values means ± S.E.M for 3-6 experiments. (Data is taken from Peachell and Pearce 1984; Pearce et al. 1981, 1983)

added cation. The relative potencies of different agonists vary strikingly in such experiments: most ligands induce some release of histamine but polyamines such as compound 48/80, peptide 401, polylysine and polymyxin are particularly effective liberators under these conditions (Table 1).

The secretion of histamine from mast cells in a calcium-free medium is normally attributed to the mobilization of membrane-bound or intracellular reservoirs of the cation. Consistently, the response is abolished by pretreatment of the cells with chelating agents, which are believed to deplete such stores (for references, see Pearce 1982a, 1984). This treatment also produces a parallel reduction in the fluorescence signal due to the calcium-chlortetracycline complex in mast cells preloaded with the antibiotic, indicative of the removal of membrane-bound calcium (White and Pearce 1983). Full reactivity is subsequently restored on reintroduction of calcium into the medium.

While the above experiments indicate the possible importance of sequestered calcium in histamine secretion, they do not define the site of these stores. Studies in our and other laboratories have suggested that these reservoirs may be located within or on the inner face of the plasma membrane (Pearce 1982a). Thus, histamine secretion in a calcium-free medium is blocked by prolonged incubation with lanthanide ions which are not thought to penetrate into the cytosol (Pearce and White 1981). Also, the response is potentiated by brief pretreatment with chelating agents (Table 1). This is in contrast to the inhibitory effect of extended incubation with these agents and, by analogy with the situation in smooth muscle, we have argued that this treatment may remove calcium from regulatory sites in the membrane, thus destabilizing the structure and facilitating the release into the cytosol of more firmly sequestered stores of the cation (Ennis et al. 1980; Pearce et al. 1981; Pearce 1982a). Conversely, supramaximal concentrations of calcium inhibit the response (Table 1), suggesting that full occupancy of the proposed regulatory sites may stabilize the membrane and prevent translocation of the cation. Recent direct measurements of calcium uptake under these conditions are in accord with this model (Ranadive and Lewis 1982).

2.3 Calcium Efflux and Histamine Secretion

As in other cells, the standing concentration of free calcium in the cytosol of the mast cell is believed to be the result of a delicate balance between the influx of the cation from the external environment, the release and uptake from intracellular or membranous stores and the extrusion into the extracellular milieu. All of these factors will contribute towards calcium homeostasis and, while influx of the cation into the mast cell has been extensively investigated (see above), the role of calcium efflux in histamine secretion has been only recently examined (Pearce and White 1984). The main mechanism for extrusion of calcium in this system appears to be through the operation of a sodium-calcium antiporter (Pearce and White 1984). Thus, the basal efflux of ^{45} calcium from mast cells preloaded with the isotope is essentially unaffected by metabolic inhibitors (thereby excluding the immediate involvement of an ATP-dependent calcium pump) but totally dependent on extracellular sodium ions. Most interestingly, antigenic stimulation of the cell causes a temporary suppression of calcium efflux (Pearce and White 1984). Such an effect would then enhance the rise in the intracellular concentration of calcium produced by influx or mobilization of the cation, and hence provide a cooperative mechanism for augmenting the secretory response.

3 Initial Events in Mast Cell Activation and Calcium Mobilization

A number of rapid biochemical changes have now been identified following immunologic or pharmacologic activation of the mast cell. The relative importance of these processes, and their relationship to calcium mobilization, is somewhat contentious, but will doubtless be clarified by further work. The main events so far characterized are summarized below.

3.1 Phosphatidylinositol Turnover

Phosphatidylinositol (PI)-turnover (Michell, this Vol.) has been observed in purified rat peritoneal mast cells following stimulation with anti-IgE, antigen, concanavalin A, chymotrypsin and compound 48/80 (Cockcroft and Gomperts 1979; Kennerly et al. 1979; Schellenberg 1980). The response to the IgE-directed ligands is potentiated by phosphatidylserine (PS), with a parallel increase in histamine release. The PI-response does not require the presence of extracellular calcium whereas the secretion of histamine is, according to the agonist, either enhanced or dependent upon the added cation. Both histamine release and PI-turnover show a similar dependence on ligand concentration. On the basis of these results, it has been argued that the PI-response is a necessary event in histamine release and is centrally involved in calcium mobilization. The exact nature of this involvement remains ill-defined but, by analogy with other systems, a number of suggestions can be made (Gil et al. 1983; Pearce 1984). Firstly, the resulting changes in membrane lipid composition might directly facilitate the activation of calcium channels or the release of bound calcium. Secondly, phosphatidic acid (PA) generated by the cycle could act as a calcium ionophore and directly transfer the cation across the cell membrane. Consistently, simple application of exogenous PA to the mast cell, albeit at high concentration, evokes histamine secretion (Pearce and Messis 1982). Thirdly, the intermediate diacylglycerol (DAG) and its cleavage products (free fatty acid and monoacylglycerol) could act as fusogens and promote the membrane changes involved in exocytosis. Alternatively, DAG might activate protein kinases to phosphorylate membrane proteins involved in calcium transport. Fourthly, the cycle could serve to generate free arachidonic acid by cleavage of DAG, arachidonyl-PI, arachidonyl-PA or arachidonyl-phosphatidylcholine (arachidonyl-PC) following transacylation. As discussed below, the further metabolism of arachidonic acid might provide intermediates involved in calcium mobilization.

More detailed studies have, however, questioned the causal role of the PI-response in histamine secretion. In particular the response is evoked, albeit in limited fashion, following stimulation of the mast cell with the calcium ionophore A23187. The latter bypasses any receptor-mediated activation steps (Kennerly et al. 1979; Schellenberg 1980). Also, ATP evokes PI-turnover and histamine secretion from the mast cell but, unlike the other ligands tested, both processes are dependent on extracellular calcium (Cockcroft and Gomperts 1980). These findings, together with results from other systems, have led to the suggestion that the PI-response may merely be an epiphenomenon of mast cell activation (Cockcroft 1981). However, the issue remains the source of argument (Michell and Kirk 1982) and the possible involvement in histamine secretion of polyphosphoinositide breakdown, recently proposed to be the initial reaction in receptor-stimulated inositol phospholipid metabolism (Michel 1983), has yet to be investigated.

3.2 Phospholipid Methylation

An alternative form of membrane activation, involving the methylation of endogenous phospholipids, has been observed in a number of cell types including the mastocyte (Hirata and Axelrod 1980). In this system, two membrane-bound enzymes (methyltransferases I and II) successively convert phosphatidylethanolamine (PE) to phosphatidyl-N-monomethylethanolamine (PME) and then to PC. The latter may be metabolized further by activation of a calcium-dependent phospholipase A_2, thus generating lyso-PC and free fatty acid. Expression of a phospholipase with a specificity for arachidonyl-PC (Crews et al. 1980, 1981; McGivney et al. 1981b) would liberate free arachidonic acid (see below).

The above sequence of reactions has now been demonstrated in stimulated rat basophil leukaemic cells and rat peritoneal mast cells (Axelrod and Hirata 1982; Crews et al. 1980, 1981; Hirata et al. 1979; Ishizaka 1982; Ishizaka et al. 1980, 1981a, 1981b). Immunologic activation of these cells induces a rapid rise and fall in methylated lipids which is followed temporally by calcium influx and histamine secretion (Crews et al. 1980; Ishizaka et al. 1980, 1981a). Inhibitors of methyltransferases block all three processes in highly correlated fashion (Crews et al. 1980; Ishizaka et al. 1980, 1981a; Morita and Siraganian 1981). These inhibitors also prevent IgE-mediated release from human basophil leucocytes and enzymically dispersed human lung mast cells (Beynon et al. 1983; Morita et al. 1981). In total, these data strongly suggest that phospholipid methylation is a primary event in immunologically induced calcium transport and histamine release. Further evidence in support of this view has come from studies with variants of the rat basophil leucocyte cell line (McGivney et al. 1981a). Sublines deficient in one or other of the methyltransferases are individually refractory to immunologic challenge, but fusion of the mutants produces hybrids with normal phospholipid methyltransferase activity and restores their ability to respond to IgE-directed ligands with calcium uptake and histamine release. It should be noted, however, that activation of the transmethylation sequence appears to be confined to IgE-mediated stimuli and does not occur in response to pharmacologic agonists including the calcium ionophore A23187, compound 48/80, polymyxin and ATP (Ishizaka et al. 1980, 1981a; Morita and Siraganian 1981; Morita et al. 1981).

The mechanism whereby lipid methylation may lead to calcium mobilization is uncertain. The methyltransferase enzymes and their products are asymmetrically distributed in the membrane: methyltransferase I is located on the cytoplasmic face and methyltransferase II is located on the exterior surface of the bilayer (Hirata and Axelrod 1980). The formation and transfer of methylated phospholipid increases the fluidity of the membrane (Hirata and Axelrod 1980) and this change in microviscosity may facilitate calcium fluxes either directly or by promoting the opening of specific calcium channels. A similar effect may be produced by the generation of the known fusogen lyso-PC, which could promote the membrane changes involved in exocytosis. The formation of arachidonic acid may also be of importance, and a number of studies have indicated that the oxidative metabolism of this compound may be directly involved in the modulation of calcium fluxes and in the control of the release process itself. As discussed above, a primary route to the generation of arachidonic acid in situ involves activation of the enzyme phospholipase A_2 and the subsequent cleavage of one or more products of the transmethylation sequence (Crews et al. 1980; McGivney et al. 1981b). Inhibitors of phospholipase A_2 block histamine secretion from rat peritoneal mast cells and human basophils (König et al. 1981; McGivney et al. 1981b; Magro 1982; Marone et al. 1981). Conversely, exogenous arachidonic acid potentiates induced histamine release from these cells (Marone et al. 1979; Sullivan and Parker 1979). Inhibitors of the lipoxygenase pathway for the metabolism of arachidonic acid, or combined inhibitors of the lipoxygenase and cyclooxygenase pathways, prevent histamine release in dose-dependent fashion, whereas inhibitors of the cyclooxygenase pathway alone have no effect or even potentiate secretion (König et al. 1981; McGivney et al. 1981b; Magro 1982; Marone et al. 1979; Nemeth and Douglas 1980, 1982; Sullivan and Parker 1979). Further work has shown that lipoxygenase inhibitors prevent histamine release evoked by IgE-directed ligands when the response is due essentially to extracellular calcium, but have no effect on polybasic agonists capable of mobilizing intracellular stores of the cation (Nemeth and Douglas 1982). These data suggest that one or more products of the lipoxygenase pathway are involved in modulating the earliest stages of stimulus-secretion coupling, possibly in the control of calcium influx into the cell. Possible candidates for this role are 5-hydroxyeicosatetraenoic acid (5-HETE), 12-hydroxyeicosatetraenoic acid (12-HETE) or 5-hydroperoxyeicosatetraenoic acid (5-HPETE) which all produce a dose-dependent enhancement of IgE-mediated histamine release (Peters et al. 1981; Stenson et al. 1980).

3.3 Changes in Cyclic Nucleotide Levels

Historically, pharmacologically induced elevations in the intracellular level of adenosine $3':5'$-cyclic monophosphate (cyclic AMP, cAMP) have been associated with inhibition of mediator release from the mast cell and basophil leucocyte (Gillespie and Lichtenstein 1975; Kaliner and Austen 1974). This parallel relationship has, however, been questioned in a number of recent publications, and it is currently not entirely clear to what extent these two parameters are directly and necessarily correlated (for discussion, see Pearce 1982a, 1984). More cogently, it has also been reported that immunologic activation of the rat peritoneal mast cell produces a rapid, transient elevation in cAMP (Burt and Stanworth 1983; Holgate et al. 1980a, b; Ishizaka 1982, Ishizaka et al. 1981 a, b; Lewis et al. 1979; Sullivan et al. 1976). There is also a late rise in cAMP and a biphasic elevation of guanosine $3:5$-cyclic monophosphate (cyclic GMP, cGMP) but these changes are abolished by preincubation with indomethacin, suggesting that they are secondary to the release of prostaglandins by the mast cell (Lewis et al. 1979). The early, indomethacin-resistant rise in cAMP is, however, considered to be a primary event in the activation sequence. The response is not blocked by β-adrenoceptor antagonists, does not appear to be mediated by inhibition of phosphodiesterase but is potentiated by guanosine triphosphate (GTP) (Ishizaka et al. 1981b). This, and other evidence (see below), suggests that the bridging of IgE receptors activates adenylate cyclase, the interaction between the receptors and the catalytic subunit being mediated, as in other systems, by combination of GTP with a specific guanine nucleotide-binding protein (Holgate et al. 1980a; Ishizaka et al. 1981b).

The relationship between changes in cAMP, phospholipid methylation, calcium influx and histamine secretion has also been studied (Ishizaka et al. 1981a, b). Temporally, the stimulated rise in cAMP parallels the increase in phospholipid methylation and precedes both calcium influx and histamine secretion. The changes in phospholipid methylation and cAMP appear to be mutually regulated. Thus, inhibitors of methyltransferase suppress the immunologically induced rise in cAMP, at least in isolated membrane preparations, and pharmacologically increased levels of the nucleotide block transmethylation (Ishizaka 1982; Ishizaka et al. 1981b). These results then indicate that IgE receptors are linked both to adenylate cyclase and methyltransferases. Stimulation of the latter enzymes may also be preceded by activation of an endogenous serine esterase, since various tryptic and chymotryptic inhibitors and substrates block both phospholipid methylation and the rise in cAMP. The possible relationship

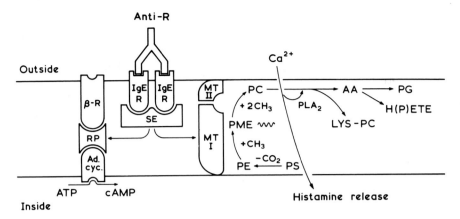

Fig. 1. Schematic representation of the relationship between IgE-receptors, methyltransferases and adenylate cyclase in the mast cell membrane. IgE receptor (*IgE-R*); anti-receptor antibody (*Anti-R*); β-adrenoceptor (*β-R*); GTP-dependent regulatory protein (*RP*); adenylate cyclase (*Ad. Cyc*); methyltransferases I and II (*MTI, MTII*); putative serine esterase (*SE*); phosphatidylserine (*PS*); phosphatidylethanolamine (*PE*); phosphatidyl-N-monomethylethanolamine (*PME*); phosphatidylcholine (*PC*); lysophosphatidylcholine (*LYS-PC*); phospholipase A$_2$ (*PLA$_2$*); arachidonic acid (*AA*); prostaglandins (*PG*); hydr(oper)oxyeicosatetraenoic acids (*H(P)ETE*); change in membrane fluidity (〰). (After Axelrod and Hirata 1982) and Ishizaka et al. 1981a).

between this enzyme, IgE receptors, methyltransferases and adenylate cyclase is represented schematically in Fig. 1.

It also remains to be decided whether the rise in cAMP following mast cell activation is involved in triggering histamine release or rather in the natural termination of the process. Evidence in favour of the former hypothesis has come from studies using adenosine and its analogues as probes either to inhibit the activation of adenylate cyclase or to stimulate the enzyme at a site independent of the immunologic reaction (Holgate et al. 1980a, b). Modulation of the induced changes in cAMP in this way produces parallel changes in histamine secretion. To explain the disparate effects of immunologically and pharmacologically induced elevations of cAMP, it has then been argued that these processes may generate the nucleotide within discrete pools in the mast cell (Holgate et al. 1980a, c, d). This may lead to the preferential activation of particular cAMP-dependent protein kinase isoenzymes involved in either the induction or suppression of the release mechanism.

It should be noted, however, that an early rise in cAMP appears to be produced only by IgE-directed and related ligands. In particular, stimulation of the mast cell with compound 48/80 or the ionophore A23187 leads to no change or to an abrupt fall, rather than an increase in the level of the nucleotide (Burt and Stanworth 1983; Sullivan et al. 1975). However, histamine release by these compounds is also potentiated by adenosine (Marquardt et al. 1978), suggesting that the mode of action of this agent may be more complex than hitherto considered. Moreover, an obligatory rise in cAMP does not occur following stimulation of the human basophil and does not appear to be a necessary part of the activation sequence in this cell type (Hughes et al. 1983; MacGlashan et al. 1983). The situation in the human lung mast cell is apparently more complex and cAMP changes following immunologic activation have been reported by some authors (Ishizaka et al. 1983) but not others (MacGlashan et al. 1983). The reasons for these discrepancies are not clear. However, in the light of these observations, and given the general inhibitory effect of cAMP in histamine release, it may be that the transient rise in the level of the nucleotide following immunologic stimulation is involved in the termination rather than the initiation of the reaction (Axelrod and Hirata 1982; Ishizaka et al. 1981a, b).

4 Calmodulin and Histamine Secretion

The exact mechanism whereby calcium links cell activation to the subsequent secretory response is not known in the mast cell or any other system. However, recent work suggests that the ubiquitous binding protein calmodulin (for general reviews, see Brostom and Wolff 1981; Cheung 1980a, b; Klee et al. 1980; Means and Dedman 1980; Means et al. 1982; Vincenzi 1981; Weiss et al. 1982) is involved in the process. This conclusion is largely based on studies of the effect of defined calmodulin antagonists on histamine release. The most complete investigation is that of Douglas and Nemeth (1982), who showed that a series of phenothiazine antipsychotic drugs and the naphthalenesulphonamide W-7 block histamine secretion from rat peritoneal mast cells activated by immunologic or pharmacologic stimuli. In the cases of compound 48/80 and the ionophore A23187, the inhibitory effects of the compounds are closely correlated with their ability to antagonize calmodulin. A greater potency and different rank order is observed in the case of antigen, suggesting an additional inhibitory action of the compunds. Neither the phenothiazines nor W-7 reduce calcium uptake in response to the ionophore A23187, indicating that they act distal to the rise in intracellular calcium following activation. Chlorpromazine sulphoxide, which shares several membrane-perturbing actions of the phenothiazines but is a weak inhibitor of calmodulin, does not prevent secretion. In total, these results suggest that the efficacy of the drugs is largely determined by their ability to inhibit calmodulin and is unlikely to be due to non-specific membrane effects. Consistently the dechloro derivative W-5, which has virtually no calmodulin-inhibiting activity, does not affect release from mast cells treated with concanavalin A or the ionophore A23187 (Suzuki et al. 1983). Inhibition of ionophore-induced histamine release by phenothiazines and the imidazole derivative R24571 has also been reported by Amellal and Landry (1983). Similar findings have been obtained with the human basophil leucocyte where phenothiazines and naphthalenesulphonamides again block secretion due to diverse stimuli (Cumella et al. 1983; Marone et al. 1983a, b). Sulphoxide derivatives and W-5 are inactive, confirming the specificity of the effect (Marone et al. 1983a, b).

Broadly similar results have been obtained in our laboratories and trifluoperazine (TFP) has been shown to inhibit histamine release in dose-dependent fashion from rat peritoneal mast cells stimulated with IgE-mediated ligands, basic secretagogues, dextran and calcium ionophores (unpublished work). The phenothiazine is generally at least equiactive, and usually considerably more active, when release is induced in the ab-

Fig. 2. Inhibition by trifluoperazine of histamine release from rat peritoneal mast cells
stimulated with compound 48/80 (0.175 μg
ml^{-1}) in the presence ($\circ$) and absence ($\triangle$) of
extracellular calcium ions (1 mM). The latter
medium contained EDTA (0.1 mM). *Points
are the means from six experiments and
vertical bars* denote S.E.M. Control releases
were ($\circ$) 41.2 ± 3.2 and ($\triangle$) 37.8 ± 3.1

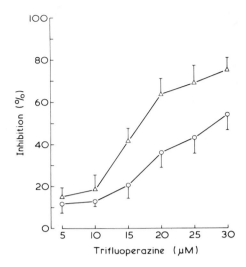

sence of added calcium and hence due to the mobilization of intracellular stores of the
cation. These findings are again consistent with an action of the compound subsequent
to calcium influx. Essentially similar conclusions have been drawn by Alm (1983).
Representative results for compound 48/80 are shown in Fig. 2.

We have further examined the effect of a range of calmodulin antagonists on histamine release induced by anti-IgE and compound 48/80 (Table 2). In general, the com-

Table 2. Effect of calmodulin antagonists on histamine secretion from rat peritoneal mast cells

Antagonist	IC_{50}^{HR} (μM) for histamine release induced by:		IC_{50}^{CaM} (μM) for inhibition of calmodulin	IC_{50}^{HR} IC_{50}^{CaM} for:	
	Anti-IgE	Compound 48/80		Anti-IgE	Compound 48/80
Chlorpromazine	13	49	42	0.31	1.17
Chlorprothixene	17	24	17	1.00	1.41
Imipramine	51	63	125	0.41	0.50
Promethazine	22	118	340	0.06	0.35
R24571	8	8	0.005	1600	1600
Thioridazine	6	17	18	0.33	0.94
Trifluoperazine	16	24	10	1.60	2.40
W-7	32	50	28	1.14	1.79

IC_{50}^{HR} is the concentration of antagonist required to produce 50% inhibition of release by the specified ligand. Concentrations of secretagogue were chosen to produce control levels of release of
ca. 35%. IC_{50}^{CaM} is the concentration of antagonist required to produce 50% inhibiton of calmodulin-dependent activation of phosphodiesterase (values taken from manufacturers information and
Weiss et al. 1982). The ratio $IC_{50}^{HR}/IC_{50}^{CaM}$ would be constant if the effects of the drugs on histamine release depended simply on their anti-calmodulin activity

pounds are seen to be rather more active against the immunologic challenge. There is a reasonable correlation between the effects of the drugs on histamine release and their reported potency as inhibitors of calmodulin, except in the case of R24571 which is much less active than would be expected on this basis and of promethazine which, particularly in the case of anti-IgE, is more active than predicted. Given that the ability of the agents to prevent histamine secretion must reflect multiple effects, including ease of penetration into the mast cell, possible interactions between drug and secretagogue and potential local membrane actions, these data may be taken as indicative of the involvement of calmodulin in the process. More firm conclusions are rendered ambiguous by the difficulties inherent in comparing data based on isolated enzyme preparations (anti-calmodulin activity) and on intact cellular systems (histamine release).

The mechanisms whereby calmodulin might regulate histamine secretion remain the source of speculation. However, considerable analogy may be drawn from other systems (Brostom and Wolff 1981; Cheung 1980a, b; Klee et al. 1980; Means and Dedman 1980; Means et al. 1982; Vincenzi 1981; Weiss et al. 1982). The enzymes whose activity is regulated by calmodulin include nucleotide phosphodiesterase, adenylate and guanylate cyclase, the calcium/magnesium-ATPase of the plasma membrane and sarcoplasmic reticulum, actomysin ATPase, phospholipase A_2, tryptophan hydroxylase, methyltransferases, myosin light chain kinase, phosphorylase b kinase, glycogen synthetase kinase, plant NAD-kinase and a variety of kinases involved in the phosphorylation of membrane proteins. This diversity provides a number of possible opportunities for the control of exocytosis. Firstly, calmodulin may regulate the activity of methyltransferases and phospholipase A_2. The importance of these enzymes in histamine release has already been discussed in detail (Sect. 3.2). Secondly, calmodulin regulates the ordered assembly-disassembly of microtubules. Various evidence indicates that these and other cytoskeletal elements are centrally involved in histamine secretion (Gillespie and Lichtenstein 1972; Kaliner 1977). Thirdly, calmodulin may control both the synthesis and degradation of cyclic nucleotides by activation of phosphodiesterase and adenylate cyclase. The stimulation of both enzymes may appear at first sight paradoxical, but the proteins may be activated in sequential fashion (Cheung 1980a, b). Thus, influx of calcium through the plasma membrane, or the release of sequestered membranous stores of the cation in response to a stimulus, may initially activate the membrane-bound adenylate cyclase, so leading to an increase in the intracellular level of cAMP. Subsequent passage of calcium into the cell would then activate the cytosolic phosphodiesterase, restoring the concentration of cAMP to its steady-state level. The net effect would be to produce a transient elevation of cAMP (Cheung 1980a, b). As discussed (Sect. 3.3), precisely such a phenomenon has been observed following stimulation of the mast cell. Finally, and perhaps most importantly, calmodulin activates a number of specific kinases capable of phosphorylating cytosolic and membrane proteins. There is now increasing evidence that such phosphorylation is a key event in stimulus-secretion coupling in a number of cell types. In particular, activation of the mast cell leads to a rapid, calcium-dependent phosphorylation of specific proteins of defined molecular weight (Sieghart et al. 1978). This phosphorylation accompanies or precedes histamine secretion. However, the participation of calmodulin in this process remains to be determined, and the effect of phenothiazines on protein phosphorylation in the mast cell would be of great interest. Thus, while the precise involvement of calcium and calmodulin in histamine secretion remains to be determined, the identification of the protein as a primary mediator of the effects of the cation and the rapid development of research in this area inspires the hope that this role may soon be clarified.

Acknowledgements. Work from the authors' laboratories was supported by grants from Fisons Pharmaceuticals Ltd, the Medical Research Council, the North Atlantic Treaty Organization, the Science and Engineering Research Council, the Wellcome Trust and the World University Service.

References

Alm PE (1983) Cyclic AMP levels during stimulation and inhibition of histamine release from rat mast cells by the calmodulin inhibitor trifluoperazine. Int Arch Allergy Appl Immunol 71:103-111

Ammellal M, Landry Y (1983) Lanthanides are transported by ionophore A23187 and mimic calcium in the histamine secretion process. Br J Pharmacol 80:365-370

Axelrod J, Hirata F (1982) Phospholipid methylation and the receptor-induced release of histamine from cells. Trends Pharmacol Sci 3:156-158

Bennet JP, Cockcroft S, Gomperts BD (1979) Ionomycin stimulates mast cell histamine secretion by forming a lipid-soluble calcium complex. Nature (Lond) 282:851-853

Bennet JP, Cockcroft S, Gomperts BD (1981) Rat mast cells permeabilized with ATP secrete histamine in response to calcium ions buffered in the micromolar range. Biochem J 317:335-345

Beynon RC, Church MK, Holgate ST (1983) Inhibitors of transmethylation decrease histamine release from rat peritoneal and human dispersed lung mast cells. Br J Pharmacol 79:242P

Brostrom CO, Wolff DJ (1981) Properties and functions of calmodulin. Biochem Pharmacol 30:1395-1405

Burt DS, Stanworth DR (1983) Changes in cellular levels of cyclic AMP in rat mast cells during secretion of histamine induced by immunoglobulin E decapeptide and ACTH (1-24) peptide. Comparison with immunological and ionophore triggers. Biochim Biophys Acta 762:458-465

Cheung WY (1980a) Calmodulin plays a pivotal role in cellular regulation. Science (Wash DC) 207:19-27

Cheung WY (ed) (1980b) Calcium and cell function, vol I. Calmodulin. Academic, New York

Cochrane DE, Distel DI (1982) Association of ^{45}calcium with rat mast cells stimulated by 48/80: effects of inactivation, calcium and metabolic inhibition. J Physiol (Lond) 330:413-427

Cockcroft S (1981) Does phosphatidylinositol breakdown control the Ca^{2+}-gating mechanism? Trends Pharmacol Sci 2:340-342

Cockcroft S, Gomperts BD (1979) Evidence for a role of phosphatidylinositol turnover in stimulus-secretion coupling. Biochem J 178:681-687

Cockcroft S, Gomperts BD (1980) The ATP^{4-} receptor of rat mast cells. Biochem J 188:789-798

Crews FT, Morita Y, Hirata F, Axelrod J, Siraganian RP (1980) Phospholipid methylation affects immunoglobulin E-mediated histamine and arachidonic acid release in rat leukemic basophils. Biochem Biophys Res Commun 93:42-49

Crews FT, Morita Y, McGivney A, Hirata F, Siraganian RP, Axelrod J (1981) IgE-mediated histamine release in rat basophilic leukemia cells: receptor activation, phospholipid methylation, Ca^{2+} flux and release of arachidonic acid. Arch Biochem Biophys 212:561-571

Cumella J, Middleton E, Drzewiecki G (1983) Effect of calmodulin (CaM) antagonists on basophil histamine release (HR). J Allergy Clin Immunol 71:94

Dahlquist R (1974) Relationship of uptake of sodium and ^{45}calcium to ATP-induced histamine release from rat mast cells. Acta Pharmacol Toxicol 35:11-22

Douglas WW, Kagayama M (1977) Calcium and stimulus-secretion coupling in the mast cell: stimulant and inhibitory effects of calcium-rich media on exocytosis. J Physiol (Lond) 270:691-703

Douglas WW, Nemeth EF (1982) On the calcium receptor activating exocytosis: inhibitory effects of calmodulin-interacting drugs on rat mast cells. J Physiol (Lond) 323:229-244

Ennis M, Truneh A, White JR, Pearce FL (1980) Calcium pools involved in histamine release from rat mast cells. Int Arch Allergy Appl Immunol 62:467-471

Ennis M, Truneh A, Pearce FL (1981) Lectin-induced histamine secretion from isolated rat and guinea pig mast cells. Biochem Pharmacol 30:2179-2181

Foreman JC, Mongar JL (1973) The action of lanthanum and manganese on anaphylactic histamine secretion. Br J Pharmacol 48:527-537

Foreman JC, Mongar JL, Gomperts BD (1973) Calcium ionophores and movement of calcium ions following the physiological stimulus to a secretory process. Nature (Lond) 245:249-251

Foreman JC, Garland LG, Mongar JL (1976) The role of calcium in secretory processes: model studies in mast cells. In: Duncan CJ (ed) Calcium in biological systems. Cambridge University Press, Cambridge, pp 193-218

Foreman JC, Hallett MB, Mongar JL (1977) The relationship between histamine secretion and ^{45} calcium uptake by mast cells. J Physiol (Lond) 271:193-214

Gil DW, Brown SA, Seeholzer SH, Wildey GM (1983) Phosphatidylinositol turnover and cellular function. Life Sci 32:2043-2046

Gillespie E, Lichtenstein LM (1972) Histamine release from human leukocytes: studies with deuterium oxide, colchicine, and cytochalasin B. J Clin Invest 51:2941-2947

Gillespie E, Lichtenstein LM (1975) Histamine release from human leukocytes: relationships between cyclic nucleotide, calcium, and antigen concentrations. J Immunol 115:1572-1576

Gomperts BD, Baldwin JM, Micklem KJ (1983) Rat mast cells permeabilized with Sendai virus secrete histamine in response to Ca^{2+} buffered in the micromolar range. Biochem J 210:737-745

Grosman N, Diamant B (1974) Studies on the role of calcium in anaphylactic histamine release from isolated rat mast cells. Acta Pharmacol Toxicol 35:284-292

Grosman N, Diamant B (1978) Binding of ^{45} calcium to isolated rat mast cells in connection with histamine release. Agents Actions 8:338-346

Hirata F, Axelrod J (1980) Phospholipid methylation and biological signal transmission. Science (Wash DC) 209:1082-1090

Hirata F, Axelrod J, Crews FT (1979) Concanavalin A stimulates methylation and phosphatidylserine decarboxylation in rat mast cells. Proc Natl Acad Sci USA 76:4813-4816

Holgate ST, Lewis RA, Austen KF (1980a) The role of cyclic nucleotides in mast cell activation and secretion. In: Fougereau M, Dausset J (eds) Immunology 80. Academic, London, pp 846-859

Holgate ST, Lewis RA, Austen KF (1980b) Role of adenylate cyclase in immunologic release of mediators from rat mast cells: agonist and antagonist effects of purine- and ribose-modified adenosine analogs. Proc Natl Acad Sci USA 77:6800-6804

Holgate ST, Lewis RA, Austen KF (1980c) 3,5-Cyclic adenosine monophosphate-dependent protein kinase of the rat serosal mast cell and its immunologic activation. J Immunol 124:2093-2099

Holgate ST, Lewis RA, Maguire JF, Roberts LJ, Oates JA, Austen KF (1980d) Effects of prostaglandin D_2 on rat serosal mast cells: discordance between immunologic mediator release and cyclic AMP levels. J Immunol 125:1367-1373

Hughes PJ, Holgate ST, Roath S, Church MK (1983) The relationship between cyclic AMP changes and histamine release from basophil-rich human leucocytes. Biochem Pharmacol 17:2557-2563

Ishizaka T (1982) Biochemical analysis of triggering signals induced by bridging of IgE receptors. Fed Proc Fed Am Soc Exp Biol 14:17-21

Ishizaka T, Hirata F, Ishizaka K, Axelrod J (1980) Stimulation of phospholipid methylation, Ca^{2+} influx and histamine release by bridging of IgE receptors on rat mast cells. Proc Natl Acad Sci USA 77:1903-1906

Ishizaka T, Hirata F, Ishizaka K, Axelrod J (1981a) Transmission and regulation of triggering signals induced by bridging of IgE receptors on rat mast cells. In: Becker EL, Simon AS, Austen KF (eds) Biochemistry of the acute allergic reactions. Liss, New York, pp 213-227

Ishizaka T, Hirata F, Sterk AR, Ishizaka K, Axelrod JA (1981b) Bridging of IgE receptors activates phospholipid methylation and adenylate cyclase in mast cell plasma membranes. Proc Natl Acad Sci USA 78:6812-6816

Ishizaka T, Conrad DH, Schulman ES, Sterk AR, Ishizaka K (1983) Biochemical analysis of initial triggering events of IgE-mediated histamine release from human lung mast cells. J Immunol 130:2357-2362

Kaliner M (1977) Human lung tissue and anaphylaxis. Evidence that cyclic nucleotides modulate the immunologic release of mediators through effects on microtubular assembly. J Clin Invest 60:951-959

Kaliner M, Austen KF (1974) Cyclic nucleotides and modulation of effector systems of inflammation. Biochem Pharmacol 23:763-771

Kano T, Cochrane DE, Douglas WW (1973) Exocytosis (secretory granule extrusion) induced by injection of calcium into mast cells. Can J Physiol Pharmacol 51:1001-1004

Kazimierczak W, Diamant B (1978) Mechanisms of histamine release in anaphylactic and anaphylactoid reactions. Progr Allergy 24:295-365

Kennerly DA, Sullivan TJ, Parker CW (1979) Activation of phospholipid metabolism during mediator release from stimulated mast cells. J Immunol 122:152-159

Klee CB, Crouch TH, Richman PG (1980) Calmodulin. Annu Rev Biochem 49:489-515

König W, Pfeiffer F, Kunau HW (1981) Effect of arachidonic acid metabolites on the histamine release from human basophils and rat mast cells. Int Arch Allergy Appl Immunol 66 (Suppl 1): 149-151

Lagunoff D, Martin TW, Read G (1983) Agents that release histamine from mast cells. Annu Rev Pharmacol Toxicol 23:331-351

Lewis RA, Austen KF (1981) Mediation of local homeostasis and inflammation by leukotrienes and other mast cell-dependent components. Nature (Lond) 293:103-108

Lewis RA, Holgate ST, Roberts LJ, Maguire JF, Oates JA, Austen KF (1979) Effects of indomethacin on cyclic nucleotide levels and histamine release from rat serosal mast cells. J Immunol 123:1663-1668

McGivney A, Crews FT, Hirata F, Axelrod J, Siraganian RP (1981a) Rat basophilic leukemia cell lines defective in phospholipid methyltransferase enzymes, Ca^{2+} influx and histamine release: reconstitution by hybridization. Proc Natl Acad Sci USA 78:6176-6180

McGivney A, Morita Y, Crews FT, Hirata F, Axelrod J, Siraganian RP (1981b) Phospholipase activation in the IgE-mediated and Ca^{2+} ionophore A23187-induced release of histamine from rat basophil leukemia cells. Arch Biochem Biophys 212:572-580

MacGlashan DW, Schleimer RP, Peters SP et al. (1983) Comparative studies of human basophils and mast cells. Fed Proc Fed Am Soc Exp Biol 42: 2504-2509

Magro AM (1982) Effect of inhibitors of arachidonic acid metabolism upon IgE and non-IgE-mediated histamine release. Int J Immunopharmacol 4:15-20

Marone G, Kagey-Sobotka A, Lichtenstein LM (1979) Effects of arachidonic acid and its metabolites on antigen-induced histamine release from human basophils in vitro. J Immunol 123:1669-1677

Marone G, Kagey-Sobotka A, Lichtenstein LM (1981) Control mechanisms of histamine release from human basophils in vitro: the role of phospholipase A_2 and of lipoxygenase metabolites. Int Arch Allergy Appl Immunol 66 (Suppl 1): 144-148

Marone G, Columbo M, Poto S, Bianco P, Torella G, Condorelli M (1983a) Possible role of calmodulin in human inflammatory reactions. Monogr Allergy 18:290-299

Marone G, Columbo M, Poto S, Conorelli M (1983b) Modulation of histamine release from human basophils by calmodulin antagonists. J Allergy Clin Immunol 71:94

Marquardt DL, Parker CW, Sullivan TJ (1978) Potentiation of mast cell mediator release by adenosine. J Immunol 120:871-878

Means AR, Dedman JR (1980) Calmodulin — an intracellular calcium receptor. Nature (Lond) 285:73-77

Means AR, Tash JS, Chafouleas JG (1982) Physiological implications of the presence, distribution, and regulation of calmodulin in eukaryotic cells. Physiol Rev 62:1-38

Metcalfe DD, Kaliner M, Donlon MA (1981) The mast cell. CRC Crit Rev Immunol 3:23-74

Michel RH (1983) Polyphosphoinositide breakdown as the initiating reaction in receptor stimulated inositol phospholipid metabolism. Life Sci 32:2083-2085

Michell RH, Kirk CJ (1982) The unknown meaning of receptor-stimulated inositol lipid metabolism. Trends Pharmacol Sci 2:140-141

Morita Y, Siraganian RP (1981) Inhibition of IgE-mediated histamine release from rat basophil leukemia cells and rat mast cells by inhibitors of transmethylation. J Immunol 127:1339-1344

Morita Y, Chiang PK, Siraganian RP (1981) Effect of inhibitors of transmethylation on histamine release from human basophils. Biochem Pharmacol 30:785-791

Nemeth EF, Douglas WW (1980) Differential inhibitory effects of the arachidonic acid analogue ETYA on rat mast cell exocytosis evoked by secretagogues utilizing cellular or extracellular calcium. Eur J Pharmacol 67:439-450

Nemeth EF, Douglas WW (1982) Lipoxygenase inhibitors exert secretagogue-specific effects on mast cell exocytosis. Eur J Pharmacol 79:315-318

Peachell PT, Pearce FL (1984) Some studies on the release of histamine from mast cells treated with polymyxin. Agents Actions 14:379-385

Pearce FL (1982a) Calcium and histamine secretion from mast cells. Prog Med Chem 19:59-109

Pearce FL (1982b) Functional heterogeneity of mast cells from different species and tissues. Klin Wochenschr 60:954-957

Pearce FL (1983) Mast cell heterogeneity. Trends Pharmacol Sci 4:165-167

Pearce FL (to be published 1984) Biochemical events involved in the release of anaphylactic mediators from mast cells. Asthma Rev

Pearce FL, Messis PD (1982) Phosphatidic acid induces histamine secretion from rat peritoneal mast cells. Int Arch Allergy Appl Immunol 68:93-95

Pearce FL, White JR (1981) Effect of lanthanide ions on histamine secretion from rat peritoneal mast cells. Br J Pharmacol 72:341-347

Pearce FL, White JR (1984) Calcium efflux and histamine secretion from rat peritoneal mast cells. Agents Actions 14:392-396

Pearce FL, Ennis M, Truneh A, White JR (1981) Role of intra- and extracellular calcium in histamine release from rat peritoneal mast cells. Agents Actions 11:51-54

Pearce FL, Barrett KE, White JR (1983) Histamine secretion from mast cells treated with chlortetracycline (aureomycin): a novel calcium ionophore. Agents Actions 13:117-122

Pepys J, Edwards AM (eds) (1979) The mast cell: its role in health and disease. Pitman, Tunbridge Wells

Peters SP, Siegel MI, Kagey-Sobotka A, Lichtenstein LM (1981) Lipoxygenase products modulate histamine release in human basophils. Nature (Lond) 292:455-457

Ranadive NS, Dhanani N (1980) Movement of calcium ions and release of histamine from rat mast cells. Int Arch Allergy Appl Immunol 61:9-18

Ranadive NS, Lewis R (1982) Extracellular effect of calcium on compound 48/80 stimulated mast cells. Can J Biochem Cell Biol 61:79-84

Samuelsson B (1983) Leukotrienes: mediators of immediate hypersensitivity reactions and inflammation. Science (Wash DC) 220:568-575

Schellenberg RR (1980) Enhanced phospholipid metabolism in rat mast cells stimulated to release histamine. Immunology 41:123-129

Sieghart W, Theoharides TC, Alper SL, Douglas WW, Greengard P (1978) Calcium-dependent protein phosphorylation during secretion by exocytosis in the mast cell. Nature (Lond) 275:329-331

Spataro AC, Bosmann HB (1976) Mechanism of action of disodium cromoglycate-mast cell calcium ion influx after a histamine-releasing stimulus. Biochem Pharmacol 25:505-510

Stenson WF, Parker CW, Sullivan TJ (1980) Augmentation of IgE-mediated release of histamine by 5-hydroxyeicosatetraenoic acid and 12-hydroxyeicosatetraenoic acid. Biochem Biophys Res Commun 96:1045-1052

Sugiyama K (1971) Significance of ATP splitting activity of rat peritoneal mast cells in the histamine release induced by exogenous ATP. Jpn J Pharmacol 21:531-539

Sullivan TJ, Parker CW (1979) Possible role of arachidonic acid and its metabolites in mediator release from rat mast cells. J Immunol 122:431-436

Sullivan TJ, Parker KL, Eisen SA, Parker CW (1975) Modulation of cyclic AMP in purified rat mast cells. II. Studies on the relationship between intracellular cyclic AMP concentrations and histamine release. J Immunol 114:1480-1485

Sullivan TJ, Parker KL, Kulczycki A, Parker CW (1976) Modulation of cylclic AMP in purified rat mast cells. III. Studies on the effects of concanavalin A and anti-IgE on cyclic AMP concentrations during histamine release. J Immunol 117:713-716

Suzuki T, Ohishi K, Uchida M (1983) Effects of calmodulin inhibitor on histamine release from rat peritoneal mast cells induced by concanavalin A and ionophore A23187. Biochem Pharmacol 14:273-275

Theoharides TC, Douglas WW (1978) Secretion in mast cells induced by calcium trapped within phospholipid vesicles. Science (Wash DC) 201:1143-1145

Truneh A, Pearce FL (1981) Characteristics of and calcium requirements for histamine release from rat peritoneal mast cells treated with concanavalin A. Int Arch Allergy Appl Immunol 66:68-75

Vincenzi FF (1981) The pharmacologcal implications of calmodulin. Trends Pharmacol Sci 2: VII-IX

Weiss B, Prozialeck WC, Wallace TL (1982) Interaction of drugs with calmodulin. Biochemical, pharmacological and clinical implications. Biochem Pharmacol 31:2217-2226

White JR, Pearce FL (1981) Role of membrane bound calcium in histamine secretion from rat peritoneal mast cells. Agents Actions 11:324-329

White JR, Pearce FL (1983) Use of chlortetracycline to monitor calcium mobilization during histamine secretion from mast cells: a cautionary note. Anal Biochem 132:1-5

Role of Calcium in Alpha-Adrenergic Regulation of Liver Function

J. H. EXTON[1]

CONTENTS

1 Introduction

The liver can be affected by the sympathetic nervous system via epinephrine and norepinephrine released into the bloodstream from the adrenal medulla or via norepinephrine released from adrenergic nerve endings within the liver. As is the case for other tissues of the body, the receptors to these catecholamines are of two major types designated α- and β-adrenergic receptors. The liver α-receptors are mainly of the α_1-subtype (Hoffman et al. 1979, 1981) and the liver β-receptors are mainly of the β_2-subtype (Morgan et al. 1983).

The relative importance of α_1- and β_2-receptors as mediators of catecholamine effects in the liver depends on a number of factors including species (Exton 1983), age (Morgan et al. 1983), sex (Studer and Borle 1982), thyroid status (Preiksaitis et al. 1982) and food intake (El-Refai and Chan 1982). Since the β-receptors are of the β_2-subtype, they are relatively unresponsive to norepinephrine (Morgan et al. 1983). Thus the effects of direct sympathetic nervous stimulation of the liver are probably mediated mainly by α_1-receptors, which are equally sensitive to epinephrine and norepinephrine (El-Refai et al. 1979). This explains why epinephrine infusion in vivo (which activates both β_2- and α_1-receptors) does not always mimic the effects of sympathetic stimulation which activates mainly α_1-receptors.

The hepatic β_2-adrenergic receptors are coupled positively to adenylate cyclase, as is the case for β-receptors in general, and their activation leads to a rise in cellular cAMP (Cherrington and Exton 1976). This in turn activates cAMP-dependent protein

[1] Laboratories for the Studies of Metabolic Disorders, Howard Hughes Medical Institute and Department of Physiology, Vanderbilt University School of Medicine, Nashville, TN 37232, USA

kinase, which phosphorylates a variety of enzymes and other cellular proteins, thus altering their activities and producing biological responses which are similar to those observed with glucagon (Cherrington and Exton 1976).

The hepatic α_1-adrenergic receptors are principally coupled to mechanisms which increase the cytosolic concentration of ionized calcium (Charest et al. 1983) and Ca^{2+}, in turn, stimulates the activities of several Ca^{2+}-dependent protein kinases and other enzymes thereby producing the cellular responses observed with α-adrenergic agonists. In mature animals, the α_1-adrenergic receptors can become coupled positively to adenylate cyclase (Morgan et al. 1983). However, the increase in cAMP is small and the catecholamine concentrations required to elicit it are five fold higher than those which increase cytosolic Ca^{2+} (Morgan et al. 1983). The concentrations needed are also above those found in the plasma, even under stress conditions. Thus it is questionable whether cAMP plays any role in hepatic α_1-adrenergic actions in vivo.

The liver possesses α_2-adrenergic receptors (Hoffmann et al. 1981) which are coupled in an inhibitory manner to adenylate cyclase (Jard et al. 1981). However, the number of α_2-receptors is low relative to α_1-receptors and, to date, there have been no demonstrations that they play a significant role in the regulation of liver cAMP levels or liver function.

2 Alpha-Adrenergic Regulation of Cell Ca^{2+}

There is much evidence that the effects of α_1-adrenergic agonists (epinephrine, norepinephrine, phenylephrine) and also vasopressin and angiotensin II in liver critically involve Ca^{2+}. Firstly, their effects are abolished or impaired in perfused livers or isolated liver cells which have been depleted of calcium by treatment with the chelator EGTA (Assimacopoulos-Jeannet et al. 1977; Van de Werve et al. 1977; Keppens et al. 1977) and are restored when calcium is replenished (Assimacopoulos-Jeannet et al. 1977). Manipulation of other divalent cations does not produce these effects. Secondly, the biological effects of these Ca^{2+}-dependent hormones can be mimicked in these preparations by the divalent cation ionophore A23187 (Assimacopuolos-Jeannet et al. 1977; Keppens et al. 1977, Blackmore et al. 1978) which causes a rapid influx of extracellular Ca^{2+} across the plasma membrane, thereby raising cytosolic Ca^{2+}. In addition, uncouplers of oxidative phosphorylation and other agents which alter the mitochondrial Ca^{2+} cycle to produce a rise in cytosolic Ca^{2+} can mimic some of the actions of the Ca^{2+}-dependent hormones. Thirdly, these hormones alter cellular Ca^{2+} fluxes as revealed by changes in the efflux and influx of $^{45}Ca^{2+}$ or $^{40}Ca^{2+}$ (Assimacopoulos-Jeannet et al. 1977; Keppens et al. 1977; Blackmore et al. 1978, 1982; Chen et al. 1978; Althaus-Salzmann et al. 1980) and by alterations in the calcium content of intracellular organelles (Blackmore et al. 1979; Babcock et al. 1979; Murphy et al. 1980). Fourthly, these hormones have recently been shown to cytosolic elevate Ca^{2+} (Murphy et al. 1980; Charest et al. 1983).

Most workers now agree that the major initial Ca^{2+} response of liver to α_1-agonists and vasopressin is cellular efflux of Ca^{2+} due to the mobilization of Ca^{2+} from intra-

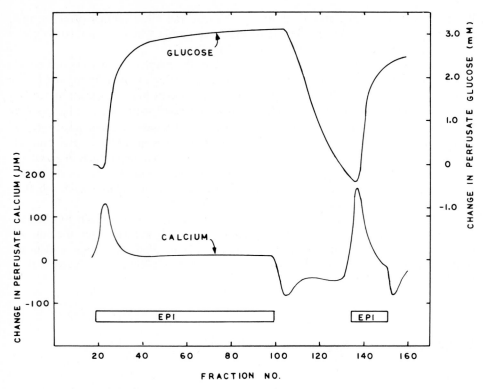

Fig. 1. Effects of two infusions of epinephrine on Ca^{2+} fluxes and glucose output in the isolated rat liver perfused with nonrecirculating medium at a constant flow rate. Epinephrine (1 μM) was infused for 25 min and, after a 10-min delay, for 5 min. Samples of effluent medium were collected every 18 s for measurement of glucose and Ca^{2+}. (Morgan et al. 1982)

cellular calcium stores. Experiments utilizing the perfused liver (Fig. 1) or isolated hepatocyte preparations and measuring Ca^{2+} by chemical means have consistenly shown a transient efflux of Ca^{2+} in response to α_1-agonists and other Ca^{2+}-dependent hormones (Blackmore et al. 1978, 1982, 1983b; Chen et al. 1978; Althaus-Salzmann et al. 1980). Net influx of Ca^{2+} is not observed until the action of the agents is terminated (Fig. 1 and Morgan et al. 1982). Although earlier studies showed a transient initial stimulation of $^{45}Ca^{2+}$ uptake (Assimacopoulos-Jeannet et al. 1977), this can now be explained on the basis of the increased intracellular Ca^{2+} pool size (Blackmore et al. 1978, 1982). Efforts to detect a rapid initial effect on Ca^{2+} uptake using chemical means have been consistently unsuccessful (Blackmore et al. 1983b).

There is other evidence that an influx of extracellular Ca^{2+} is neither required nor involved (Blackmore et al. 1978, 1982; Charest et al. 1984) in the initial rise in cytosolic Ca^{2+} induced by Ca^{2+}-dependent hormones. For example, studies with isolated hepatocytes have shown that rapid depletion of extracellular Ca^{2+} by EGTA to levels of micromolar or lower does not initially impair the ability of the agents to raise cytosolic Ca^{2+} or activate phosphorylase (Fig. 2 and Blackmore et al. 1978, 1982; Charest et al.

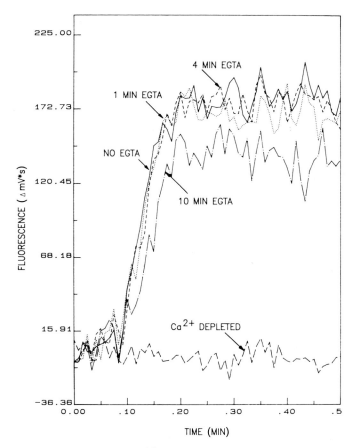

Fig. 2. Effect of short-term chelation of extracellular Ca^{2+} with *EGTA* on the ability of vasopressin to increase cytosolic Ca^{2+} in isolated rat hepatocytes. Hepatocytes were loaded with the fluorescent Ca^{2+} indicator Quin-2 as described by Charest et al. (1983). The Ca^{2+} in the incubation medium was 1 mM, and 2 mM *EGTA* was added prior to vasopressin (10^{-8} M) for the various periods of time indicated. Ca^{2+}-depleted hepatocytes were prepared as described elsewhere (Assimacopoulos-Jeannet et al. 1977). The values shown are the differences between the fluorescence changes in Quin-2 loaded and nonloaded cells

1985). On the other hand, prolonged EGTA treatment abolishes these responses (Fig. 2), since it causes depletion of the intracellular calcium pools which are mobilized by the hormones (Assimacopoulos-Jeannet et al. 1977; Blackmore et al. 1982; Charest et al. 1985). Furthermore, there is no evidence that α_1-agonists and other Ca^{2+}-dependent hormones stimulate the influx of Ca^{2+} at earlier times since the initial rate at which Ca^{2+} added to calcium-depleted liver cells increases cytosolic Ca^{2+} is unaffected by these agents (R. Charest and J. H. Exton, unpublished observations).

As will be discussed below, although there is no evidence that Ca^{2+} influx plays a role in the initial effects of α_1-agonists and other Ca^{2+}-dependent agents on liver, i.e., those occurring within 1 min, there is much evidence that extracellular Ca^{2+} is required to sustain the effects of these agents beyond 1 min (Charest et al. 1985). This is because it permits them to maintain an elevated cytosolic Ca^{2+} level (Charest et al. 1984).

As noted above, the major initial Ca^{2+} response induced by α_1-agonists in liver is Ca^{2+} efflux from the cells. This appears to be secondary to the rise in cytosolic Ca^{2+} and presumably reflects the increased activity of the plasma membrane $(Ca^{2+}\text{-}Mg^{2+})$ ATPase pump due to this rise. The sources of the intracellular Ca^{2+} released into the cytosol by hormone action remain a matter of some dispute. Most investigators agree that mitochondria are a major source in liver, based on measurements of the calcium content of these organelles following hormone stimulation (Blackmore et al. 1979; Babcock et al. 1979; Dehaye et al. 1980, 1981; Murphy et al. 1980; Barritt et al. 1981). Other intracellular sources (endoplasmic reticulum, plasma membrane) have been suggested (Exton 1983), but no consistent changes in their calcium content have been measured.

The mechanism(s) by which α_1-agonists release Ca^{2+} from mitochondria remains unknown. Since the α_1-adrenergic receptors are located exclusively on the plasma membrane (Clarke et al. 1978; Guellaen et al. 1978; El-Refai et al. 1979), it has been proposed that the communication between the receptors and the mitochondria involves an intracellular signal or second messenger. Although a large number of candidates have been proposed (Exton 1983), the most likely compound is inositol-1,4,5-P_3 which is generated by phosphatidylinositol-4,5-P_2 breakdown (see Sect. 4).

The speed with which mitochondrial Ca^{2+} release is stimulated by α_1-agonists is not known with certainty. Mitochondrial parameters such as respiration and pyruvate dehydrogenase activity are increased after 10-20 s (Assimacopoulos-Jeannet et al. 1983; Blackmore et al. 1983a, 1983b) and cellular pyridine nucleotides are reduced after 5 s (Blackmore et al. 1983b), but a rise in cytosolic Ca^{2+} is detectable within 2 s (Charest et al. 1983). These considerations have led to the proposal that there is a more immediate source of mobilized Ca^{2+} than mitochondria, and the plasma membrane has been suggested as a likely site (Blackmore et al. 1983b). However, as alluded to above, direct evidence that α_1-agonists or other Ca^{2+}-dependent hormones induce the release of Ca^{2+} from liver plasma membranes is lacking.

The release of Ca^{2+} from mitochondria or other organelles induced by Ca^{2+}-dependent hormones is obviously limited by the magnitude of the calcium stores in these organelles. For this reason, the rise in cytosolic Ca^{2+} induced by these hormones in hepatocytes incubated in a low Ca^{2+} medium (30 μM) is transient (Fig. 3). However, if the medium contains Ca^{2+} in a more physiological concentration range (500 μM), the increase in cytosolic Ca^{2+} is sustained for at least 10 min. These findings indicate that the hormones have an additional action on cell Ca^{2+} which is dependent upon extracellular Ca^{2+}. Although several possibilities exist, one which is supported by experimental data is that the hormones affect transmembrane Ca^{2+} flux through inhibition of the plasma membrane $(Ca^{2+}\text{-}Mg^{2+})$ATPase or Ca^{2+} pump (Lin et al. 1983; Prpic et al. 1984). Analysis of the data of Fig. 3 indicates that the dependence of the hormone-induced increase in cytosolic Ca^{2+} upon extracellular Ca^{2+} is only evident after about 1 min of hormone exposure (Charest et al. 1985). Similarly, hormone effects on the liver plasma membrane Ca^{2+} pump are not significant before 1 min (Fig. 4). The relative slowness of these effects would explain why variations in extracellular Ca^{2+} do

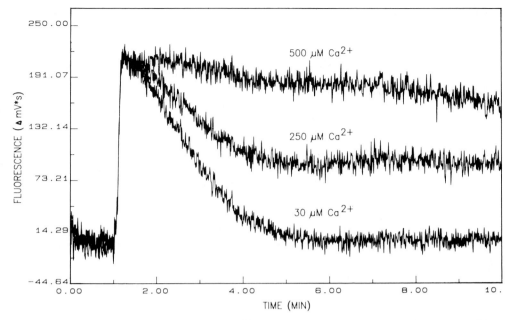

Fig. 3. Effects of various extracellular Ca^{2+} concentrations on the increase in cytosolic Ca^{2+} induced in hepatocytes by vasopressin. Control and Quin-2-loaded hepatocytes (prepared as described by Charest et al. 1983) were washed and resuspended in medium containing 30 μM, 250 μM or 500 μM Ca^{2+} for 6 min prior to addition of 10^{-8} M vasopressin (at 1 min). The data are presented as in Fig. 2

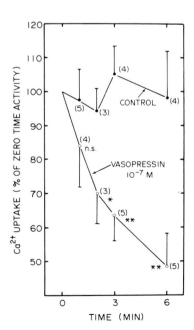

Fig. 4. Ca^{2+} uptake by plasma membrane vesicles prepared from rat livers perfused with saline or vasopressin (10^{-7} M) for various times. Full experimental details are given in Prpic et al. (1984)

not affect the initial (30-60 s) changes in cytosolic Ca^{2+} induced by Ca^{2+}-dependent hormones (Fig. 3).

3 Intracellular Actions of Ca^{2+}

Almost all of the α_1-adrenergic effects that have been examined in the liver have been shown to be Ca^{2+}-dependent (Exton 1983). However, the precise role of Ca^{2+} in these effects is not known except for very few cases. The most clear-cut case is the activation of glycogen phosphorylase and resulting breakdown of glycogen and release of glucose. The relationship between the α_1-adrenergic activation of phosphorylase and elevation of cytosolic Ca^{2+} has been established by the high correlation between the two effects with respect to agonist dose-dependence (Fig. 5 and Charest et al. 1983, 1985) and temporal sequence (Charest et al. 1983, 1985). The target for cytosolic Ca^{2+} is almost certainly the calmodulin-containing enzyme phosphorylase kinase (Chrisman et al. 1982) which phosphorylates phosphorylase b thereby converting it to the active form phosphorylase a. Phosphorylase kinase is stimulated allosterically by increases in Ca^{2+} (Chrisman et al. 1982) within the cytosolic range (Charest et al. 1983). The major, if not exclusive, role of phosphorylase kinase in mediating the increase in phosphorylase a induced by Ca^{2+}-dependent hormones is indicated by the fact that in livers of rats with a genetic deficiency of this kinase, these hormones are unable to activate phosphorylase or break down glycogen (Blackmore and Exton 1981).

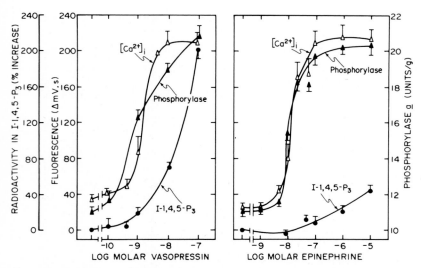

Fig. 5. Dose responses of vasopressin and epinephrine on phosphorylase a activity, cytosolic Ca^{2+} and inositol-1,4,5-P_3 level in hepatocytes. All measurements were made after incubation with hormone for 15 s. Cytosolic Ca^{2+} was measured as described by Charest et al. (1983), phosphorylase a was assayed according to Hutson et al. (1976) and inositol-1,4,5-P_3 was measured according to Berridge (1983)

Another enzymatic change for which an explanation currently exists is the inactivation of glycogen synthase by Ca^{2+}-dependent hormones. Two Ca^{2+}-dependent protein kinases have been shown to phosphorylate and inactivate this enzyme, namely phosphorylase kinase (Chrisman et al. 1982) and a calmodulin-dependent glycogen synthase kinase (Payne and Soderling 1980). In rat liver, it appears that phosphorylase kinase is the predominant enzyme which phosphorylates and inactivates glycogen synthase in a Ca^{2+}-dependent manner (Fig. 6 and Imazu et al. 1984). However, the calmodulin-dependent kinase may play a larger role in rabbit liver (Payne and Soderling 1980). Compared with phosphorylase, glycogen synthase is a poor substrate for phosphorylase kinase (Chrisman et al. 1982), and it is possible that the Ca^{2+} regulation of glycogen synthase in the liver cell may mainly involve phosphorylase kinase only indirectly. This is because experiments utilizing liver filtrates have indicated that an increase in phosphorylase a induced by Ca^{2+} stimulation of phosphorylase kinase inhibits glycogen synthase phosphatase (Strickland et al. 1983), thus permitting endogenous protein kinases to phosphorylate and inactivate glycogen synthase (Imazu et al. 1984).

Garrison and coworkers (Garrison 1978; Garrison et al. 1979; Garrison and Wagner 1982) have shown taht, in addition to stimulating the phosphorylation of phosphorylase and glycogen synthase, Ca^{2+}-dependent hormones, including the α_1-agonists, promote the incorporation of phosphate into several other liver cytosolic proteins. Some of these proteins correspond to known enzymes, e.g., pyruvate kinase and phenylalanine hydroxylase, whereas others are unknown. Pyruvate kinase has been shown to be phosphorylated and inactivated by α-agonists in other studies utilizing hepatocytes (Chan and Exton 1978; Foster and Blair 1978; Garrison et al. 1979; Nagano et al. 1980; Steiner et al. 1980), but the Ca^{2+}-sensitive protein kinase (or phosphoprotein phosphatase) involved is unknown. Phosphorylase kinase and calmodulin-dependent glyco-

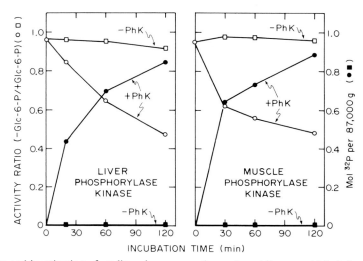

Fig. 6. Phosphorylation and inactivation of rat liver glycogen synthase a by rat liver or rabbit skeletal muscle phosphorylase kinase (PhK). Glycogen synthase was incubated at 30°C with $[\gamma^{-32}P]ATP$, Mg^{2+}, glycogen, albumin, buffer, and the respective phosphorylase kinase. Phosphorylation is presented as mol ^{32}P incorporated per 87,000 g synthase subunit. Other experimental details are given in Imazu et al. (1984)

gen synthase kinase do not phosphorylate pyruvate kinase in vitro (Payne et al. 1983 and unpublished findings by T. D. Chrisman, Vanderbilt University).

The inactivation of pyruvate kinase by α_1-agonists probably contributes to the stimulation of gluconeogenesis caused by these agents (Tolbert et al. 1973; Hutson et al. 1976; Kneer et al. 1974), but other factors are involved. One is a stimulation of mitochondrial pyruvate carboxylation (Garrison and Borlund 1979) and another is a stimulation of the mitochondrial oxidation of cytosolic NAD(P)H (Kneer et al. 1979). The mechanism(s) of the increase in pyruvate carboxylation is unknown, but could involve increased pyruvate entry into the mitochondria or alterations in the concentrations of effectors of pyruvate carboxylase (Thomas and Halestrap 1981). The stimulation of NAD(P)H oxidation could be the result of increased activity of glycerol-3-P dehydrogenase. This enzyme is stimulated by an increase in cytosolic Ca^{2+} (Wernette et al. 1981), and participates in the transfer of NADH from the cytosol to the mitochondria, which can be rate-limiting for gluconeogenesis from reduced substrates (Yip and Lardy 1981).

Contrary to Richards et al. (1981), Hue et al. (1981) were unable to observe a Ca^{2+}-dependent reduction in the concentration of fructose-2,6-P_2 with α_1-adrenergic agonists or vasopressin in hepatocytes. This hexose bisphosphate can control gluconeogenesis by stimulating phosphofructokinase and inhibiting fructose 1,6-bisphosphatase (see Pilkis et al. 1982). Likewise, the report (Furuya et al. 1982) that phosphorylase b kinase phosphorylates and inactivates 6-phosphofructo-2-kinase (which converts fructose-6-P to fructose-2,6-P_2) has not been confirmed (Pilkis et al. 1983). It thus seems unlikely that Ca^{2+}-dependent hormones stimulate gluconeogenesis at the fructose-1,6P_2-fructose-6-P substrate cycle through inhibition of fructose-2,6-P_2 formation.

α-Adrenergic agonists have been reported to inhibit hepatic fatty acid synthesis (Ly and Kim 1981) and to stimulate ketogenesis slightly (Kosugi et al. 1983), although these findings have not been universally observed (Assimacopoulos-Jeannet et al. 1981). The inhibition of lipogenesis has been attributed to phosphorylation and inactivation of acetyl-CoA carboxylase (Ly and Kim 1981), but the possible role of cAMP has not been rigorously excluded. Acetyl-CoA carboxylase and ATP-citrate lyase have been shown to be substrates for calmodulin-dependent glycogen synthase kinase from muscle (Woodgett et al. 1983), but it is not known whether they are substrates for the liver enzyme. Sugden and her associates have reported that α-adrenergic stimulation increases the oxidation of fatty acids to CO_2 (Sugden et al. 1980) and have attributed this to Ca^{2+} activation of NAD^+-linked isocitrate dehydrogenase and 2-oxoglutarate dheydrogenase complex (Sugden and Watts 1983). Pyruvate dehydrogenase is also stimulated by α_1-adrenergic agonists in liver (Denton et al. 1981; Blackmore et al. 1983b; Assimacopoulos-Jeannet et al. 1983). The stimulation is not observed in Ca^{2+}-depleted cells and is mimicked by the Ca^{2+} ionophore A23187 (Blackmore et al. 1983b; Assimacopoulos-Jeannet et al. 1983).

α-Adrenergic agonists transiently increase the $NAD(P)H/NAD(P)^+$ ratio in hepatocytes (Fig. 7) and the perfused liver (Blackmore et al. 1983a; Buxton et al. 1982) and part, at least, of this change is due to an alteration in the mitochondrial oxidation-reduction state as reflected by an increase in the β-hydroxybutyrate/acetoacetate ratio (Buxton et al. 1982). There is also an associated transient inhibition of branched chain

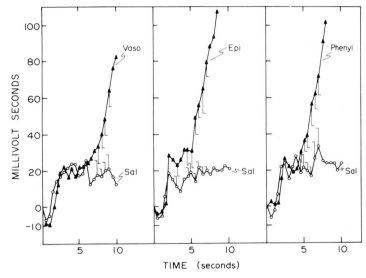

Fig. 7. Effects of vasopressin (10^{-8} M), epinephrine (10^{-6} M) and phenylephrine (10^{-5} M) on NAD(P)$^+$ reduction in hepatocytes. NAD(P)H fluorescence was measured fluorimetrically as described in Blackmore et al. (1983b)

α-ketoacid oxidation (Buxton et al. 1982). These changes are apparently dependent on alterations in mitochondrial Ca^{2+} since they are reduced or abolished by removing Ca^{2+} from the medium (Buxton et al. 1982).

Hepatic respiration is stimulated by α-adrenergic agonists (Jakob and Diem 1975; Dehaye et al. 1981; Blackmore et al. 1983a) in a Ca^{2+}-dependent manner (Blackmore et al. 1983a). The stimulation follows in time the increase in NAD(P)H/NAD(P)$^+$ ratio and therefore may be secondary to it. As noted above, several mitochondrial dehydrogenases may also be stimulated by a rise in free Ca^{2+} in the matrix.

Some workers have reported a stimulation of hepatic ureogenesis by α-adrenergic activation (Titheradge and Haynes 1980; Corvera and Garcia-Sainz 1981) but the mechanisms involved are unclear.

Catecholamines markedly affect K$^+$ fluxes in the perfused liver, but the direction of the net change depends on the species. In guinea pig, α-adrenergic stimulation promotes net K$^+$ efflux (Haylett and Jenkinson 1972; Burgess et al. 1981) and this has been attributed to the opening of Ca^{2+}-sensitive K$^+$ channels in the plasma membrane (Weiss and Putney 1978; Burgess et al. 1981). However in rat, α-agonists stimulate a transient net uptake of K$^+$ (Jakob and Diem 1975; Blackmore et al. 1979). This has been attributed to a paucity of Ca^{2+}-sensitive K$^+$ channels in this species (Burgess et al. 1981) and a stimulation of the plasma membrane (Na$^+$-K$^+$)ATPase (Berthon et al. 1983). The mechanism(s) by which Ca^{2+}-dependent agonists stimulate the (Na$^+$-K$^+$) ATPase is unknown.

4 Mechanisms by Which Alpha-Adrenergic Agonists Could Increase Cytosolic Ca^{2+}

As described above, the two mechanisms by which α-agonists increase cytosolic Ca^{2+} in liver are by initially mobilizing intracellular Ca^{2+} stores and by later inhibiting Ca^{2+} efflux from the cell mediated by the plasma membrane $(Ca^{2+}$-$Mg^{2+})$ATPase. No clear evidence has been obtained for an initial stimulation of Ca^{2+} influx across the plasma membrane, although this might occur in other cell types.

A current hypothesis for the coupling of α_1-adrenergic vasopressin receptors to cellular Ca^{2+} fluxes is that this involves the breakdown of phosphatidylinositol-4,5-P_2 in the plasma membrane (Kirk et al. 1981; Creba et al. 1983; Berridge et al. 1983). Originally it was proposed that phosphatidylinositol breakdown to 1,2-diacylglycerol and inositol monophosphates fulfilled this role (Michell 1975, 1979), but this breakdown was found to occur too slowly in hepatocytes (Prpic et al. 1982) to play a role in the changes in Ca^{2+}, which are detectable within 2 s (Charest et al. 1983, 1985). Further work revealed an earlier reaction, namely phosphatidylinositol-4,5-P_2 breakdown to 1,2-diacylglycerol and inositol-1,4,5-P_3. Stimulation of this reaction by epinephrine or vasopressin in hepatocytes was detectable at 5 or 15 s (Rhodes et al. 1983; Charest et al. 1985; Creba et al. 1983; Thomas et al. 1983; Litosch et al. 1983). Thus it may be as fast as the rise in cytosolic Ca^{2+}. Measurements of the concentration dependencies for the effects of epinephrine and vasopressin on inositol-1,4,5-P_3 accumulation, phosphorylase activation and cytosolic Ca^{2+} illustrate that maximal increases in phosphorylase a and cytosolic Ca^{2+} can be achieved at 15 s with an increase in inositol-1,4,5-P_3 of only 20% with epinephrine, but 40% with vasopressin (Fig. 5 and Charest et al. 1985). For these data to be compatible with the hypothesis that phosphatidylinositol-4,5-P_2 breakdown is mechanistically involved in the coupling of α_1-adrenergic and vasopressin receptors to cellular Ca^{2+} fluxes in liver (Creba et al. 1983; Thomas et al. 1983), it seems that there would have to be more than one pool of phosphatidylinositol-4,5-P_2 and that an extremely small increase in breakdown would be sufficient to trigger Ca^{2+} changes.

The increases in inositol-P_3 provoked by epinephrine and vasopressin can be markedly potentiated by addition of Li^+ ions without any corresponding potentiation of Ca^{2+} mobilization or phosphorylase activation (Charest et al. 1985). Unless Li^+ increases another isomer, these findings would seem to be contrary to the proposal that inositol-1,4,5-P_3 functions as a second messenger for Ca^{2+}-dependent hormones by releasing Ca^{2+} from internal pools (Berridge 1983; Streb et al. 1983). However, support for this proposal comes from observations that additional of a preparation of inositol-1,4,5-P_3 to hepatocytes or paratid cells made permeable with saponin causes the release of Ca^{2+} (Streb et al. 1983; Berridge 1984; Putney 1984; Williamson et al. 1984). In summary, evidence is increasing in support of the hypothesis that inositol-1,4,5-P_3 is the intracellular messenger of the Ca^{2+}-dependent hormones in vivo.

1,2-Diacylglycerol, the other product of phosphatidylinositol-4,5-P_2 breakdown, has been found to be an activator of protein kinase C, a specific protein kinase which requires Ca^{2+} and a phospholipid (e.g., phosphatidylserine) for activity (Takai et al. 1979). Diacylglycerol activates the enzyme by reducing its Ca^{2+} requirement to the

cytosolic range (Kishimoto et al. 1980), and its action can be mimicked by tumor-promoting phorbol esters (Castagna et al. 1982). Addition of these esters to hepatocytes
leads to increased phosphorylation of three cytosolic proteins of unknown function
(Garrison 1983). These proteins are also phosphorylated, in addition to others, by
vasopressin and norepinephrine, but not by A23187 (Garrison 1978, 1983; Garrison
et al. 1979; Garrison and Wagner 1982). It has also been reported that these esters
inactivate glycogen synthase in hepatocytes without affecting glycogen phosphorylase
(Roach and Goldman 1983). These findings suggest a role for diacylglycerol and protein kinase C in the regulation by Ca^{2+}-dependent hormones of certain cytosolic
proteins through phosphorylation.

5 Summary

The effects of $alpha_1$-adrenergic agonists, vasopressin and angiotensin II on the liver
and certain other tissues fundamentally involve Ca^{2+} ions. These agents rapidly raise
cytosolic Ca^{2+} by initially mobilizing Ca^{2+} from intracellular stores and later by promoting net Ca^{2+} influx from the extracellular milieu. In the case of liver, mitochondria are a major site from which Ca^{2+} is mobilized, and net Ca^{2+} influx is enhanced
through inhibition of the plasma membrane (Ca^{2+}-Mg^{2+})ATPase pump (Fig. 8).

   The rise in cytosolic Ca^{2+} induced by the Ca^{2+}-dependent hormones alters the activities of several Ca^{2+}-sensitive enzymes or other proteins resulting in some of the
physiological responses observed (Fig. 8). Some of the Ca^{2+}-sensitive enzymes are
protein kinases and contain or require calmodulin. However, for many of the responses, the mechanisms have not been clarified.

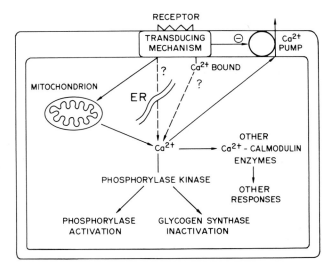

Fig. 8. Proposed scheme by which α_1-adrenergic agonists, vasopressin, and angiotensin II act in liver. Ca^{2+} may also be mobilized from endoplasmic reticulum (*ER*) or from binding sites on membranes (Ca^{2+}-bound). Inhibition of the plasma membrane Ca^{2+} pump may be more indirect than shown

Cell-surface receptors for the Ca^{2+}-dependent hormones have been defined in many tissues using radioligand binding. However, the effector systems to which these receptors are coupled in the plasma membrane remain unknown. There is much evidence that breakdown of phosphatidylinositol-4,5-P_2 is a very early event following receptor activation, but the mechanism of this breakdown is still unclear. The breakdown products are inositol-1,4,5-P_3 and 1,2-diacylglycerol. Inositol-1,4,5-P_3 rapidly accumulates in hepatocytes incubated with Ca^{2+}-dependent hormones. It also releases intracellular Ca^{2+} from permeabilized liver and parotid cells. For these reasons it has been postulated to act as an intracellular message for Ca^{2+} mobilization. Unsaturated diacylglycerol activates a specific Ca^{2+}-phospholipid-dependent protein kinase in many tissues. This enzyme appears to be responsible for the phosphorylation of some, but not all, of the proteins phosphorylated in hepatocytes in response to Ca^{2+}-dependent hormones.

It is clear that many key questions remain concerning the mechanism(s) of action of α_1-adrenergic agonists and other Ca^{2+}-dependent hormones. The major ones are: definition of the mechanism of phosphatidylinositol-4,5-P_2 breakdown, determination of the primary effector system, elucidation of its coupling to the receptor, determination of the mechanism(s) for mobilization of intracellular Ca^{2+} and for the control of plasma membrane Ca^{2+} flux, definition of the Ca^{2+}-sensitive enzymes or other proteins involved in the biological responses.

References

Althaus-Salzmann M, Carafoli E, Jakob A (1980) Ca^{2+}, K^+ redistributions and α-adrenergic activation of glycogenolysis in perfused rat livers. Eur J Biochem 106:241-248

Assimacopoulos-Jeannet F, Denton RM, Jeanrenaud B (1981) Stimulation of hepatic lipogenesis and acetyl-coenzyme A carboxylase by vasopressin. Biochem J 198:485-490

Assimacopoulos-Jeannet F, McCormack JG, Jeanrenaud B (1983) Effect of phenylephrine on pyruvate dehydrogenase activity in rat hepatocytes and its interaction with insulin and glucagon. FEBS Lett 159:83-88

Assimacopoulos-Jeannet FD, Blackmore PF, Exton JH (1977) Studies on the α-adrenergic activation of hepatic glucose output. III. Studies on the role of calcium in the α-adrenergic activation of phosphorylase. J Biol Chem 252:2662-2669

Babcock DF, Chen J-LJ, Yip BP, Lardy HA (1979) Evidence for mitochondrial localization of the hormone-responsive pool of Ca^{2+} in isolated hepatocytes. J Biol Chem 254:8117-8120

Barritt GJ, Parker JC, Wadsworth JC (1981) A kinetic analysis of effects of adrenaline on calcium distribution in isolated rat liver parenchymal cells. J Physiol 312:29-55

Berridge MJ (1983) Rapid accumulation of inositol trisphosphate reveals that agonists hydrolyse polyphosphoinositides instead of phosphatidylinositol. Biochem J 212:849-858

Berridge MJ (1984) Inositol trisphosphate and calcium mobilization. In: Bleasdale J (ed) Proceedings of Chilton conference on inositol and phosphoinositides. Humana Press, Clifton (In press)

Berridge MJ, Dawson RMC, Downes CP, Heslop JP, Irvine RF (1983) Changes in the levels of inositol phosphates after agonist-dependent hydrolysis of membrane phosphoinositides. Biochem J 212:473-82

Berthon B, Burgess GM, Capiod T, Claret M, Poggioli J (1983) Mechanism of action of noradrenaline on the sodium-potassium pump in isolated rat liver cells. J Physiol 341:25-40

Blackmore PF, Exton JH (1981) α_1-Adrenergic stimulation of calcium mobilization without phosphorylase activation in hepatocytes from phosphorylase b kinase-deficient *gsd/gsd* rats. Biochem J 198:379-383

Blackmore PF, Brumley FT, Marks JL, Exton JH (1978) Studies on α-adrenergic activation of hepatic glucose output: relationship between α-adrenergic stimulation of calcium efflux and activation of phosphorylase in isolated rat liver parenchymal cells. J Biol Chem 253:4851-4858

Blackmore PF, Dehaye J-P, Exton JH (1979) Studies on α-adrenergic activation of hepatic glucose output: the role of mitochondrial calcium release in α-adrenergic activation of phosphorylase in perfused rat liver. J Biol Chem 254:6945-6950

Blackmore PF, Hughes BP, Shuman EA, Exton JH (1982) α-Adrenergic activation of phosphorylase in liver cells involves mobilization of intracellular calcium without influx of extracellular calcium. J Biol Chem 257:190-197

Blackmore PF, Hughes BP, Exton JH (1983a) Time course of α-adrenergic and vasopressin effects in isolated hepatocytes. In: Harris RA, Cornell NW (eds) Isolation, characterization and use of hepatocytes. Elsevier/North Holland Biomedical, Press, Amsterdam, pp 433-438

Blackmore PF, Hughes BP, Charest R, Schuman EA, IV, Exton JH (1983b) Time course of α_1-adrenergic and vasopressin actions on phosphorylase activation, calcium efflux, pyridine nucleotide reduction, and respiration in hepatocytes. J Biol Chem 258:10488-10494

Burgess GM, Claret M, Jenkinson DH (1981) Effects of quinine and apamin on calcium-dependent potassium permeability of mammalian hepatocytes and red cells. J Physiol 317:67-90

Buxton D, Barron L, Olson MS (1982) The effects of α-adrenergic agonists on the regulation of the branched chain α-ketoacid oxidation in the perfused rat liver. J Biol Chem 257:14318-14323

Castagna M, Takai Y, Kaibuchi K, Sano K, Kikkawa U, Nishizuka Y (1982) Direct activation of calcium-activated, phospholipid-dependent protein kinase by tumor-promoting phorbol esters. J Biol Chem 257:7847-7851

Chan TM, Exton JH (1978) Studies on α-adrenergic activation of hepatic glucose output: studies on α-adrenergic inhibition of hepatic pyruvate kinase and activity of gluconeogenesis. J Biol Chem 253:6393-6400

Charest R, Blackmore PF, Berthon B, Exton JH (1983) Changes in free cytosolic Ca^{2+} in hepatocytes following α_1-adrenergic stimulation. J Biol Chem 258:8769-8773

Charest R, Prpic V, Exton JH, Blackmore PF (1985) Stimulation of *myo*-inositol trisphosphate formation in hepatocytes by vasopressin and epinephrine and its relationship to changes in cytosolic free Ca^{2+}. Biochem J (in press)

Chen J-LJ, Babcock DF, Lardy HA (1978) Norepinephrine, vasopressin, glucagon, and A23187 induce efflux of calcium from an exchangeable pool in isolated rat hepatocytes. Proc Natl Acad Sci USA 75:2234-2238

Cherrington AD, Exton JH (1976) Studies on the role of cAMP-dependent protein kinase in the actions of glucagon and catecholamines on liver glycogen metabolism. Metabolism 25:1351-1354

Chrisman TD, Jordan JE, Exton JH (1982) Purification of rat liver phosphorylase kinase. J Biol Chem 257:10798-10804

Clarke WR, Jones LR, Lefkowitz RJ (1978) Hepatic α-adrenergic receptors. Identification and subcellular localization using [^{3}H]dihydroergocryptine. J Biol Chem 253:5975-5979

Corvera S, Garcia-Sainz JA (1981) α_1-Adrenoceptor activation stimulates ureogenesis in rat hepatocytes. Eur J Pharmacol 72:387-390

Creba JA, Downes CP, Hawkins PT, Brewster G, Michell RH, Kirk CJ (1983) Rapid breakdown of phosphatidylinositol 4-phosphate and phosphatidylinositol 4,5-bisphosphate in rat hepatocytes stimulated by vasopressin and other Ca^{2+}-mobilizing hormones. Biochem J 212:733-747

Dehaye J-P, Blackmore PF, Venter JC, Exton JH (1980) Studies on the alpha-adrenergic activation of hepatic glucose output. alpha-Adrenergic activation of phosphorylase by immobilized epinephrine. J Biol Chem 255:3905-3910

Dehaye J-P, Hughes BP, Blackmore PF, Exton JH (1981) Insulin inhibition of α-adrenergic actions in liver. Biochem J 194:949-956

Denton RM, McCormack JG, Oviasu OA (1981) Short-term regulation of pyruvate dehydrogenase activity in the liver. In: Hue L, Van de Werve G (eds) Short-term regulation of liver metabolism. Elsevier/North Holland Biomedical. Press, Amsterdam, pp 159-174

El-Refai MR, Chan TM (1982) Effects of fasting on hepatic catecholamine receptors. FEBS Lett 146:397-402

El-Refai MF, Blackmore PF, Exton JH (1979) Evidence for two α-adrenergic binding sites in liver plasma membranes. Studies with [^{3}H]epinephrine and [^{3}H]dihydroergocryptine. J Biol Chem 254:4375-4386

Exton JH (1983) α-Adrenergic agonists and Ca^{2+} movement. In: Cheung WY (ed) Calcium and cell function vol 4. Academic Press, London New York, pp 63-97

Foster JL, Blair JB (1978) Acute hormonal control of pyruvate kinase and lactate formation in isolated rat hepatocytes. Arch Biochem Biophys 189:263-276

Furuya E, Yokoyama M, Uyeda K (1982) Regulation of fructose-6-phosphate 2-kinase by phosphorylation and dephosphorylation: possible mechanism for coordinated control of glycolysis and gluconeogenesis. Proc Natl Acad Sci USA 79:325-329

Garrison JC (1978) The effects of glucagon, catecholamines, and the calcium ionophore A23187 on the phosphorylation of rat hepatocyte cytosolic proteins. J Biol Chem 253:7091-7100

Garrison JC (1983) Role of Ca^{2+}-dependent protein kinases in the response of hepatocytes to α-agonists, angiotensin II and vasopressin. In: Harris RA, Cornell NW (eds) Isolation, characterization, and use of hepatocytes. Elsevier/North Holland Biomedical Press, Amsterdam New York, pp 551-559

Garrison JC, Borland MK (1979) Regulation of mitochondrial pyruvate carboxylation and gluconeogenesis in rat hepatocytes via an α-adrenergic, adenosine 3',5'-monophosphate-independent mechanism. J Biol Chem 254:1129-1133

Garrison JC, Wagner JD (1982) Glucagon and the Ca^{2+}-linked hormones angiotensin II, norepinephrine, and vasopressin stimulate the phosphorylation of distinct substrates in intact hepatocytes. J Biol Chem 257:13135-13143

Garrison JC, Borland MK, Florio VA, Twible DA (1979) The role of calcium ion as a mediator of the effects of angiotensin II, catecholamines, and vasopressin on the phosphorylation and activity of enzymes in isolated hepatocytes. J Biol Chem 254:7147-7156

Guellaen G, Yates-Aggerbeck M, Vauquelin G, Strosberg D, Hanoune J (1978) Characterization with [^{3}H]dihydroergocryptine of the α-adrenergic receptor of the hepatic plasma membrane. J Biol Chem 253:1114-1120

Haylett DG, Jenkinson DH (1972) Effects of noradrenaline on potassium efflux, membrane potential and electrolyte levels in tissue slices prepared from guinea-pig liver. J Physiol 225:721-750

Hoffman BB, DeLean A, Wood CL, Schocken DD, Lefkowitz RJ (1979) Alpha-adrenergic receptor subtypes: quantitative assessment by ligand binding. Life Sci 24:1739-1746

Hoffman BB, Dukes DF, Lefkowitz RJ (1981) Alpha-adrenergic receptors in liver membranes: delineation with subtype selective radioligands. Life Sci 28:265-272

Hue L, Blackmore PF, Exton JH (1981) Fructose 2,6-bisphosphate: hormonal regulation and mechanism of its formation in liver. J Biol Chem 256:8900-8903

Hutson NJ, Brumley FT, Assimacopoulos FD, Harper SC, Exton JH (1976) Studies on the α-adrenergic activation of hepatic glucose output. I. Studies on the α-adrenergic activation of phosphorylase and gluconeogenesis and inactivation of glycogen synthase in isolated rat liver parenchymal cells. J Biol Chem 251:5200-5208

Imazu M, Strickland WG, Chrisman TD, Exton JH (1984) Phosphorylation and inactivation of liver glycogen synthase by liver protein kinases. J Biol Chem 259:1813-1821

Jakob A, Diem S (1975) Metabolic responses of perfused rat livers to alpha- and beta-adrenergic agonists, glucagon and cyclic AMP. Biochim Biophys Acta 404:57-66

Jard S, Cantau B, Jakobs KH (1981) Inhibition of rat liver adenylate cyclase angiotensin II and α-adrenergic agonists. J Biol Chem 256:2603-2606

Keppens S, Vandenheede JR, DeWulf H (1977) On the role of calcium as second messenger in liver for the hormonally induced activation of glycogen phosphorylase. Biochim Biophys Acta 496:448-457

Kirk CJ, Creba JA, Downes CP, Michell RH (1981) Hormone-stimulated metabolism of inositol lipids and its relationship to hepatic receptor function. Biochem Soc Trans 7:377-379

Kishimoto A, Takai Y, Mori T, Kikkawa U, Nishizuka Y (1980) Activation of calcium and phospholipid-dependent protein kinase by diacylglycerol, its possible relation to phosphatidylinositol turnover. J Biol Chem 255:2273-2276

Kneer NM, Bosch AL, Clark MG, Lardy HA (1974) Glucose inhibition of epinephrine stimulation of hepatic gluconeogenesis by blockade of the α-receptor function. Proc Natl Acad Sci USA 71:4523-4527

Kneer NM, Wagner MJ, Lardy HA (1979) Regulation by calcium of hormonal effects on gluconeogenesis. J Biol Chem 254:12160-12168

Kosugi K, Harano Y, Nakano T, Suzuki M, Kashiwagi A, Shigeta Y (1983) Mechanism of adrenergic stimulation of hepatic ketogenesis. Metabolism 32:1081-1087

Lin S-H, Wallace MA, Fain JN (1983) Regulation of Ca^{2+}-Mg^{2+}-ATPase activity in hepatocyte plasma membranes by vasopressin and phenylephrine. Endocrinology 113:2268-2275

Litosch I, Lin SH, Fain JN (1983) Rapid changes in hepatocyte phosphoinositides induced by vasopressin. J Biol Chem 258:13727-13732

Ly S, Kim KH (1981) Inactivation of hepatic acetyl-CoA carboxylase by catecholamine and its agonists through the α-adrenergic receptors. J Biol Chem 256:11585-11590

Michell RH (1975) Inositol phospholipids and cell surface receptor function. Biochim Biophys Acta 415:81-147

Michell RH (1979) Inositol phospholipids in membrane function. Trends Biochem Sci 4:128-131

Morgan NG, Shuman EA, Exton JH, Blackmore PF (1982) Stimulation of hepatic glycogenolysis by α_1- and β_2-adrenergic agonists. J Biol Chem 257:13907-13910

Morgan NG, Blackmore PF, Exton JH (1983) Age-related changes in the control of hepatic cyclic AMP levels by α_1- and β_2-adrenergic receptors in male rats. J Biol Chem 258:5103-5109

Murphy E, Coll K, Rich TL, Williamson JR (1980) Hormonal effects on calcium homeostasis in isolated hepatocytes. J Biol Chem 255:6600-6608

Nagano M, Ishibashi H, McCully V, Cottam GL (1980) Epinephrine-stimulated phosphorylation of pyruvate kinase in hepatocytes. Arch Biochem Biophys 203:271-281

Payne ME, Soderling TR (1980) Calmodulin-dependent glycogen synthase kinase. J Biol Chem 255:8054-8056

Payne ME, Schworer CM, Soderling TR (1983) Purification and characterization of rabbit liver calmodulin-dependent glycogen synthase kinase. J Biol Chem 258:2376-2382

Pilkis SJ, El-Maghrabi MR, McGrane M, Pilkis J, Fox E, Claus TH (1982) Fructose 2,6-bisphosphate: a mediator of hormone action at the fructose 6-phosphate/fructose 1,6-bisphosphate substrate cycle. Mol Cell Endocrinol 25:245-266

Pilkis SJ, Chrisman TD, El-Maghrabi R, Colosia A, Fox E, Pilkis J, Claus TH (1983) The action of insulin on hepatic fructose 2,6-bisphosphate metabolism. J Biol Chem 258:1495-1503

Preiksaitis HG, Kan WH, Kunos G (1982) Decreased α_1-adrenoceptor responsiveness and density in liver cells of thyroidectomized rats. J Biol Chem 257:4321-4327

Prpic V, Blackmore PF, Exton JH (1982) Phosphatidylinositol breakdown induced by vasopressin and enephrine in hepatocytes is calcium-dependent. J Biol Chem 257:11323-11331

Prpic V, Green K, Blackmore PF, Exton JH (1984) Vasopressin-, angiotensin II- and α_1-adrenergic-induced inhibition of Ca^{2+} transport by rat liver plasma membrane vesicles. J Biol Chem 259:1382-1385

Putney JW (1984) Messages of the phosphoinositide effect. In: Bleasdale J (ed) Proceedings of Chilton conference on inositol and phosphoinositides. Humana Press, Clifton (In press)

Rhodes D, Prpic V, Exton JH, Blackmore PF (1983) Stimulation of phosphatidylinositol 4,5-bisphosphate hydrolysis in hepatocytes by vasopressin. J Biol Chem 258:2770-2773

Richards CS, Furuya E, Uyeda K (1981) Regulation of fructose 2,6-P_2 concentration in isolated hepatocytes. Biochem Biophys Res Commun 100:1673-1679

Roach PJ, Goldman M (1983) Modification of glycogen synthase activity in isolated rat hepatocytes by tumor-promoting phorbol esters: evidence for differential regulation of glycogen synthase and phosphorylase. Proc Natl Acad Sci USA 79:7170-7172

Steiner KE, Chan TM, Claus TH, Exton JH, Pilkis SJ (1980) The role of phosphorylation in the alpha-adrenergic-mediated inhibition of rat hepatic pyruvate kinase. Biochim Biophys Acta 632:366-374

Streb H, Irvine RF, Berridge MJ, Schulz I (1983) Release of Ca^{2+} from a nonmitochondrial intracellular store in pancreatic acinar cells by inositol-1,4,5-trisphosphate. Nature 306:67-69

Strickland WG, Imazu M, Chrisman TD, Exton JH (1983) Regulation of rat liver glycogen synthase. Roles of Ca^{2+}, phosphorylase kinase, and phosphorylase a. J Biol Chem 5490-5497

Studer RK, Borle AB (1982) Differences between male and female rats in the regulation of hepatic glycogenolysis: the relative role of calcium and cAMP in phosphorylase activation by catecholamines. J Biol Chem 257:7987-7993

Sugden MC, Watts DI (1983) Stimulation of [I-^{14}C]loleate oxidation to $^{14}CO_2$ in isolated rat hepatocytes by the catecholamines, vasopressin and angiotensin. Biochem J 212:85-91

Sugden MC, Tordoff AFC, Ilic V, Williamson DH (1980) α-Adrenergic stimulation of [I-^{14}C]oleate oxidation to $^{14}CO_2$ in isolated rat hepatocytes. FEBS Lett 120:80-84

Takai Y, Kishimoto A, Iwasa Y, Kawahara Y, Mori T, Nishizuka Y (1979) Calcium-dependent activation of a multifunctional protein kinase by membrane phospholipids. J Biol Chem 254: 3692-3695

Thomas AP, Halestrap AP (1981) The role of mitochondrial pyruvate transport in the stimulation by glucagon and phenylephrine of gluconeogenesis from L-lacate in isolated rat hepatocytes. Biochem J 198:551-564

Thomas AP, Marks JS, Coll KE, Williamson JR (1983) Quantitation and early kinetics of inositol lipids changes induced by vasopressin in isolated and cultured hepatocytes. J Biol Chem 258:5716-5725

Titheradge MA, Haynes RC Jr (1980) The hormonal stimulation of ureogenesis in isolated hepatocytes through increases in mitochondrial ATP production. Arch Biochem Biophys 201:44-55

Tolbert MEM, Butcher FR, Fain JN (1973) Lack of correlation between catecholamine effects on cyclic adenosine 3':5'-monophosphate and gluconeogenesis in isolated rat liver cells. J Biol Chem 248:5686-5692

Van de Werve G, Hue L, Hers H-G (1977) Hormonal and ionic control of the glycogenolytic cascade in rat liver. Biochem J 162:135-142

Weiss SJ, Putney JW Jr (1978) Does calcium mediate the increase in potassium permeability due to phenylephrine or angiotensin II in the liver. J Pharmacol Exp Ther 207:669-676

Wernette ME, Ochs RS, Lardy HA (1981) Ca^{2+}-stimulation of rat liver mitochondrial glycerophosphate dehydrogenase. J Biol Chem 256:12767-12771

Williamson JR, Thomas AP, Joseph SK (1984) The role of inositol trisphosphate in hormone-induced Ca^{2+} mobilization in liver. In: Bleasdale J (ed) Proceedings of Chilton conference on inositol and phosphoinositides. Humana Press, Clifton (In press)

Woodgett JR, Davison MT, Cohen P (1983) The calmodulin-dependent glycogen synthase kinase from rabbit skeletal muscle. Purification, subunit structure and substrate specificity. Eur J Biochem 136:481-487

Yip B, Lardy HA (1981) The role of calcium in the stimulation of gluconeogenesis by catecholamines. Arch Biochem Biophys 212:370-377

Calcium and Platelet Function

M. B. FEINSTEIN, S. P. HALENDA and G. B. ZAVOICO[1]

CONTENTS

Abbreviations: PtdIns, 1-(3-sn-phosphatidyl)-L-myo-inositol); PtdIns4P, 1-(3-sn-phosphatidyl-inositol)-L-myo-inositol-4-phosphate; PtdIns4,5P$_2$, I-(3-sn-phosphatidylinositol)-L-myo-inositol 4,5-bisphosphate; Quin-2, the tetraanion of Quin-2-tetra (acetoxymethyl) ester; EGTA, ethylene-glycol bis β-aminoethyl ether)-N,N,N',N'-tetraacetic acid; TXA2, thromboxane A2; PGE$_1$, PGD$_2$, PGI$_2$, prostaglandins E$_1$, D$_2$ and I$_2$ (prostacyclin); PS, phosphatidylserine; OAG, 1-oleoyl-2-acetyl-glycerol; DG, 1,2-diacylglycerol; PA, phosphatidic acid; IP3, inositol 1,4,5-trisphosphate.

[1] Department of Pharmacology, University of Connecticut Health Center, Farmington, CT 06032, USA

1 Introduction

The platelet plays a central role in the hemostatic processes that protect the body against the consequences of traumatic damage to blood vessels. Bleeding is arrested by platelets which form a hemostatic plug at the site of vascular injury, promote coagulation to stabilize the plug, and release potent vasoactive substances. The platelet is both a contractile and a secretory cell. An extraordinary variety of biologically active molecules are released from several types of secretory vesicles. The initial responses to stimulation result in changes in cell shape which increase surface area and provide a surface that promotes adhesion to the subendothelium and activation of coagulant factors. Activated platelets also stick to each other to form aggregates, elaborate biologically potent lipid metabolites, and release the contents of their secretory granules, including agglutinin activities on the platelet surface and into the surrounding medium. In common with other secretory and contractile cells, Ca^{2+} may play the role of a second messenger in platelets that links receptor activation by extracellular agonists to response (Detwiler et al. 1978; Gerrard et al. 1981; Feinstein et al. 1981). Platelet functions can also be controlled by cyclic AMP which serves as a second messenger mediating the inhibition of activation (Haslam et al. 1978; Feinstein et al. 1981).

The interaction of stimulating agents (e.g., collagen, thrombin, platelet-activating factor) with receptors on the platelet produces the activated state that is characterized by specific morphologic and biochemical responses; enhanced metabolism of membrane phospholipids and arachidonic acid, phosphorylation of proteins, disassembly of microtubules, and enhanced assembly of cytoskeleton microfilaments and their association with membrane glycoproteins that are involved in aggregation. Many of these responses are regulated by calcium ions, and in fact a change in the state or localization of intracellular Ca^{2+} caused by the stimulus has long been assumed to provide the impetus for the initiation of many biochemical events (Gerrard et al. 1981; Feinstein et al. 1981). Some of the biochemical effects of Ca^{2+} are exerted through the activation of two enzymatic pathways for protein phosphorylation; i.e., Ca^{2+}-calmodulin-dependent myosin light chain kinase (Hathaway and Adelstein 1979) and Ca^{2+}-PS-dependent protein kinase C (Kawahara et al. 1980). In addition, the activity of phospholipase C (Hofmann and Majerus 1982; Rittenhouse 1983) and A2 (Ballou and Cheung 1983) may be stimulated by Ca^{2+} (Fig. 1).

Since the discovery that the Ca^{2+} ionophore A23187 could elicit aggregation and the release reaction (Feinman and Detwiler 1974) it has been widely assumed that Ca^{2+} was the principal if not sole mediator of activation-response coupling in platelets. Nevertheless, important gaps exist in our understanding of the role of Ca^{2+} in platelet function. The last several years have seen the emergence of several significant methodological developments that have enhanced and modified our understanding of the role of Ca^{2+} in platelet functions. Previously the major impediments to progress in the study of the role of Ca^{2+} in platelet function were the lack of a good experimental model to control cytoplasmic Ca^{2+}, and the inability to monitor intracellular free calcium in intact cells. Two developments have largely overcome these obstacles. One was the development of the technique of applying intense electrical fields to cell suspensions that generate small "holes" in the surface membrane, thereby rendering the cell

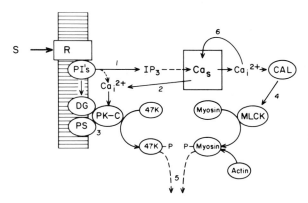

Fig. 1. Proposed sequence of initial events in the activation of platelets. *1* the binding of stimulatory agonists (*S*), such as thrombin, to membrane receptors (*R*) initiates the hydrolyis and subsequent resynthesis of phosphatidylinositides (*PI*s) and the increase of $[Ca^{2+}]_i$. PtdIns4,5P$_2$ is probably the first phospholipid broken down by phosphodiesterase activity (phospholipase C) to produce 1,2-diacylglycerol (DG) and inositol trisphosphate (IP3). This reaction has been suggested to (a) release lipid-bound Ca^{2+}, and (b) to regulate membrane Ca^{2+}-permeability; *2* Ca^{2+} is mobilized from an internal source by mechanisms that are unknown at present. IP3 has been suggested to be a second-messenger that releases Ca^{2+} from a non-mitochondrial source; *3* the increase of $[Ca^{2+}]_i$ acting in concert with 1,2-DG and PS activates protein kinase C, which phosphorylates a cytosolic M_r 47,000 protein; *4* $[Ca^{2+}]_i$ also binds to calmodulin (CAL) which can then activate myosin light chain kinase (MLCK) to phosphorylate the M_r 20,000 light chains of myosin. The ATPase activity of phosphorylated myosin can then be activated by actin; *5* Actomyosin and M_r 47,000 protein may both participate in reactions that enable secretion to occur. *6* $[Ca^{2+}]_i$ can be resequestered within the platelet. Speculative reactions are shown as *dashed lines*

interior accessible to small molecules like Ca^{2+} and ATP without significant loss of protein constituents ($> M_r$ 80,000) of the cytoplasm (Knight and Scrutton 1980; Knight et al. 1982). The other advance was the synthesis of a fluorescent calcium chelator molecule termed Quin-2 that can easily be introduced into the cell cytoplasm to monitor $[Ca^{2+}]_i$ (Tsien et al. 1982). Based in part on information gained from these methods the theory that Ca^{2+} is the second messenger linking receptor-activation to response has undergone an intensive reevaluation. One important finding was that a synthetic 1,2-diacylglycerol (OAG) and phorbol esters could cause secretion and aggregation of platelets in the absence of any rise in $[Ca^{2+}]_i$ detectable by Quin-2 (Rink et al. 1983). This was correlated with the discovery that DG and phorbol esters stimulate the Ca^{2+}-PS-dependent protein kinase C which phosphorylates a cytosolic protein associated with secretion (Kawahara et al. 1980).

Thus, the necessity for a stimulus to mobilize Ca^{2+} has been seriously questioned (Rink et al. 1982, 1983). An alternative hypothesis postulates that 1,2-diacylglycerol is the second messenger that links receptor stimulation to response by activating the enzyme protein kinase C (Kaibuchi et al. 1982), and it has been suggested that this reaction can occur without a rise in cytoplasmic free Ca^{2+} (Rink et al. 1983). Stimulation with extracellular agonists such as thrombin and collagen normally generates DG rapidly from membrane phosphoinositides (Rittenhouse-Simmons 1979). In this review we will discuss in more detail the changes in $[Ca^{2+}]_i$ that occur in response to platelet stimulation, the relationship between the mobilization of Ca^{2+} and concurrent changes in lipid metabolism, the importance of cooperative interactions between Ca^{2+} and diacylglycerol, and the mechanisms by which cyclic AMP regulates $[Ca^{2+}]_i$. Many excellent reviews of various aspects of platelet biology have appeared in recent years, so that our emphasis will be limitied in scope and will concentrate on the most recent work.

2 Relationships Between Ca^{2+}, Diacylglycerol, and Protein Kinases

The activation of platelets initiates two Ca^{2+}-dependent parallel pathways for protein phosphorylation, one is the Ca^{2+}-calmodulin-dependent myosin light chain kinase that phosphorylates myosin and stimulates actin-activated myosin MgATPase activity (Daniel et al. 1981; Adelstein et al. 1981), and the other is the Ca^{2+}-phosphatidylserine-dependent protein kinase C (Kaibuchi et al. 1982) that phosphorylates a M_r 47,000 cytosolic polypeptide of unknown function (Imaoka et al. 1983). Other significant roles for protein kinase C remain to be defined. One possibility is an involvement in Ca^{2+} transport, since Ca^{2+}-ATPase of sarcoplasmic reticulum was reported to be stimulated by the enzyme (Limas 1980). In cardiac sarcolemma a crude enzyme preparation phosphorylated several proteins including a polypeptide believed to be phospholamban, which stimulates Ca^{2+} transport (Iwasa and Hosey 1984).

In addition to requiring Ca^{2+} and PS, protein kinase C activity (using soluble proteins as substrates) is also stimulated by 1,2-diacylglycerol, which is rapidly and transiently formed in activated platelets from membrane phosphoinositides (Rittenhouse-Simmons 1979). Thus, DG also appears to act as a second messenger linking receptor stimulation by extracellular agonists to activation of protein kinase C (Kaibuchi et al. 1982; Rink et al. 1983). Both Ca^{2+} mobilization and DG production are consequences of the action of the first messenger, and both may be coupled to phosphoinositide turnover (see below). The elucidation of this enzymatic pathway provided a vital link to the early lipid changes that characterize receptor-linked platelet activation. DG may also promote its own production because it lowers the Ca^{2+} requirement for phospholipase C (Hofmann and Majerus 1982) and increases the enzyme's ability to hydrolyze membrane phosphoinositide (Dawson et al. 1983).

DG acts by increasing the affinity of the enzyme system for PS and Ca^{2+}; the K_m for activation of protein kinase C by Ca^{2+} is decreased to the range of 0.1 μM by the presence of diolein plus PS (Takai et al. 1981). Furthermore the maximum activity of the enzyme is increased about four fold. Therefore, the activity of protein kinase C is normally a function of the concentration of DG and the phospholipid composition of the membrane, and could conceivably be activated at resting levels of $[Ca^{2+}]_i$ provided sufficient DG is produced. Certain phorbol esters can also stimulate protein kinase C *in vitro* in the same way as DG (Castagna et al. 1982). Interestingly, the phosphorylation of membrane proteins, although dependent on Ca^{2+} and PS was actually inhibited by diolein (Iwasa and Hosey 1984). Variable or inhibitory effects of diolein have also been reported for liver protein kinase C (Kiss and Mhina 1982; Wu et al. 1982). No evidence for similar effects in platelets has been reported, but the possibility that DG could inhibit protein phosphorylation involved in Ca^{2+} transport represents an interesting mechanism for facilitating the cooperative actions of Ca^{2+} and DG.

The protein kinase C pathway for protein phosphorylation can be activated in intact platelets by adding OAG, a synthetic DG, or phorbol ester, as indicated by phosphorylation of the M_r 47,000 protein (Kaibuchi et al. 1982). Both OAG and phorbol esters cause aggregation and the release reaction (Kaibuchi et al. 1983; Yamanishi et al. 1983b), without a detectable increase in $[Ca^{2+}]_i$ (Rink et al. 1983) or phosphoinositide breakdown. Similar effects were observed in neutrophils (Sha'afi et al. 1983). Secretion due to OAG or phorbol esters is unusual in several respects: it is much slower than secretion evoked by thrombin or A23187, and it is often less than maximal (Kaibuchi et al. 1982; Yamanishi et al. 1983b; Kajikawa et al. 1983; Rink et al. 1983). Phorbol ester causes coalescence of alpha-granules and their fusion with the SCCS, but the granule contents are not expelled from the SCCS, which suggested that a contractile event was lacking (White and Estensen 1974). This is consistent with recent findings that the amount of myosin phosphorylation induced by PMA is substantially less than that caused by thrombin (Kaibuchi et al. 1982). The explanation for these apparently calcium-independent effects is that the Ca^{2+} affinity of the protein kinase C-lipid complex is greatly enhanced, into the range of the resting $[Ca^{2+}]_i$, by the added DG or phorbol esters (Takai et al. 1981)[2]. Indeed, it has been suggested that if sufficient DG is produced by stimulation with agents such as collagen and thrombin, responses may occur without the need to increase $[Ca^{2+}]_i$ beyond its normal resting level (Rink et al. 1982).

A concept that harmonizes the conflicting theories about the requirement for Ca^{2+} in platelet responses is based on the fact that a small increase in $[Ca^{2+}]_i$, caused by a calcium ionophore, produces a striking synergistic increase in the rate and extent of secretion elicited by OAG or phorbol esters (Kaibuchi et al. 1982; Rink et al. 1983). This response is accompanied by increased phosphorylation of myosin light chains (M_r 20,000) that is necessary for actin-activated myosin ATPase activity (Kajikawa et al. 1983; Kaibuchi et al. 1983). Kaibuchi et al. (1982, 1983) concluded from these experiments that the diacylglyerol/protein kinase C pathway was a prerequisite for secretion, but not sufficient by itself. Similar conclusion can be drawn from experiments in adrenal chromaffin cells "permeabilized" by an intense electric field. In these cells Knight and Baker (1983) observed that phorbol ester caused a left shift of the concentration-response (secretion) curve for free Ca^{2+}. Thus, Ca^{2+} was not dispensable in the presence of the protein kinase C activator, only less of it was needed for secretion.

The formation of DG and its modulation of the sensitivity of protein kinase C to activation by Ca^{2+} can be viewed as part of a supplementary messenger system that allows for maximal activation of Ca^{2+}-dependent reactions without the necessity to increase $[Ca^{2+}]_i$ to a level that is potentially toxic to the cell. Our investigations indicate that low concentrations of thrombin which do not mobilize Ca^{2+} also do not cause secretion. Therefore, not enough DG could have been produced to independently initiate secretion (see below). Furthermore, it has been proposed that the mobilization of Ca^{2+} occurs as a result of receptor-linked hydrolysis of phosphoinositides (Berridge 1981; Downes and Michell 1982; Streb et al. 1983), suggesting that both second messengers are generated simultaneously. The simultaneous increase of $[Ca^{2+}]_i$

and DG by the stimulus, as normally occurs with thrombin, ensures a mechanism to synergistically activate protein kinase C.

Recently protein kinase C was shown to phosphorylate the M_r 20 K light chains of myosin *in vitro* at a different site than the Ca^{2+}-calmodulin-dependent light chain kinase (Endo et al. 1982; Naka et al. 1983; Nishikawa et al. 1983). Phosphorylation of this site did not occur in platelets stimulated with 0.1 Uml^{-1} thrombin, but it took place slowly in phorbol ester-stimulated platelets (Naka et al. 1983). Another interesting finding in this study was that a transient phosphorylation of the calmodulin-dependent site also occurred in phorbol ester-stimulated platelets (Naka et al. 1983). The extent of phosphorylation was about 15% of that due to 0.1 Uml^{-1} thrombin. A possible interpretation of this result is that phorbol esters cause some degree of Ca^{2+} mobilization, despite the inability of Quin-2 to detect it, thereby slightly activating myosin light chain kinase. A redistribution of intracellular Ca^{2+} by phorbol ester is likely since it causes some fall in $[Ca^{2+}]_i$ that is reflected by Quin-2 fluorescence (Rink et al. 1983; Sha'afi et al. 1983). These findings prompt the following questions: (1) is the slow rate of secretion caused by OAG or phorbol ester due to the low extent of myosin phosphorylation at the calmodulin-dependent site? and (2) is secretion possible at all if myosin is not phosphorylated at the calmodulin-dependent site? No functional effect of the protein kinase C-dependent phosphorylation of platelet myosin has yet been described. Smooth muscle myosin can also be phosphorylated by this enzyme, and in that case the Ca^{2+}-calmodulin-stimulated, actin-activated, myosin ATPase activity is reduced by 50% (Nishikawa et al. 1983). This may be a mechanism to reduce the energy expenditure of contraction without affecting sustained maintenance of force (Dillon et al. 1981).

3 Proposed Mechanisms for Mobilization of Calcium

One of the most vexing problems in platelet physiology is to determine how stimuli mobilize platelet Ca^{2+} (Gerrard et al. 1981). Both recruitment of internal calcium stores as well as influx from the external media are believed to occur, but how this is accomplished is not understood and the source(s) of internal activator Ca^{2+} has not been identified. Receptor-agonist complexes could alter plasma membrane permeability to ions by opening associated channels, or the activated receptor itself could function as a channel for extracellular Ca^{2+}. No evidence exists for the operation of voltage-dependent membrane Ca^{2+} channels, however, we find that the presence of Na^+ in the medium does increase the rise in $[Ca^{2+}]_i$ caused by thrombin. Increased Na^+ influx is known to occur when platelets are stimulated by ADP (Feinberg et al. 1977), and may play a role in the recruitment of intracellular calcium.

3.1 Possible Lipid Mediators of Calcium Mobilization

Several biochemical reactions involving membrane lipids have also been proposed to be involved in mobilization of calcium; i.e., hydrolysis of inositol phosphatides, formation of phosphatidic acid, and the generation of products of arachidonic acid metabolism. PtdIns4,5P2 present in the membrane surface facing the cytosol is thought to provide a potential source of intracellular calcium that could be released by receptor-activated hydrolysis of the lipid by phospholipase C (Vickers et al. 1982b). It was calculated that the amount of PtdIns4,5P2 present in platelet membranes could potentially provide about 35 μM Ca^{2+}. This is at least several fold below the total amount of Ca^{2+} that is likely to be mobilized by a maximal stimulus (see below), but it could provide "trigger" Ca^{2+} to induce the release of a larger additional pool of calcium; i.e., so-called calcium-induced calcium release (Fabiato and Fabiato 1977). Given the known affinity of Ca^{2+} for PtdIns4,5P2 it is also uncertain that the lipid could be a significant binding site for Ca^{2+} at equilibrium with 10^{-7}M free cytoplasmic Ca^{2+} in the presence of much larger amounts of Mg^{2+} (Hendrickson and Reinertsen 1971; Downes and Michell 1982). Nevertheless, this interesting hypothesis deserves further investigation.

The intracellular formation of compounds with ionophore-like activity has commonly been suggested as a mechanism for the mobilization of platelet calcium. Gerrard et al. (1978, 1979), Gerrard and Carroll (1981) and Imai et al. (1982) proposed that thromboxane A2 acts like a Ca^{2+} ionophore, but with more restricted effects than A23187. Another view is that thromboxane A2 acts through specific receptors (Le Breton et al. 1979). Although the cellular location of such presumptive receptors has not been determined conclusively, it is thought that they reside in the dense tubule system, which is also believed to be the site of formation of thromboxane A2. Although not essential for thrombin-induced responses thromboxane A2 and/or PG endoperoxides probably contribute somewhat to calcium mobilization by thrombin, since aspirin partially inhibits the rise in $[Ca^{2+}]_i$ measured by Quin 2 (Rink et al. 1982; Zavoico et al. 1984). Exogenous arachidonic acid also increases $[Ca^{2+}]_i$ and this effect is completely blocked by aspirin (Zavoico et al. 1984).

PA has been proposed to act as an ionophore for the transport of Ca^{2+} (Serhan et al. 1982), but this has recently been disputed and the effect attributed to lipid impurities such as fatty acid oxidation products (Holmes and Yoss 1983). We also find that the time course of the appeareance of PA is slow after thrombin stimulation so that it is unlikely to be responsible for the initial rise in $[Ca^{2+}]_i$ (Fig. 2). Furthermore, PA levels are maximal when $[Ca^{2+}]_i$ has fallen back to its control level. LysoPA is also unlikely to be responsible for initial Ca^{2+} mobilization since it appears in platelets even later than PA and was not detected at a thrombin concentration of 0.25 Uml^{-1} (Billah and Lapetina 1982) which in our studies produces a maximal increase in $[Ca^{2+}]_i$ (Zavoico et al. 1984). The evidence in favor of a role for PG endoperoxides and/or TXA2 in the mobilization of Ca^{2+} is more convincing than that for PA. In fact PA production caused by exogenous arachidonic acid requires an intact cyclooxygenase pathway (Siess et al. 1983).

Another hypothesis currently receiving great attention holds that IP3 generated by the phosphodiesteratic cleavage of PtdIns4,5P2 is a second messenger for the mobili-

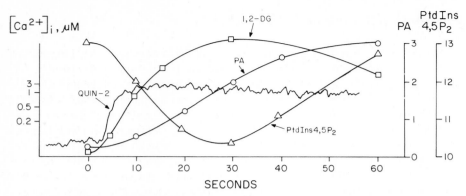

Fig. 2. The temporal relationships between Ca^{2+}_i mobilization and lipid metabolism. Washed human platelets were loaded with Quin-2 and the phospholipids were labeled with ^{32}P-phosphate or ^{3}H-arachidonic acid. The platelets were then stimulated with thrombin (1 Uml^{-1}). Quin-2 fluorescence measured cytoplasmic free Ca^{2+}, $[Ca^{2+}]_i$. Concurrently the platelet lipids were extracted and analyzed for ^{3}H-1,2-diacylglycerol (1,2-DG), and ^{32}P-labeled PtdIns4,5P$_2$ and phosphatidic acid (PA) by thin-layer chromatography and liquid scintillation counting (dpm/10^8 platelets). Temperature 23°C

zation of Ca^{2+} (Berridge 1983). Recent experiments have shown that added IP3 can release a nonmitochondrial source of intracellular Ca^{2+} in permeabilized pancreatic acinar cells (Streb et al. 1983), providing the first experimental verification of this hypothesis. Since the hydrolysis of PtdIns4,5P2 produces IP3 and DG a mechanism exists by which a single receptor-linked enzymatic reaction could simultaneously generate second messengers that mediate the two pathways for Ca^{2+}-dependent protein phosphorylation. The level of IP3 has been shown to increase about two fold in 5 s after stimulation by a high concentration of thrombin (Agranoff et al. 1983). However, the role of IP3 in platelet calcium economy remains to be established.

3.2 Relationships Between Ca^{2+} and Phosphoinositides

The metabolism of phosphoinositides and other phospholipids in platelets has been extensively reviewed recently (Rittenhouse 1982; Lapetina 1983) and will not be the subject of this discussion. However, there are certain important relationships between lipid metabolism and Ca^{2+} that are germane to our discussion. From the standpoint of the role of Ca^{2+} in platelet function, the phosphoinositides are of especial interest for the following reasons: (a) the metabolic turnover of phosphoinositides has been proposed to be linked to Ca^{2+}-gating; i.e., the regulation of receptor-linked membrane Ca^{2+} permeability; (b) PtdIns4,5P2 has been suggested to bind Ca^{2+}, so that its hydrolysis would release Ca^{2+} into the cytosol; (c) PA and TXA2 are reputed to be calcium ionophores; (d) 1,2-diacylglycerol can promote membrane fusion, stimulate Ca^{2+}-PS-dependent protein kinase C and phospholipase C, and facilitate the hydrolysis of membrane lipids by phospholipase C and A2.

The relationship between phosphoinositide metabolism and Ca^{2+} has been a subject of intense investigation since Michell in 1975 first suggested that PtdIns metabolism regulated receptor-linked Ca^{2+}-gating in the plasma membrane. Subsequently attention has turned towards PtdIns4,5P2 (Downes and Michell 1982) and this phospholipid is probably the first one broken down in platelets in response to receptor stimulation (Vickers et al. 1982a, b; Rendu et al. 1983a, b; Perret et al. 1983; Imai et al. 1983). IP3 produced by the hydrolysis of PtdIns4,5P2 has been proposed to be a second messenger that is responsible for mediating receptor-linked mobilization of nonmitochondrial intracellular stores of Ca^{2+} (Streb et al. 1983). We find that maximal mobilization of Ca^{2+} occurs about 15 s prior to the measured peak of ^{32}P-labeled PtdIns4,5P2 breakdown and ^{3}H-DG production (Fig. 2). (Halenda et al. 1984). The maximal mobilization of Ca^{2+} also occurs at thrombin concentrations well below that necessary to produce maximal breakdown of ^{32}P-labeled PtdIns4,5P2 (Lapetina 1983) or maximal production of DG (Rittenhouse-Simmons 1981) and PA (Billah and Lapetina 1982). These results suggest that the lipid metabolism may not be the direct cause of Ca^{2+} mobilization. However, the crucial factor may be the production of IP3 which could be sufficient at a much earlier time. Furthermore, the amount of inositol 1,4,5-trisphosphate produced may greatly exceed the amount of fall in radioactivity present in PtdIns4,5P2 because the latter may be rapidly replenished from the pool of PtdIns (Downes and Wusteman 1983). The validity of the theory proposing IP_3 as a mediator for Ca^{2+} mobilization will therefore require more detailed information about the relationship between $[Ca^{2+}]_i$ and the production of the watersoluble inositol phosphates from polyphosphoinositides (Agranoff et al. 1983).

A major premise of the Michell theory is that the lipid metabolism mobilizes Ca^{2+} and the reactions involved should therefore be independent of a requirement for increased $[Ca^{2+}]_i$ (Downes and Michell 1982). Therefore, the stimulation of platelet phospholipase C through receptor activation should occur without a rise in $[Ca^{2+}]_i$, but definite proof of this premise in platelets, as in many other cell types, has been elusive and controversial (see Symposium in Life Sciences Vol. 32 pp. 2043-2085). Ca^{2+} (K_m 1 μM) is required for the in vitro activity of phospholipase C (Hofmann and Majerus 1982) which hydrolyzes all the phosphoinositides, including PtdIns4,5P2 (Rittenhouse 1983). Nevertheless, merely increasing platelet $[Ca^{2+}]_i$, which can be accomplished by applying ionophores, does not produce as extensive a breakdown of phosphoinositides, or production of DG, as caused by thrombin. In fact, much of the PtdIns metabolism induced by A23187 is mediated by phospholipase A2 rather than phospholipase C (Rittenhouse-Simmons 1981). Therefore, receptor activation appears to be a necessary precondition to render phosphoinositides susceptible to phospholipase C activity. It has been claimed that breakdown of PtdIns4,5P2 is also independent of Ca^{2+}, whereas its resynthesis may be increased in the presence of elevated $[Ca^{2+}]_i$ (Lapetina 1983). In support of this thesis it was shown that PGI$_2$ could inhibit ^{32}P incorporation into PtdIns4,5P2 (resynthesis) but did not prevent its breakdown. However, we find that dibutyryl cyclic AMP, PGD$_2$, and PGI$_2$ all inhibit Ca^{2+} mobilization, as well as breakdown of PtdIns4,5P2 and the formation of 1,2-diacylglycerol (Halenda et al. 1984) (Fig. 3). Ca^{2+} mobilization and PtdIns4,5P2 metabolism appear to be closely related events, but it remains to be seen if there is a cause-and-effect

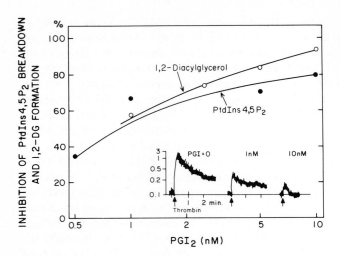

Fig. 3. Inhibition of Ca_i^{2+} mobilization, PtdIns4,5P$_2$ breakdown and 1,2-diacylglycerol formation by PGI$_2$. Conditions were the same as in Fig. 1

relationship, or whether they are parallel events that are both evoked by receptor occupancy.

PtdIns4,5P2 may be involved in the regulation of $[Ca^{2+}]_i$ in another way. PtdIns4,5P2 was a more effective stimulator of purified Ca^{2+}-pumping ATPase from rat brain synaptosome membranes than calmodulin (Penniston 1982; see also discussion on pp. 326-327 of same ref.). Thus, the breakdown and subsequent resynthesis of PtdIns4,5P2 could lead to a transient inhibition of Ca^{2+} pumping that would allow the optimal development of a transient Ca^{2+} signal. Although it is not known what effect PtdIns4,5P2 has on platelet Ca^{2+}-pumps, this hypothesis would be consistent with our data showing that PtdIns4,5P2 breakdown reaches a peak about 15 s after the mobilization of Ca^{2+}_i is maximal. Resynthesis of PtdIns4,5P2 thereafter occurs rapidly, and is followed by a fall of $[Ca^{2+}]_i$ back to normal. It is also possible that a product of PtdIns4,5P2 hydrolysis inhibits Ca^{2+} transport. This hypothesis holds that the breakdown of PtdIns4,5P$_2$ is not the direct cause of the mobilization of Ca^{2+}, but rather a mechanism to potentiate and briefly prolong the rise in $[Ca^{2+}]_i$.

4 Relationship Between $[Ca^{2+}]_i$ and Platelet Responses

The relationship between platelet responses and $[Ca^{2+}]_i$ have been measured by two types of method. In one case platelets have been permeabilized by intense electrical fields, or by treatment with digitonin. The other method utilizes intact platelets that have been loaded with Quin-2, the fluorescent indicator for cytoplasmic Ca^{2+}. The most important direct evidence that Ca^{2+} causes secretion comes from work of Knight et al. (1982), who used platelets permeabilized by an intense electric field (20,000 $V cm^{-1}$) that were then incubated in Ca^{2+}-EGTA buffers of varying free Ca^{2+} activity.

This method has also been used by several groups to study the Ca^{2+} dependency of secretion in chromaffin cells. The threshold $[Ca^{2+}]$ for 5-HT and acid hydrolase secretion from electrically permeabilized platelets was obove 0.5 μM, and was maximal at 10 μM. The phosphorylation of myosin, which may be necessary for secretion, was measured in digitonin-permeabilized platelets (Daniel et al. 1982); phosphorylation of the M_r 20,000 light chains occurred over a range of free Ca^{2+} from 0.3 μM to 4 μM; $K_{0.5} = 1$ μM (Daniel et al. 1982).

It is most revealing to compare these data with that from Quin-2-loaded intact platelets. The fluorescent indicator for Ca^{2+}, Quin-2, has yielded much new and valuable information concerning cellular regulation of $[Ca^{2+}]_i$. Quin-2 is distributed uniformly in the cytoplasm and does not appear to bind to membranes, or enter mitochondria or secretory granules (Tsien et al. 1982). Quin-2 has been used to measure $[Ca^{2+}]_i$ in lymphocytes (Tsien et al. 1982), neutrophils (Sha'afi et al. 1983), adrenal medullary cells (Knight and Kestenen 1983), hepatocytes (Charest et al. 1983), as well as in platelets. However, certain of its properties make it less than ideal for physiological studies. The quantum yield is relatively low, so that high concentrations of the indicator must be attained intracellularly to produce usable signals at $[Ca^{2+}]_i$ levels from 0.1 to 1-10 μM. Excessively high concentrations (several millimolar) can cause intracellular shifts in calcium distribution, or they can buffer changes in $[Ca^{2+}]_i$ and affect Ca^{2+}-dependent rections. There is already evidence that Quin-2 can inhibit secretion in adrenal cells, depending on the extent of loading of the cells with the Ca^{2+} indicator (Knight and Kestenen 1982). Another factor to be considered is that the Ca^{2+}-Quin-2 complex has a Kd of about 115 nM under physiological conditions, which although ideal to measure free Ca^{2+} in the range of $[Ca^{2+}]_i$ normally present in unstimulated cells (i.e., 10^{-7}M) is poor for resolving changes above 1 μM, at which concentration the fluorescence response is about 90% saturated.

Because Quin-2 measures the average overall $[Ca^{2+}]_i$, highly localized regional changes in $[Ca^{2+}]_i$ could escape detection, especially if changes in opposite directions occur at different regions of the cell. Despite these problems it is possible to obtain important data with due attention to the inherent limitations of the method, and by comparison with other methods that can provide information about the relationship between $[Ca^{2+}]_i$ and responses.

The resting $[Ca^{2+}]_i$ in platelets determined with Quin-2 is about 0.1 μM (Rink et al. 1981, 1982; Rink and Smith 1983; Yamanishi et al. 1983a). In response to certain agonists such as thrombin, and the ionophore A23187 and ionomycin, $[Ca^{2+}]_i$ rises rapidly to > 2-3 μM (Rink et al. 1981; Yamanishi et al. 1983a). Employing ionomycin as the stimulus Rink et al. (1981) observed that the $[Ca^{2+}]_i$ threshold for shape change was between 0.4 and 0.6 μM, whereas aggregation required a further increase to about 2 μM. Secretion of 5-HT induced by ionomycin had a threshold of about 0.7-1.0 μM $[Ca^{2+}]_i$; at 3 μM $[Ca^{2+}]_i$ secretion was 68% of maximum. These values correspond well with data for the secretion of 5-HT as a function of $[Ca^{2+}]_i$ in electrically permeabilized platelets (Knight et al. 1982).

The responses to thrombin in intact platelets occur at lower concentrations of $[Ca^{2+}]_i$ than were required in electrically permeabilized platelets, or were measured in ionophore-stimulated platelets. Secretion of 5-HT in Quin-2 loaded platelets was 90% maximal when $[Ca^{2+}]_i$ increased to only about 0.4 μM, and 65% release occurred at 0.2 μM $[Ca^{2+}]_i$ (Rink et al. 1982). In further experiments Rink et al. (1982) found that it was possible to obtain secretion with thrombin concentrations as high as 0.5 Uml^{-1} with an increase in $[Ca^{2+}]_i$ of no more than 100 nM. The conditions for these experiments were quite specific, either thrombin was added in aspirin-treated platelets in EGTA, or after a priming dose of ionomycin had first produced a partial increase in $[Ca^{2+}]_i$ to about 200 nM. These results prompted Rink et al. (1982) to propose that a factor other than Ca^{2+} was involved in eliciting secretion.

A major revelation from the experiments of Rink et al. (1983) was the finding of specific conditions under which secretion of ATP or 5-HT could occur without a detectable rise in $[Ca^{2+}]_i$. A slow secretion of ATP, ranging in extent from 57%-79% of that produced by thrombin, could be elicited when 20 nM phorbol ester or 60 μgml^{-1} OAG were added to aspirin-treated platelets, both in the absence or presence of $[Ca^{2+}]_o$. The appearance of β-thromboglobulin in the medium indicated that alpha-granule release had also occurred. Collagen (10 μg ml^{-1}) also did not increase $[Ca^{2+}]_i$ in aspirin-treated platelets, but an attenuated secretory response occurred which resembled, in rate and extent, that produced by phorbol ester or OAG (Rink et al. 1983). The fact that exogenous DG can elicit secretion without a detectable rise in $[Ca^{2+}]_i$ suggests that the DG that is normally produced in response to thrombin (Rittenhouse-Simmons 1979) may be itself be responsible for eliciting secretion. DG production was detectable when as little as 0.05 U thrombin per 10^8 platelets was added: however, the thrombin concentration was actually 0.5 Uml^{-1}, since the platelet concentration was 10^9 ml^{-1} (Rittenhouse-Simmons 1979; Kaibuchi et al. 1982). From our experiments a maximal increase of $[Ca^{2+}]_i$ would occur at that concentration of thrombin. At issue then is whether low concentrations of thrombin can mobilize sufficient DG to initiate secretion without an increase in $[Ca^{2+}]_i$. To approach this problem it was necessary to obtain more detailed information about changes in platelet $[Ca^{2+}]_i$ in response to thrombin and its relationship to secretion and the production of DG.

5 Mobilization of Calcium and its Regulation

5.1 Dose-Response to Thrombin

We have further investigated the dose-response relationship for thrombin with respect to both $[Ca^{2+}]_i$ and secretion. Many experiments were conducted at 21-23°C because one of the major objects of our work was to determine the temporal relationships between various platelet responses to stimulation and the concentration of cytoplasmic free Ca^{2+}. At room temperature the typical platelet responses remain intact, but their time courses are sufficiently slowed, permitting a more satisfactory temporal resolution of biochemical events which often appear coincidental at 37°C (e.g., see Hofmann and Majerus 1982). Quin 2-loaded platelets in Ca^{2+}-containing (1 mM) or Ca^{2+}-free (+ 1 mM EGTA) buffers were stimulated with thrombin over a concentration range of from 0.01 Uml^{-1} to 2.0 Uml^{-1} (0.1 nM to 20 nM). The basal level of $[Ca^{2+}]_i$

in resting platelets was normally between 70 and 130 nM, and increased to about 2-3 μM^3 when maximally stimulated with thrombin, as previously described by Rink et al. (1981, 1982) and Rink and Smith (1983). The rate and extent of dense granule secretion (Ca^{2+} and ATP), and the rise of $[Ca^{2+}]_i$ was increased as a function of thrombin concentration. A clearly detectable rise in $[Ca^{2+}]_i$ above the basal level was observed with as little as 0.015 Uml^{-1} thrombin (0.015 $U/10^8$ platelets) and the maximal response occurred at a thrombin concentration of about 0.25 Uml^{-1} (Fig. 4). This range is in close agreement with the data of Knight et al. (1982) for the secretion of 5-HT induced by thrombin in intact platelets at 20°C. Even at low concentrations of thrombin secretion was never detected without some prior elevation of $[Ca^{2+}]_i$. In fact a rise in $[Ca^{2+}]_i$ was detectable at concentrations of thrombin below the threshold to produce secretion.

It is clear from our results that the secretory response to thrombin occurs along a $[Ca^{2+}]_i$ curve that is shifted significantly to the left of that in electrically permeabilized cells. An obvious conclusion, in agreement with the work of Rink et al. (1982), is that factors other than $[Ca^{2+}]_i$, such as DG, come into play when intact platelets are stimulated with thrombin. However, it is also likely that normal responses to thrombin probably involve the cooperative action of both mediators. DG production increases progressively as a function of thrombin concentration as high as 5 Uml^{-1} (Rittenhouse-Simmons 1981). The amount of DG produced at 0.25 Uml^{-1} thrombin (which elicits maximum Ca^{2+} mobilization) was only 25% of that at 5 Uml^{-1}, nevertheless it was equivalent to the amount of DG ($\sim 5 \mu M$) that fully activated protein kinase C in vitro (Kaibuchi et al. 1982).

5.2 Kinetics of Response to Thrombin

After the addition of maximally effective concentrations of thrombin there was a minimum lag phase of 2.0-2.5 s until the onset of the rise in $[Ca^{2+}]_i$; about 1.0-1.5 s of this interval was accounted for by the mixing time. $[Ca^{2+}]_i$ reached its peak level 8-12 s after adding thrombin, and then spontaneously declined slowly to its original basal level, usually over a period varying from 6-15 min for platelets from different donors. The decline of $[Ca^{2+}]_i$ in any batch of platelets was more rapid in Ca^{2+}-free medium, suggesting that influx of Ca^{2+} from the extracellular medium contributed to the sustained elevated cytosolic levels. Most of the rise in $[Ca^{2+}]_i$ occurred prior to the onset of secretion, which is an agreement with our previous experiments using the fluorescent probe chlortetracycline to measure the rate of mobilization of intracellular membrane-bound calcium (Feinstein 1980).

Since Ca^{2+} and DG interact importantly, the temporal relationships between them are of interest. The production of DG produced was reported to peak at 15 s after adding thrombin at 37°C (Rittenhouse-Simmons 1981). At 25°C the first rise in DG was detected at 25 s by Prescott and Majerus (1983), and the peak occurred at about 60 s. We find the generation of DG to be more rapid. At 1 Uml^{-1} thrombin at 23°C the formation of DG lags slightly behind the rise in $[Ca^{2+}]_i$ and reaches a peak at about 30 s in Quin-2-loaded platelets (Fig. 2). Both Ca^{2+} and DG are therefore entirely, or substantially, mobilized prior to the secretory response (see below) and in sufficient amounts to qualify as mediators of the reactions that enable secretion to occur.

5.3 Evidence for Intracellular Mechanisms for Mobilization and Sequestration of Calcium

The extent to which influx of Ca^{2+} from the medium contributes to the stimulus-induced rise in $[Ca^{2+}]_i$ is not clear although secretion induced by thrombin occurs in the absence of $[Ca^{2+}]_o$. In the absence of $[Ca^{2+}]_o$ (+ EGTA) Rink et al. (1981) and Rink and Smith (1982) reported that the rise in $[Ca^{2+}]_i$ produced by thrombin or ionomycin was reduced from about 3 μM to only 0.2-0.3 μM; although it was stated that due to buffering of Ca^{2+} by Quin-2 the rise may have been to as much as 0.5 μM (Rink et al. 1982). Thus, it was concluded that most of the increase in $[Ca^{2+}]_i$ is normally due to influx of Ca^{2+} from the medium. We find that in Ca^{2+}-free medium thrombin rapidly increases $[Ca^{2+}]_i$ up to a maximal level of 1-1.5 μM. Although the highest level of $[Ca^{2+}]_i$ attained was less than in the presence of 1 mM $[Ca^{2+}]_o$ the dose-response curve to thrombin was not significantly shifted, and mobilization of intracellular Ca^{2+} was still detectable at 0.015 Uml^{-1} thrombin (Fig. 4). These responses in Ca^{2+}-free media were significantly larger than those previously observed by Rink et al. (1981) and Rink and Smith (1983) at 37°C. This may be due to the fact that in our experiments exposure to Ca^{2+}-free media was very brief prior to stimulation so as to minimize depletion of calcium from the platelets. It remains to be proven whether a large influx of Ca^{2+} actually occurs, or whether the absence of $[Ca^{2+}]_o$ depletes membrane calcium and thereby some reaction necessary for the mobilization of intracellular Ca^{2+}. Another possibility is that extracellular Ca^{2+} suppresses some mechanism for resequestration of released Ca^{2+}, which would in effect increase $[Ca^{2+}]_i$. Most of the Ca^{2+} could come from intracellular sources that would be substantially buffered by calcium-binding proteins (e.g., calmodulin). Under such conditions a relatively small further influx of Ca^{2+} from the medium would produce a disproportionately larger increase in $[Ca^{2+}]_i$. Measurements of $^{45}Ca^{2+}$ uptake in response to thrombin (Imai et al. 1982) do not settle these issues, because the experimental conditions cannot distinguish between net influx of Ca^{2+} as opposed to merely enhanced $[Ca^{2+}]_o/[Ca^{2+}]_i$ exchange that could occur as a result of the increase in $[Ca^{2+}]_i$.

5.4 Effects of Receptor Occupancy on $[Ca^{2+}]_i$

Holmsen et al. (1981) reported that certain platelet responses to thrombin at 37°C

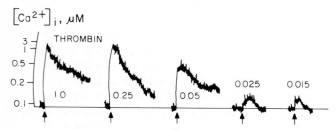

Fig. 4. Increase of platelet cytoplasmic Ca^{2+} in response to thrombin. Quin-2 loaded platelets in Ca^{2+}-free medium (+ 2 mM EGTA). Temperature 23°C

appear to require continuous occupancy of the receptors: i.e., phosphatidic acid production and secretion of acid hydrolases, but not ATP secretion or PtdIns breakdown. We therefore assessed the role of receptor occupancy on $[Ca^{2+}]_i$ by employing the potent thrombin-inactivator hirudin, which rapidly strips thrombin from its receptors (Tam et al. 1979). When excess hirudin was added at the peak of the response to thrombin (10-15 s) the spontaneous decline of $[Ca^{2+}]_i$ was greatly accelerated, even in Ca^{2+}-free medium. Thus, the rate at which Ca^{2+} could be removed from the cytoplasm was much greater when thrombin receptors were unoccupied. The concentration of free cytosolic Ca^{2+} in an activated platelet probably represents a steady-state determined by the rate of Ca^{2+} mobilization and the rate of Ca^{2+} sequestration (e.g., pumping, binding). The accelerated fall in $[Ca^{2+}]_i$ caused by hirudin was most probably due to the rapid inactivation of the mechanism for Ca^{2+} mobilization caused by the decline in receptor occupancy, leaving the mechanisms for Ca^{2+} removal (e.g., pumping) unopposed.

When receptor occupancy was maintained the spontaneous rate of decline of $[Ca^{2+}]_i$ was a much slower process. Under these conditions the fall of $[Ca^{2+}]_i$ would involve the desensitization of thrombin receptors, the rate of depletion of a Ca^{2+}-mobilizing messenger, the inactivation of a Ca^{2+}-channel, or some combination of these processes. Platelets are in fact known to become rapidly desensitized to thrombin, with respect to their ability to aggregate and secrete, but the mechanism is unknown (Shuman et al. 1979; McGowan and Detwiler 1983). The Ca^{2+}-mobilizing action of thrombin also becomes desensitized. This desensitization is apparent when thrombin is removed from its receptors by hirudin and then the platelets are restimulated with excess thrombin. A possible mechanism for desensitization is that some step in lipid metabolism that effects Ca^{2+} mobilization, such as the breakdown of PI-4,5P2, might be exhausted. However, stimulation of platelets with trypsin, after initial treatment with thrombin, usually evoked normal mobilization of $[Ca^{2+}]_i$, and even a greater breakdown of PtdIns-4,5P2 and formation of DG (Fig. 5). Phosphatidic acid production was also increased. Therefore, it appears that desensitization to thrombin is more likely to involve the thrombin receptor itself, or Ca^{2+}-mobilizing reactions (e.g., inactivation of Ca^{2+} channels) that are uniquely linked to the thrombin receptor. However, another possibility worth considering is that thrombin and trypsin interact with different receptors and that each receptor has a finite pool of PtdIns-4,5P2 that can be exhaused by maximal stimulation. The absence of specific receptor antagonists for the proteases makes his hypothesis difficult to test at present.

5.5 Recycling of Intracellular Calcium

Regardless of the extent to which influx of Ca_o^{2+} occurs it is quite apparent that a major source of Ca_i^{2+} mobilized by thrombin comes from an intracellular source. Not only is the rise of Ca_i^{2+} in the absence of Ca_o^{2+} substantial, but we have also been able to establish that recycling of intracellular Ca^{2+} can occur.

As described above, stimulation by thrombin caused a cyclic change in $[Ca^{2+}]_i$. The spontaneous fall of $[Ca^{2+}]_i$ must be due to expulsion of cytosolic Ca^{2+} from the cell across the plasma membrane (or SCCS) or its resequestration within the platelet. To

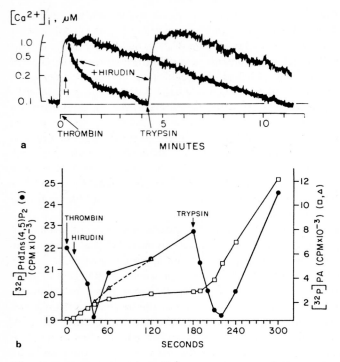

Fig. 5 a. Recycling of internal Ca^{2+} measured by Quin-2 fluorescence. Same experimental protocol as in Fig. 3. A control response to thrombin shows the spontaneous rate of decline of $[Ca^{2+}]_i$ over 10-11 min. The addition of hirudin 15 s after thrombin caused a more rapid fall of $[Ca^{2+}]_i$. Subsequent addition of trypsin evoked a second mobilization of Ca^{2+}. Temperature 23°C, Ca^{2+}-free medium plus 2 mM EGTA. **b** Sequential responses of platelets to thrombin and trypsin. Washed platelets loaded with Quin-2 and lipids labeled with ^{32}P-phosphate were stimulated with thrombin ($1.0 \ Uml^{-1}$). After 15 s hirudin ($3 \ Uml^{-1}$) was added to strip thrombin from its receptors. At 3 min trypsin ($0.2 \ \mu M$) was added. Breakdown and resynthesis of $PtdIns4,5P_2$ occurred in response to each stimulus. The first increase in PA, in response to thrombin, was abbreviated by the addition of hirudin. Temperature 23°C

assess these possibilities, we studied the reactivity of the platelets after one cycle of stimulation. The addition of another agonist like trypsin, post-thrombin stimulation, when $[Ca^{2+}]_i$ had fallen to its basal level, evoked a second wave of increase of $[Ca^{2+}]_i$. The magnitude of this response was variable, but it was often equivalent to, or greater than, the first response to thrombin. This sequential pattern of responses could be produced even when the platelets were in Ca^{2+}-free medium (+ EGTA) (Fig. 5). Thrombin itself could evoke a second cycle of calcium release. In these experiments platelets were first stimulated with thrombin (0.5-$1.0 \ Uml^{-1}$), which was then inactivated with a small excess of hirudin. When thrombin was added a second time in large excess over hirudin $[Ca^{2+}]_i$ again increased, although the second response to thrombin was smaller, indicating that some type of desensitization to thrombin had occurred. Under the same conditions (i.e., after inactivation of thrombin by hirudin) the response to trypsin was normal. These experiments indicate that in Ca^{2+}-free medium the spontaneous fall of $[Ca^{2+}]_i$ after stimulation was largely due to resequestration of calcium within the platelets, and furthermore suggest that Ca^{2+} was probably restored to those sites from which it was originally released. If activator Ca^{2+} had mainly been extruded from the cell after the first stimulation it would have been chelated by the large excess of EGTA. As a result, the pool of available intracellular activator Ca^{2+} would become exhausted, thereby attenuating the second cycle of response to trypsin. Since this did not occur (i.e., the response to trypsin was normal) the activator pool of Ca^{2+} was largely retained within the cell. The fact that trypsin evokes the same changes in phosphoinositide and diacylglycerol metabolism as thrombin strongly suggests that both proteases act on receptors that are similarly linked to both Ca^{2+} mobilization and lipid metabolism. In this regard trypsin is clearly different from Ca^{2+} ionophores which mobilize platelet calcium with much less stimulation of phosphoinositide metabolism (Rittenhouse-Simmons 1981; Lapetina et al. 1981; Rittenhouse 1982).

The sources for intracellular mobilization and resequestration of calcium have not been identified. However, mitochondria do not seem to be importantly involved because neither the rise in Ca^{2+} due to stimulation, nor the resequestration of Ca^{2+} was affected significantly by FCCP, which prevents transport by dissipating the mitochondrial transmembrane potential. Another distinct possibility for a pool of releasable Ca^{2+} is the DTS. White (1972) first called attention to the association of DTS and SCCS membranes in platelets and their resemblance to that of sarcotubules of embryonic muscle, and suggested that the DTS might play a role analogous to that of the sarcoplasmic reticulum of muscle as a site for the release of Ca^{2+}. The endoplasmic reticulum of other nonmuscle cells has been shown to actively accumulate Ca^{2+}.

6 Regulation of $[Ca^{2+}]_i$ by Cyclic AMP

One of the mechanisms by which cyclic AMP regulates platelet functions is by controlling the level of Ca^{2+} in the cytosol. Although this has long been assumed to be so, no direct evidence was available until the introduction of Quin-2 made such experiments feasible. In Quin-2-loaded platelets the increase in $[Ca^{2+}]_i$ caused by thrombin is suppressed by PGD_2 and PGI_2 (and PGE_1) which act respectively on different receptors linked to adenylate cyclase (Miller and Gorman 1979; Schafer et al. 1979), and by forskolin which acts directly on the catalytic unit of the enzyme (Seamon and Daly 1981) and/or on the stimulatory guanine nucleotide regulatory protein, Ns (Durfler et al. 1982). Dibutyryl cyclic AMP also prevents the mobilization of Ca^{2+}, which provides the most direct evidence that cAMP-dependent reactions can regulate $[Ca^{2+}]_i$ in intact platelets. Cyclic AMP-dependent processes exerted two types of effects on $[Ca^{2+}]_i$; (a) they inhibited the rise of Ca^{2+} that would normally occur in response to platelet stimulation by thrombin (Rink and Smith 1983; Feinstein et al. 1983a; Yamanishi et al. 1983a), and (b) they stimulated resequestration of Ca^{2+} that had already been released into the cytosol by thrombin (Feinstein et al. 1983; Yamanishi et al. 1983a). The possible mechanisms responsible for these effects will be discussed below.

6.1 Inhibition of Calcium Mobilization by Cyclic AMP

Owen and LeBreton (1981) using chlortetracycline as a probe for membrane-bound Ca^{2+} obtained evidence that increased cyclic AMP stabilized internal Ca^{2+} stores that were susceptible to release by epinephrine, A23187 and the PG- endoperoxide analog U46619. Subsequently the effects of adenylate cyclase stimulants were investigated by employing Quin-2 to measure $[Ca^{2+}]_i$. Pretreatment of platelets with PGI_2, PGE_1 (Rink et al. 1983; Feinstein et al. 1983), PGD_2 or forskolin (Feinstein et al. 1983a) inhibited the rate and extent of rise in $[Ca^{2+}]_i$. Dibutyryl cAMP also produced a time- and concentration-dependent inhibition of calcium mobilization. PGI_2 was by far the most potent antagonist. The I50 for inhibition of increase in $[Ca^{2+}]_i$ due to 1.0 Uml^{-1} thrombin was about 0.5-1.0 nM PGI_2, and 95% inhibiton was attained at 10 nM PGI_2. The dose-response curve to PGI_2 was not affected by the lack of Ca_o^{2+}, suggesting that cyclic AMP had no selective action on Ca^{2+} influx. The effects of low to moderate concentrations of prostaglandins on $[Ca^{2+}]_i$ were substantially potentiated by the cyclic AMP-phosphodiesterase inhibitor theophylline. Phosphodiesterase inhibitors similarly potentiate the inhibitory effects of prostaglandins on platelet functions (Mills and Smith 1971), an effect that is attributable to the fact that platelets possess high cAMP-phosphodiesterase activity that can be stimulated indirectly by prostaglandins which increase cyclic AMP (Alvarez et al. 1981). It has been suggested that the antithrombotic effectiveness of certain agents observed clinically may be

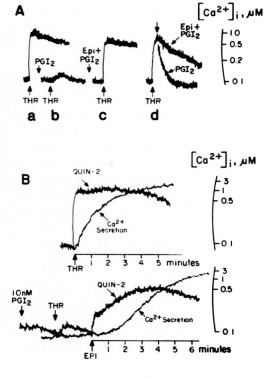

Fig. 6 A, B. Antagonism of PGI_2 and epinephrine. **A** $[Ca^{2+}]_i$ measured with Quin-2. Control response to thrombin (1.0 Uml^{-1}), inhibition by 10 nM PGI_2, and reversal of PGI_2 effect by 1 μM epinephrine. *Last panel* shows rapid fall of $[Ca^{2+}]_i$ when PGI_2 is added at peak of response to thrombin (10-15 s) and the inhibition of PGI_2 by epinephrine. **B** Simultaneous measurement of $[Ca^{2+}]_i$ with Quin-2 and dense granule secretion with an extracellular Ca^{2+} electrode. *Upper panel* is control response to 1.0 Uml^{-1} thrombin. Lower panel shows inhibition of Ca_i^{2+} mobilization and secretion by PGI_2, and the reversal of both effects by epinephrine (1 μM). Note that the rise in $[Ca^{2+}]_i$ after epinephrine was slower and less than the control response. Furthermore, secretion occurred after a much longer lag period and was much slower than normal, although the same total amount of calcium was secreted. Temperature 23°C, extracellular Ca^{2+} was 50 μM

attributable to their antiphosphodiesterase activity which enhances the action of prostacyclin, released from blood vessel walls, on platelet cAMP levels (Weksler 1982).

6.2 Stimulation of $[Ca^{2+}]_i$ Resequestration by Cyclic AMP

Stimulation of adenylate cyclase might prevent the thrombin-induced rise in $[Ca^{2+}]_i$ either by inhibiting the initial receptor-linked reactions that mobilize Ca^{2+} stores, or by enhancing Ca^{2+} sequestration. Cyclic AMP can inhibit phospholipase C activity and therefore can prevent the hydrolysis of the phosphoinositides that have been proposed to be involved in Ca^{2+} mobilization (Rittenhouse-Simmons 1979). Therefore, in order to test the hypothesis that cyclic AMP can stimulate Ca^{2+} sequestration platelets were first stimulated with thrombin before adenylate cyclase was stimulated. This procedure allowed mobilization of Ca^{2+} and DG and the breakdown of PtdIns4,5P2 to occur normally. The addition of PGI$_2$ (or PGD$_2$, PGE$_1$, forskolin) after $[Ca^{2+}]_i$ reached its peak caused the elevated $[Ca^{2+}]_i$ to rapidly fall back to the normal pre-stimulus level at either 37°C (Feinstein et al. 1983a) or 23°C (Zavoico et al. 1984) (Fig. 6). A similar effect with PGE$_1$ was subsequently reported by Yamanishi et al. (1983a). PGI$_2$ also caused a rapid fall of $[Ca^{2+}]_i$ that had been induced by low concentrations of thrombin which did not cause detectable breakdown of PtdIns4,5P2. PGI$_2$ was more potent and caused a faster rate of fall of $[Ca^{2+}]_i$ than PGD$_2$. This is in keeping with the relative potency of PGI$_2$ as a stimulator of adenylate cyclase, and its ability to produce much higher maximal levels of cyclic AMP (Alvarez et al. 1981; Gorman et al. 1977). These experiments indicate that cyclic AMP activates powerful processes for the resequestration of $[Ca^{2+}]_i$ or that it terminates the production or action of a mediator of Ca^{2+} mobilization.

6.3 Antagonism of Adenylate Cyclase Stimulators by Epinephrine and ADP

In intact platelets epinephrine and ADP can reduce the concentration of cAMP that has previously been elevated by adenylate cyclase-stimulants (Mills 1974). They also inhibit basal and hormone-stimulated adenylate cyclase activity in platelet particulate fractions (Jakobs et al. 1978; Cooper and Rodbell 1979). Both antagonists act through specific receptors coupled to guanine nucleotide-binding proteins (Ni) that inhibit the enzyme (Sabol and Nirenberg 1979; Smith and Limbird 1982; Motulsky et al. 1982). Therefore, in the simultaneous presence of prostaglandins and epinephrine (or ADP), the rate of cyclic AMP production should represent the net balance between the effects of the various receptors on the stimulatory (Ns) and inhibitory (Ni) (Murayama and Ui 1983; Jakobs et al. 1981; Rodbell 1980) GTP-binding protein regulators of adenylate cyclase activity.

This dynamic balance between the opposing forces acting on the enzyme was clearly reflected in the changes in $[Ca^{2+}]_i$ levels that were produced in stimulated platelets. Epinephrine and ADP antagonized the effects of PGI$_2$, PGD$_2$ and forskolin on $[Ca^{2+}]_i$ in thrombin-stimulated platelets, but they were unable to reverse the effects of dB-cAMP, presumably because the cyclic AMP analog can bypass adenylate cyclase to directly activate cAMP-dependent kinases. Similarly epinephrine could not overcome the inhibition of platelet aggregation by dibutyryl cyclic AMP (Cox et al. 1984). Epinephrine and ADP by themselves had little or no effect on $[Ca^{2+}]_i$ when added to washed platelets at 23°C in the absence of thrombin or prostaglandins; sometimes a transient rise of $[Ca^{2+}]_i$ of < 100 nM was observed. The inhibition of Ca^{2+}-mobilization by the adenylate cyclase stimulants was largely prevented if epinephrine (or ADP) was added prior to prostaglandins (or forskolin), and the platelets then were stimulated with thrombin (Fig. 6). The ability of PGD_2 and PGI_2, added after stimulation by thrombin, to rapidly restore elevated $[Ca^{2+}]_i$ to its basal level was also counteracted by epinephrine and ADP (Fig. 6).

Epinephrine and ADP can also reverse the effects of PGI_2, PGD_2 or forskolin. Platelets were first incubated with PGD_2 or PGI_2 and then stimulated with thrombin. After the abortive rise in $[Ca^{2+}]_i$ had subsided the addition of epinephrine (or ADP) caused a further increase in $[Ca^{2+}]_i$ up to levels ranging from 300 nM to about 700 nM (Fig. 6). Both the rate and extent of rise of $[Ca^{2+}]_i$ under these conditions were substantially less than the response of untreated platelets to thrombin. This was probably due to the onset of desensitization to thrombin by the time epinephrine (or ADP) was added. The partial rise in $[Ca^{2+}]_i$ was accompanied by secretion, albeit at a lower rate than normal for the amount of thrombin present (Fig. 6). Epinephrine has also been shown to reverse the inhibition by PGI_2 of arachidonic acid-induced cytoskeleton assembly and M_r 47,000 protein phosphorylation (Cox et al. 1984).

All of the effects of epinephrine on $[Ca^{2+}]_i$ were blocked by yohimbine, an alpha$_2$-adrenoreceptor antagonist, but not by corynanthine, which is a highly selective alpha$_1$-antagonist (Weitzell et al. 1979). Alpha-adrenergic receptors mediate two types of physiological responses: (1) inhibition of adenylate cyclase (Fain and Garcia-Sainz 1980), and (2) increasing $[Ca^{2+}]_i$ (Charest et al. 1983) and increased turnover of phosphoinositides (Fain and Garcia-Sainz 1980). Different subsets of alpha-receptors mediate these effects, since alpha$_1$-receptors are linked to the Ca^{2+} and lipid effects and alpha$_2$-receptors have been suggested to act solely on adenylate cyclase (Fain and Garcia-Sainz 1980). Platelet aggregation appears to be mediated by alpha$_2$-receptors since it is much more effectively inhibited by the alpha$_2$-antagonist yohimbine than by prazosin, an alpha$_1$-antagonist (Grant and Scrutton 1979). Although platelets may contain about 10-15% alpha$_1$-receptors their presence and content varies among donors. The significance of this subtype in platelets is not known (Scrutton and Wallis 1981).

Epinephrine and ADP were able to increase $[Ca^{2+}]_i$ even in the absence of Ca^{2+} in the medium, but intact phosphodiesterase activity was necessary since the effect was inhibited by theophylline. Although ADP and epinephrine can reduce elevated cyclic AMP back to basal levels, the presence of a phosphodiesterase inhibitor slows the fall, and the new steady-state level of cyclic AMP attained remains well above normal (Mills 1974). The response to epinephrine (and ADP) also depended upon the occupancy of thrombin receptors, since no effects on $[Ca^{2+}]_i$ were elicited if thrombin was first removed by hirudin. These experiments imply that the mechanisms for mobilizing calcium, although inhibited by a cAMP-dependent reaction, remain potentially operant for some time as long as thrombin continues to be associated with its receptor.

6.4 Stimulation of Ca^{2+} Transport by Cyclic AMP

$[Ca^{2+}]_i$ in cells is controlled in part by transport systems in the surface membrane, the smooth endoplasmic reticulum and mitochondria. Several laboratories have described the presence in platelets of a Ca^{2+}-Mg^{2+}-ATPase and ATP-driven calcium uptake system in isolated microsomal membrane fractions (Robblee et al. 1973; Statland et al. 1969). Although several smooth ER marker enzymes are enriched in these membrane fractions, the presence of surface membranes has not been completely eliminated (Kaser-Glanzmann et al. 1978). In common with Ca^{2+} pumps located in sarcoplasmic reticulum or endoplasmic reticulum of other cells, Ca^{2+} uptake is stimulated by oxalate (Robblee et al. 1973). In a detailed kinetic analysis of the platelet Ca^{2+} uptake system Javors et al. (1982) found that the $K_{0.5}$ for the initial rate of uptake was 0.145 μM free Ca^{2+}. Maximal uptake occurred at 1-5 μM Ca^{2+} and the Vmax was 8 nmol min^{-1} mg^{-1} protein. These data indicate that the Ca^{2+} transport system of platelets has the capability to regulate free Ca^{2+} in the precise range found to occur in resting and stimulated cells.

How adenylate cyclase stimulators increase resequestration of $[Ca^{2+}]_i$ is not known. One possibility is that the production of a messenger necessary for the mobilization of Ca^{2+} is terminated by a cyclic AMP-dependent reaction, thereby allowing the normal transport processes to act unimpeded to restore $[Ca^{2+}]_i$ to its basal level. The finding that dibutyryl cyclic AMP inhibits phospholipase C activity (Rittenhouse-Simmons 1979; Billah et al. 1979), and therefore phosphoinositide metabolism, is consistent with this type of mechanism. If cyclic AMP were to inhibit PtdIns4,5P2 breakdown the formation of IP3, the reputed messenger for mobilization of Ca^{2+} would be reduced. Other findings suggest that the dense tubule system may contain a calcium transport system which may may be directly stimulated by platelet cyclic AMP-dependent protein kinases (Kaser-Glanzmann et al. 1977, 1979). It was previously suggested that the association of the dense tubule membrane system with the surface-connected open-canalicular system of platelets resembles the sarcoplasmic reticulum-T tubule system of embryonic muscle (White 1972). By analogy with the SR the dense tubule system might, therefore, provide regions that are specialized for Ca^{2+} transport as well as stimulus-induced Ca^{2+}-release. The existence of intracellular structures for the sequestration and release of Ca^{2+} is supported by the evidence of internal cycling of Ca^{2+}.

Cyclic AMP along with calmodulin can regulate the Ca^{2+} pump ATPase in cardiac sarcolemma and cardiac sarcoplasmic reticulum and may have similar effects in platelets. However, LePeuch et al. (1983) found no effect of calmodulin on Ca^{2+} transport by microsomal membranes. The rapid fall of $[Ca^{2+}]_i$, in thrombin-stimulated platelets, caused by the addition of adenylate cyclase stimulators could be due to enhancement of a Ca^{2+} pump. Cytochemical studies reveal that the dense tubule system, which has been proposed to act as a calcium "sink", contains Ca^{2+}-activated ATPase (Cutler et al. 1978, 1981), as well as adenylate cyclase activity that is stimulated by PGE_1 (Cutler et al. 1978), PGI_2, PGD_2 and forskolin, and is inhibited by epinephrine (Cutler et al. 1984). Under certain conditions cyclic AMP-dependent protein kinase has been reported to stimulate calcium uptake by platelet microsomal membrane vesicles, and promote the phosphorylation of a M_r 22,000 membrane polypeptide (Kaser-Glanzmann et al. 1979). It was suggested that this polypeptide might be phospholamban, a known activator of Ca^{2+} transport in cardiac sarcoplasmic reticulum. A polypeptide of the same molecular weight is phosphorylated when intact platelets are treated with prostaglandins that stimulate adenylate cyclase (Haslam et al. 1979). Le Peuch et al. (1983) have provided strong evidence that the M_r 22,000 polypeptide is not identical with cardiac muscle phospholamban. Furthermore, they dispute the purported regulation of Ca^{2+} transport by cyclic AMP. Their experiments were conducted with oxalate in the medium, which results in the precipitation of accumulated Ca^{2+} as insoluble calcium oxalate within the membrane vesicles and maximizes Ca^{2+} uptake. It is possible that under such conditions the modulating effect of cyclic AMP on Ca^{2+} transport is obscured. Further work on the properties of platelet calcium pumps is necessary to understand the mechanism by which cyclic AMP controls calcium homeostasis in this cell.

Other cells with calcium pumps in the smooth endoplasmic reticulum or SR also transport Ca^{2+} via surface membranes that is mediated by Ca^{2+} ATPases or Na^+/Ca^{2+} exchange. Indeed the SCCS of the platelet resembles muscle T-tubules in its association with dense tubules (White 1972), and T-tubules of muscle fibers have recently been shown to contain a Ca^{2+} ATPase and Ca^{2+}-transport activity (Rosenblatt et al. 1981). With cytochemical methods the SCCS of platelets, but not the plasma membrane, also exhibits Ca^{2+} ATPase activity (Cutler et al. 1978, 1981). It is not known if membranes of the SCCS are present in the isolated microsomal membrane vesicles that take up Ca^{2+}. A Na^+-Ca^{2+} exchange mechanism may be absent, or of low activity, since the spontaneous and PGI_2-stimulated decrease in $[Ca^{2+}]_i$ were not inhibited by lack of Na^+ in the medium. On the contrary PGI_2 was more effective in the absence of Na^+. The participation of the mitochondria in the regulation of platelet $[Ca^{2+}]_i$ is even less well understood, although a mitochondrial uncoupler (e.g., FCCP) did not inhibit the spontaneous, hirudin-stimulated, or PGI_2-stimulated fall of $[Ca^{2+}]_i$ (Zavoico et al. 1984).

6.5 Effects of Ca^{2+} and Cyclic AMP on Cytoskeleton and Contractile Proteins

The cytoskeleton of cells is involved in the regulation of cell shape, motility, pseudopod formation, and the lateral mobility of intrinsic membrane proteins. Cytoskeleton assembly and turnover is under the control of many proteins; i.e., actin length regulating and capping (e.g., villin, gelsolin, platelet cytochalasin-like protein) (Grumet and Lin 1980; Tellam and Frieden 1982), actin depolymerizing proteins (e.g., profilin), actin crosslinking proteins (e.g., ABP, filamin, alpha-actinin, caldesmon) and proteins involved in tubulin assembly and disassembly (MAP's and tau protein) (see reviews by Stossel 1978; Craig and Pollard 1982; Harris 1981). Immunofluorescence has revealed the presence of myosin, actin, tropomyosin, alpha-actinin, tubulin and filamin (actin-binding protein, ABP) in platelet cytoskeleton (Debus et al. 1981). Stimulation of platelets causes the disassembly of microtubules, and the assembly of the cytoskeleton apparatus composed of actin, myosin, actin-binding protein and other proteins (Carlsson et al. 1979; Jennings et al. 1981; Rosenberg et al. 1981). These reactions are normally associated with platelet shape change, secretion and aggregation. The cytoskeleton is a highly dynamic structure that can undergo cyclical changes in organization in response to variations in the extent of platelet activation and inhibition (Carlsson et al. 1979; Jennings et al. 1981; Rosenberg et al. 1981; Fox and Philipps 1982; Feinstein et al. 1983a, b; Cox et al. 1984; Pribluda and Rotman 1982). An important concept to be kept in mind comes from the work of Carroll et al. (Carroll et al. 1982), who found that specific pools of actin are probably involved in different platelet functions; i.e., pseudopod formation and cytoplasmic contraction. This also implies that the various actin pools are susceptible to some degree of independent regulation, perhaps by highly localized changes in Ca^{2+}, DG, cyclic nucleotides and protein kinase activity. For example, phorbol esters activate pseudopod formation, but not cytoplasmic contraction, and the assembly of cytoskeletons containing actin and actinbinding protein, but lacking myosin (Carroll et al. 1982). The fact that these changes are caused by phorbol ester implies some specific role for protein kinase C. However, a localized fall in $[Ca^{2+}]_i$ (Yassin et al. 1984), and/or the phosphorylation of a vinculin-like protein (Werth et al. 1983) caused by phorbol esters may also be involved.

In addition to the structural cytoskeletal elements the platelet contains a system of contractile proteins that resembles smooth muscle in some of its biochemical properties. Extracellular messengers acting on platelet receptors effect the cytoskeleton and contractile proteins by causing the elaboration of second-messengers, such as Ca^{2+} and cyclic AMP, that act in turn on membrane and cytosolic proteins. Some of the effects of Ca^{2+} on platelet functions and cytoskeleton and contractile proteins are exerted through the Ca^{2+}-binding protein calmodulin. Calmodulin and various calmodulin-binding proteins (e.g., spectrin-like membrane proteins, tau protein, caldesmon, and a M_r 135,000 protein that may be myosin light chain kinase) are importantly involved in the control of microtubule assembly-disassembly, actin-myosin interactions, and cytoskeleton-membrane interactions (Kakiuchi and Sobue 1983; Harris 1981). The calmodulin-binding proteins appear to interact in a flip-flop mechanism, controlled by Ca^{2+}, with either actin or tubulin and calmodulin (Kakiuchi and Sobue 1983). By virtue of their shifting associations with cytoskeleton proteins the calmodulin-binding proteins appear to regulate both structure and function. Several examples of this mechanism have been described. For example, high molecular weight MAP2 and low molecular weight tau protein can bind specifically to calmodulin that contains only 1-2 bound Ca^{2+} (Lee and Wolff 1984). Their affinity for calmodulin is weak compared to enzymes such as phosphodiesterase, but the shifts of these regulatory proteins between tubulin and calmodulin provides a mechanism for the control of microtubule assembly that is sensitive to Ca^{2+} (Lee and Wolff 1984). Myosin light chain kinase is a high affinity calmodulin-binding protein, which in association with protein phosphatases regulates actomyosin ATPase activity. The enzyme is found associated with stress fibers in fibroblasts (DeLanerolle et al. 1981), and lymphocyte cytoskeletons (Bourguignon et al. 1982), and the enzyme from smooth muscle binds to actin (Dabrowska et al. 1982). These findings suggest that the kinase might be located on actin filaments in resting cells, and it has been suggested to flip-flop from actin to calmodulin when $[Ca^{2+}]_i$ increases, and from calmodulin back to actin when $[Ca^{2+}]_i$ is low (Kakiuchi and Sobue 1983). In smooth muscle, caldesmon, another protein that binds to actin at low $[Ca^{2+}]_i$ shifts to calmodulin at micromolar $[Ca^{2+}]_i$, and is believed to act as an on-off switch for actin-myosin interaction, providing a dual mechanism for control involving both actin and myosin filaments (Kakuichi and Sobue 1983). Because of the many similarities between the contractile systems in smooth muscle and platelets, these findings may be relevant to platelet physiology. Smooth muscle contraction may be regulated by yet other proteins (Small and Sobieszek 1980; Ebashi et al. 1982) and there is no reason to believe that the regulation of platelet actomyosin will prove to be any less complicated.

Early experiments employing the calmodulin antagonists TFP and W-7 indicated that calmodulin was involved in platelet aggregation and secretion. Furthermore, secretion was inhibited without a reduction in Ca^{2+} mobilization. However, it was soon demonstrated that these reagents were also effective inhibitors of protein kinase C (Takai et al. 1981), and they blocked phosphorylation of its major substrate, M_r 47 K protein, in intact platelets (Feinstein and Hadjian 1982). In view of these findings, it is not possible at present to selectively dissect out the role of calmodulin and myosin phosphorylation in platelet responses using these pharmacological inhibitors.

Another mechanism for biochemical control of cytoskeleton structure and function of cells is exerted through variation in the state of phosphorylation of certain proteins, such as actin-binding protein (Wallach et al. 1978; Carroll et al. 1982), myosin (Adelstein et al. 1981) and vinculin (Werth et al. 1983). Actin-binding protein is phosphorylated by a cyclic AMP-dependent reaction (Carroll et al. 1982), and vinculin, which is present at focal adhesion plaques and may link actin bundles to the membrane, is phosphorylated by protein kinase C (Werth et al. 1983). The phosphorylation of myosin by Ca^{2+}-calmodulin-dependent myosin light chain kinase (Adelstein et al. 1981) promotes the binding of myosin to actin (Michnicka et al. 1982; Fox and Philipps 1982), stimulates actin-activated myosin ATPase activity (Adelstein et al. 1981) and stabilizes the formation of myosin filaments (Smith et al. 1983).

Cyclic AMP is also an important regulator of cytoskeleton organization and contractile protein function. Agonists that stimulate adenylate cyclase inhibit thrombin-induced protein phosphorylation and cytoskeleton assembly (Haslam et al. 1978; Fox and Philipps 1982; Feinstein et al. 1983a, b; Cox et al. 1984). Stimulators of adenylate cyclase also reverse these responses after they have been produced by thrombin (Feinstein et al. 1983a, b) or arachidonic acid (Cox et al. 1984). Prostaglandins and forskolin may inhibit cytoskeleton assembly and protein phosphorylation by suppression of increases in $[Ca^{2+}]_i$. The effects of cyclic AMP on phospholipid and DG metabolism may also be involved, especially with regard to the influence of lipids on membrane proteins that interact with the cytoplasmic cytoskeleton elements (Schick et al. 1983; Sheetz 1983; Philipps et al. 1980). In addition cyclic AMP-dependent phosphorylation can directly inhibit myosin light chain kinase in vitro (Hathaway et al. 1981), but this reaction has not yet been demonstrated in intact cells. This would in effect shift the sensitivity of the myosin kinase reaction to require higher levels of Ca^{2+}. Preexisting phosphorylation of myosin kinase by cyclic AMP-dependent kinase would be an effective mechanism for inhibition of platelet functions dependent on actomyosin. However, calmodulin/Ca^{2+} binding prevents phosphorylation of the critical site on myosin kinase. Therefore, reversal of myosin phosphorylation after it has occurred (Feinstein et al. 1983a) may be more likely to be caused by the fall in $[Ca^{2+}]_i$ due to cyclic AMP than to phosphorylation of myosin kinase.

The effects of platelet-aggregating agents and cyclic AMP on actin polymerization in intact platelets are more difficult to interpret since so many factors affect that process (Craig and Pollard 1982). However, some of the proteins involved in nucleation and polymerization of actin are affected by cyclic AMP or Ca^{2+}. While cyclic AMP can have direct effects on regulatory proteins it may also influence conditions indirectly by its effects on $[Ca^{2+}]_i$ (Feinstein et al. 1983b).

Kaibuchi et al. (1982) showed that a calcium ionophore synergistically promoted secretion initiated by DG or phorbol esters. Since this response was associated with an increase in myosin phosphorylation, it is reasonable to suppose that a contractile process may have enhanced the rate of granule release (Carroll et al. 1982). There is some evidence that different types of granule are released at distinct membrane regions (Skaer 1981; White 1973; Allen et al. 1979), which suggests that specific recognition sites are involved. It is possible that the slow rate of secretion induced by DG and phorbol esters is attributable to a low probability of random associations of granules with release sites. The normal process of secretion evoked by other agonists (e.g., thrombin) is much more rapid, perhaps because in the presence of elevated $[Ca^{2+}]_i$ myosin provides the motive force for propulsion of granules, which with the aid of proper orientation of cytoskeletal fibers, efficiently directs the granules to the release sites. In this process the rate-limiting factor may be the availability of Ca^{2+}. In digitonin-permeabilized platelets myosin phosphorylation was stimulated at a free Ca^{2+} threshold of about 0.3 μM and was maximal at about 4 μM; $K_{0.5} = 1$ μM (Daniel et al. 1982). Platelet actomyosin ATPase activity had a threshold of 0.1 μM free Ca^{2+} and was maximal at about 1.0 μM (Cohen and DeVries 1973). From the measurements of $[Ca^{2+}]_i$ by Quin-2 it is evident that agonists such as thrombin can mobilize sufficient Ca^{2+} to activate myosin phosphorylation and contraction of actomyosin.

7 Addendum

Since the observation by Streb et al. (1983) that IP_3 can release stored Ca^{2+} in permeabilized pancreatic acinar cells, several groups have reported similar findings in detergent-permeabilized hepatocytes (Joseph et al. 1984; Burgess et al. 1984) and smooth muscle cells (Suematsu et al. 1984). In each case, the mitochondria were excluded as the source of IP_3-releasable Ca^{2+} and endoplasmic reticulum was proposed as the subcellular site of action for IP_3. These studies were extended by Prentki et al. (1984), who found that IP_3 could elicit the release of Ca^{2+} stored in a microsomal fraction, but not from mitochondria or secretory granules prepared from insulinoma cells. Also, in platelets, a subcellular fraction apparently derived from the dense tubular system has been found to release stored Ca^{2+} in response to IP_3 (O'Rourke et al. 1985).

These recent findings provide strong support for a direct link between polyphosphoinositide breakdown and Ca^{2+} mobilization in cells responding to receptor stimulation.

References

Adelstein RS, Pato MD, Conti MA (1981) The role of phosphorylation in regulating contractile proteins. Adv Cyclic Nucleotide Res 14: 361-373

Agranoff BW, Murthy P, Seguin EB (1983) Thrombin-induced phosphodiesteratic cleavage of phosphatidylinositol bisphosphate in human platelets. J Biol Chem 258: 2076-2078

Allen RD, Zacharski LR, Widirstky ST, Rosenstein R, Zaitlin LM, Burgess DR (1979) Transformation and motility of human platelets. Details of shape change and release reaction observed by optical and electron microscopy. J Cell Biol 83: 126-142

Alvarez R, Taylor A, Fazzari JJ, Jacobs JR (1981) Regulation of cyclic AMP metabolism in human platelets. Sequential activation of adenylate cyclase and cyclic AMP phosphodiesterase by prostaglandins. Mol Pharmacol 20: 302-309

Ballou LR, Cheung WY (1983) Marked increase of human platelet phospholipase A_2 activity in vitro and demonstration of an endogenous inhibitor. Proc Natl Acad Sci USA 80: 5203-5207

Berridge M (1983) Rapid accumulation of inositol trisphosphate reveals that agonists hydrolyse polyphosphoinositides instead of phosphatidylinositol. Biochem J 212: 849-858

Berridge MJ (1981) Phosphatidylinositol hydrolysis and calcium signaling. Adv Cyclic Nucleotide Res 14: 289-299

Billah MM, Lapetina EG (1982) Evidence for multiple metabolic pools of phosphatidylinositol in stimulated platelets. J Biol Chem 257: 11856-11859

Billah MM, Lapetina EG, Cuatrecasas P (1979) Phosphatidylinositol specific phospholipase-C of platelets: association with 1,2-diacylglycerol-kinase and inhibition by cyclic-AMP. Biochem Biophys Res Commun 90: 92-98

Bourguignon LYW, Nagpal ML, Balazovich K, Guierriero V, Means AR (1982) Association of myosin light chain kinase with lymphocyte membrane-cytoskeleton complex. J Cell Biol 95: 793-797

Burgess GM, Godfrey PP, McKinney JS, Berridge MJ, Irvine RF, Putney JW Jr (1984) The second messenger linking receptor activator to internal Ca release in liver. Nature 309: 63-66

Carlsson L, Markey F, Blikstad I, Persson I, Lindberg V (1979) Reorganization of actin in platelets stimulated by thrombin as measured by the DNAase I inhibition assay. Proc Natl Acad Sci USA 76: 6376-6380

Carroll RC, Butler RG, Morris PA, Gerrard JM (1982) Separable assembly of platelet pseudopodal and contractile cytoskeletons. Cell 30: 385-393

Castagna M, Takai Y, Kaibuchi K, Sano K, Kikkawa U, Nishizuka Y (1982) Direct activation of calcium-activated, phospholipid-dependent protein kinase by tumor-promoting phorbol esters. J Biol Chem 257: 7847-7851

Charest R, Blackmore PF, Berthon B, Exton JH (1983) Changes in free cytosolic Ca^{2+} in hepatocytes following α_1-adrenergic stimulation. Studies on Quin-2-loaded hepatocytes. J Biol Chem 258: 8769-8773

Cohen I, DeVries A (1973) Platelet contractile regulation in an isometric system. Nature 246: 36-37

Cooper DMF, Rodbell M (1979) ADP is a potent inhibitor of human platelet adenylate cyclase. Nature 282: 517-518

Cox AC, Carrol RC, White JG, Rao GH (1984) Recycling of platelet phosphorylation and cytoskeleton assembly. J Cell Biol 98: 8-15

Craig SW, Pollard TD (1982) Actin-binding proteins. Trends Biochem Sci 7: 88-92

Cutler LS, Feinstein MB, Rodan GA, Christian CP (1981) Cytochemical evidence for the segregation of adenylate cyclase, Ca^{2+}-, Mg^{2+}-ATPase, K^+-dependent p-nitrophenyl phosphatase in separate membrane compartments in human platelets. Histochem J 13: 547-554

Cutler LS, Rodan G, Feinstein MB (1978) Cytochemical localization of adenylate cyclase and of calcium ion, magnesium ion-activated ATPases in the dense tubular system of human blood platelets. Biochim Biophys Acta 542: 357-371

Dabrowska R, Hinkins S, Walsh MP, Hartshorne DJ (1982) The binding of smooth muscle myosin light chain kinase to actin. Biochem Biophys Res Commun 107: 1524-1531

Daniel JC, Purdon AD, Molish IR (1982) Platelet myosin phosphorylation as an indicator of cellular calcium concentration. Fed Proc 41: 1118

Daniel JL, Molish IR, Holmsen H, Salganicoff L (1981a) Phosphorylation of myosin light chain in intact platelets: Possible role in platelet secreation and clot retraction. Cold Spring Harbor Conf Cell Proliferation, Vol. 8, Protein Phosphorylation, pp 913-928

Daniel JL, Molish IR, Holmsen H (1981b) Myosin phosphorylation in intact platelets. J Biol Chem 256: 7510-7514

Dawson RMC, Hemington NL, Irvine RF (1983) Diacylglycerol potentiates phospholipase attack upon phospholipid bilayers: possible connection with cell stimulation. Biochem Biophys Res Commun 117: 196-201

Debus E, Weber K, Osborn M (1981) The cytoskeleton of blood platelets viewed by immuno-fluorescence microscopy. Eur J Cell Biol 24: 45-52

DeLanerolle P, Adelstein RS, Feramisco JR, Burridge K (1981) Characterization of antibodies to smooth muscle myosin kinase and their use in localizing myosin kinase in non-muscle cells. Proc Natl Acad Sci USA 78: 4738-4742

Detwiler TC, Charo IF, Feinman RD (1978) Evidence that calcium regulates platelet function. Thromb Haemostasis 40: 207-211

Dillon PF, Askoy MO, Driska SP, Murphy RA (1981) Myosin phosphorylation and the cross-bridge cycle in arterial smooth muscle. Science 211: 495-497

Downes P, Michell RH (1982) Phosphatidylinositol 4-phosphate and phosphatidylinositol 4,5-bis-phosphate: lipids in search of a function. Cell Calcium 3: 467-502

Downes CP, Wusteman MM (1983) Breakdown of polyphosphoinositides and not phosphatidyl-inositol accounts for muscarinic agonist-stimulated inositol phospholipid metabolism in rat parotid glands. Biochem J 216: 633-640

Durfler FJ, Mahan LC, Koachman AM, Insel PA (1982) Stimulation by forskolin of intact S49 lymphoma cells involves the nucleotide regulatory protein of adenylate cyclase. J Biol Chem 257: 11901-11907

Endo T, Naka M, Hidaka H (1982) Ca^{2+}-phospholipid-dependent phosphorylation of smooth muscle myosin. Biochem Biophys Res Commun 105: 942-948

Ebashi S, Nomongura Y, Hirata M (1982) Mode of calcium binding to smooth muscle contractile system. In: Kakiuchi S, Hidaka M, Means AR (eds) Calmodulin and intracellular Ca^{2+} receptors. Plenum, New York, pp. 393-401

Fabiato A, Fabiato F (1977) Calcium release from the sarcoplasmic reticulum. Circ Res 40: 119-129

Fain JN, Garcia-Sainz JA (1980) Role of phosphatidylinositol turnover in alpha 1 and of adenylate cyclase inhibition in alpha 2 effects of catecholamines. Life Sci 26: 1183-1194

Feinberg H, Sandler WC, Scorer M, LeBreton GC, Grossman and Born GVR (1977) Movement of sodium into human platelets induced by ADP. Biochim Biophys Acta 470: 317-324

Feinman RD, Detwiler TC (1974) Platelet secretion induced by divalent cation ionophores. Nature 249: 172-173

Feinstein MB (1980) Release of intracellular membrane-bound calcium precedes the onset of stimulus-induced exocytosis in platelets. Biochem Biophys Res Commun 93: 593-600

Feinstein MB (1982) The role of calmodulin in hemostasis. In: Spaet TH (ed) Progress in hemostasis and thrombosis Vol 6. Grune and Stratton, New York, pp. 25-61

Feinstein MB, Hadjian R (1982) Effects of the calmodulin antagonist trifluoperazine on stimulus-induced calcium mobilization, aggregation, secretion, and protein phosphorylation in platelets. Mol Pharmacol 21: 422-431

Feinstein MB, Egan JJ, Sha'afi RI, White J (1983) The cytoplasmic concentration of free calcium in platelets is controlled by stimulators of cyclic AMP production (PGD_2, PGE_1, forskolin). Biochem Biophys Res Commun 113: 598-604

Feinstein MB, Egan JJ, Opas EE (1983) Reversal of thrombin-induced myosin phosphorylation and the assembly of cytoskeletal structures in platelets by the adenylate cyclase stimulants prosta-glandin D_2 and forskolin. J Biol Chem 258: 1260-1267

Feinstein MB, Rodan GA, Cutler LS (1981) Cyclic AMP and calcium in platelet function. In: Gordon IL (ed) Platelets in biology and pathology-2. Elsevier/North-Holland, Amsterdam, pp 437-472

Fox JEB, Phillips DR (1982) Role of phosphorylation in mediating the association of myosin with the cytoskeletal structures of human platelets. J Biol Chem 257: 4120-4126

Gerrard JM, Carroll RC (1981) Stimulation of platelet protein phosphorylation by arachidonic acid and endoperoxide analogs. Prostaglandins 22: 81-94

Gerrard JM, Kindom SE, Peterson DA, Peller J, Krantz KE, White JG (1979) Lysophosphatidic acids. Influence on platelet aggregation and intracellular calcium flux. Am J Pathol 96: 423-438

Gerrard JM, Peterson DA, White JG (1981) Calcium mobilization. In: Gordon IL (ed) Platelets in biology and pathology-2. Elsevier/North-Holland, Amsterdam, pp 407-436

Gerrard JM, White JG, Peterson DA (1978) The platelet dense tubular system: its relationship to prostaglandin synthesis and calcium flux. Thromb Haemostasis 40: 224-231

Gorman RR, Bunting S, Miller OV (1977) Modulation of human platelet adenylate cyclase by prostacyclin (PGX). Prostaglandins 13: 377-388

Grant JA, Scrutton MB (1979) Novel α_2-adrenoreceptors primarily responsible for inducing human platelet aggregation. Nature 277: 659-661

Grumet M, Lin S (1980) A platelet inhibitor protein with cytochalasin-like activity against actin polymerization in vitro. Cell 21: 439-444

Halenda S, Zavoico GB, Chester D, Feinstein MB (1985) Interrelationship between Ca^{2+} mobilization, phosphoinositide metabolism and diacylglycerol formation in thrombin-stimulated platelets. Submitted for publication.

Harris H (1981) Regulation of motile activity in blood platelets. In: Gordon IL (ed) Platelets in biology and pathology-2. Elsevier/North-Holland, Amsterdam, pp 473-500

Haslam RJ, Davidson MML, Davies T, Lynham JA, McClenaghan MD (1978) Regulation of blood platelet function by cyclic nucleotides. Adv Cyclic Nucleotide Res 9: 533-552

Haslam RJ, Lynham JA, Fox JEB (1979) Effects of collagen, ionophore A23187 and prostaglandin E_1 on the phosphorylation of specific proteins in blood platelets. Biochem J 178: 397-406

Hathaway DR, Adelstein RS (1979) Human platelet myosin light chain kinase requires the calcium-binding protein calmodulin for activity. Proc Natl Acad Sci USA 76: 1653-1657

Hathaway DR, Eaton CR, Adelstein RS (1981) Regulation of human platelet myosin light chain kinase by the catalytic subunit of cyclic AMP-dependent protein kinase. Nature 291: 252-254

Hendrickson HS, Reinertsen JL (1971) Phosphoinositide interconversion: A model for control of Na^+ and K^+ permeability in the nerve axon membrane. Biochem Biophys Res Commun 44: 1258-1264

Hofmann SL, Majerus PW (1982) Modulation of phosphatidylinositol-specific phospholipase C activity by phospholipid interactions, diglycerides and calcium ions. J Biol Chem 257: 14359-14363

Holmes RP, Yoss NL (1983) Failure of phosphatidic acid to translocate Ca^{2+} across phosphatidylcholine membranes. Nature 305: 637-638

Holmsen H, Dangelmaier CA, Holmsen HK (1981) Thrombin-induced platelet responses differ in requirement for receptor occupancy. Evidence for tight coupling of occupancy and compartmentalized phosphatidic acid formation. J Biol Chem 256: 9393-9396

Imai A, Ishizuka Y, Kawai K, Nozawa Y (1982) Evidence for coupling of phosphatidic acid formation and calcium influx in thrombin-activated human platelets. Biochem Biophys Res Commun 108: 752-759

Imai A, Nakashima S, Nozawa Y (1983) The rapid polyphosphoinositide metabolism may be a triggering event for thrombin-mediated stimulation of human platelets. Biochem Biophys Res Commun 110: 108-115

Imaoka T, Lynham JA, Haslam RJ (1983) Purification and characterization of the 47,000-dalton protein phosphorylated during degranulation of human platelets. J Biol Chem 258: 11404-11414

Iwasa Y, Hosey MM (1984) Phosphorylation of cardiac sarcolemma proteins by the calcium-activated phospholipid-dependent protein kinase. J Biol Chem 259: 534-540

Jakobs KH, Aktories K, Schultz J (1981) Inhibition of adenylate cyclase by hormones and neurotransmitters. Adv Cyclic Nucleotide Res 14: 173-187

Jakobs KH, Saur W, Schultz G (1978) Inhibition of platelet adenylate cyclase by epinephrine requires GTP. FEBS Lett 85: 167-170

Javors MA, Bowden CL, Ross DH (1982) Kinetic characterization and substrate requirement for the Ca^{2+}-uptake system in platelet membrane. Biochim Biophys Acta 691: 220-226

Jennings LK, Fox JEB, Edwards HH, Phillips DR (1981) Changes in the cytoskeletal structure of human platelets following thrombin activation. J Biol Chem 256: 6927-6932

Joseph SKL, Thomas AP, Williams RJ, Irvine RF, Williamson JR (1984) Myo-inositol 1,4,5-tris-phosphate. A second messenger for the hormonal mobilization of intracellular Ca^{2+} in liver. J Biol Chem 259: 3077-3081

Kaibuchi K, Sano K, Hoshijima M, Takai Y, Nishizuka Y (1982) Phosphatidylinositol turnover in platelet activation; calcium mobilization and protein turnover. Cell Calcium 3: 323-335

Kaibuchi K, Takai Y, Sawamura M, Hoshijima M, Fujikura T, Nishizuka Y (1983) Synergistic functions of protein phosphorylation and calcium mobilization in platelet activation. J Biol Chem 258: 6701-6704

Kajikawa N, Kaibuchi K, Matsubara T, Kikkawa U, Takai Y, Nishizuka Y (1983) A possible role of protein kinase C in signal-induced lyosomal enzyme release. Biochem Biophys Res Commun 116: 743-750

Kakuichi S, Sobue K (1983) Control of the cytoskeleton by calmodulin and calmodulin-binding proteins. Trends Biochem Sci 8: 59-62

Kaser-Glanzmann R, Gerber E, Luscher EF (1979) Regulation of the intracellular calcium level in human blood platelets: cyclic adenosine 3',5'-monophosphate-dependent phosphorylation of a 22,000 Dalton component in isolated Ca^{2+}-accumulating vesicles. Biochim Biophys Acta 558: 344-347

Kaser-Glanzmann R, Jakabova George JN, Luscher EF (1977) Stimulation of calcium uptake in platelet membrane vesicles by adenosine 3',5'-cyclic monophosphate and protein kinase. Biochim Biophys Acta 466: 429-440

Kawahara Y, Takai Y, Minakuchi R, Sano K, Nishizuka Y (1980) Phospholipid turnover as a possible transmembrane signal for protein phosphorylation during human platelet activation by thrombin. Biochem Biophys Res Commun 97: 309-317

Kiss Z, Mhina Y (1982) Rat liver plasma membranes contain a lipid-dependent protein kinase activity. FEBS Lett 148: 131-134

Knight DE, Baker PF (1983) The phorbol ester TPA increases the affinity of exocytosis for calcium in "leaky" adrenal medullary cells. FEBS Lett 160: 98-100

Knight DE, Kestenen NT (1983) Evoked transient intracellular free Ca^{2+} changes and secretion in isolated bovine adrenal medullary cells. Proc R Soc Lond B 218: 177-199

Knight DE, Scrutton MB (1980) Direct evidence for a role for Ca^{2+} in amine storage granule secretion by human platelets. Thrombos Res 20: 437-446

Knight DE, Hallam TJ, Scrutton MC (1982) Agonist selectivity and second messenger concentration in Ca^{2+}-mediated secretion. Nature 296: 256-257

Lapetina EG (1983) Metabolism of inosities and the activation of platelets. Life Sci 32: 2069-2082

Lapetina EG, Billah MM, Cuatrecasas P (1981) The phosphatidylinositol cycle and the regulation of arachidonic acid production. Nature 292: 367-369

LeBreton GC, Venton DL, Enke SE, Haluska PV (1979) 13-Azaprostanoic acid: A specific antagonist of the human blood platelet thromboxane/endoperoxide receptor. Proc Natl Acad Sci USA 76: 4097-4101

Lee YC, Wolff J (1984) Calmodulin binds to both microtubule-associated protein 2 and proteins. J Biol Chem 259: 1226-1236

Le Peuch CJ, Le Peuch DAM, Katz S, DeMaille JG, Hincke MD, Bredoux R, Enouf J, Levy-Toledano S, Caen J (1983) Regulation of calcium accumulation and efflux from platelet vesicles. Possible role for cyclic-AMP-dependent phosphorylation and calmodulin. Biochim Biophys Acta 731: 456-464

Limas CJ (1980) Phosphorylation of cardiac sarcoplasmic reticulum by a calcium-activated, phospholipid-dependent protein kinase. Biochem Biophys Res Commun 96: 1378-1383

McGowan EB, Detwiler TC (1983) Characterization of the thrombin induced desensitization of platelet activation by thrombin. Thromb Res 31: 297-304

Michnicka M, Kasman K, Kakol I (1982) The binding of actin to phosphorylated and dephosphorylated myosin. Biochim Biophys Acta 704: 470-475

Miller OV, Gorman RR (1979) Evidence for distinct prostaglandin I_2 and D_2 receptors in human platelets. J Pharmacol Exp Ther 210: 134-140

Mills DCB (1974) Factors influencing the adenylate cyclase system in human blood platelets. In: Sherry S, Scriabine A (eds) Platelets and thrombosis. Univ Park Press, Baltimore, pp 45-67

Mills DCB, Smith JB (1971) The influence on platelet aggregation of drugs that affect the accumulation of adenosine 3':5'-cyclic monophosphate in platelets. Biochem J 121: 185-196

Motulsky HJ, Hughes RJ, Brickman AS, Farfel Z, Bourne HR, Insel PA (1982) Platelets of pseudo-hypoparathyroid patients: Evidence that distinct receptor-cyclase coupling proteins mediate stimulation and inhibition of adenylate cyclase. Proc Natl Acad Sci USA 79: 4193-4197

Murayama T, Ui M (1983) Loss of the inhibitory function of the guanine nucleotide regulatory component of adenylate cyclase due to its ADP ribosylation by islet-activating protein, pertussis toxin, in adipocyte membranes. J Biol Chem 258: 3319-3326

Naka M, Nishikawa M, Adelstein RS, Hidaka H (1983) Phorbol ester induced activation of human platelets is associated with protein kinase C phosphorylation of myosin light chains. Nature 306: 490-492

Nishikawa M, Hidaka H, Adelstein RS (1983) Phosphorylation of smooth muscle heavy meromyosin by calcium-activated, phospholipid-dependent protein kinase. The effect on actin-activated Mg-ATPase activity. J Biol Chem 258: 14069-14072

O'Rourke F, Halenda SP, Zavoico GB, Feinstein MB (1985) Inositol 1,4,5-trisphosphate releases Ca^{2+} from a Ca^{2+} transporting membrane vesicle fraction derived from human platelets. In press, J Biol Chem

Owen NE, LeBreton GL (1981) The involvement of calcium in epinephrine or ADP potentiation of human platelet aggregation. Am J Physiol 241: H613-H619

Penniston JT (1982) Plasma membrane Ca^{2+}-pumping ATPases. Ann NY Acad Sci 402: 296-303

Perret BP, Plantavid M, Chap H, Douste-Blazy L (1983) Are polyphosphoinositides involved in platelet activation? Biochem Biophys Res Commun 110: 660-667

Phillips DR, Jennings LK, Edwards HH (1980) Identification of membrane proteins mediating the interaction of human platelets. J Cell Biol 86: 77-86

Prentki M, Biden TJ, Janjic D, Irvine RF, Berridge MJ, Wollheim CB (1984) Rapid mobilization of Ca^{2+} from rat insulinoma microsomes by inositol-1,4,5-trisphosphate. Nature 309: 562-564

Prescott SM, Majerus PW (1983) Characterization of 1,2-diacylglycerol hydrolysis in human platelets. J Biol Chem 258: 764-769

Pribluda V, Rotman A (1982) Dynamics of membrane-cytoskeleton interactions in activated blood platelets. Biochemistry 21: 2825-2832

Rendu F, Marche P, Maclouf J, Girard A, Levy-Toledano S (1983) Triphosphoinositide breakdown and dense body release as the earliest events in thrombin-induced activation of human platelets. Biochem Biophys Res Commun 116: 513-519

Rink TJ, Smith SW (1983) Inhibitory prostaglandins suppress Ca^{2+} influx, the release of intracellular Ca^{2+} and the responsiveness to cytoplasmic Ca^{2+} in human platelets. J Physiol 338: 66-67P

Rink TJ, Sanchez A, Hallam TJ (1983) Diacylglycerol and phorbol ester stimulate secretion without raising cytoplasmic free calcium in human platelets. Nature 305: 317-319

Rink TJ, Smith WS, Tsien RY (1981) Intracellular free calcium in platelet shape change and aggregation. J Physiol 324: 53-54P

Rink TJ, Smith SW, Tsien RY (1982) Cytoplasmic free Ca^{2+} in human platelets: Ca^{2+} thresholds and Ca-independent activation for shape-change and secretion. FEBS Letts 148: 21-26

Rittenhouse SE (1982) Inositol lipid metabolism in the responses of stimulated platelets. Cell Calcium 3: 311-322

Rittenhouse SE (1983) Human platelets contain phospholipase C that hydrolyzes polyphospho-inositides. Proc Natl Acad Sci USA 80: 5417-5420

Rittenhouse-Simmons S (1981) Differential activation of platelet phospholipases by thrombin and ionophore A23187. J Biol Chem 256: 4153-4155

Rittenhouse-Simmons S (1979) Production of diglyceride from phosphatidylinositol in activated platelets. J Clin Invest 63: 580-587

Robblee LS, Shepro D, Belamarich FA (1973) Calcium uptake and associated adenosine triphosphate activity of isolated platelet membranes. J Gen Physiol 61: 462-481

Rodbell M (1980) The role of hormone receptors and GTP-regulatory proteins in membrane transduction. Nature 284: 17-22

Rosenberg S, Stracher A, Lucas R (1981) Isolation and characterization of actin and actin-binding protein from human platelets. J Cell Biol 91: 201-211

Rosenblatt M, Hidalog C, Vergara C, Ikemoto N (1981) Immunological and biochemical properties of transverse tubule membranes isolated from rabbit skeletal muscle. J Biol Chem 256: 8140-8148

Sabol SL, Nirenberg M (1979) Regulation of adenylate cyclase of neuroblastoma x glioma hybrid cells by α-adrenergic receptors. I. Inhibition of adenylate cyclase mediated by α receptors. J Biol Chem 254: 1913-1920

Schafer AI, Cooper B, O'Hara D, Handin RI (1979) Identification of platelet receptors for prostaglandin I_2 and D_2. J Biol Chem 254: 2914-2917

Schick PK, Tuszynski GP, Vander Voort PW (1983) Human platelet cytoskeletons: specific content of glycolipids and phospholipids. Blood 61: 163-166

Scrutton MC, Wallis RB (1981) Catecholamine receptors. In: Gordon IL (ed) Platelets in biology and pathology-2. Elsevier/North-Holland, Amsterdam, pp 179-210

Seamon KB, Daly JW (1981) Forskolin: a unique diterpene activator of cyclic AMP-generating systems. J Cyclic Nucleotide Res 7: 201-224

Serhan CN, Fridovich J, Goetzl EJ, Dunham PB, Weissmann G (1982) Leukotriene B_4 and phosphatidic acid are calcium ionophores. Studies employing arsenazo III in liposomes. J Biol Chem 257: 4746-4752

Sheetz MP (1983) Membrane skeletal dynamics: role in modulation of red cell deformability, mobility of transmembrane proteins, and shape. Semin Hematol 20: 175-188

Shuman MA, Botney M, Fenton II JW (1979) Thrombin-induced platelet secretion. Further evidence for a specific pathway. J Clin Invest 63: 1211-1218

Siess W, Cuatrecasas P, Lapetina EG (1983) A role for cyclooxygenase products in the formation of phosphatidic acid in stimulated human platelets. Differential mechanisms of action of thrombin and collagen. J Biol Chem 258: 4683-4686

Skaer RI (1981) Platelet degranulation. In: Gordon IL (ed) Platelets in biology and pathology-2. Elsevier/North-Holland, Amsterdam, pp 321-348

Small JV, Sobieszek A (1980) The contractile apparatus of smooth muscle. Int Rev Cytol 64: 241-306

Smith RC, Cande WZ, Craig R, Tooth PJ, Scholey JM, Kendrick-Jones J (1983) Regulation of myosin filament assembly by light chain phosphorylation. In: Perry SV, Cohen P (eds) Biological roles of protein phosphorylation. The Roy Soc London, London, pp 73-82

Smith SK, Limbird LE (1982) Evidence that human platelet α-adrenergic receptors coupled to inhibition of adenylate cyclase are not associated with the subunit of adenylate cyclase ADP-ribosylated by cholera toxin. J Biol Chem 257: 10471-10478

Somlyo AV, Gonzalez-Serratos H, Shuman H, McClellan G, Somlyo AP (1981) Calcium release and ionic changes in the sarcoplasmic reticulum of tetanized muscle: an electron probe study. J Cell Biol 90: 577-594

Statland BD, Heagan BM, White JG (1969) Uptake of calcium by platelet relaxing factor. Nature 223: 521-522

Stossel TP (1978) Contractile proteins in cell structure and function. Annu Rev Med 29: 427-457

Streb H, Irvine RF, Berridge MJ, Schulz I (1983) Release of Ca^{2+} from a nonmitochondrial intracellular store in pancreatic acinar cells by inositol-1,4,5-trisphosphate. Nature 306: 67-68

Suematsu E, Hirata M, Hashimoto T, Kuriyama H (1984) Inositol 1,4,5-trisphosphate releases Ca^{2+} from intracellular store sites in skinned single cells of procine coronary artery. Biochem Biophys Res Commun 120: 481-485

Takai Y, Kaibuchi K, Sano K, Nishizuka Y (1982) Counteraction of calcium-activated, phospholipid-dependent protein kinase activation by adenosine 3',5'-monophosphate and guanosine 3',5'-monophosphate in platelets. J Biochem 91: 403-406

Takai Y, Kishimoto A, Kawahara Y, Minakuchi R, Sano K, Kikkawa Y, Mori T, Yu B, Kaibuchi K and Nishizuka Y (1981) Calcium and phosphatidylinositol turnover as signalling for transmembrane control of protein phosphorylation. Adv Cyclic Nucleotide Res 14: 301-313

Tam SW, Fenton II JW, Detwiler TC (1979) Dissociation of thrombin from platelets by hirudin. Evidence for receptor processing. J Biol Chem 254: 8723-8725

Tellam R, Frieden C (1982) Cytochalasin D and platelet gelsolin accelerate actin polymer formation. A model for regulation of the exent of actin polymer formation in vivo. Biochemistry 21: 3207-3214

Tsien RY, Pozzan T, Rink TJ (1982) Calcium homeostasis in intact lymphocytes: cytoplasmic free calcium monitored with a new, intracellularly trapped fluorescent indicator. J Cell Biol 94: 325-334

Vickers JD, Knilough-Rathbone RL, Mustard JF (1982a) The effect of prostaglandins E_1, I_2 and $F_{2\alpha}$ on the shape and phosphatidyl-4,5-bisphosphate metabolism of washed rabbit platelets. Thromb Res 28: 731-740

Vickers JD, Kinlough-Rathbone RL, Mustard JF (1982b) Changes in phosphatidylinositol-4,5-bisphosphate 10 seconds after stimulation of washed rabbit platelets with ADP. Blood 60: 1247-1250

Wallach D, Davies PJA, Pastan J (1978) Cyclic AMP-dependent phosphorylation of filamin in mammalian smooth muscle. J Biol Chem 253: 4739-4745

Weitzell R, Tanaka T, Starke K (1979) Pre- and postsynaptic effects of yohimbine stereoisomers on noradrenergic transmission in the pulmonary artery of the rabbit. Nauyn-Schmiedebergs Arch Pharmakol. 308: 127-136

Weksler B (1982) Prostacyclin. In: Spaet TH (ed) Progress in hemostasis and thrombosis. Vol 6. Grune and Stratton, New York, pp 113-118

Werth DK, Niedel JE, Pastan I (1983) Vinculin, a cytoskeletal substrate of protein kinase C. J Biol Chem 258: 11423-11426

White JG (1972) Interaction of membrane systems in blood platelets. Amer J Pathol 66: 295-312

White JG (1973) Identification of platelet secretion in the electron microscope. Ser Haematol 67: 429-459

White JG, Rao GHR, Estensen RD (1974) Investigation of the release reaction in platelets exposed to phorbol myristate acetate. Am J Pathol 75: 301-314

Wu WC-S, Walaas SI, Nairn AC, Greengard P (1982) Calcium/phospholipid regulates phosphorylation of a M_r "87K" substrate protein in brain synaptosomes. Proc Natl Acad Sci USA 79: 5249-5253

Yamanishi J, Kawahara Y, Fukuzaki H (1983) Effect of cyclic AMP on cytoplasmic free calcium in human platelets stimulated by thrombin: direct measurement with Quin2. Thrombosis Res 32: 183-188

Yamanishi J, Takai Y, Kaibuchi K, Sano K, Castagna M, Nishizuka Y (1983) Synergistic functions of phorbol ester and calcium in serotonin release from human platelets. Biochem Biophys Res Commun 112: 778-786

Yassin R, Shefcyk J, Tao W, White JR, Molski TFP, Naccache PH, Sha'afi RI (1984) Actin association with the cytoskeleton in human and rabbit neutrophils. Fed Proc 43: 1507

Zavoico GB, Feinstein MB (1984) Cytoplasmic Ca^{2+} in platelets is controlled by cyclic AMP: Antagonism between stimulators and inhibitors of adenylate cyclase. Biochem Biophys Res Commun 120: 579-585, 1984

Zavoico GB, Halenda S, Chester D, Feinstein MB (1984) Control of Ca^{2+} mobilization and polyphosphoinositide metabolism in platelets by prostacyclin. In: Bailey JM (ed) Prostaglandins and leukotrienes. Plenum, New York (in Press)

Zawalich W, Brown C, Rasmussen H (1983) Insulin-secretion: Combined effects of phorbol ester and A23187. Biochem Biophys Res Commun 117: 448-455

Footnotes to pages 349 and 357:

[2] This effect resembles that of cyclic AMP on skeletal muscle phosphorylase b kinase. Cyclic AMP-dependent phosphorylation of the Ca^{2+}-dependent enzyme increases its sensitivity to Ca^{2+} so that basal $[Ca^{2+}]_i$ may be sufficient to activate.

[3] Although the rise in $[Ca^{2+}]_i$ is about 2-3 μM (i.e., 2-3 $nmol ml^{-1}$) the actual amount of Ca^{2+} released into the cytoplasm is probably much larger. Ca^{2+} must act by binding to its activator sites such as calmodulin. From the approximate calmodulin content of platelets (Feinstein 1982) we calculate that at least 100-200 $nmol ml^{-1}$ calcium would be required to saturate its binding sites. In activated muscle cells $[Ca^{2+}]_i$ also rises to about several micromolar, but the total amount of Ca^{2+} mobilized is equivalent to at least several hundred micromolar (Somlyo et al. 1981).

Protein Kinase C and Polyphosphoinositide Metabolites: Their Role in Cellular Signal Transduction

D. MARME[1] [2] and S. MATZENAUER[1]

CONTENTS

The important role of cyclic nucleotides and Ca^{2+} ions as intracellular messengers has been generally accepted. Cyclic AMP and cyclic GMP achieve their regulatory function through the activation of cyclic AMP-dependent and cyclic GMP-dependent protein kinases. Calcium ions bind to intracellular calcium-binding proteins such as troponin C and calmodulin, and thus interfere with cellular processes either directly or — as in the case of cyclic nucleotides — by stimulating Ca^{2+}, calmodulin-dependent protein kinases. Both signal-transducing systems have been extensively reviewed over the past several years (Klee and Vanaman, 1982; Beavo and Mumby, 1982; Kuo and Shoji, 1982; Nestler and Greengard, 1983).

Two recent discoveries have led to another important system which is involved in cellular signal transduction: the coordinated action of protein kinase C and polyphosphoinositide metabolites. Nishizuka and his colleagues have identified protein kinase C as a Ca^{2+}-dependent and phospholipid-dependent protein kinase which appeared to be activated by diacylglycerol (for review: Nishizuka, 1984). Michell in 1975 had hypothesized that stimulated inositol lipid metabolism is essential to a process by which receptor stimulation leads to a rise in cytosolic Ca^{2+} concentration and hence to a variety of cellular responses (Michell, 1975). Only recently the mechanism by which this is achieved has been clarified (Burgess et al., 1984). It is the aim of this chapter to describe this novel signal-transducing system and to outline its physiological significance.

[1] Institute of Biology III, University of Freiburg, Schänzlestr. 1, 7800 Freiburg, FRG
[2] Department of Biophysics, Gödecke Research Institute, Mooswaldallee 1-9, 7800 Freiburg, FRG

1 Protein Kinase C

Covalent modification of proteins such as phosphorylation has been shown to be an appropriate mechanism to control cellular metabolism and physiology. Among those enzymes which catalyze phosphorylation, the cyclic AMP-dependent and the cyclic GMP-dependent protein kinases exist as virtually one type. The calcium-dependent protein kinases can be classified as those which are activated in conjunction with calmodulin (i.e., phosphorylase kinase, myosin light chain kinase, protein kinase I, protein kinase II; for review see Nestler and Greengard, 1983), and those which are activated by phosphatidylserine and other lipids. This latter has been termed protein kinase C (for review see Nishizuka, 1984).

Protein kinase C, which was discovered by Nishizuka and his colleagues (Inoue et al., 1977), is widely distributed in animals. It occurs in many tissues and organs within the same organism and sometimes exceeds the activity of the cyclic AMP-dependent protein kinase. It consists of one polypeptide with a relative molecular weight of 77,000. The enzyme can be cleaved by Ca^{2+}-dependent thiol-proteases into two functional different domains. One is hydrophilic and contains the catalytic center, the other is hydrophobic and may represent the domain which interacts with membrane phospholipids. Protein kinase C can be activated by diacylglycerol (Takai et al., 1979) which is transiently produced from inositol lipids upon stimulation by extracellular signals (Fig. 1). The association with diacylglycerol leads to a dramatic alteration in the Ca^{2+} activation profile of the protein kinase C. Without diacylglycerol even in the presence of phosphatidylserine the enzyme is only slightly activated by high Ca^{2+} concentrations (10^{-5} M

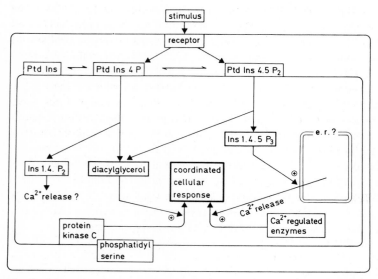

Fig. 1. Schematic representation of polyphosphoinositide breakdown, Ins 1,4,5 P_3-induced Ca^{2+} release from nonmitochondrial stores (possibly endo- or sarcoplasmic reticulum = e.r.) and the co-ordinated action of protein kinase C and Ca^{2+}-regulated enzymes

to 10^{-4} M). However, in the presence of diacylglycerol the enzyme is activated at about 10^{-7} M free Ca^{2+}. Thus the generation of diacylglycerol leads to a sensitivity shift of the protein kinase C: the Ca^{2+}-dependent enzyme can be activated at Ca^{2+} concentrations slightly greater than those found in the nonactivated cell and those found in the activated cell after the stimulus-dependent initial and transient increase of cytoplasmic calcium (see also: Rasmussen et al., p. 1 this Vol.). Diacylglycerol is the link between receptor-mediated polyphosphoinositide metabolism and the protein kinase C (Fig. 1).

2 Polyphosphoinositide Metabolism

Phosphatidylinositol (Ptd Ins) is present as a few percent of the total phospholipids in all eukaryotic cells. The polyphosphoinositides phosphatidylinositol 4-phosphate (Ptd Ins 4P) and phosphatidylinositol 4,5-bisphosphate (Ptd Ins 4,5 P_2) also appear to be ubiquitous components of eukaryotic cells. Since the pioneering work of Hokin and Hokin (1953), it is now well established that a number of extracellular stimuli cause a very rapid increase of the inositol phospholipid metabolism. Some of these receptor-mediated effects are summarized in Table 1.

The network of metabolic pathways involved in the stimulus-dependent degradation and biosynthesis is rather complex and has been assembled from experiments with many different cell types. Figure 1 shows the pathways for degradation of the various inositol phospholipids (for details see Downes and Michell, 1982). Two features of this degradation scheme are important: *first*, Ptd Ins can be converted into Ptd Ins 4P and Ptd Ins 4,5 P_2; *second*, Ptd Ins 4P and Ptd Ins 4,5 P_2 are degraded by appropriate phosphodiesterases to give inositol 1,4 biphosphate (Ins 1,4 P_2) and inositol 1,4,5 trisphosphate (Ins 1,4,5 P_3) and diacylglycerol. The two reaction products, Ins 1,4,5 P_3 and diacylglycerol, are of special interest with regard to their important role in intracellular Ca^{2+} release (Ins 1,4,5 P_3) and protein kinase C activation (diacylglycerol). They are therefore the key substances for the coordination of receptor-coupled inositol phospholipid metabolism-dependent *and* at the same time, Ca^{2+}-dependent mechanisms (Fig. 1). As Rasmussen et al. (this chap. Rasmussen et al. p. 1, Vol.) have pointed out, the combination of these two regulatory pathways could possibly explain a long-lasting Ca^{2+} effect which persists even after the cellular transient Ca^{2+} increase has disappeared.

3 Inositol 1,4,5-Trisphosphate-Dependent Ca^{2+} Release

A wide variety of agonists act on cell-surface receptors to cause an increase of cytosolic free Ca^{2+}. Another feature of this receptor stimulation is the hydrolysis of polyphosphoinositides. It has been suggested by Michell (1975) that this inositol phospholipid breakdown is essential to couple receptor stimulation to the release of Ca^{2+} from intracellular stores. The primary lipid substrates for the receptor-mediated degradation

Table 1. Receptor-mediated effects

Responsive tissues or cells	Type of stimulus	Cellular response[a]	Reference
Hepatocytes	Epinephrine, angiotensin	PI, Ca^{2+}	Billah et al. (1979)
Hepatocytes	Vasopressin	PI	Rhodes et al. (1983)
Platelets	Thrombin	PI, Ca^{2+}	Kaibuchi et al. (1983)
Platelets	Collagen, thrombin	PI	Kaibuchi et al. (1982)
Parotid gland	Carbachol, substance P	PI, Ca^{2+}	Berridge et al. (1983)
Insect salivary gland	5-Hydroxytryptamine	PI, Ca^{2+}	Berridge (1982)
Pancreas	Carbachol, caerulein	PI, Ca^{2+}	Putney et al. (1983)
Iris smooth muscle	Acetylcholin	PI	Abdel-Latif et al. (1977)
Brain	Corticotropin	PI	Jolles et al. (1981)

[a]PI = inositol phospholipid breakdown
Ca^{2+} = increase of cytosolic Ca^{2+}

mechanism seem to be the two polyphosphoinositides Ptd Ins 4P and Ptd Ins 4,5 P_2 which are hydrolysed very rapidly. It has been shown recently that 5 s after stimulation of rat parotid gland slices with carbachol a rapid and large increase of Ins 1,4 P_2 and Ins 1,4,5 P_3 could be observed (Downes and Wusteman, 1983). The ATP dependence of the appearance of Ins 1,4 P_2 and Ins, 1,4,5 P_3 and the fact that only minor amounts of inositol 1-phosphate are released (the latter probably due to the hydrolysis of Ins 1,4 P_2 and not Ptd Ins) suggested to these authors that muscarinic agonists stimulate a polyphosphoinositide-specific phospholipase C and that these lipids are continuously replenished from the Ptd Ins pool (see Fig. 1). Similar results were obtained by Berridge et al. (1983) for insect salivary glands and parotid glands upon stimulation with 5-hydroxytryptamine. From these findings one may conclude that it is not the Ptd Ins breakdown that initiates Ca^{2+} release as originally suggested by Michell (1975) but the polyphosphoinositide breakdown.

Recently it has been examined whether the two reaction products resulting from polyphosphoinositide hydrolysis, Ins 1,4 P_2 and Ins 1,4,5 P_3, are capable to induce intracellular Ca^{2+} release. Streb et al. (1983) could show that Ins 1,4,5 P_3 at micromolar concentrations caused a release of Ca^{2+} from a nonmitochondiral intracellular Ca^{2+} store in pancreatic acinar cells. Ins 1,4 P_2 as well as inositol 1-phosphate were ineffective. Similar results were obtained by Burgess et al. (1984), who could demonstrate that addition of Ins 1,4,5 P_3 to permeabilized guinea pig hepatocytes did induce a rapid Ca^{2+} release most probably from the endoplasmic reticulum. Furthermore they demonstrated that α-adreno-receptor stimulation caused a rapid increase of Ins 1,4,5 P_3, as well as a rapid Ca^{2+} release from these permeabilized cells. Suematsu et al. (1984) found Ca^{2+} release from nonmitochondrial stores of vascular smooth muscles upon addition of microsomal Ins 1,4,5 P_3.

From these data a reaction sequence as represented in Fig. 1 can be deduced: receptor stimulation initiates the breakdown of Ptd Ins 4P and Ptd Ins 4,5-P_2 to yield Ins 1,4 P_2 and Ins 1,4,5 P_3. Subsequently Ins 1,4,5 P_3 causes a Ca^{2+} release from an intracellular store (probably endoplasmic reticulum). The resulting increase in free cytoplasmic Ca^{2+} initiates, together with the diacylglycerol-activated protein kinase C, the coordinated physiological response.

4 Coordinated Regulation of Cellular Functions

A large number of different hormones and neurotransmitters have more than one type of receptor that can be separated pharmacologically. One receptor type seems to be coupled to the cAMP-generating system. The other receptor type seems to have multifunctional properties: hydrolysis of polyphosphoinositides, mobilization of Ca^{2+}, generation of cGMP, formation of arachidonic acid metabolites (e.g., prostaglandines, thromboxanes, prostacyclin, leukotrienes). Pharmacological studies have clearly revealed that receptor-mediated hydrolysis of inositol phospholipids is associated with the generation of Ca^{2+} signals. In the preceding paragraph preliminary evidence has been presented that Ca^{2+} release is mediated by Ins 1,4,5 P_3, the stimulus-dependent reaction product of Ptd Ins 4,5 P$_2$ breakdown. Furthermore, the second reaction product resulting from polyphosphoinositide metabolism, diacyclglycerol has an important regulatory function: the stimulation of the protein kinase C at Ca^{2+} concentrations occurring in the nonexcited cell.

From these experimental findings it becomes evident that receptor-mediated poly-phosphoinositide breakdown initiates two responses in the cell: generation of Ca^{2+} releasing Ins 1,4,5 P$_3$ and protein kinase C activating diacylglycerol. These two cellular signals obviously can initiate a coordinated response (Fig. 1). Evidence that this is indeed the case comes from experiments with platelets (for review see: Nishizuka, 1984). Upon stimulation by thrombin, collagen, or platelet-activating factor it was shown that diacylglycerol increases as well as cytosolic Ca^{2+} (whether this Ca^{2+} in-crease is a function of Ins 1,4,5 P$_3$ has not yet been demonstrated). As a consequence two polypeptides become phosphorylated a 40 K protein of yet unknown function and a 20 K protein which has been identified as myosin light chain. These two phospho-proteins are the endproducts of the two chains of events which result in the coordinated response, serotonin release. The 40 K phosphoprotein has been shown to be generated by the diacylglycerol-activated protein kinase C without elevation of Ca^{2+}. The 20 K protein becomes phosphorylated solely upon increase of cytosolic Ca^{2+} by the cal-modulin-dependent myosin light chain kinase. Only prolonged incubation with TPA (see below) leads to phosphorylation of the 20 K protein by the protein kinase C (Naka et al. 1983).

However, the physiological role of the protein kinase C is still poorly understood. Table 2 summarizes a few substrates of the protein kinase C and gives some indications of a possible physiological function. Before we are able to understand the underlying molecular mechanisms of this regulatory system more substrates of the protein kinase C with known functions have to be identified. In addition, one should keep in mind that the kinetics of the Ca^{2+}-dependent and diacylglycerol-dependent mechanisms may be different and may differ from cell to cell (transient versus sustained response; see Rasmussen et al. p XX, this Vol.).

5 Protein Kinase C and Tumor Promotion

Recently Nishizuka and his colleagues could show that several phorbol esters such as 12-0-tetradecanoylphorbol-13-acetate (TPA) are able to stimulate protein kinase C ac-tivity in vitro and in vivo. They found a correlation between the ability of individual phorbol esters to activate the enzyme and to induce tumor promotion (for review see: Nishizuka, 1984). TPA which has a diacylglycerol-like structure can stubstitute for dia-cylglycerol and shifts the Ca^{2+} dependence of the protein kinase C below the micro-molar range. Binding of the ^{3}H-labeled phorbol-12,13-dibutyrate (PDBu) to the enzyme has an absolute requirement for Ca^{2+} and phospholipid (Kikkawa et al., 1983). The apparent dissociation constant K$_d$ = 8nM corresponds with the activation constant for the enzyme. The K$_d$ for PDBu binding is also comparable to those found for specific tumor-promotor binding sites in various tissues. Experiments using phorbol ester photo-

Table 2. Some substrates of the protein kinase C

Tissue or cell type	Substrate Protein	Possible physiological function	Reference
Platelets	40 K protein	Serotonin release	Sano et al. (1983)
Platelets	Myosin light chain	?	Naka et al. (1983)
Turkey gizzard	Myosin light chain	Inhibition of Mg^{2+}-ATPase activity	Nishikawa et al. (1983)
Chicken gizzard	Vinculin, Filamin	Filament organization	Kawamoto et al. (1984) Werth et al. (1983)
Human epidermoid carcinoma cells	Epidermal growth factor receptor	Inhibition of the EGF receptor tyrosine kinase activity	Cochet et al. (1984)
Artemia salina	Ribosomal S 6 protein	?	Le Peuch et al. (1983)
Rat cerebral cortex	Myelin basic protein	Prevention of experimental allergic encephalomyelitis	Turner et al. (1982)
Rat hepatocytes	16 K protein	?	Cooper et al. (1984)
Rabbit reticulocytes	Initiation factor 2	Protein synthesis	Schatzman et al. (1983)

affinity probes have shown that phorbol ester may interact indirectly — via the phospho-lipids — with protein kinase C (Delclos et al. 1983). As a consequence of these findings, one may anticipate that perturbations of the plasmamembrane could cause patho-physiological changes involving deregulation of the protein kinase C system.

As Nishizuka states in his review (Nishizuka 1984) the results obtained with tumor promotors raise the possibility that these agents use a physiological system (the pro-tein kinase C which is under normal conditions *reversibly* activated by diacylglycerol) for a *long-lasting* expression of their pleiotropic effects such as cell proliferation. There-fore it will be of great interest to investigate the pathophysiology of tumor promotion, supposing protein kinase C activation as a possible mode of action.

Recently some very interesting aspects concerning the relation of the inositol lipid metabolism, protein kinase C activation and enzymatic activity expressed by oncogenes have been outlined by Michell (1984). Based on the finding that two oncogene products can act as inositol lipid kinases — *first:* purified $pp^{60v\text{-}src}$ phosphorylates Ptd Ins and Ptd Ins 4P (Sugimoto et al. 1984) and *second:* $p^{68v\text{-}ros}$ phosphorylates Pdt Ins (Macara et al. 1984) — Michell raises the possibility that there might be some sort of threshold of signal intensity that has to be exceeded before proliferation is triggered in competent cells. In this interpretation activation of cell proliferation would require a combination of high rate synthesis of Ptd Ins 4,5 P_2 (by oncogene products) *and* a signal that activ-ates Ptd Ins 4,5 P_2 breakdown. Any oncogene product that could enhance the activity of either of these pathways might act at a proliferative signal.

References

Abdel-Latif AA, Akhtar RA, Hawthrone JN (1977) Acetylcholine increases breakdown of triphospho-inositide of rabbit iris muscle prelabeled with phosphate-P-32. Biochem J 162: 61-73

Beavo JA, Mumby MC (1982) Cyclic AMP-dependent protein phosphorylation. In: Nathanson JA, Kebabian JW (eds) Cyclic Nucleotides, Part I: Biochemistry. Springer, Berlin-Heidelberg-New York (Handbook of experimental pharmacology) 58: 363-392

Berridge MJ (1982) 5-Hydroxytryptamine stimulation of phosphatidylinositol hydrolysis and cal-cium signalling in the blowfly salivary gland. Cell Calcium 3: 385-397

Berridge MJ, Dawson RMC, Downes CP, Heslop JP, Irvine RF (1983) Changes in the levels of ino-sitol phosphates after agonist-dependent hydrolysis of membrane phosphoinositides. Biochem J 212: 473-482

Billah MM, Michel RH (1979) Phopshatidylinositol metabolism in rat hepatocytes stimulated by glycogenolytic hormones. Biochem J 1982: 661-668

Burgess GM, Godfrey PP, McKinney JS, Berridge MJ, Irvine RF, Putney Jr. JW (1984) The second messenger linking receptor activation to internal calcium release in liver. Nature 309: 63-66

Cochet C, Gill GN, Meisenhelder J, Cooper JA, Hunter T (1984) C-kinase phosphorylates the epi-dermal growth factor receptor and reduces its epidermal growth factor-stimulated tyrosine pro-tein kinase activity. J Biol Chem 259: 2553-2558

Cooper RH, Kobayashi K, Williamson JR (1984) Phosphorylation of a 16 K Da protein by diacyl-glycerol-activated protein kinase C in vitro and by vasopressin in intact hepatocytes. FEBS Lett 166: 125-130

Delcos KB, Yeh E, Blumberg PM (1983) Specific labeling of mouse brain membrane phospholipids with [20-^{3}H] phorbol 12-p-azidobenzoate 13-benzoate, a photolabile phorbol ester. Proc Natl Acad Sci USA 80: 3054-3058

Downes CP, Michell RH (1982) Phosphatidylinositol 4-phosphate and phosphatidylinositol 4,5-bis-phosphate: lipids in search of a function. Cell Calcium 3: 467-502

Downes CP, Wusteman MM (1983) Breakdown of polyphosphoinositides and not phosphatidyl-inositol accounts for muscarinic agonist-stimulated inositol phospholipid metabolism in rat parotid glands. Biochem J 216: 633-640

Hokin MR, Hokin LE (1953) Enzyme secretion and the incorporation of P^{32} into phospholipids of pancreas slices. J Biol Chem 203: 967-977

Inoue M, Kishimoto A, Takai Y, Nishizuka Y (1977) Studies on a cyclic nucleotide-independent protein kinase and its proenzyme in mammalian tissues. J Biol Chem 252: 7610-7616

Jolles J, Zwiers H, Dekker A, Wirtz KWA, Gispen WH (1981) Corticotropin-(1-24)-tetracosapeptide affects protein phosphorylation and polyphosphoinositide metabolism in rat brain. Biochem J 194: 283-291

Kaibuchi K, Sano K, Hoshijima M, Takai Y, Nishizuka Y (1982) Phosphatidylinositol turnover in platelet activation; calcium mobilization and protein phosphorylation. Cell Calcium 3: 323-335

Kaibuchi K, Takai Y, Sawamura M, Hoshijima M, Fujikura T and Nishizuka Y (1983) Synergistic functions of protein phosphorylation and calcium mobilization in platelet activation. J Biol Chem 258: 6701-6704

Kawamoto S, Hidaka H (1984) Calcium-activated, phospholipid-dependent protein kinase catalyzes the phosphorylation of actin-binding proteins. Biochem Biophys Res Commun 118: 736-742

Kikkawa U, Takai Y, Tanaka Y, Miyake R, Nishizuka Y (1983) Protein kinase C as a possible receptor protein of tumor-promoting phorbol esters. J Biol Chem 258: 11442-11445

Klee CB, Vanaman TC (1982) Calmodulin. In: Advances in protein chemistry, vol. 35. Academic Press, New York, 213-321

Kuo JF, Shoji M (1982) Cyclic GMP-dependent protein phosphorylation. In: Nathanson JA, Kebabian JW (eds) Cyclic Nucleotides, part I: Biochemistry. Springer, Berlin-Heidelberg-New York (Handbook of experimental pharmacology) 58: 393-424

Le Peuch ChJ, Ballester R, Rosen OM (1983) Purified rat brain calcium- and phospholipid-dependent protein kinase phosphorylates ribosomal protein S 6. Proc Natl Acad Sci USA 80: 6858-6862

Macara IG, Marinetti GV, Balduzzi PC (1984) Transforming protein of avian sarcoma virus UR2 is associated with phosphatidylinositol kinase activity: Possible role in tumorigenesis. Proc Natl Acad Sci USA 81: 2728-2732

Michell RH (1975) Inositol phospholipids and cell surface receptor function. Biochem Biophys Acta 415: 81-147

Michell RH (1984) Oncogenes and inositol lipids. Nature 308: 770

Naka M, Nishikawa M, Adelstein RS, Hidaka H (1983) Phorbol ester-induced activation of human platelets is associated with protein kinase C phosphorylation of myosin light chains. Nature 306: 490-492

Nestler EJ, Greengard P (1983) Protein phosphorylation in the brain. Nature 305: 583-588

Nishikawa M, Hidaka H, Adelstein RS (1983) Phosphorylation of smooth muscle heavy meromyosin by calcium-activated, phospholipid-dependent protein kinase. J Biol Chem 258: 14069-14072

Nishizuka Y (1984) The role of protein kinase C in cell surface signal transduction and tumor promotion. Nature 308: 693-698

Putney JW, Burgess GM, Halenda SP, McKinney JS, Rubin RP (1983) Effects of secretagogues on [^{32}P] phosphatidylinositol 4,5-bisphosphate metabolism in the exocrine pancreas. Biochem J 212: 483-488

Rhodes D, Prpic V, Exton JH, Blackmore PF (1983) Stimulation of phosphatidylinositol 4,5-bis-phosphate hydrolysis in hepatocytes by vasopressin. J Biol Chem 258: 2770-2773

Sano K, Takai Y, Yamanishi J, Nishizuka Y (1983) A role of calcium-activated, phospholipid-dependent protein kinase in human platelet activation; comparison of thrombin and collagen actions. J Biol Chem 258: 2010-2013

Schatzman RC, Grifo JA, Merrick WC, Kuo JF (1983) Phospholipid-sensitive calcium-dependent protein kinase phosphorylates the β subunit of eukaryotic initiation factor 2 (eIF-2). FEBS Lett 159: 167-170

Streb H, Irvine RF, Berridge MJ, Schulz I (1983) Release of calcium from a nommitochondrial intracellular store in pancreatic acinar cells by inositol-1,4,5-trisphosphate. Nature 306: 67-69

Suematsu E, Hirata M, Hashimoto T, Kuriyama H (1984) Inositol 1,4,5-Trisphosphate releases calcium from intracellular store sites in skinned single cells of porcine coronary artery. Biochem Biophys Res Commun 120: 481-485

Sugimoto Y, Whitman M, Cantley LC, Erikson RL (1984) Evidence that the Roussarcoma virus transforming geneproduct phosphorylates phosphatidylinositol and diacylglycerol. Proc Natl Acad Sci USA 81: 2117-2121

Takai Y, Kishimoto A, Kikkawa U, Mori T, Nishizuka Y (1979) Unsaturated diacylglycerol as a possible messenger for the activation of calcium-activated, phospholipid-dependent protein kinase system. Biochem Biophys Res Commun 91: 1218-1224

Turner RS, Chou CHJ, Kibler RF, Kuo JF (1982) Basic protein in brain myelin is phosphorylated by endogenous phospholipid-sensitive calcium-dependent protein kinase. J Neurochem 39: 1397-1404

Werth DK, Niedel JE, Pastan I (1983) Vinculin, a cytoskeletal substrate of protein kinase C. J Biol Chem 258: 11423-11426

Subject Index